Lecture Notes in Physics

Edited by J. Ehlers, München, K. Hepp, Zürich
R. Kippenhahn, München, H. A. Weidenmüller, Heidelberg
and J. Zittartz, Köln
Managing Editor: W. Beiglböck, Heidelberg

112

Imaging Processes and Coherence in Physics

Proceedings of a Workshop,
Held at the Centre de Physique,
Les Houches, France, March 1979

Edited by
M. Schlenker, M. Fink, J. P. Goedgebuer,
C. Malgrange, J. Ch. Viénot and R. H. Wade

Springer-Verlag
Berlin Heidelberg New York 1980

Editors

M. Schlenker
Institut Max von Laue-Paul Langevin
and Laboratoire Louis Néel du CNRS
B.P. 166
F-38042 Grenoble Cedex

J. P. Goedgebuer
Lab. d'Optique
Faculté des Sciences et Techniques
F-25030 Besançon

J. Ch. Viénot
Lab. d'Optique
Faculté des Sciences et Techniques
F-25030 Besançon

M. Fink
Lab. Mécanique Physique
Univ. Paris VI
2, Place de la Gare de Ceinture
F-78210 Saint-Cyr l'Ecole

C. Malgrange
Lab. Minéralogie Cristallographie
Tour 16, Univ. P & M Curie
4, Pl. Jussieu
F-75230 Paris Cedex 05

R. H. Wade
Physique du Solide
DRF-G
CENG, 85X
F-38041 Grenoble Cedex

ISBN 3-540-09727-9 Springer-Verlag Berlin Heidelberg New York
ISBN 0-387-09727-9 Springer-Verlag New York Heidelberg Berlin

Library of Congress Cataloging in Publication Data
Main entry under title:
Imaging processes and coherence in physics.
(Lecture notes in physics; 112)
Bibliography: p.
Includes index.
1. Coherence (Optics)--Congresses. 2. Coherence (Nuclear physics)--Congresses.
3. Image processing--Congresses. I. Schlenker, Michel, 1940- II. Series.
QC476.C6I42 539.2 79-27215
ISBN 0-387-09727-9

This work is subject to copyright. All rights are reserved, whether the whole or part
of the material is concerned, specifically those of translation, reprinting, re-use of
illustrations, broadcasting, reproduction by photocopying machine or similar means,
and storage in data banks. Under § 54 of the German Copyright Law where copies
are made for other than private use, a fee is payable to the publisher, the amount
of the fee to be determined by agreement with the publisher.

© by Springer-Verlag Berlin Heidelberg 1980
Printed in Germany

Printing and binding: Beltz Offsetdruck, Hemsbach/Bergstr.
2153/3140-543210

INTRODUCTION

The techniques developed over the last two decades or so have considerably widened the range of particles used as imaging probes. The present volume, concerning imaging with photons (including hard and soft X-rays, visible, infrared and microwave regions), electrons, neutrons, ions (protons) and phonons, is a collection of the lectures given during a two week workshop held in Les Houches, France in March 1979 and attended by 80 invited participants from 11 countries.

The information concerning imaging with different probes tends to remain enclosed within the frontiers of each specialized field. In response to the clear need for a collection and synthesis of information about the numerous, very distinct approaches to imaging which have been developed for different particles, the workshop aimed to bring together people from different fields so that each could contribute to, and benefit from, a common pool of knowledge. Personal contacts can help to throw bridges across the gaps separating neighbouring activities. This book aims to materialize these bridges by making this unique collection of material available to all physicists, biologists, materials scientists and others interested in "imaging".

Some papers in the book deal with more general aspects of the physics involved in imaging processes and especially with coherence and the theoretical formalism common to different particle probes. The bulk of the book describes sources, beam characteristics, interaction with matter, image formation, detection and processing. Finally some recent techniques (NMR imaging, nuclear scattering radiography, channeling) are described separately.

We have tried to make it as easy as possible for the reader to find his way into and around the many different subjects dealt with in this book by providing three means of access :

1) a table of contents in the publication order which closely follows the order of presentation at the workshop. This is based on the physical sequence in any probe experiment, (the *source* produces a *beam* which propagates through space then *interacts with matter*, etc.)

2) a table of contents by probe particle

3) an analytical index based on keywords.

Remembering that Les Houches is a beautiful mountain resort in the French Alps we have allowed ourselves the fantasy of including the text of a talk on the "Coherent Theory of Skiing". This talk encouraged many participants to ski for the first time and others to attempt to improve their existing capacities.

We would like to thank J. Joffrin, D. Thoulouze and M.Th. Beal-Monod who initiated the idea of the workshop, the staff at the Centre de Physique des Houches for so pleasantly looking after the material side of communal life, all the participants for making the meeting so friendly and interesting and, of course, particularly the authors of contributions appearing in this book.

Naturally an enterprise such as this meeting and this book cannot come about within a financial vacuum. We would like to express our gratitude to the following organizations for their support :

Mission à la Recherche du Ministère des Universités
Centre National de la Recherche Scientifique
Délégation à la Recherche Scientifique et Technique
Direction des Recherches Etudes et Techniques.

M. Schlenker, M. Fink, J.P. Goedgebuer, C. Malgrange, J.Ch. Vienot, R.H. Wade

TABLE OF CONTENTS

F.T. ARECCHI: Laser Sources...1
P. CEREZ: Lasers as Stable Frequency Sources...............................17
G. FONTAINE: Electron Sources..28
J.B. FORSYTH: Intense Sources of Thermal Neutrons for Research.............37
R. COISSON: X-Ray Sources..51
A. ZAREMBOVITCH: Ultrasonic Waves and Optics...............................57
A. MARECHAL: Imaging Processes and Coherence in Optics.....................64
A. STEYERL: Propagation of Neutron Beams...................................77
H.A. FERWERDA: Coherence of Illumination in Electron Microscopy............85
S.M. CANDEL: Propagation of Sound and Ultrasound in Non-Homogeneous Media..95
J.M. LEVY-LEBLOND: Classical Apples and Quantum Potatoes..................106
F. MEZEI, G. JOUBERT: Outline of a Coherent Theory of Skiing..............113
J. SIVARDIERE: Wave-Matter Interactions: A General Survey.................121
L. GERWARD: Basic X-Ray Interactions with Matter..........................131
F. COHEN-TENOUDJI: Scattering Experiments in Ultrasonic Spectroscopy......138
F. MICHARD: Acousto-Optical Interactions..................................147
J.P. MANEVAL: H.F. Phonon Transmission as a Probe of Condensed Matter.....157
A. HADNI: Infrared Detectors, Infrared Imaging Systems....................164
B.F. BUXTON: Elastic Scattering of Fast Electrons.........................172
C. COLLIEX, C. MORY, P. TREBBIA: Inelastic Electron Scattering............182
G.L. SQUIRES: Interaction of Thermal Neutrons with Matter.................198
V. LUZZATI: Structure Information Retrieval from Solution X-Ray and Neutron
  Scattering Experiments..206
Y. LEVY: Guided Wave Propagation and Integrated Optics....................213
D. SAYRE: Prospects for Long-Wavelength X-Ray Microscopy and Diffraction..226
A. AUTHIER: Kinematical and Dynamical Diffraction Theories................233
F. BALIBAR: Dynamical Theory of X-Ray Propagation in Distorted Crystals...248
P. SKALICKY: Polarization Phenomena in X-Ray Diffraction..................258
A. ZEILINGER: Perfect Crystal Neutron Optics..............................267
F. MEZEI: Coherent Approach to Neutron Beam Polarization..................279
W. GRAEFF: X-Ray and Neutron Interferometry...............................293
C. MALGRANGE: X-Ray Topography: Principles................................299
A. MATHIOT: Examples of X-Ray Topographic Results.........................307
M. SCHLENKER, J. BARUCHEL: Neutron Topography.............................317
M. HART: Crystal Diffraction Optics for X-Rays and Neutrons...............322
R.H. WADE: Electron Imaging Techniques....................................333
F. RODDIER: Speckle and Intensity Interferometry. Applications to Astronomy......344
J.P. GOEDGEBUER, J. CH. VIENOT: Holography and λ-Coding...........357
J. ROGERS: Electron Holography..362
V.V. ARISTOV: Prospects of X-Ray Holography...............................368

Y. QUERE: Imaging by Means of Channelled Particles...................................373
A.K. FREUND: Neutron Optics Using Non-Perfect Crystals.............................378
C. BRUNEEL: Ultrasonic Real-Time Reconstruction Imaging...........................393
B. JOUFFREY: What Can Be Done with High Voltage Electron Microscopy?.............399
P.W. HAWKES: Transfer Functions and Electron Microscope Image Formation..........412
B. NONGAILLARD: Acoustic Microscopy..420
J. BRUNOL: Use of Coded Apertures in Gamma-Imaging Techniques.....................428
M. FINK: Principles and Techniques of Acoustical Imaging..........................434
I.L. PYKETT, P.MANSFIELD, P.G. MORRIS, R.G. ORDIDGE, V.BANGERT: N.M.R. Imaging...449
D. GARRETA: Nuclear Scattering Radiography..459
R. KLAUSZ: Computerized Tomography Scanners.......................................469
J.J. CLAIR: Recording Materials and Transducers in Optics.........................477
D. CHARRAUT: Some Applications in Hybrid Image Processing.........................490
J. MILTAT: X-Ray Image Detectors..499
E. ZEITLER: Electron Detectors..509
J.E. MELLEMA: Image Processing of Regular Biological Objects......................519
A.R. LANG: Cathodoluminescence Topography...529
O. KÜBLER: Image Processing in Electron Microscopy: Non-Periodic Objects.........541
Y. EPELBOIN: Interpretation of X-Ray Topography...................................551
J. JOFFRIN: Coherence is Beautiful..560

TABLE OF CONTENTS BY SUBJECT

Visible Light

F.T. ARECCHI: Laser Sources..1
P. CEREZ: Lasers as Stable Frequency Sources..17
A. MARECHAL: Imaging Processes and Coherence in Optics..............................64
F. MICHARD: Acousto-Optical Interactions...147
A. HADNI: Infrared Detectors, Infrared Imaging Systems.............................164
Y. LEVY: Guided Wave Propagation and Integrated Optics.............................213
F. RODDIER: Speckle and Intensity Interferometry. Applications to Astronomy.......344
J.P. GOEDGEBUER, J.CH. VIENOT: Holography and λ-Coding.....................357
J.J. CLAIR: Recording Materials and Transducers in Optics..........................477
D. CHARRAUT: Some Applications in Hybrid Image Processing.........................490
A.R. LANG: Cathodoluminescence Topography...529

Acoustics

A. ZAREMBOVITCH: Ultrasonic Waves and Optics.......................................57
S.M. CANDEL: Propagation of Sound and Ultrasound in Non-Homogeneous Media..........95
F. COHEN-TENOUDJI: Scattering Experiments in Ultrasonic Spectroscopy..............138
F. MICHARD: Acousto-Optical Interactions..147
J.P. MANEVAL: H.F. Phonon Transmission as a Probe of Condensed Matter.............157
C. BRUNEEL: Ultrasonic Real-Time Reconstruction Imaging...........................393
B. NONGAILLARD: Acoustic Microscopy...420
M. FINK: Principles and Techniques of Acoustical Imaging..........................434

Electrons

G. FONTAINE: Electron Sources..28
H.A. FERWERDA: Coherence of Illumination in Electron Microscopy....................85
B.F. BUXTON: Elastic Scattering of Fast Electrons.................................172
C. COLLIEX, C. MORY, P. TREBBIA: Inelastic Electron Scattering....................182
R.H. WADE: Electron Imaging Techniques..333
J. ROGERS: Electron Holography..362
B. JOUFFREY: What Can be Done With High Voltage Electron Microscopy?..............399
P.W. HAWKES: Transfer Functions and Electron Microscope Image Formation..........412
E. ZEITLER: Electron Detectors..509
J.E. MELLEMA: Image Processing of Regular Biological Objects......................519
A.R. LANG: Cathodoluminescence Topography...529
O. KÜBLER: Image Processing in Electron Microscopy: Non-Periodic Objects.........541

X-Rays

R. COISSON: X-Ray Sources..51
L. GERWARD: Basic X-Ray Interactions with Matter...........................131
V. LUZZATI: Structure Information Retrieval from Solution X-Ray and Neutron
    Scattering Experiments...206
D. SAYRE: Prospects for Long-Wavelength X-Ray Microscopy and Diffraction...226
A. AUTHIER: Kinematical and Dynamical Diffraction Theories.................233
F. BALIBAR: Dynamical Theory of X-Ray Propagation in Distorted Crystals....248
P. SKALICKY: Polarization Phenomena in X-Ray Diffraction...................258
W. GRAEFF: X-Ray and Neutron Interferometry................................293
C. MALGRANGE: X-Ray Topography: Principles.................................299
A. MATHIOT: Examples of X-Ray Topographic Results..........................307
M. HART: Crystal Diffraction Optics for X-Rays and Neutrons................322
V.V. ARISTOV: Prospects of X-Ray Holography................................368
R. KLAUSZ: Computerized Tomography Scanners................................469
J. MILTAT: X-Ray Image Detectors...499
Y. EPELBOIN: Interpretation of X-Ray Topography............................551

Neutrons

J.B. FORSYTH: Intense Sources of Thermal Neutrons for Research..............37
A. STEYERL: Propagation of Neutron Beams....................................77
G.L. SQUIRES: Interaction of Thermal Neutrons with Matter..................198
V. LUZZATI: Structure Information Retrieval from Solution X-Ray and Neutron
    Scattering Experiments...206
A. AUTHIER: Kinematical and Dynamical Diffraction Theories.................233
A. ZEILINGER: Perfect Crystal Neutron Optics...............................267
F. MEZEI: Coherent Approach to Neutron Beam Polarization...................279
W. GRAEFF: X-Ray and Neutron Interferometry................................293
M. SCHLENKER, J. BARUCHEL: Neutron Topography..............................317
M. HART: Crystal Diffraction Optics for X-Rays and Neutrons................322
A.K. FREUND: Neutron Optics Using Non-Perfect Crystals.....................378

General

J.M. LEVY-LEBLOND: Classical Apples and Quantum Potatoes...................106
J. SIVARDIERE: Wave-Matter Interactions: A General Survey..................121
J. JOFFRIN: Coherence is Beautiful...560

Sundry

P. CEREZ: Lasers as Stable Frequency Sources...17
F. MEZEI, G. JOUBERT: Outline of a Coherent Theory of Skiing.....................113
Y. QUERE: Imaging by Means of Channelled Particles...............................373
J. BRUNOL: Use of Coded Apertures in Gamma-Imaging Techniques....................428
I.L. PYKETT, P. MANSFIELD, P.G. MORRIS, R.G. ORDIDGE, V. BANGERT: N.M.R. Imaging..449
D. GARRETA: Nuclear Scattering Radiography.......................................459

Medical Applications

J. BRUNOL: Use of Codes Apertures in Gamma-Imaging Techniques....................428
M. FINK: Principles and Techniques of Acoustical Imaging.........................434
I.L. PYKETT, P. MANSFIELD, P.G. MORRIS, R.G. ORDIDGE, V. BANGERT: N.M.R. Imaging..449
D. GARRETA: Nuclear Scattering Radiography.......................................459
R. KLAUSZ: Computerized Tomography Scanners......................................469

ALPHABETICAL SUBJECT INDEX

Absorption, Neutron 198, 318
Absorption, X-ray 135
Acoustic Activity 59, 154
Acoustic Attenuation 422
Acoustic Lens 422, 437
Acoustic Microscopy 420
Acoustic Scattering Center 434
Acoustic Waves, Polarized 59
Acoustical Echography 440
Acoustical Holography 438
Acoustics, Non-linear 59
Acousto-electronic Lens 442
Acousto-optical Deflector 151
Acousto-optical Interactions 147, 393
Acousto-optical Modulator 151
Activity, Acoustic 59, 154
Aerial Photograph Processing 494
Alpha Particles, Imaging with 374
Amplitude, Scattering 121, 132
Angular Amplification 246, 273
Anomalous Dispersion Corrections 134, 296, 328
Anomalous Scattering 328
Anomalous Transmission, Neutron 268
Anomalous Transmission, X-ray (Borrmann Effect) 136, 244
Atomic Displacement by Electrons 531
Atomic Scattering Factor 132
Attenuation in Optical Fibers 216
Attenuation, Acoustic 422
Available Thickness in H.v.e.m. 401
Beam Modulation 327
Bethe Theory for Electron Inelastic Scattering 191
Biological Material, Electron Energy Loss Spectrum for 189
Biological Objects, Imaging in the E.m. 520
Bloch Waves 177, 239
Boersch Effect Energy Broadening of Electron Beams 35
Bonse-Hart Interferometer 293, 327
Booster, Fission 45
Born Approximation 122, 174, 199
Borrmann Effect or Anomalous Transmission 136, 243, 295
Bragg Diffraction 233
Bragg or Reflection Case 244, 322
Bragg Scattering in Acoustics 152
Bragg's Law 235, 322, 381
Bremsstrahlung 51, 531
Bright Field Electron Microscopy 417
Brightness of Electron Source 29
Burgers' Vector Determination 302, 554
Cathodoluminescence Topography 529
Channel Electron Multiplier 515
Channelling 373
Characteristic Length 249, 302
Characteristic Parameters in X-ray Dynamical Theory 248
Characteristic X-ray Lines 51
Chemical Analysis Using Electron Microscopy 408
Classical Apples 106
Classical Treatment of Electron Scattering 403

The editors are happy to thank Y. Siret, head of the scientific computing service at I.L.L, who kindly wrote the FORTRAN program for producing this index.

Coded Aperture Tomography 428
Coherence and Holography 359
Coherence of Electron Beam 28
Coherence, Degree of 67, 345
Coherence, Image Formation and 71, 414
Coherence, Laser 9
Coherence, Partial 64
Coherence, Spatiotemporal 70, 85, 86, 416
Coherent Generation of Phonons 160
Coherent Image Processing 490
Coherent Scattering, X-ray 132
Coherent States 279, 563
Coherent Theory of Skiing 113
Collision Processes, Elastic and Inelastic 182
Collision Theory, Formalism of 121
Comparison of Light and Electron Microscopy 85
Complementarity Principle 109
Compton Scattering 136
Computerized Axial Tomography 469
Conical Refraction Effect 59
Contrast in Electron Microscopy 338, 399, 412
Contrast in X-ray Image Detection 501, 504, 553
Conventional Transmission Electron Microscope (Ctem) 335
Correlation Function 201
Creation of New Wavefields 257, 303
Critical Voltage Effect in Electron Scattering 404
Cross-over from Electron Gun 32
Cross-section, Scattering 121, 132, 191, 227, 402
Cross-sections, X-ray Reaction 131, 226
Crystal Defects, Investigation of 252, 294, 299, 309, 318, 373
Crystal Diffraction ..233, 248, 258, 267, 293, 299, 307, 317, 322, 368, 378, 551
Ctem and Stem, Comparison of 94
Ctem and Stem, Reciprocity between 93, 341
Ctem, Image Formation in 89, 339
Dark Field Electron Microscopy 340, 416
Dechannelling by Defects 373
Detection of Photons, Quantum Noise 355
Detective Quantum Efficiency (Dqe) 510
Detectors, Electron 509
Detectors, Infrared (Quantum, Thermal) 164
Detectors, Optical 477
Detectors, X-ray 499
Dichroism, X-ray 261
Diffraction Focusing 271, 370
Direct Image 301, 555
Dislocation Luminescence 536
Dispersion Corrections, Anomalous 134, 296, 328
Dispersion Surface 177, 242, 248, 262
Double Crystal X-ray Topography 305
Duality Principle 109
Dynamical Diffraction Theory 175, 233, 258, 267, 293, 404
Earthquake Detection by Laser 23
Echo 568
Echo Planar Imaging, N.m.r. 450
Echography, Acoustical 440
Echography, B-mode 398, 443
Effective Source in Electron Microscopy 89
Eikonal Approximation 174, 252
Elastic and Inelastic Collision Processes 182
Elastic Scattering in High Energy Electron Diffraction .. 172

Elastic Scattering, Schroedinger Equation for 121,172
Electrochromic Transducers 486
Electron Beam, Coherence of 28
Electron Detectors 509
Electron Diffraction 173
Electron Dose Effects on Structure of Biological Specimens 523
Electron Energy Loss Spectroscopy 195
Electron Energy Loss Spectrum for Biological Material 189
Electron Gun, Cross-over from 32
Electron Gun, Effective Source 86
Electron Holography 364
Electron Imaging, Transfer Theory of 414
Electron Inelastic Scattering 193
Electron Lenses 334
Electron Lenses, Wave Aberration Function for 90
Electron Micrographs, Fourier Analysis of 521,546
Electron Micrographs, Optical Diffractograms of 524
Electron Microscope Specimens 337
Electron Microscope(Stem), Scanning Transmission 336
Electron Microscope, Conventional Transmission 335
Electron Microscope, Description of 412
Electron Microscope, High Voltage 399
Electron Microscopy, Contrast in 338,412
Electron Microscopy, Dark Field 340,416
Electron Microscopy, Effective Source in 89
Electron Microscopy, Resolution in 333,342,399,416,542
Electron Multiplier, Channel 515
Electron Penetration in Solids 401,530
Electron Radiation Damage 342
Electron Scattering 126,172,182,338,402,511
Electron Scattering and Image Contrast 338
Electron Scattering by a Coulomb Potential 126
Electron Scattering, Dynamical Theory of 404
Electron Scattering, Optical Potential in 172
Electron Sources 28
Electronic Devices, Topographic Study 311
Electrons, Field and Thermal Emission of 32
Electrons, Helmholtz Equation for 86
Electrons, Image Intensifiers for 516
Electrons, Photographic Emulsion for 517
Electrons, Scintillation Counters for 516
Electrons, Solid State Detector for 515
Electrons, Thermionic Emission of 30
Electrooptical Image Detectors 504
Emulsions, X-ray Photographic 553
Energy Broadening of Electron Beams, Boersch Effect 35
Energy Loss Processes 182,402,531
Ensemble Average 280
Environmental Cells in Electron Microscopy 407
EXAFS 135
Expectation Value 279
Exposure Time for X-ray Topographs 308
Extinction, or Pendelloesung, Length 249,302
Eye, Sensitivity of 552
Faraday Cup 514
Faraday Effect, Acoustical 59
Fericon Transducers 487
Fermi Golden Rule 123
Fermi Pseudopotential 199
Ferroelectrics As Optical Recording Material 480

Field and Thermal Field Emission of Electrons 32
Fission 39
Fission Booster 45
Fission Fragments, Imaging with 375
Focusing in Diffraction by Imperfect Crystals 378
Focusing, Diffraction 271,370
Formalism of Collision Theory 121
Forward Scattering Approximation in Electron Diffraction .. 173
Fourier Analysis of Electron Micrographs 521,546
Fourier Transform of Helical Projection 521
Fraunhofer Electron Holography 365
Frequency Stabilized Lasers 17
Fresnel Diffraction 394
Fresnel Focusing 446
Fresnel Transform Device, Analogical 397
Geometric Aberrations 422
Geometrical Approximation To Wave Propagation 99
Geometrical Diffraction Theory 234
Grain Boundaries 374
Grain Boundary Dislocations, Topographic Study 310
Gravitational Effect on Neutrons 81,297
Gravity Measurement by Laser 24
Green's Function 255,435
Growth Defects, Topographic Study 308,318
Guided Mode Propagation in Fibers 214
Guide-tube, Neutron 80,318
Guinier-Preston Zones 377
Handwriting, Optical Processing of 493
Harmonic-free Monochromator 325
Helical Projection, Fourier Transform of 521
Helmholtz Equation for Electrons 86
High Energy Electron Diffraction, Elastic Scattering in .. 172
High Voltage Electron Microscopy (Hvem) 399
Holographic Memory 478,572
Holographic Reconstruction 363
Holography, Acoustical 438
Holography, Electron 364
Holography, Optical 357
Holography, X-ray 368
Howie-Whelan Equations in Electron Diffraction 175
Hybrid Image Processing 492
Hydrogen Density Distribution 460
Image Formation and Coherence 71,414
Image Formation in Ctem 89,339
Image Formation in Stem 92,340
Image Intensifiers for Electrons 516
Image Processing 490,541
Image Reconstruction 393
Image Restoration 543
Image Treatment 507
Imaging in Three Dimensions, X-ray 231
Imaging Parameters for X-rays, Neutrons, Electrons ... 302,400
Imaging Systems, Infrared 168
Imaging with Partially Coherent Illumination 74,416
Imaging, Rose Condition for 542
Imaging, X-ray 226,299,307,551
Incoherent Scattering 132,200
Inelastic Electron Scattering 182,182,531
Inelastic Scattering, Electron 193
Inelastic Scattering, Neutron 204,289,380

Infrared Detectors, Noise (Johnson, Thermal, Background) in	165
Infrared Imaging Systems	168
Infrared(Quantum, Thermal) Detectors	164
integrated Intensity in X-ray Diffraction	237
integrated Optical Circuits	219
Interactions with Matter, X-ray	131
Interference	279
Interferometer, Bonse-Hart	293, 327
Interferometry, Applications in Astronomy	351
Interferometry, Intensity and Speckle	349
Interferometry, X-ray and Neutron	293, 370
inverse Scattering Problem in One Dimension	98
Jellium Model of Electron Inelastic Scattering	193
Karhunen-Loeve Transform Image Processing	495
Kinematical Diffraction Theory	234, 381
Kirchhoff Approximation	139
Lang's Method	249, 300
Larmor Precession	81, 281
Lasers, Frequency Stabilized	17
Laser Coherence	9
Laser Modes	2
Laser Mode Locking	6
Laser Pulse	6
Laser Stability	17
Laser Stimulated Photon Emission	1
Lattice Parameter, Absolute Determination	295
Laue or Transmission Case	244, 322
Lean in Skiing	117
Lens, Acoustic	422, 437
Linac	45
Line-scan Imaging, N.m.r.	449
Liquid Crystal Optical Transducers	484
Lithium Niobate As Recording Material	482
Low-energy Electron Scattering	180
Macromolecules, Scattering by	206
Magnetic Domains and Walls, Topographic Study	311, 319
Magnetic Resonance, Imaging by Nuclear	449
Magnetic Scattering, Neutron	202, 319
Magnetooptic Recording Material	481
Majorana Flipper	288
Malus' Law, Acoustic	58
Mean Free Paths for Electrons	134
Memory, Holographic	478, 572
Microanalysis Using Electron Energy Loss	195
Microradiography	229
Microscopy, Acoustic	420
Microscopy, Comparison of Light and Electron	85
Microscopy, X-ray	226, 371
Mode Locking, Laser	6
Moderation, Neutron	41
Momentum Space	381
Monochromators, Neutron	380
Mosaic Spread	381
Multi-slice Method in Electron Diffraction	175
N.m.r. Imaging	449
Negative Staining Technique	520
Neutron Absorption	198, 318
Neutron Bottle	79
Neutron Guide-tube	80, 318
Neutron Incoherent Scattering	200

Neutron Inelastic Scattering 204,289,380
Neutron Interferometry 293
Neutron Moderation 41
Neutron Monochromators 380
Neutron Polarization 79,279
Neutron Sources 37
Neutron Spin 280
Neutron Topography 317
Neutrons,Ultracold 79
Noise (Johnson,Thermal,Background) in Infrared Detectors 165
Noise in Gamma-ray Tomography 431
Noise,Effect on Phase Contrast Imaging 541
Non-linear Acoustics 59
Non-linear Optical Processes 8
Nuclear Scattering of Protons 460
Nuclear Scattering,Neutron 198
One Diffusion Process Approximation 435
Optical Detectors 477
Optical Diffractograms of Electron Micrographs 524
Optical Fibers 213
Optical Holography 357
Optical Potential in Electron Scattering 172
Optical Potential,Neutron 78
Optics,Ultrasonic 57
Parabolic Equation Method 98
Parametric Arrays 61
Partial Waves and Phase Shifts 123
Partially Coherent Illumination,Imaging with 74,416
Particle Concept 106
Pendelloesung 294
Pendelloesung Fringes 244,262,300,314
Pendelloesung or Extinction,Length 249,302
Penetration in Solids,Electron 401,530
Phase and Amplitude Transfer Functions 413
Phase Contrast Topography 297
Phase Filter 543
Phase Measurements 327,369
Phase Object Approximation in Electron Diffraction 174
Phase Operator 562
Phase Shifts,Born Approximation for 124
Phonon Echo 571
Phonons 157
Photochromic Recording Material 481
Photographic Detector for Charged Particles 374
Photographic Emulsion for Electrons 517
Photographic Emulsion for X-rays 501
Photon Noise 504
Photosensitive Materials,X-ray 229,507
Phototitus Transducers 488
Pivoting,Ski 116
Planar Defect 252,304
Plastic Deformation,Topographic Study 315
PLZT As Recording Material 482
Pockels Effect 59
Polarization,Neutron 79,279
Polarization,X-ray 258
Polarized Acoustic Waves 59
Polarized Neutron Topography 319
Polarizer,X-ray 328
Poynting Vector 243,303

```
PROM Transducers .............................................. 488
Proton Channelling ............................................ 374
Proton N.m.r. Imaging ......................................... 450
Proton Radiography ............................................ 459
Protons,Nuclear Scattering of ................................. 460
Pseudopotential,Fermi ......................................... 199
Pulsed Plasma Sources .................................... 54,226
Pulsed Reactor ................................................. 45
Pulse,Laser ..................................................... 6
Quantons ...................................................... 106
Quantum Mechanics of Electron Inelastic Collisions ............ 189
Quantum Potatoes .............................................. 106
Quantum Transducers,Phonon .................................... 159
Radiation Damage in Electron Microscopy ....................... 523
Radiation Damage in High Voltage Electron Microscopy .......... 407
Radiative Processes ........................................... 532
Radiography,Proton ............................................ 459
Radius of Gyration ............................................ 208
Rainbow Holography ............................................ 359
Ray Theory for a Perfect Crystal .............................. 250
Ray Theory in Perfect Crystals ................................ 241
Rayleigh Scattering in Optical Fibers ......................... 216
Reactor,Fission ................................................ 39
Reciprocal Space, Focusing in ................................. 378
Reciprocity Between Ctem and Stem ......................... 93,341
Reconstruction,Holographic .................................... 363
Recording Materials in Optics ................................. 477
Recrystallization,Topographic Study ........................... 310
Reference Wave in Holography .................................. 362
Reflection (Bragg) Case in Bragg Diffraction .............. 244,322
Reflection (Bragg) Topography ................................. 299
Reflection Acoustic Microscope ................................ 424
Refraction,X-ray .............................................. 323
Refractive Index in Scattering Processes ...................... 128
Refractive Index,X-ray and Neutron ................... 77,241,296
Relativistic Effects in Electron Scattering ................... 172
Relaxation Time ............................................... 567
Resolution Function in Neutron Scattering ..................... 385
Resolution in Electron Microscopy ............ 333,342,399,416,542
Resolving Power ............................................... 503
Restoration of Electron Images ................................ 543
Rocking Curve ................................................. 235
Rose Condition for Imaging .................................... 542
Rotatory Power,X-ray .......................................... 261
Scanner ....................................................... 469
Scanning Transmission Electron Microscope(Stem) .......... 336,417
Scattering Amplitude ................................... 121,132,198
Scattering and Energy Loss Effect on Electron Detection ....... 511
Scattering by Randomly Rough Surfaces ......................... 141
Scattering Cross-section ................... 121,132,191,227,402
Scattering Length ......................................... 198,296
Scattering of Slow Particles .................................. 125
Scattering Potential,Neutron ................................... 78
Schroedinger Equation for Elastic Scattering .............. 121,172
Schroedinger Equation for Electrons ............................ 85
Scintillation Counters for Electrons .......................... 516
Section Topograph ........................................ 301,319,553
Sensitivity of Eye ............................................ 552
Side Cut in Skis .............................................. 114
Simulation of X-ray Topographs ................................ 556
```

Single Loss Electron Inelastic Scattering	186
Single Side-band Electron Holography	365
Ski Boots and Skis	114
Skiing, Coherent Theory of	113
Slow Particles, Scattering of	125
Small Angle Scattering	206
Solid State Detector for Electrons	515
Sources, Electron	28
Sources, Neutron	37
Sources, Ultrasonic	57
Sources, X-ray	51
Spallation Neutron Source	46
Spatiotemporal Coherence	70, 85, 86, 416
Spatiotemporal Holography	360
Speckle, Applications in Astronomy	353
Spectroscopic Applications of Stabilized Lasers	24
Spectroscopic Instruments, X-ray	324
Spectroscopy, Ultrasonic	138
Spherical Wave	245
Spin Echo, Neutron	290, 569
Spin, Neutron	280
Spinor Rotation Experiments	297
Statistical Analysis in Image Processing	495
Steering of Skis	116
Stem, Image Formation in	92, 340
Stereo Images	301, 555
Stern-Gerlach Experiment	284
Stimulated Phonon Emission	160
Stimulated Photon Emission, Laser	1
Storage Ring, Electron	54
Storage Ring, Neutron	83
Structure Factor	134, 235
Superfluidity	564
Supermirror for Neutrons	79
Surface Waves, Ultrasonic	61
Synchrotron Radiation	52, 226, 264, 296, 305, 499
Takagi's Equations	255, 556
Thermal Diffuse Scattering	204
Thermal Radiations, Statistical Properties	344
Thermal Transducers, Phonon	158
Thermionic Emission of Electrons	30
Thermoplastic Recording Material	481
Three-dimensional Imaging, X-ray	231
Three-dimensional Structure from Electron Micrographs	522
Tomography	231, 428, 469
Topography, Cathodoluminescence	529
Topography, Interpretation of, X-ray	551
Topography, Neutron	317
Topography, Phase Contrast	297
Topography, X-ray	299, 307
Total Reflection, Neutron	79
Transducers, Optical	483
Transducer Array	437
Transfer Functions, Phase and Amplitude	413
Transfer Theory of Electron Imaging	414
Transmission (Laue) Case in Bragg Diffraction	244, 322
Transmission (Laue) Topography	299, 307, 317, 551
Transmission Acoustic Microscope	421
Transmission Function for Electron Transparent Objects	90, 414
Traverse Topograph	301, 553

Tubes, X-ray	51
Turn in Skiing	119
Twins	374
Ultracold Neutrons	79
Ultrasonic Optics	57
Ultrasonic Sources	57
Ultrasonic Spectroscopy	138
Ultrasonic Surface Waves	61
Undulators	54
Vibronic Spectra	535
Videodisc	479
Volume Holography	359
Wave Aberration Function for Electron Lenses	90
Wave Concept	106
Wave-matter Interactions	121
Wavefield	239, 254, 303
Wavelength Standards	22
White Light Holography	359
Wiener Filter	543
Wigglers	54
X-ray Absorption	135
X-ray Detectors	499
X-ray Holography	368
X-ray Image Detectors	499
X-ray Imaging in Three Dimensions	231
X-ray Interactions with Matter	131
X-ray Interferometry	293
X-ray Microscopy	226, 371
X-ray Photographic Emulsions	553
X-ray Photon Fluxes	51, 499
X-ray Photosensitive Materials	229, 507
X-ray Polarization	258
X-ray Polarizer	328
X-ray Propagation in Distorted Crystals	248, 303
X-ray Propagation in Perfect Crystals	241
X-ray Reaction Cross-sections	131, 226
X-ray Scattering by Electrons	126
X-ray Sources	51
X-ray Spectroscopic instruments	324
X-ray Topography	299, 307
X-ray Topography, Interpretation of	551
X-ray Tubes	51
Zone Mirror, Neutron	83

LASERS SOURCES [*]

F. T. Arecchi
University and
Istituto Nazionale di Ottica, Firenze

Since 1960 the Laser has become a useful device in many areas of physics and technology. A description of the laser can be done from two points of view, namely:
  i) physics of the stimulated emission processes;
  ii) coherence and cooperative phenomena in radiation-matter interaction.

We shall discuss the two aspects in sequence, defining the terms and giving the orders of magnitude.

1. Physics of the stimulated emission processes

If the e.m. cavity where we are considering the radiation-atom interaction is a rectangular cavity of sides X_1, X_2, X_3, volume $V = X_1 X_2 X_3$, then the solution of the wave equation, with periodic boundary conditions, yields the plane wave expansion for the field

$$E(x,y,z,t) = \sum_{\vec{k}} E(\vec{k},t) e^{i(k_1 x + k_2 y + k_3 z)} \qquad (1)$$

where $k_i = n_i \cdot 2\pi/X_i$ ($i = 1,2,3$; $n_i = 1,2,\ldots$).
For each set of k_i we have a different field configuration, or mode.

The dispersion relation imposes a constraint between frequency ω and amplitude $k = (k_1^2 + k_2^2 + k_3^2)^{1/2}$ of the k vector

$$\omega = c k \qquad (2)$$

In k space each mode occupies an elementary volume

$$\delta^3 k = (2\pi)^3/V. \qquad (3)$$

In a spherical shell of radius k and thickness Δ k there

---

[*] Workshop on "IMAGING PROCESSES AND COHERENCE IN PHYSICS"
Les Houches - 12-23 March 1979

are

$$\Delta M = 2 \frac{4\pi k^2 \Delta k}{\delta^3 k} = \frac{k^2 \Delta k}{\pi^2} V = \frac{8\pi \nu^2}{c^3} \Delta \nu V \qquad (4)$$

modes. The extra-factor 2 accounts for the two possible polarizations for each k vector. If the cavity contains radiators (atoms on the walls or inside) in thermal equilibrium at a temperature T, then the electromagnetic energy density in the cavity is given by Planck's blackbody formula

$$\frac{dW}{d\nu} = \frac{dM}{d\nu} \cdot h\nu \cdot \bar{n}_1(\nu) = \frac{8\pi \nu^2}{c^3} V h\nu \frac{1}{e^{h\nu/kT} - 1} \qquad (5)$$

Here
$$\bar{m}_1(\nu) = (e^{h\nu/kT} - 1)^{-1}$$

is the average photon number for <u>each</u> mode whose k vector lies on the spherical surface of radius k .
The distinction between <u>spontaneous</u> and <u>stimulated</u> emission came in the 1917 Einstein's derivation of Eq. (6), as follows. Consider two relevant levels of an atom separated by the energy $\Delta E = h\nu$ and coupled by an optical transition. Each time the atom goes up (<u>absorption</u>) or down (<u>emission</u>), this is a one-photon exchange process. (fig. 2)

The emission or decay process can be spontaneous (i.e. not triggered by photons) as well as stimulated (i.e., proportional to the photon number $\bar{n}$ at the frequency ν).
If there is an ensemble of N atoms in thermal equilibrium, with N_2 in the upper state and $N_1 = N - N_2$ in the lower, then we have

$$N_1/N_2 = e^{\Delta E/kT} \qquad (6)$$

and equating the rates of absorption and emission

$$N_1 B' \bar{n} = N_2 B \bar{n} + N_2 A \qquad (7)$$

From this latter equation

$$\bar{n} = \frac{A/B}{\frac{B'}{B}\frac{N_1}{N_2} - 1} = \bar{n}_1(\nu)\Delta\nu.$$

By use of Eq. (6) and comparison with (5), we get

$$\frac{A}{B} = \frac{8\pi\nu^2}{c^3} V \Delta\nu = \Delta M \qquad (8)$$

$$B' = B. \qquad (9)$$

This can be interpreted by representing the degrees of freedom of the e.m. field as boxes and the excited atom as linked to all of them as in Fig. 3.

With the probabilities there indicated, the stimulated emission probability into the mode with n photons is larger than the total spontaneous emission over the empty ΔM modes (all those within the linewidth of the atomic emission) when

$$n > \Delta M. \qquad (10)$$

Let us call ΔN the atomic population difference between upper and lower states, P the rate of excitation (pump), n the photon number in the laser mode and

$$T_c = \frac{L}{c}\frac{1}{\theta} \qquad (11)$$

the decay time of photons in the cavity made of two facing mirrors separated by a length L. T_c is equal to $1/\theta$ transit times, since the limited mirror transmittivity $\theta = 1-\rho < 1$ increases the number of transits. Condition (11) stems from considering photons as particles. It is a necessary, but not sufficient condition. Indeed, if we account for wave propagation and phase matching between forward and backward waves, the cavity is resonant for those frequencies corresponding to the standing wave condition (fig. 7)

$$m\lambda/2 = L \qquad \text{(m integer)}$$

which amounts to a minimum frequency separation

$$\Delta\nu_{m,m+1} = c/2L \qquad (12)$$

Only for these resonances the escape time is given by (11), otherwise it is much faster (just one transit time L/c).

The rate equations for photons and a population inversion are then

$$\frac{dn}{dt} = B \cdot \Delta N n - n/T_c$$

$$\frac{d\Delta N}{dt} = P - B \cdot \Delta N n \qquad (13)$$

where we have neglected spontaneous processes. Solving them at equilibrium, the first gives

$$\Delta N = \frac{1}{BT_c} \qquad (14)$$

and the second

$$P = B \cdot \Delta N n \qquad (14)'$$

Combining the two with (10), the pump rate must be

$$P > \Delta M/T_c . \qquad (15)$$

Let us now introduce the concept of atomic cross section σ per atom. The stimulated emission rate Bn can be written as

$$Bn = \sigma \phi \qquad (16)$$

where $\phi = cn/V$ is the photon flux and hence

$$\sigma = \frac{BV}{c}$$

When the atomic line is broadened only by spontaneous emission process, then, we can put $\Delta \nu = A$ in eq. (4) and have

$$\Delta M = \frac{8\pi}{\lambda^2} \frac{VB}{c} \Delta M$$

Hence

$$\sigma = \frac{\lambda^2}{8\pi} \qquad (17)$$

If there is an extra broadening $\Delta \nu_a > A$ for collision process or other decays, σ reduces as

$$\sigma = \frac{\lambda^2}{8\pi} \frac{A}{\Delta \nu_a} . \tag{17}'$$

Cross section (17) holds for a bound electron, while for a free electron it is much smaller (fig. 4) since it is given by the square of the classical electron radius $r_0 \approx 10^{-13}$ cm.

On the other hand, writing the volume as $V = S.L$, condition (14) can be rewritten as

$$P > \theta \frac{S}{\sigma} \Delta \nu \tag{15}'$$

which shows that the excitation rate is proportional to the ratio between the laser beam cross section S and the atomic cross section. Fig. 4 shows why bound electrons are better than free electrons. However nowadays using high energy ($\sim$ 1 GeV) free electrons in a storage ring one can produce laser action down to $\lambda \sim 1 \mu m$. Fig. 5 summarizes the different interactions and the spectral regions covered by lasers.

Once $n > \Delta M$ is fulfilled, that is, once the privileged mode has enough photons to neglect spontaneous decay channels, we must also take care for the cavity losses, and by (14) request that

$$B \Delta N \geqslant 1/T_c . \tag{16}$$

This condition is represented in Fig. 6 for two different ΔN. In the first case only one mode is above threshold, hence we have a single monochromatic frequency. In the second case we may have emission at three frequencies. Here we must introduce the fundamental difference between <u>homogeneous</u> and <u>inhomogeneous</u> linewidth. In the former case a monochromatic transition is broadened by circumstances which are <u>equal</u> for all atoms in the cavity (as spontaneous lifetime broadening in a gas, phonon interaction in a solid matrix). All atoms can contribute over the <u>whole</u> linewidth. Hence, once the mode nearest to the peak has been excited, as the associated field "sweeps" the cavity, it will "eat" all atomic contributions, forbidding the other modes from going above threshold. In the standing wave case this frequency picture is not sufficient and one should also consider the space pattern. As sketched in fig. 7 two different modes have nodes and maxima in different positions, hence they will "exploit" different atoms, releasing the competition. The simultaneous laser action over many modes is then possible.

The inhomogeneous line broadening corresponds to different frequency locations of different atoms. This can be due, e.g., to Doppler shift in a gas where thermal agitation gives a distribution of velocities.

Another inhomogeneity occurs in a crystal where active ions are exposed to a crystal field which changes from site to site.

For an inhomogeneous line, different modes can go above threshold even without a standing wave pattern.

In general, if c/2 L is much smaller than the atomic line width $\Delta\nu_a$ there are many independent laser lines, without phase relations.

In Fig. 8 we have shown the scheme of mode locking operation. In that figure, the several parts have the following meaning:

a) frequency picture of a many-mode laser
b) if the different laser fields have fixed phase relations, they act as the different Fourier components of a train of pulses, each lasting $1/\Delta\nu_a$ and separated by 2 L/c.
b) is the Fourier transform of the amplitude spectrum a), provided the phases are all equal
c) practical scheme of a many-mode laser.
Besides the three main ingredients (active medium, incoherent light to excite the atoms at the upper level, mirrors) there is also a saturable dye which becomes transparent at a critical light intensity l_c (see d) . All the standing waves of the different modes will self-adjust their phases to have a maximum when the dye is transparent. Transparency is then lost with a decay time $\tau_{dye} \ll$ 2 L/c and then recovered after a transient 2 L/c. This corresponds to having a narrow pulse bouncing forth and back between two mirrors. Notice that, from b) the pulse duration is $1/\Delta\nu_a$.

In the scheme of Fig. 9 we have thus explained how to lock in phase the laser lines, in order to make short pulses (as short as the uncertainty relation permits, i.e. $1/\Delta\nu_a$).

By using a Doppler broadened atomic line in a gas (like in

a He-Ne, or in an A^+ laser), then

$$\Delta \nu_a \sim \frac{\nu}{c}\sqrt{\frac{kT}{M}} \sim 10^9 \text{ Hz}$$

hence

$$t_{pulse} \sim 1 \text{ n s.}$$

Using ions of a transition element embedded in a crystal or glass matrix, as Cr^{3+} in Al_2O_3 (ruby), or Nd^{3+} in glass, one may have large $\Delta \nu_a$. A large $\Delta \nu_a$ can also be achieved in the case of complex dye molecules in a liquid solution because of the overlapping among many vibrational and rotational levels.
It is nowadays easy to achieve

$$\Delta \nu_a \sim 10^{13} \text{ Hz}$$

and hence

$$t_{pulse} \sim 0.1 \text{ p sec}$$

Notice that the range of picosecond times can be attained only by techniques as in Fig. 8, and <u>not</u> by electronic shutters.

2. Stimulated emission and nonlinear optics (NLO)

We have seen (fig. 3) that the transition rate for an emission process is B . 1 if spontaneous, or B (n+1) if stimulated.

Also in higher order processes as those studied in NLO we can have a spontaneous and a stimulated version. Take a parametric process implying the annihilation of one quantum $\hbar \omega_1$, and the creation of two quanta $\hbar \cdot \omega_2$ and $\hbar \cdot \omega_3$ as in Fig. 9. The transition rate is B for spontaneous emission in the field 2 or B $(n_2 + 1)$ for stimulated emission in the field 2.

In the first case, we look at 90°, and we collect point-like processes, having to satisfy the conservation of energy:

$$\omega_1 = \omega_2 + \omega_3$$

In the second case, we look in a direction (forward or backward) almost collinear with the impinging beam. Here, in order to add coherently the field contributions, the momentum matching condition

$$\vec{k}_1 = \vec{k}_2 + \vec{k}_3$$

has also to be satisfied.

In a similar way we may describe usual light propagation in a transparent medium as an elastic two-photon process. Since the scattered contributions sum in phase, it is more convenient to speak of a linear polarization

$$P_i = \varepsilon_0 \chi_{ij}^{(2)} E_j \qquad (17)$$

rather than stimulated emission in the scattered channel. Similarly, there are 4 photon processes leading to "self-actions" in the propagation of a large e.m. field, that is, self focusing, self-defocusing, self modulation in phase (self-broadening) and amplitude (self-steepening). Thses non linearities on the same light beam are described by a nonlinear polarization index as

$$P_i = \chi^{(4)}_{ijk\ell} E_j E_k E_\ell \qquad (18)$$

The nonlinear refraction index can be written in the isotropic case as

$$n = n_0 + n_2 \cdot |E|^2 \qquad (19)$$

In a liquid of anisotropic molecules, self actions stem from orientation of the molecules due to interaction with the induced dipole moments (high frequency Kerr effect). In a liquid of isotropic molecules, or in solids and gasses, self actions are due to distortion of the electron cloud.

In Table 1 we show some examples of NLO processes.

T A B L E   1

Nature	of the quanta	Name of the process
2	3	
light	molecular vibrations	Raman
light	optical phonons in solids	Raman
light	acoustical phonons in solids	Brillouin
light	sound waves in liquids	Brillouin
light	light	parametric conversion (sum or difference of frequency, second harmonic generation, etc.)

3. Coherence and cooperative phenomena

As shown in Fig. 3, stimulated emission explains mode selection, that is, a narrowing in the frequency spectrum and in the spectrum of possible directions (monochromaticity and directionality). This amounts to increasing the spectral purity, and use can be made of it in physics and technology (linear spectroscopy, holography, plasma production and compression by powerful pulses). But all this has very little to do with <u>coherence</u>.

Each mode has still a harmonic oscillator dynamics, that is, it is like a particle in a parabolic potential well, with an equilibrium statistical distribution given by Maxwell-Boltzmann, that is, peaked at the minimum energy. To have a sizeable amount of energy $|E_o|^2$ one has to increase the "temperature" i.e. the excitation, thus broadening the distribution and increasing the entropy as well (Fig. 10).

However as the field E increases, one must consider high order processes, besides the one photon emission, as e.g. the three-photon process of Fig. 11 which gives a cubic polarization

$$P = G_o E - \beta |E|^2 E \tag{20}$$

and hence a quartic free energy

$$W = -P \cdot E = -\frac{G_o}{2}|E|^2 + \frac{\beta}{4}|E|^4 \tag{21}$$

As E increases, the quartic potential well becomes steeper and steeper (Fig. 12), so that a useful E_o can be reached with a little amount of spread, or statistical fluctuations, around it.

We call <u>coherent</u> this highly excited field state <u>without noise</u>. The field can be described with very good approximation by a complex number with constant amplitude and phase. Such a field can bring the induced atomic dipoles to a coherent motion in which the phase relations among atomic wave functions are kept for long times.

This is the basis for <u>coherent nonlinear spectroscopy</u> which sheds information on fine properties of atoms and molecules.

4. Practical limitations to coherence

The above picture however is misleading if we aim to long distance interferometry. In fact high order (nonlinear) corrections regard the photon number, that is, the square of the field amplitude and not its phase. Even when the laser amplitude is stabilized by the above nonlinearities, there is a residual phase noise giving an ultimate linewidth as (Townes formula)

$$\Delta\nu = \frac{\pi h \nu}{2P}(\Delta\nu_c)^2$$

where $\Delta\nu_c$ is the cavity width (that is, the resolution of the empty laser interferometer) and P the power output. For $P \sim 1 mW$ it would yield $\Delta\nu \ll 10^{-2} Hz$.

A more stringent limitation is imposed by practical features as the thermal noise on the cavity or mechanical fluctuations in the laboratory room. Practically, today one can achieve a long term stability around 10 k Hz.

Figure Captions

1 - Spherical shell in K-space.

2 - Radiative transitions in a two-level atom.

3 - Decay channels of an excited atom into different field modes.

4 - Radiative electron cross-section versus frequency.

5 - Map of coherent radiation emission mechanisms.

6 - Scheme of two standing waves in a Fabry-Perot cavity, and interplay between cavity resonances and atomic gain line.

7 - Intensity distributions of two standing waves.

8 - Mode locking operation.

9b- Angular relation in a non-linear optical process.

9a- Non-linear process.

10 - Harmonic potential well and equilibrium statistical distribution for a linear field.

11 - Third-order radiative process in a two-level atom.

12 - Quartic potential well and statistical distribution for a laser field above threshold.

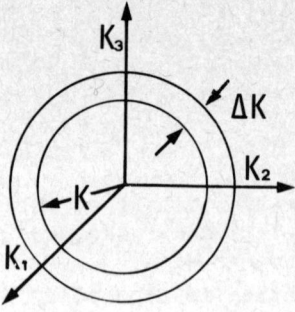

Fig. 1

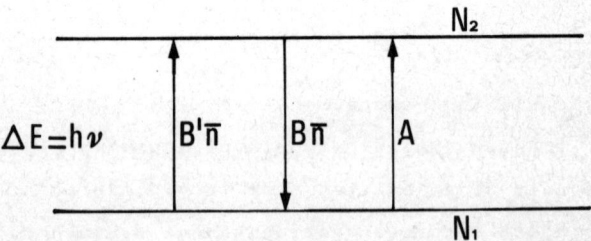

Fig. 2

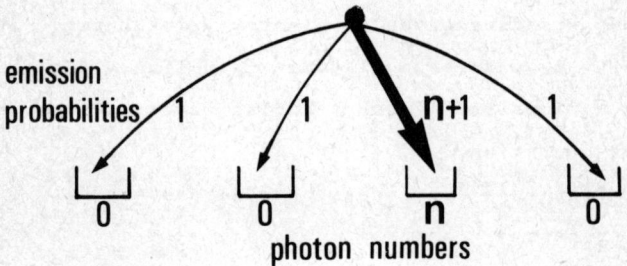

Fig. 3

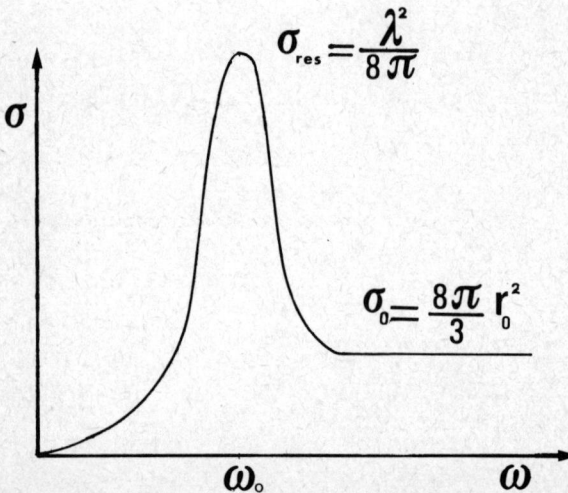

Fig. 4

LASERS

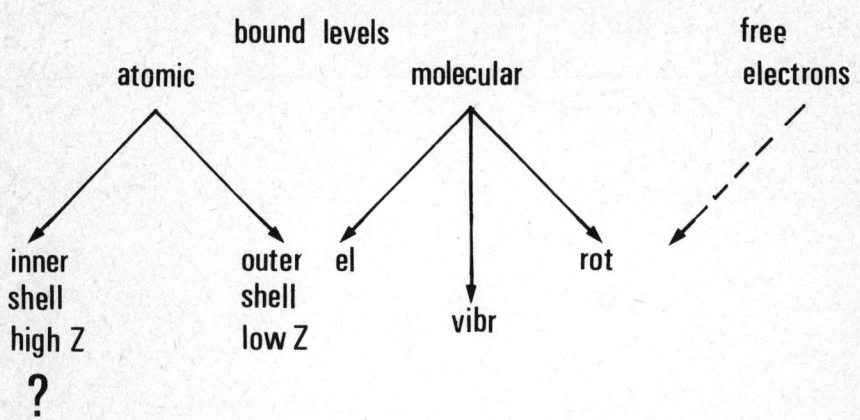

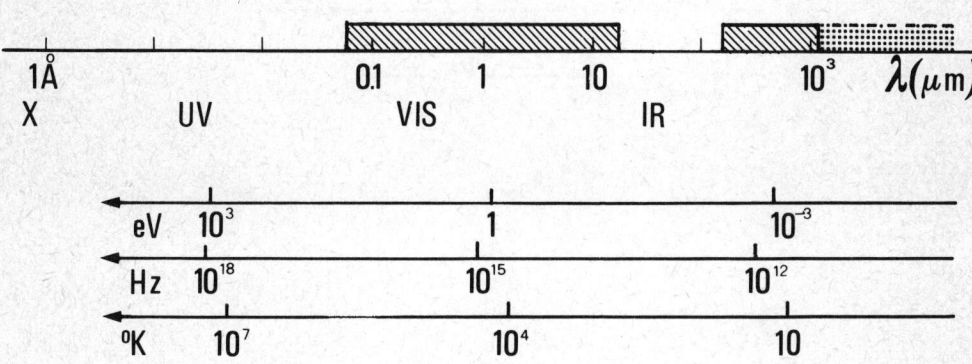

Fig. 5

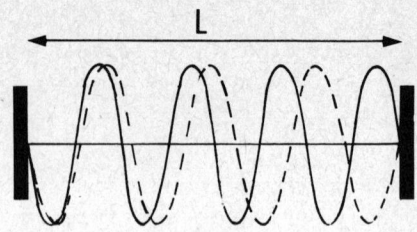

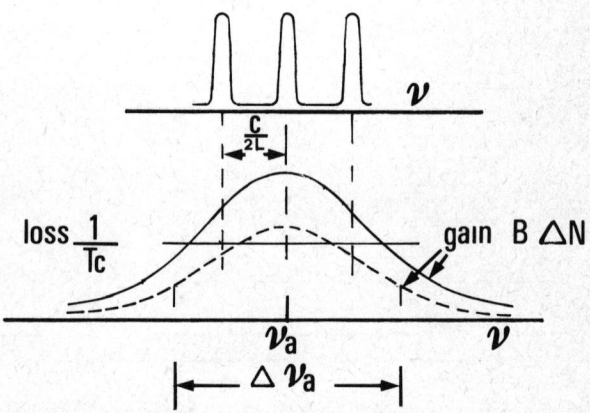

Fig. 6

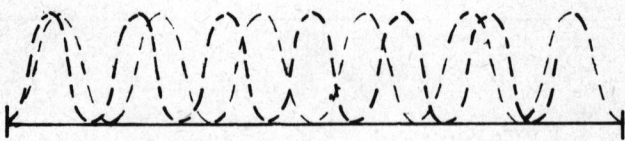

Fig. 7

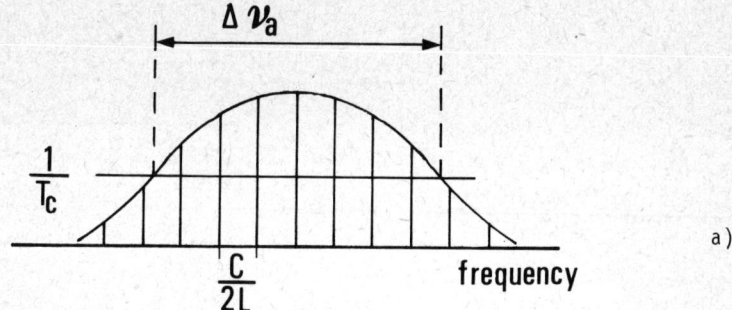

a)

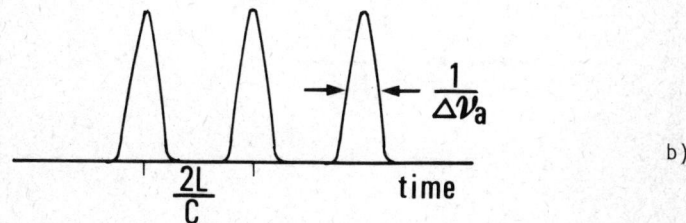

b)

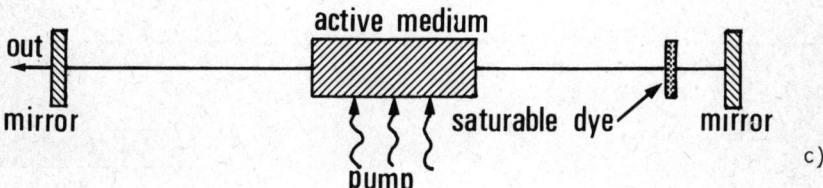

c)

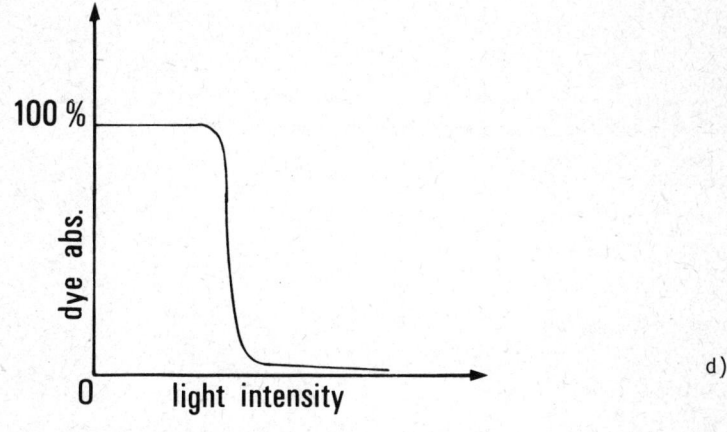

d)

Fig. 8

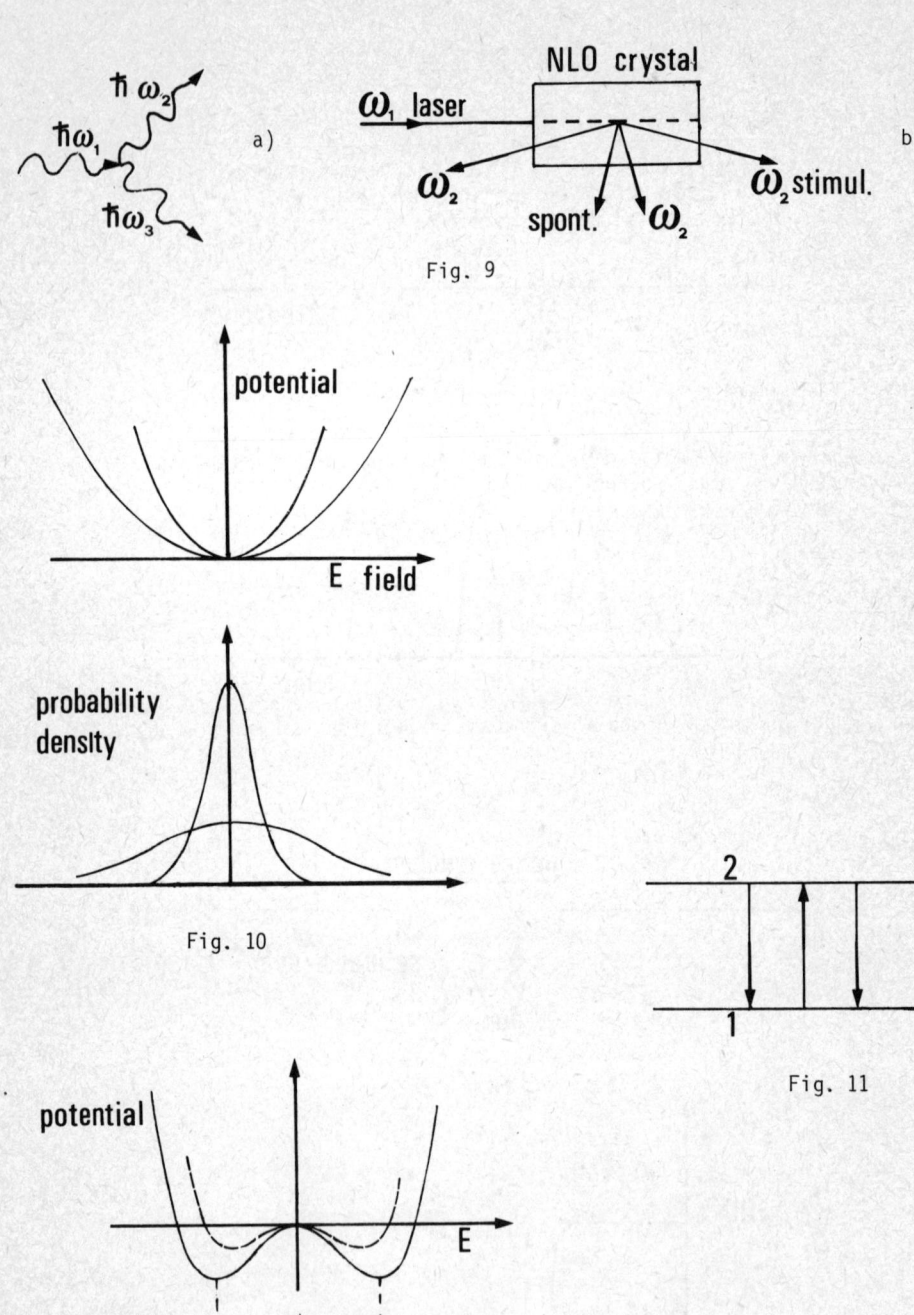

Fig. 9

Fig. 10

Fig. 11

Fig. 12

LASERS AS STABLE FREQUENCY SOURCES

P. Cérez
Laboratoire de l'Horloge Atomique
Equipe de Recherche du CNRS, associée à l'Université Paris-Sud
Bât. 221 - Université Paris-Sud
91405 Orsay - France

Abstract

The development of low noise gas lasers and the advent of Doppler free spectroscopy techniques has opened up the possibility of building frequency and length standards in the optical domain. The problems encountered, the present state of the art and the application field of stabilized lasers are briefly reviewed.

I. Generalities

The oscillation frequency ν of a single mode free running gas laser is primarily determined by the length L of its resonant cavity, according to the relation $\nu = \frac{nc}{2L}$, when n is an integer, and C is the speed of light. Such a laser when placed in good environmental conditions can exhibit a very high short term frequency stability [1]. However, the oscillation frequency can lie anywhere within the amplification bandwidth of the active medium (Fig.1) and

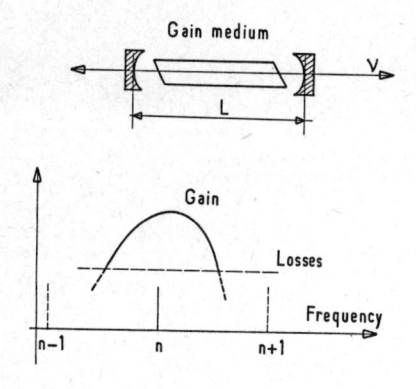

Fig.1 Scheme of a single mode free running laser

no significant accuracy can be defined for this device. To obtain a frequency standard, it is therefore necessary to lock the output frequency ν to some absolute frequency reference ν_{ref}, the best of which is the center of a suitable atomic or molecular absorption line. It is customary to characterize a stabilized oscillator by its :
- <u>stability</u> (the degree to which it will produce the same frequency over a period of time in continuous operation)
- <u>reproducibility</u> (the degree to which it will produce the same frequency from unit to unit, of from one run to another)
- <u>accuracy</u> (the degree to which the difference between the delivered frequency and the frequency of the corresponding unperturbed reference line can be evaluated)

Stability and reproducibility call for the narrowest possible linewidth $\Delta \nu$ of the reference line compatible with other constraints such as signal to noise ratio.

The absorption linewidth in low pressure molecular gas of interest here is essentially determined by the Doppler effect due to molecular thermal motion. In the optical domain, $\frac{\Delta \nu}{\nu_{ref}} \simeq 10^{-6}$ which is too large to obtain any interesting figures for the metrological qualities listed above.

It is thus necessary to have a means of observing absorption lines without any Doppler broadening. Three techniques have been investigated up to now for this purpose :
- saturated absorption (S.A.) [2]
- use of molecular beams excited by a light beam perpendicular to it [3]
- two photons absorption [4].

All three techniques lead to absorption lines with a relative linewidth $\sim 10^{-9}$ suitable for frequency stabilization. However, saturated absorption has been the most studied and we will mainly discuss this case.

2. The Saturated Absorption Technique

We assume that a Doppler-broadened transition between two levels a, b ($E_a - E_b = h \nu_{ref}$) is submitted to a monochromatic standing light wave (Fig.2).

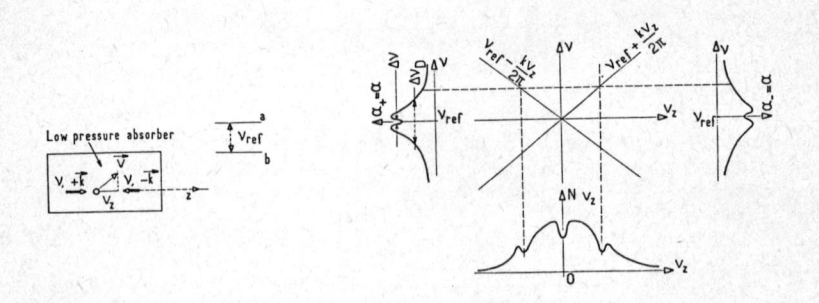

Fig.2 The saturated absorption technique in a strong monochromatic standing light wave, and the shape of the absorption coefficient of the medium for light waves propagating along the + z and - z directions.

The field interacts, resonantly with molecules whose velocity satisfy the condition

$$\nu = \nu_{ref} \pm \frac{k\,v_z}{2\pi} \qquad (1)$$

where ν is the frequency of the laser field, $\pm v_z$ the projection of the velocity of the molecule on the direction of propagation of each of the travelling waves forming the standing wave. It the amplitude of the field is sufficient to change the level population (to saturate), two holes are then burned in the population difference distribution $N(v_z)$. When the field frequency is tuned, the holes overlap. Both travelling waves change the population difference with $kv_z = 0$. The degree of saturation is twice as large in this case. The dip which appears in the center of the Doppler absorption contour $\alpha_+ = \alpha_- = \alpha$ is called the saturated absorption dip. The ratio of its width compared to the Doppler width is typically 10^{-2} or 10^{-3}.

The saturated absorption signal can be observed directly when the absorber is outside the laser source (Fig.3).

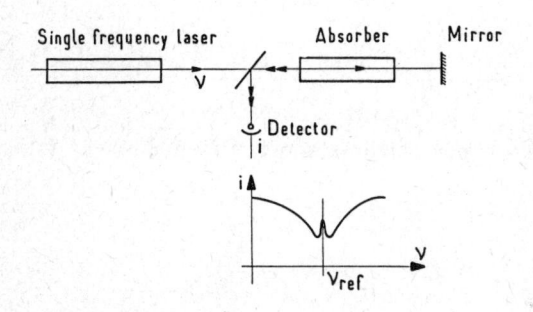

Fig.3 External cell saturated absorption set-up

The observation of narrow resonances of saturated absorption can also be made when the absorber is placed inside the laser resonator. When such a situation occurs, it is obvious that the amplifying medium must satisfy the following conditions :
1) the absorption line must coincide with the gain line of the active medium or lie inside the gain profile.
2) the laser emission intensity must be sufficient to saturate the absorber.

Figure 4 shows that the dip in the absorption line leads to a maximum of transmission. The "inverted Lambdip" which appears on the laser output power profile at $\nu = \nu_{ref}$ is suitable for frequency stabilization.

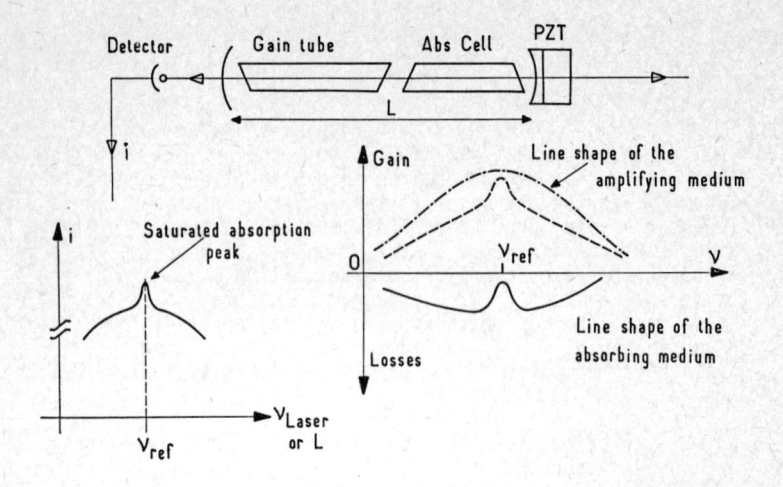

Fig.4 Laser with a non linearly absorbing medium inside the resonator

3. Principle of the Servo System Used to Stabilize the Laser Frequency

The experimental set-up enabling such a stabilization is depicted on Fig.5. A small frequency modulation is applied to the laser. This modulation

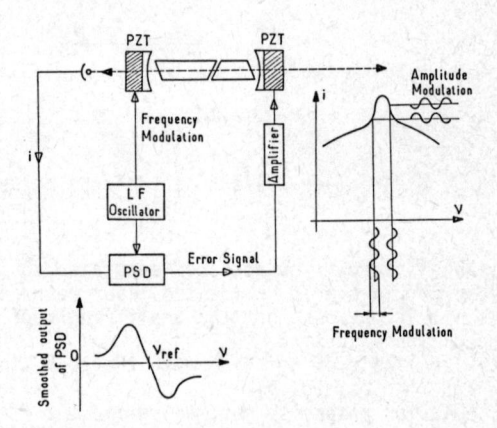

Fig.5 Principle of the servo system used to stabilize the laser frequency.

results in an intensity modulation of the laser output power. The intensity modulation is fed in a phase sensitive detector and suitably smoothed. A D C error signal is obtained. A servo system controls the laser length so as to maintain zero PSD voltage. The laser frequency is continually referred to the centre frequency of the saturated peak and perturbations of the optical frequency will automatically be compensated for.

4. Performances

Several practical optical frequency standards based on this scheme have been studied. Up to now, they use low noise single mode gas lasers and non linear absorber molecules whose absorption frequency coincides accidentally with the laser gain line. They are :
  - the infrared He-Ne laser stabilized to a saturated absorption line of CH_4 at 3.39 μm [5],
  - the visible He-Ne laser stabilized to S.A.L. of iodine at 0.612 μm and 0.633 μm [6],
  - the visible Ar^+ laser stabilized to S.A.L. of iodine at 0.5145 μm [7],
  - the infrared CO_2 laser stabilized to S.A.L. of CO_2, SF_6, OsO_4 at 10.6 μm [8].

4.1. Frequency Stability

Frequency stability as a function of the observation time has been determined for these standards in different laboratories using a beat frequency method between two stabilized lasers operating under similar conditions. Fig.6 is a plot of the results in the form of Allan variance curves. The long term

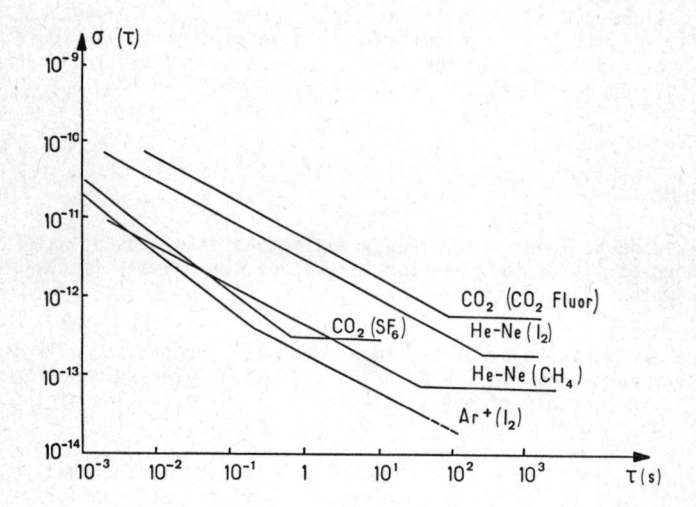

Fig.6 Relative frequency stability curves as a function of the observation time τ.

frequency stability is reaching parts in 10^{13} for SF_6 [9] and Iodine [10], and even parts in 10^{14} for CH_4 [11].

4.2. Reproducibility

The main effects which limit the reproducibility (and also accuracy) of frequency stabilized lasers are the following :
. Some are clearly of technological origin, such as electronic offsets in the servo system or modulation distorsions [12].
. Others have a physical origin depending on the system under considerations such as frequency shifts due to molecular collisions (Iodine) or to an unresolved hyperfine structure (CH_4) [13].

The present reproducibility of these optical frequency standards is $\simeq 10^{-11}$. Let us mention that it does not reach the fundamental limitation ($< 10^{-12}$) which are linked to the molecular photon interaction. With the saturated absorption technique, these fundamental limitations are [14] :
- the residual first order Doppler effect not completely eliminated by the saturated absorption technique [15],
- the second order Doppler effect [16],
- the recoil effect which gives rise to a splitting of the saturated absorption lines into two close peaks unresolved in conventional devices [17].

4.3. Accuracy

At the present time, the accuracy of small Iodine conventional saturated absorption devices may be estimated at about a few 10^{-11}. In the case of small CH_4 devices, it is even unsuitable to give a significant figure because the emitted frequency does not correspond to a given transition between two CH_4 energy levels.

5. Applications

5.1. Metrology

The stabilized He-Ne lasers are already applied as standards of wavelength. Their reproducibility is far superior to that of the primary length standard, the Krypton lamp.

The absolute vacuum wavelength of the Methane and Iodine stabilized He-Ne lasers have been measured by direct interferometric comparison with the [86]Kr primary standard line [18]. A by product of these precise measurements has reveal an uncertainty of $\pm 4 \times 10^{-9}$ in the 1960 definition of the meter.

By measuring both the frequency and the wavelength of the stabilized He-Ne laser at 3.39 µm [19], one obtains an experimental value for the speed of light recommended to be $C = \lambda f = 299\ 792\ 458$ m/s in 1973 by CCDM.

The main source of error is due to the uncertainty in the definition of the meter which, when reported on the speed of light, gives an error of ± 1.2 m/s. This value is 100 fold smaller than the former determination of C. One can predict that the speed of light and stabilized lasers will play a central role in length metrology. May be they will contribute to a new definition for the meter.

5.2. Geophysics

Geophysical research uses stabilized lasers as light sources for strain meters where the movement of rocks over long periods are studied [20]. In this experiment, a slave He-Ne laser is frequency locked to the transmission maxima of a long Fabry Perot interferometer. The frequency of the laser is therefore related to the length

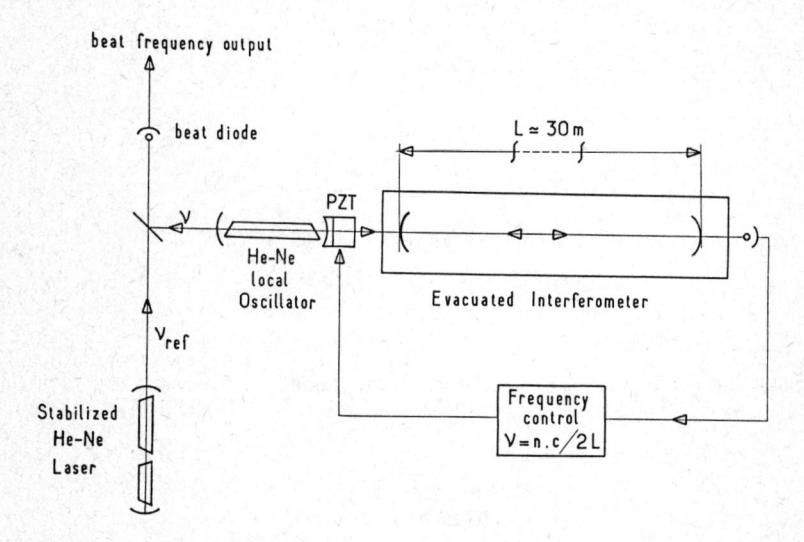

Fig.7 A Strain Meter Design

of the interferometer by $f = n\frac{c}{2L}$ where n is an integer, C is the velocity of light and L is the length of the interferometer. Then $\frac{\Delta f}{f} = -\frac{\Delta L}{L}$. This laser is compared with a reference stabilized He-Ne laser and the resulting beat frequency variations are a direct measure of the earth strain variations. A beat frequency shift of 10 MHz between two visible He-Ne lasers in an interferometer 30 m long correspond to a strain of 2.10^{-8}. This apparatus has been used to detect earthquakes [21].

Figure 8 illustrates the sensitivity of the method. We can see the initial arrival of a burst of seismic energy and subsequent arrivals after the pulse has travelled around the surface of the earth. Pulses up to the fifth transit are visible on the record.

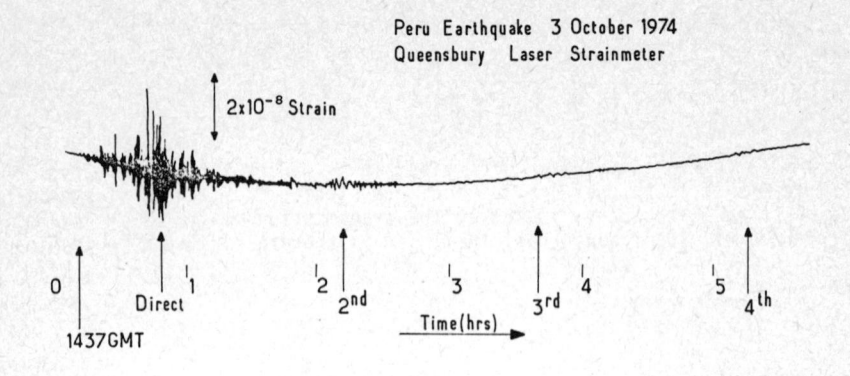

Fig.8 Record of the earth strain variations during an earthquake. From [21].

5.3. Gravity measurements

Exact knowledge of gravitational acceleration is important for many measurements (barometry, geological and mining prospection, determination of daily and secular g variations.

The principle of the absolute measurement of g is the following : we use the fundamental law of dynamics $e = \frac{1}{2} g t^2$ which applies for a body falling in vacuum. e represents the distance, while t is the duration of the fall. The distance measured by a Michelson interferometer is expressed by the number of interference fringes of a laser beam reflecting upon an optical corner which constitutes the moving body. The time is measured by mean of a high stability clock (Rb clock). The accuracy of g measurement is 10^{-9} [22].

5.4. Relativity tests

New versions of the fundamental tests of relativity based on the Michelson-Morley [23] and Kennedy Thorndike [24] experiments can be realized using frequency metrology with stabilized lasers instead of interferometry. A recent version of the Michelson-Morley experiment demonstrates a gain in sensitivity by a factor of 12 000 [25].

5.5. Frequency stabilized CW dye lasers for spectroscopic applications

The tunable CW dye laser has become an increasingly important laboratory source of light for spectroscopic and even metrological purposes. Sources suitable for sub-Doppler atomic spectroscopy where one is dealing with linewidths of $\sim$ 10 MHz are commercially availables. But the much narrower and stable dye source needed for molecular high resolution spectroscopy requires many man hours of technology [26]. The spectrum of the frequency fluctuations of a dye laser turns out to be essentially conditionned by the perturbations occuring in the jet stream region. The low frequency fluctuations (0 - 1 KHz) can be ascribed to thermal drifts, drafts, jet stream thickness fluctuations due to vibrations in the dye circulator. They result in a frequency jitter of a few MHz and a drift of $\sim$ 1 MHz/minute, provided elementary precautions are taken for vibrations and acoustical isolation. In an intermediate frequency range (tens of KHz) the mechanical resonances of the nozzle can induce a line broadening of 150 KHz. We have also high frequency noise which we think to be due to surface waves in the dye stream.

In order to obtain a high short and long term frequency stability together with some tunability of the laser, it is necessary to use 3 successives servo loop whose principle is depicted in Fig.9.

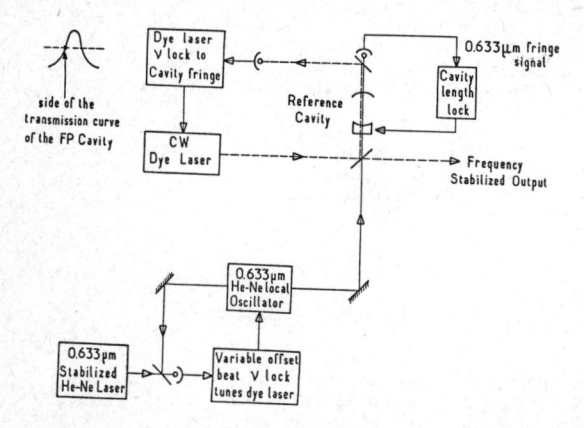

Fig.9 Experimental set-up used to stabilize the frequency of a CW dye laser

The first fast loop acting on the cavity output mirror PZT locks the dye laser frequency to the side of the transmission curve of an external reference cavity, thereby contributing to the short term frequency stability of the dye laser. To achieve good long term frequency stability of the laser, the length of the reference external cavity is locked to the emission line of a single frequency He-Ne transfer oscillator. The frequency of the local oscillator is stabilized by offset locking it to a iodine stabilized He-Ne laser using standard techniques.

Such CW dye lasers exhibit linewidths of a few tens of KHz whereas the long term frequency stability is reaching a few parts in 10^{11} for observation times longer than 100 s.

5.6. Lambdameter

The recent development of high resolution laser spectroscopy has created, in many laboratories, the need for accurate optical wavemeters. For instance, the study of heavy molecules or of atomic Rydberg states requires a fast and reliable way to measure the wavelength of the laser with a precision better than a tenth of a wavenumber (or one Doppler width). In two photon spectroscopy where very narrow ($<< 10^{-3}$ cm^{-1}), but sometimes very weak signals are to be expected, one eventually wishes an accuracy of the order of a milliwavenumber, which is, by far, not achievable by any monochormator.

We have realized a highly performant and easy to operate lambdameter. The basic set-up is described in Fig.10 [27]. Fringes of an unknown and a reference wavelengths are simultaneously generated through the motion af a moving double corner cube reflector in a Michelson type interferometer. An electronic device compare the two wavelengths to within one hundredth of a fringes and numerically display the result.

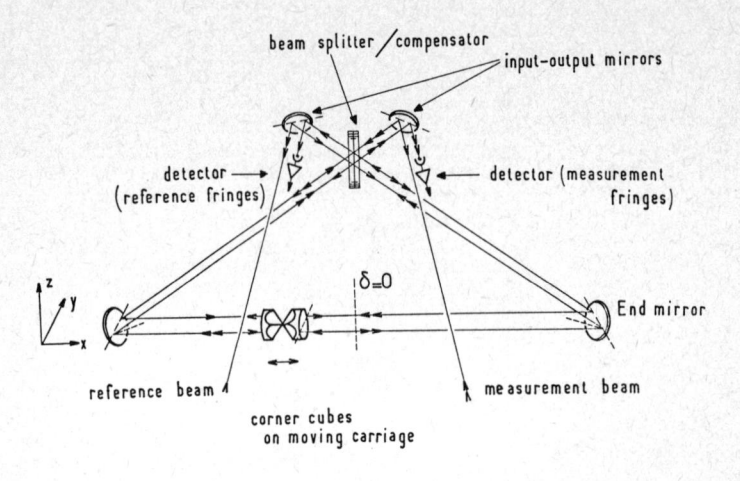

Fig.10 Design of an improved lambdameter

Most of the systematic errors are eliminated by the total superposition of the optical paths of the reference and the measured beams ensuring a high potential accuracy. As an illustration we have measured with this device the wavelength of the $\lambda = 612$ nm radiation generated by a He-Ne laser stabilized on a saturated absorption peak of iodine. The results exhibit a statistical dispersion of $\pm 6 \; 10^{-9}$ about a mean value which agrees to within 2×10^{-9} with a more exact result obtained at BIPM [28].

Conclusion

The developments in stabilized laser physics over the past few years have led to great improvements in frequency stability and reproducibility. Important applications in metrology, geophysics, and super high resolution spectroscopy offer to physics new tools which hold the promise of exciting new observational techniques and measurement capabilities.

References

1. T.S. Jaseja, A. Javan. C.H. Townes, Phys. Rev. Lett. 10,165,(1963)
2. P.H. Lee, M.L. Skolnick, Appl. Phys. Lett. 10,303,(1967)
3. S. Ezekiel and R. Weiss, Phys. Rev. Lett. 20,91 (1968)
4. L.S. Vasilenko, V.P. Chebotaev, A.S. Shishaev, JETP Lett. 12,113,(1970)
   B. Cagnac, G. Grynberg, F. Biraben, J. Phys. 34,845,(1973)
5. R.L. Barger and J.L. Hall, Phys. Rev. Lett. 22, 4-8,(1969
   A. Brillet, P. Cérez, H. Clergeot, IEEE Journal of Quantum Electronics, QE-10, 526-528,(1974)
6. G.R. Hanes, C.E. Dahlstrom, Appl. Phys. Lett. 14, 362-364, (1969)
   P. Cérez, A. Brillet, F. Hartmann, IEEE Trans. on Instr. and Meas. IM-23,526, (1974)
   A.J. Wallard, J. Phys. E 5,926,(1972)
   J. Helmcke, F. Bayer-Helms, IEEE Trans. on Instr. and Meas IM-23, 529, (1974)
   J.M. Chartier, J. Helmcke, A.J. Wallard, IEEE Trans. on Instr. and Meas. IM-25, 450, (1976)
   P. Cérez, S.J. Bennett, IEEE Trans. on Instr. and Meas. IM-27, n° 4, 396, (1978)
   P. Cérez, S.J. Bennett, Applied Optics, 18, 1079-1083, (1979)

7. F. Spieweck, IEEE Trans. on Instr. and Meas. IM-27,398-400,(1978)
   G. Camy, B. Decomps, J.L. Gardinat, C.J. Bordé, Metrologia 13,145-148,(1977)
8. C. Freed, O'Donnell, R.G. Ross, IEEE Trans. on Instr. and Meas. IM-25,431-437, (1976)
   J.J. Jimenez, F.R. Petersen, P. Plainchamp, C. Sallot, X. Drago, Bull. d'information du BNM 9,19-27,(1978)
9. A. Clairon, L. Henry, C.R. Acad. Sci. Paris B 279,419,(1974)
   V.M. Gusev, O.N. Kompanets, A.R. Kukudzhanov, V.S. Letokhov, E.L. Mikhailov, Sov. J. Quant. Elec. 4,1370,(1975)
10. P. Cérez, S.J. Bennett, C. Audoin, C.R. Acad. Sci. Paris B53, t.286,(1978)
11. J.L. Hall, S.J. Smith and G.K. Walters, Ed. Plenum, 615, (1972)
12. P. Cérez, A. Brillet, Metrologia 13,29,(1977)
13. J.L. Hall, C.J. Bordé, Phys. Rev. Lett. 30,1101,(1973)
14. F. Hartmann, 18th URSI General Assembly, Lima Pérou - August 1975
15. C.J. Bordé, J.L. Hall, C.V. Kunasz, D.G. Hummer, Phys. Rev. 14,236,(1976)
16. S.N. Bagaev, V.P. Chebotaev, JETP Lett. 16,433,(1972)
17. A.P. Kol'Chenko, S.G. Rautian, R.I. Sokolovskii, JETP 28,986,(1969)
18. A.J. Wallard, J.M. Chartier, J. Hamon, Metrologia 11,89-95,(1975)
    H.P. Layer, R.D. Deslattes, W.G. Schweitzer, Applied Optics 15,734,(1976)
19. K.M. Evenson, J.S. Wells, F.R. Petersen, B.L. Danielson, G.W. Day, R.L. Barger, J.L.Hall, Phys. Rev. Lett. 29,1346,(1972)
20. J. Levine, J.L. Hall, J. Geophys. Res. 77,2595,(1972)
21. Goulty, King, Wallard, Geophys. J. Roy. Astr. Soc. 39,269-282,(1974)
22. A. Sakuma, NBS Special publications 343,447-456,(1971)
23. A.A. Michelson, E.W. Morley, Am. J. Sci. 34,333,(1887)
24. R.J. Kennedy, E.M. Thorndike, Phys. Rev. 42,400,(1932)
25. A. Brillet, J.L. Hall, Phys. Rev. Lett. 42,549,(1979)
26. R.L. Barger, J.B. West, T.C. English, Appl. Phys. Lett. 27,31-33,(1975)
    J.L. Hall, S.A. Lee, Springer 1976, p. 361-366, Editor Mooradian
    C.N. Man, P. Cérez, A. Brillet, F. Hartmann, J. de Physique Lettres 38,287,(1977)
27. J.L. Hall, S.A. Lee, Appl. Phys. Lett. 29,367,(1976)
    J. Cachenaut, C. Man, P. Cérez, A. Brillet, F. Stoeckel, A. Jourdan, F. Hartmann, Rev. de Physique Appliquée 14,685,(1979)
28. S.J. Bennett, P. Cérez, J. Hamon, A. Chartier, To be published in the July 1979 issue of Metrologia .

ELECTRON SOURCES

G. FONTAINE

Département de Physique des Matériaux
(associé au C.N.R.S.)
Université LYON I   69621 - VILLEURBANNE   (France)

1 - NUMBER OF MODES IN AN ELECTRON BEAM, COHERENCE

If an electron beam is limited by an aperture of size dS perpendicular to the mean ray and if v is the speed of the electrons, the wave front sweeps over the volume $V = vdSdt$ in time dt. The number g of modes available in the phase volume $d\omega p^2 dpV$ ($E = p^2/2m$) results from Fermi-Dirac statistics (a factor 2 for the spin) :

$$g = 2 \frac{p^2 dp d\omega v dS dt}{h^3} = 2 \frac{d\omega dS}{\lambda^2} \frac{dt dE}{h} = \frac{4m}{h^3} dS d\omega E dE dt \tag{1}$$

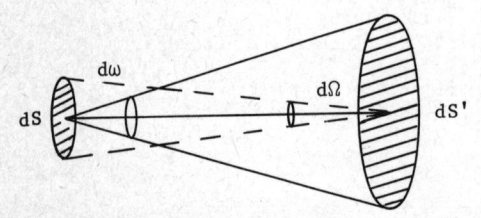

Fig.1 Extent of an electron beam

By analogy with optics[1] we can write the second equality of equation (1) as :

$$g = 2 \frac{d^2 U}{d^2 U_1} \frac{dt}{\tau} \qquad d^2 U = d\omega dS = d\Omega dS'$$

$d^2 U_1 = \lambda^2$ and $\tau = h/dE$ define respectively the extent of the spatial and temporal coherence. Typically $E = 100$ keV in electron microscopy ($\lambda = 0.037$ Å) and $dE \simeq 1$ eV which gives $d^2 U_1 = 1.37 \, 10^{-3}$ Å^2 and $\tau = 4.14 \, 10^{-15}$ s. Total spatial coherence can only be realized in a beam of circular cross section $dS = \pi d_\perp^2/4$ and half aperture α ($d\omega = \pi \alpha^2$) if $d^2 U_1 / d^2 U \gg 1$, which means :

$$\alpha \ll \frac{2}{\pi} \frac{\lambda}{d_\perp} \quad (\alpha < 10^{-4} \text{ rad for } d_\perp = 240 \text{ Å})$$

Associated with τ is the longitudinal coherence length $d_{/\!/} = v\tau = 6800$ Å with $E = 100$ keV and $dE = 1$ eV.

The third equality in equation (1) determines in fact the maximum "theoretical" intensity of the beam ($i_{max} = eg/dt$) or the theoretical limit in brightness (luminance)

$$B_{max} = \frac{i_{max}}{dSd\omega} = \frac{4em}{h^3} E.dE = 5.2 \; 10^9 \; VdV \; (A \; cm^{-2} \; sr^{-1})$$

if V is expressed in volts. This is an enormous value ($5.2 \; 10^{14}$ with V = 100 kV and dV = 1 V) as compared to the actual best experimental realisations using field emission sources ($< 10^{10}$ A cm^{-2} sr^{-1}). This means that the occupation numbers of the available phase cells are in fact very small (one in 10^5 or so in the extreme experiments) hence as noticed many years ago by D. Gabor[2] the "almost complete identity of light optics and electron optics, in spite of the extreme difference between Einstein-Bose and Fermi-Dirac statistics".

2 - BRIGHTNESS

Whatever the emission process at a metallic cathode (as thermionic, field emission T.F. emission, ...) it is easy to compute the number of electrons $G(E_n, E_t) \, dE_n \, dE_t$ emitted by unit area of a plane cathode by unit time whose normal (to the cathode) and transverse energies are in the range dE_n and dE_t respectively. From the distribution G it is possible to compute:

- mean energies $<E_n>$ and $<E_t>$

- density of emitted current $J_c = \iint G(E_n, E_t) \, dE_n \, dE_t$

- axial brightness of the beam $B_c = \lim \frac{dJc}{d\omega}$ at the cathode

$$\begin{cases} dS \to 0 \\ d\omega \to 0 \end{cases}$$

In the limit $dS, d\omega \to 0$, it is easy to see that $d\omega = \frac{\pi dE_t}{E_n}$ and $E_t \to 0$ so that:

$$B_c = \frac{e}{\pi} \int G(E_n, 0) \, E_n dE_n = \int B_c(E_n) \, dE_n$$

where $B_c(E_n)$ represents the contribution to the cathode brightness of electrons whose normal energies are in the range dE_n.

Consider now a "cross-over" where at some distance M from the cathode the beam converges due to some lens effects (Figure 2).

The electrons are coming from a zone dS_c, $d\omega_c$ at the cathode. From Liouville's theorem:

$$(dx_1 \, dx_2 \, dx_3 \, dp_1 \, dp_2 \, dp_3)_M = (dx_1 \, dx_2 \, dx_3 \, dp_1 \, dp_2 \, dp_3)_C$$

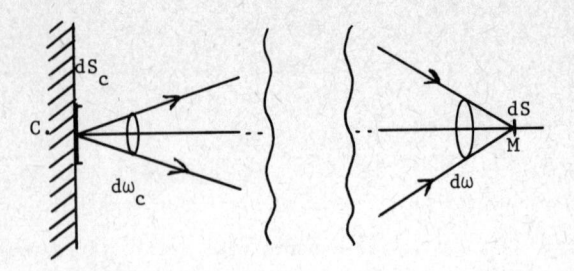

Fig. 2 Relation between cathode C and cross-over M

In an electron gun the beam is accelerated up to a potential V at M (relative to the cathode) so that $(E_n)_M = E_n + eV$ or $p_3 \, dp_3$ = constant. We also have $dx_3 \sim p_3 dt$ and $d\omega = dp_1 \, dp_2/p_3^2$ so that Liouville's theorem gives :

$$dS_M \, d\omega_M \, (E_n)_M = dS_M \, d\omega_M \, (E_n + eV) = dS_c \, d\omega_c \, E_n$$

Hence the differential brightness at M :

$$\frac{B_A(E_n)}{B_c(E_n)} = \frac{(di/dSd\omega)_M}{(di/dSd\omega)_C} = \frac{E_n + eV}{E_n}$$

A result first deduced by Gabor[3]. Finally we get the total brightness B at A[4]

$$B = \int B_A(E_n) \, dE_n = \frac{e}{\pi} \int (E_n + eV) \, G(E_n, 0) \, dE_n \qquad (2)$$

As soon as the beam enters an equipotential space after the gun, V = constant and the brightness of the beam is conserved ($dSd\omega$ = constant).

3 - THERMIONIC EMISSION

Thermionic emission is the most extensively used in electron microscopy. It's characteristics are well known and can be deduced easily from the preceding discussion.

Since the work function ϕ (4.5 volts for W) $\gg kT$, it is easy to get :

$$G(E_n, E_t) = \frac{4\pi}{h^3} m \exp - \frac{(\phi + E_n + E_t)}{kT}$$

$$<E_n> = <E_t> = kT$$

$$J_c = \frac{4\pi em}{h^3} (kT)^2 \, e^{-\frac{\phi}{kT}} \quad \text{(for W : 1 A cm}^{-2}\text{ at 2650 K)}$$

$$B = \frac{J_c}{\pi} (1 + \frac{eV}{kT}) \quad (3)$$

a result first deduced by Langmuir[5].

It is also easy to determine the normal energy distribution $P(E_n)$, $P(E_n) dE_n$ giving the fraction of electrons whose normal energy is the range dE_n.

$$P(E_n) = \frac{\int G(E_n E_t) dE_t}{\int G(E_n E_t) dE_n dE_t} = \frac{1}{kT} e^{-\frac{E_n}{kT}}$$

The total energy distribution $P(E) dE$ whose total energy is in the range dE :

$$P(E) = \frac{E}{(kT)^2} e^{-\frac{E}{kT}} \quad (4)$$

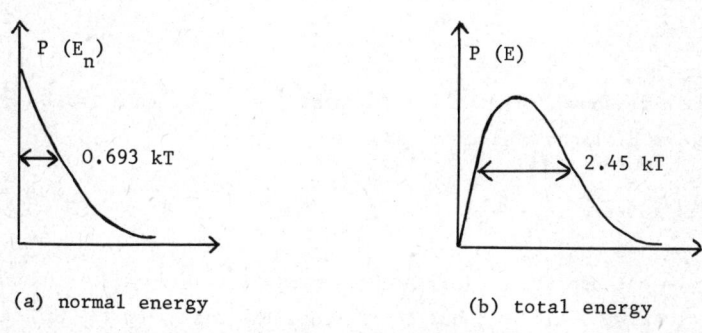

(a) normal energy

(b) total energy

Fig. 3 Energy distributions for thermionic emission

The half width of $P(E)$ is equal to 2.45 kT. In an electron microscope the thermionic cathode is usually made from tungsten wire bent into the shape of a hairpin and emitting only at the tip. From these cathodes heated to 2650 K current densities $J_c = 1$ A cm^{-2} can be drawn from an area of 0.1 mm diameter. The emitted current is controlled by a grid, namely the Wehnelt cylinder surrounding the cathode and biased negatively some - 200, - 300 V relative to the cathode. The electron beam is then accelerated by the anode to a potential V relative to the cathode.

Electrons proceed approximatively parallel to the axis, spreading due to the tangential componant of emission energies. As the Wehnelt has a repulsive action on the emitted electrons, the first part of the gun has a strong converging action. The last part near the anode and its aperture acts as a diverging lens (Figure 4).

Each point of emission gives electron bundles whose peripheral rays leave the cathode surface at grazing angles. The different bundles converge as shown in the figure into a disc of least confusion which is called the cross-over. Each point of the cross-over emits a pencil of rays so that it is the cross-over itself which must be considered as the electron source.

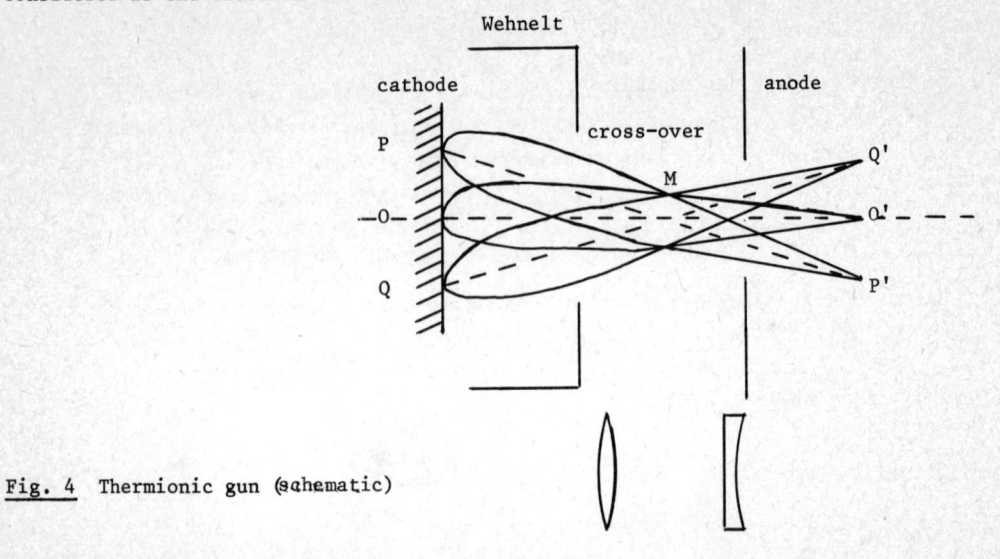

Fig. 4 Thermionic gun (schematic)

It is possible to show[4] that the brightness decreases in the cross-over exponentially with increasing distance r from the axis as :

$$B(r) = B \exp\left(-\frac{r^2}{M^2 a^2}\right)$$

where a is the cathode tip radius, M the gun magnification and B the axial brightness given by (3), $B \sim eVJ_c/\pi kT$ in usual conditions. The radius of the cross-over (Ma) is usually around 25 μm. The maximal axial brightness is around 10^5 A cm^{-2} sr^{-1} at an accelerating voltage V = 100 kV. The average directional beam intensity defined as di/dω is around 10^{-2} A sr^{-1} in these conditions.

4 - FIELD AND THERMAL FIELD EMISSION (T.F.)

High electric fields applied to the cathode allow electrons to tunnel directly from energies near the Fermi Level into the vacuum. Typically the field F must be as high as 4.10^7 V cm^{-1} which is only possible with a small tip radius (around 1000 Å). The emissive power J_c is now in the range 10^4 A cm^{-2} and is very sensitive to field fluctuations. Typically a 1 % increase in F produces a 10 % increase of J_c and a 1 % increase in the work function φ produces a 15 % decrease of J_c. These fluctuations result from adsorption/desorption processes from residual gases and desorption of the extracting anode under electron bombardment. For this reason the tip is very often heated to a moderate temperature (1300 K) to stabilise these effects and regulate the stability of the emission. We now have[6] :

$$G(E_n E_t) = \frac{4\pi m}{h^3} \frac{\exp\{-c + (E_n - E_f)/d\}}{1 + \exp\{(E_n + E_t - E_f)/kT\}}$$

That is to say the Fermi-Dirac distribution multiplied by the probability of penetration of the barrier (the numerator). The energy d corresponds as we shall see to the mean transverse energy : $d = \langle E_t \rangle$; d varies varies with the field F as :

$$d(eV) = 9.76 \cdot 10^{-9} F(volts)/\phi^{1/2}$$

typically $d = 0.25$ eV for a field $F = 5 \cdot 10^7$ V cm^{-1}

$$\langle E_t \rangle = d \tag{5}$$

$$\langle E_n \rangle = E_F - d - \pi kT \cot g \left(\frac{\pi kT}{d}\right)$$

$$J_c = \frac{4\pi em}{h^3} d^2 e^{-c} \frac{\frac{\pi kT}{d}}{\sin\left(\frac{\pi kT}{d}\right)} = J_{co} \frac{\frac{\pi kT}{d}}{\sin\left(\frac{\pi kT}{d}\right)} \tag{6}$$

(for W, $J_{co} = 10^4$ A cm^{-2} for $F = 4 \cdot 10^7$ V cm^{-1} and 10^5 A cm^{-2} for $F = 4.5 \cdot 10^7$ V.cm^{-1})

$$B_c = \frac{J_c}{\pi} \left\{ \frac{E_F}{d} - \frac{\pi kT}{d} \cot g\left(\frac{\pi kT}{d}\right) \right\}$$

$$B = \frac{J_c}{\pi} \frac{eV}{d} + B_c \simeq \frac{J_c}{\pi} \frac{eV}{d} \text{ in practical situations.} \tag{7}$$

At 0°K the energy distributions take a simple form :

$$P(E_n) = \frac{(E_F - E_n)}{d^2} e^{(E_n - E_F)/d} \quad \text{and} \quad P(E) = \frac{1}{d} e^{(E - E_F)/d} \tag{8}$$

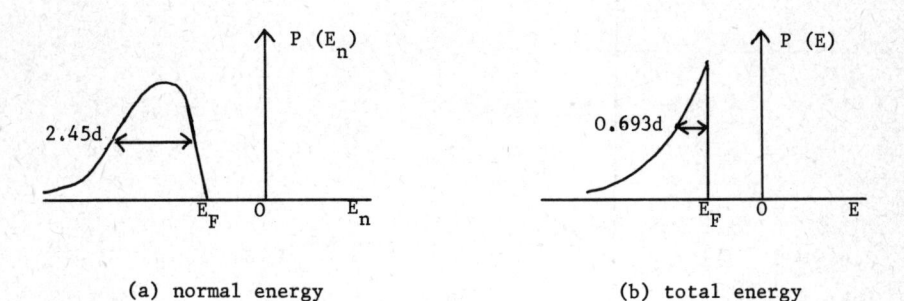

(a) normal energy

(b) total energy

Fig. 5 Energy distributions for field emission

Thus d in field emission plays the role of kT in thermionic emission. Total energy distribution P(E) in field emission has the same form and width as the normal energy distribution $P(E_n)$ in thermionic emission[7].

At a finite temperature the total current density emitted with total energy between E and E + dE may be written (J_c and J_{co} as defined in equation (6)) :

$$J_c(E) \, dE = P(E) \, J_c \, dE = \frac{J_{co}}{d} \, dE \cdot \frac{\exp(E - E_F)/d}{1 + \exp(E - E_F)/kT}$$

As compared to the total current density emitted in thermionic emission :

$$J_c^{th}(E) \, dE = \frac{4\pi em}{h^3} \cdot E \cdot dE \, \exp{-(E + \phi)/kT}$$

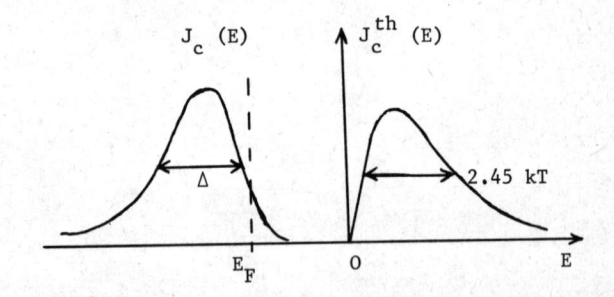

Fig. 6 Total current density at a cathode for field (left) and thermionic emission (right)

Values of the half width Δ/d are given[8] as a function of kT/d. The analytical treatment given here is exact only if $kT/d < 0.8$ which means $T < 1800$ K for W with a field $F = 4.10^7$ V cm^{-1}. The condition is thus always satisfied in practical situations.

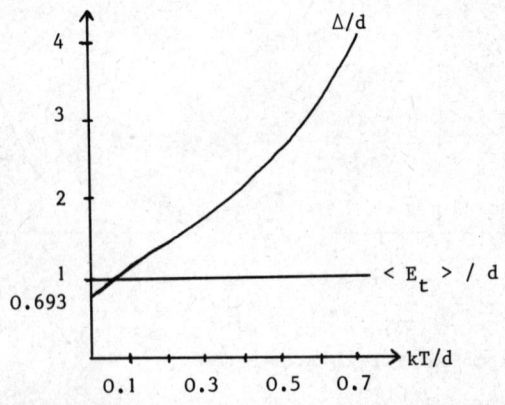

Fig. 7 Half width (Δ) and mean transverse energy ($<E_t>$ of field emission as a function of temperature

Typically with a field $F = 4 \cdot 10^7$ V cm^{-1}, $d = 0.18$ eV ; $\Delta = 0.13$ eV at 0°K ; 0.23 eV at 300 K and 0.72 eV at 1300 K a temperature often used to stabilize the emission. For comparison $2.45 \, kT = 0.59$ eV for half width of thermionic emission at 2750 K. In practice a field emission gun often works with a pointed cathode of small radius a ($\simeq$ 1000 Å) at a small distance from an extracting anode held at some positive potential $V_1 \simeq a \, F$ where F is the electric field at the cathode. The cross-over radius is[9]

$$r_c \simeq a \left(\frac{<E_t>}{V_1}\right)^{1/2} \simeq a^{1/2} \left(\frac{<E_t>}{F}\right)^{1/2} \tag{9}$$

Typical values are $<E_t> = 0.2$ eV, $F = 4.10^7$ V/cm, $a = 1000$ Å and $r_c \simeq 22$ Å. Source size is thus approximately four orders of magnitude lower than conventional thermionic one. Similarly axial brightness is approximately four orders of magnitude higher. One difficulty however is the poor directivity of field emission, 10^{-4} A.sr^{-1} at the best as compared to 10^{-2} A.sr^{-1} or so for thermionic emission so confinement processes are often used (oxygen processing[10], build-up[11], ...).

5 - ENERGY BROADENING IN ELECTRON BEAMS

Boersch[12] was the first to observe an increase in the energy spread of an electron beam at high current densities. Simultaneously, the Maxwellian energy distribution changes to an apparent Gaussian distribution. These observations have been verified repeatedly, generally with thermionic guns and recently with a field emission cathode[8]. In this later case, for instance with an emission current of 0.3 µA an energy half width of 0.24 eV was measured. The half width reaches 0.4 eV for an emission current of 4 µA (at room temperature). A theory by Loeffler[13] was based upon an analysis of electron-electron collisions at a cross-over and predicts a broadening :

$$\Delta E \propto \frac{1}{\alpha} \left(\frac{I}{r}\right)^{1/2} \tag{10}$$

where α is the angle of convergence, I the current and r the radius of the beam at the cross-over. However, it seems that such a law is not experimentally satisfied. Recently Crewe[14] proposed for ΔE a double-valued function of α with a maximum effect at some critical angle of convergence. This law has however not been compared to experimental results mainly due to difficulties in defining cross-over parameters in the electron gun. This situation is rather unfortunate since the energy spread of the illuminating beam attenuates the contrast transfer function in electron microscopy. Improvement of resolution limit can thus be reached only at low beam current not exceeding a few µA. Recently Troyon[15] measured the contrast attenuation due to the different envelope of the transfer function and deduced from these measurements the energy spread half width ΔE of a field emission electron microscope. At about 5 µA $\Delta E = 1.4 \pm 0.3$ eV and at about 100 µA $\Delta E = 3 \pm 0.6$ eV, the corresponding brightness being at 75 kV respectively 7×10^7 A cm^{-2} or $^{-1}$ and 2.5×10^8 A cm^{-2} or $^{-1}$.

References

1   G. Bruhat, A. Kastler : Thermodynamique (1962) Masson éditeur (p. 666 à 670)

2   D. Gabor : (1961) Progress in optics I, 109

3   L.M. Myers : (1939, Electron optics p. 498 (New York : Van Nostrand)

4   J. Worster : (1969), J. Phys. D. 2, 2, 457 and 889

5   D.B. Langmuir : (1937), Proc. I.R.E. 25, 977

6   R.H. Good, E.W. Müller : (1956), Handb. d. Phys., 21, 176

7   R.D. Young : (1959), Phys. Rev., 113, 110

8   K.H. Gaukler, R. Speidel, F. Vorster : (1975), Optik, 42, 391

9   J.C. Wiesner, T.E. Everhart : (1973), J. Appl. Phys., 44, 2140

10 L.H. Veneklasen, B.M. Siegel : (1972), J. Appl. Phys., 43, 1600

11 S. Ranc, M. Pitaval, G. Fontaine : (1976), Surface Science 57, 667

12 H. Boersch : (1954), Zeit. Phys. 139, 115

13 K.H. Loeffler : (1969), Z. Angew. Phys. 27, 145

14 A.V. Crewe : (1978), Optik, 50, 205

15 M. Troyon : (1979), Optik, to be published

INTENSE SOURCES OF THERMAL NEUTRONS FOR RESEARCH

J B Forsyth
Neutron Beam Research Unit
Science Research Council's Rutherford Laboratory
Chilton Didcot Oxon England

1   INTRODUCTION

This lecture will cover the production of neutrons and their subsequent moderation to epithermal and thermal energies in volumes from which they can be extracted in the form of a neutron beam. Only those devices which are capable of yielding the most intense beams will be described since it is these beams which are required for many of the most interesting experiments in physics, chemistry and biology. It is salutory to recall that the highest thermal neutron fluxes currently available are still orders of magnitude weaker than the photon fluxes produced by a modest x-ray tube, whereas the cross-sections for coherent scattering of the two radiations are roughly similar.

It is obvious from the historical survey of neutron production presented in Figure 1 that the most common technique - the steady-state fission reactor

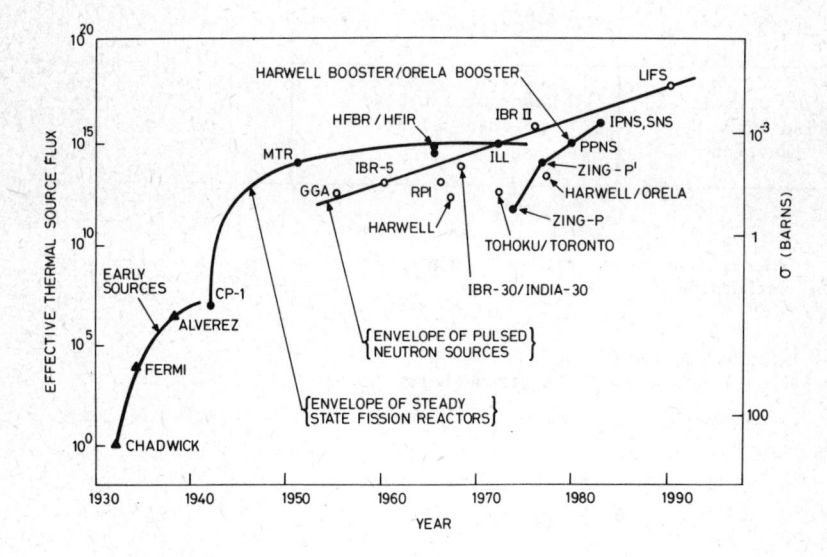

Figure 1   A historical survey of neutron production (Carpenter, 1977). The right hand scale has been added to indicate the typical scattering cross section (in barns) which can be studied with a given source flux.

– is rapidly nearing the limit of its potential performance. High flux reactors are expensive, some $10-100M, so even a linear relationship between cost and intensity would be bad enough; unfortunately, the real difficulty in achieving high reactor fluxes is more fundamental. Table 1 lists the energy deposition per neutron produced for a number of different mechanisms, including fission. Although the latter process has by no means the highest energy deposition, heat removal from the core is the biggest technological limitation to achieving continuous fluxes of thermal neutrons significantly in excess of 10^{15} n/cm^2/sec. It is therefore not surprising that in the quest for higher fluxes attention has now turned to the spallation process, since the feasibility of controlled thermonuclear reation (CTR) devices has yet to be demonstrated.

Process	Example	Energy Deposition
Fission	^{235}U (n,f)	200 MeV/n
Photonuclear	100 MeV electrons on ^{238}U (e$^-$,γ,n)	2000 MeV/n
Spallation	800 MeV protons on ^{238}U (p,n)	55 MeV/n
Deuterium, tritium controlled thermo-nuclear reaction	Laser or ion-beam imploded pellet	3 MeV/n

Table 1 Energy deposition per neutron produced for a number of different mechanisms.

Figure 2 indicates the ranges of neutron energy to be associated with the various descriptive terms such as cold, fast, epithermal etc. The relationship between a neutron's energy, E, and its wavelength, λ, or velocity, v, is given by

$$\lambda = \frac{h}{mv} = \frac{3.96}{v} = \frac{0.286}{\sqrt{E}} \text{ Å}$$

where v is expressed in km sec^{-1} and E is in eV.

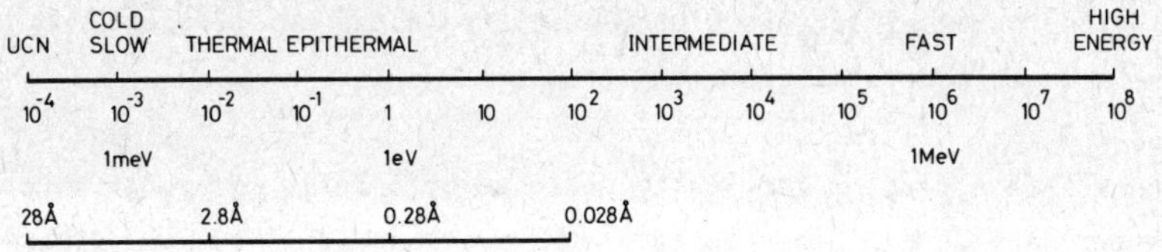

Figure 2   The ranges of neutron energy and their corresponding wavelengths.

The energy and time of flight corresponding to several different wavelengths are:

λ (Å)	Energy (meV)	(Velocity)$^{-1}$ (μs/m)
0.5	~ 327	~ 126
1.0	~ 82	~ 253
4.0	~ 5	~ 1011

Neutron scattering experiments normally require an input beam of defined energy and this is achieved by selecting the neutron wavelength by Bragg scattering from a monochromator crystal or by selecting the neutron velocity.

2   THE REACTOR

A fission reactor is a device in which the fuel contains a fissile isotope of one of the heavy elements. If sufficient of the neutrons released at fission interact with other nuclei in the fuel to induce further fission, then the reaction can be made self sustaining.

The fission cross section of the ^{235}U nucleus is much higher for low energy (meV) slow neutrons than for the high energy (MeV) fast neutrons emitted in the fission process. The fuel elements are therefore surrounded by a moderating material which slows down the energetic neutron by inelastic collisions. The energy loss per collision is largest if the mass of the moderator nuclei is the same as the mass of the neutron, so hydrogen or deuterium make the best moderators. After a number of such collisions, the energy of the neutron comes into equilibrium with the thermal energy of the moderator and the neutrons are said to be thermalised. The thermal flux per energy interval dE at energy E is given by

$$n_f(E)dE = 2n_f \left(\frac{E}{kT}\right)^2 \exp\left(\frac{-E}{kT}\right) dE$$

The peak of this Maxwellian distribution occurs in the neighbourhood of 0.03eV for a water moderator operating at about 60°C.

The whole core of a reactor is usually enclosed by a reflector whose function is to scatter back some of the fast neutrons which would otherwise lose the chance of being moderated and of inducing further fission. Beryllium, heavy water and graphite are good reflectors.

Most research reactors produce a continuous flux of neutrons whose intensity is controlled by the insertion of neutron-absorbing rods in between the array of fuel elements.

The fission process is also accompanied by the release of high energy γ-radiation and the whole assembly of core, moderator and reflector must be well shielded to reduce the escaping radiations to biologically acceptable levels. Light water and concrete containing steel or barytes are commonly used, the former to moderate the fast neutrons and the latter to attenuate the γ-radiation. Thermal neutron beams are extracted from the reactor through holes pierced in the massive shielding and penetrating to the regions of highest flux. These beams are contaminated to a greater or lesser extent with γ-rays and fast neutrons, both of which have their peak intensities at the positions of the fuel elements. The situation is worst in light-water moderated reactors, where the thermal flux is also peaked at the fuel elements and falls off rapidly with distance due to the short moderation length of fast neutrons in H_2O and also to the appreciable absorption cross section of hydrogen. There is, therefore, little alternative to arranging the beam tubes radially to the core, thereby achieving the highest fluxes of thermal neutrons, fast neutrons and γ-rays. Heavy water is less absorbing for neutrons and beam tubes may be placed tangentially to the core, thus lowering the beam contamination without a severe loss of thermal neutron flux.

The highest thermal fluxes are achieved by using an undermoderated core of restricted physical size. The thermal flux peaks in a D_2O reflector which surrounds the core and the core is moderated and cooled by a flow of D_2O under pressure. This design was first used for the Brookhaven High Flux Reactor, in which the maximum thermal flux is 7.5×10^{14} n cm^{-2}sec^{-1} (Kouts 1963). The flux at the HFR, Oak Ridge, and at the reactor of the Institut Laue-Langevin, Grenoble, is some 1.2×10^{15} n cm^{-2}sec^{-1}. The latter reactor first reached its full thermal power of 57 megawatts towards the end of 1971 and its large complement of instruments makes it one of the best research reactors in the world. The reactor is powered by a single fuel element containing 8.5 kg of uranium which is enriched to 93% ^{235}U. Figure 3 illustrates the construction of the element whose overall dimensions are 1476 mm x 418 mm in diameter.

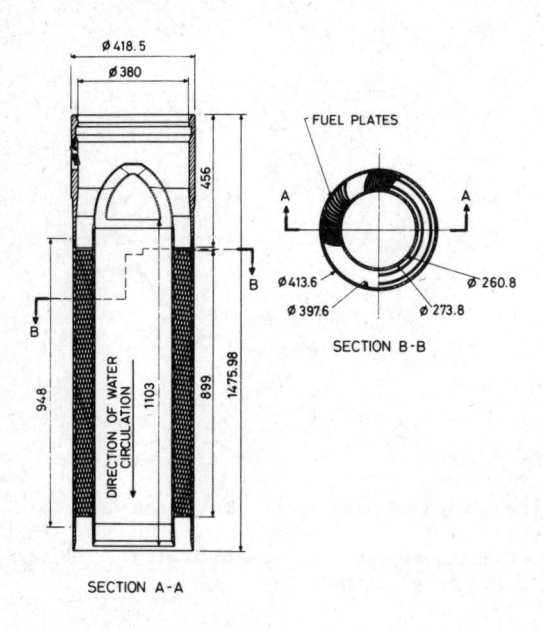

Figure 3   The fuel element of the HFR, Institut Laue-Langevin (ILL), Grenoble, France. The dimensions are in mm and the fuel plates occupy the hatched annulus in the section A-A.

Heavy water acts as coolant, moderator and reflector, and high pressure pumps force D_2O between the 280 curved fuel plates with a flow rate of 2140 m^3/hour. The D_2O then passes through heat exchangers where it loses its heat to the nearby river Drac. The normal cycle of operation is 44 days, followed by a 12 day shutdown to change the fuel element in which some 30% of the ^{235}U has undergone fission.

Figure 4 shows the arrangement of beam tubes, none of which points directly at the fuel element. The effect of the undermoderated core on the spatial distribution of the thermal flux is illustrated in Figure 5. The peak in its Maxwellian distribution is around 1.2Å, but this is modified for certain beam tubes by the inclusion of two different volumes of moderator. The hot source (10 dm^3 of graphite at 2000K) and the cold source (25 dm^3 of liquid deuterium at 25K) give enhanced neutron intensities in the wavelength ranges $0.4 < \lambda < 0.8$Å and $\lambda > 4.0$Å respectively, as shown in Figure 6.

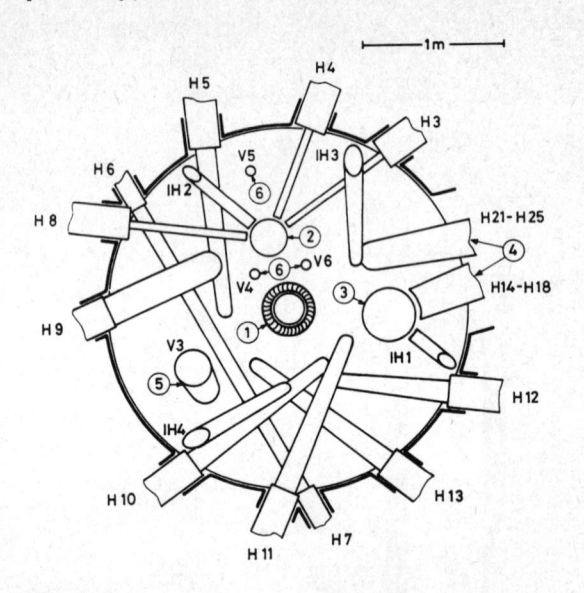

Figure 4   The arrangement of beam tubes (H) at the HFR, ILL. The positions of the core (1), hot (2) and cold source (3), the guides (4) and the vertical beam tube (5) are indicated. The vertical channels (6) are for irradiations.

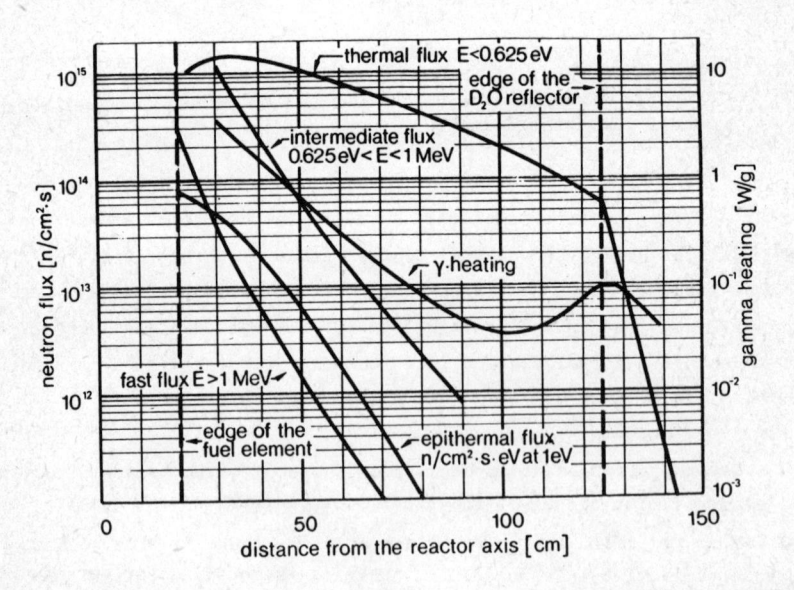

Figure 5   Spatial distribution of the fast, thermal and γ-fluxes at the HFR, ILL. It can be seen that the thermal flux peaks outside the radius of the fuel element.

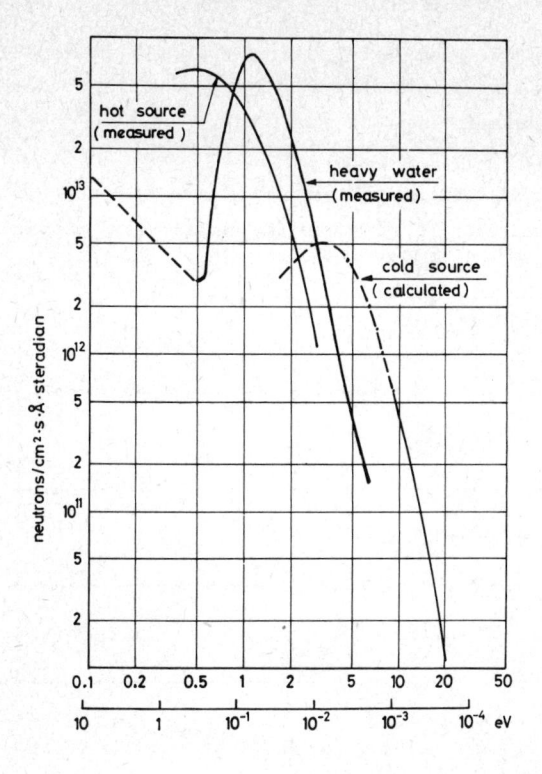

Figure 6   Flux spectra at the HFR, ILL. The shifts in the flux distribution
             introduced by the hot and cold sources relative to the thermal
             flux in the heavy water moderator are clearly seen.

3   PULSED SOURCES

Pulsed neutron sources are able to provide momentarily higher neutron fluxes
than those produced by steady state reactors. Since the neutrons are only
generated for a small fraction of the total time, the average power density is
lower and the problem of cooling is reduced. Table 2 summarises three
different types of pulsed sources: those based on the pulsed reactor, the
electron linear accelerator and the proton synchrotron. All these devices
produce fast neutrons ($\sim$ 1 MeV) and unwanted radiation and heat. Moderators
are again required to optimise the performance at a given wavelength; the
source must be shielded and collimators inserted through the shielding to
direct the pulsed beams onto the spectrometers. The pulsed nature of these
sources enables the energy of the neutrons to be determined by measuring
their arrival time at detectors placed some 10-100 m from the moderator. In
many experimental arrangements, the effective flux is equivalent to that from

a steady state source working at the peak of the pulse flux, but providing neutron beams which have been either monochromated by crystal reflection or chopped to define their incident energy.

The pulse length $\Delta t (\lambda)$ defines one element of the resolution function for time-of-flight apparatus. If $n_o (\lambda)$ is the neutron flux per unit wavelength at

	Pulsed Reactor (or Booster)	Electron Accelerator	Proton Accelerator
Production Reaction	fission	e-γ-n (+e-γ-f)	Spallation + fission
Typical Energy	~ 0.1 MeV	50 MeV	800 MeV
Typical Yield	1 n fission^{-1}	5.10^{-2} n e^{-1}	30 n proton^{-1}
Energy Produced	200 MeV n^{-1}	2000 MeV n^{-1}	55 MeV n^{-1}
Typical Pulse Time	70 μs	2 μs	0.5 μs

Table 2 Typical characteristics of three different types of pulsed neutron source.

the moderator of area A, then the flux at a sample position, distance L_o from the moderator, will be

$$n_L (\lambda) = n_o (\lambda) \frac{A}{4\pi L_o^2} \qquad (1)$$

The flight time of the neutrons is $t_o = \frac{m}{h} \lambda L_o$ and therefore the fractional wavelength resolution from the moderator is

$$R = \frac{\Delta \lambda_o}{\lambda_o} = \frac{\Delta t(\lambda)}{t_o} = \frac{h}{m} \frac{\Delta t(\lambda)}{\lambda L_o} \qquad (2)$$

$\frac{h}{m} = 3.95603$ for t in μs, λ in Å and L_o in mm.

If we assume that the incident path L_0 may be selected at will to produce a given resolution, we can eliminate it from equations (1) and (2) above and write

$$n_L(\lambda) = \frac{\lambda^2 R^2}{4\pi \left(\frac{h}{m}\right)^2} \left[\frac{An_0(\lambda)}{\Delta t(\lambda)^2}\right]$$

For a given λ and R the term in the square brackets defines a figure of merit and it can be seen that a short pulse length is very important. Returning to Table 2, it may be noted that the longer pulse times associated with the pulsed reactor will adversely affect its performance as a source for time-of-flight experiments.

3.1 The Pulsed Reactor or Fission Booster

A series of pulsed reactors has been developed at Dubna, USSR. The prototype IBR was completed in 1960 and had a mean power of 3KW. The IBR-30 (1969) had 25KW mean power and 100MW pulsed. The IBR-II will operate with a mean thermal power of 4MW (2.10^{17} fn s^{-1}). The instantaneous thermal neutron flux will be 10^{16} n s^{-1} cm^{-2} pulsing at 5/sec but the duration of the power pulse is 90μs. After moderation, the thermal pulse has a full width at half height of some 120μs. However, the IBR-II may be operated as a booster with a 200KW, 30 MeV electron accelerator and the pulse width then drops to 5μs with a pulse repetition frequency of 50 Hz.

3.2 The Electron Linear Accelerator (LINAC)

The electron linear accelerator is essentially a series of tuned microwave cavities fed by klystons. Electrons are accelerated by surfriding on a travelling wave and stability considerations require that the electrons are bunched at the stable position along the wave. The new Linac at AERE, Harwell, operates at an r.f. frequency of 1300 MHz with a power of 45 KW (Windsor 1978). The pulse repetition frequency is 150 Hz **and the pulse leng**th is 5 μs. Operation is expected to begin towards the end of 1979 and the fast neutron flux will be 2.10^{14} n s^{-1}. The electron beam can be switched to four different cells: low energy, booster, fast neutron, and the condensed matter cell. The latter has a target consisting of a stack of zircalloy-clad plates of natural uranium and will be used for neutron scattering experiments. Water cooling is used to remove the 90 KW generated in the target and the uranium gives a gain in neutron production of a factor of two due to fission. Figure 7 shows a schematic view of the layout of this target cell and beam tubes. Both low temperature and ambient moderators are provided and 'slab' and 'wing' geometries are used. The former provides good flux but poor γ and fast neutron backgrounds, whereas these are better for the wing geometry. Unfortunately, the flux is lower in this case and the solid angle over which the moderator can be viewed is reduced by the presence of the reflector.

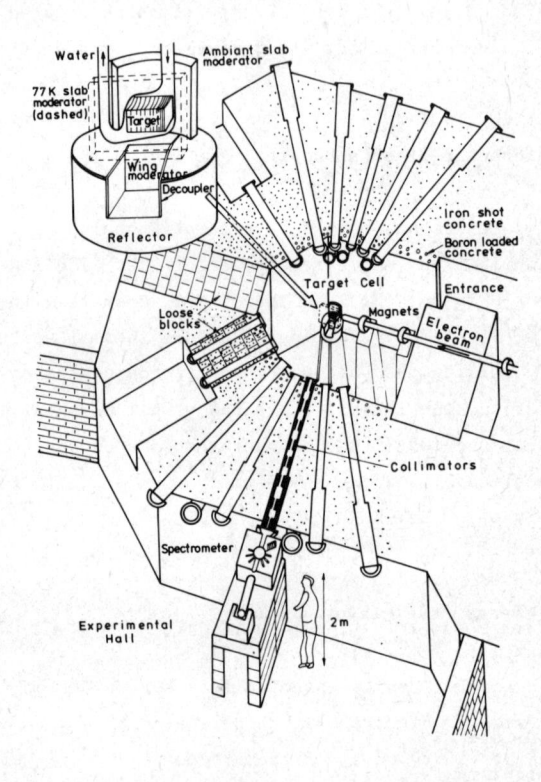

Figure 7   Schematic view of the layout of the condensed matter target cell, moderators and beam tubes at the new Harwell linac (Windsor 1978).

3.3 Proton Spallation Sources

Carpenter (1977) has recently reviewed the problem of providing enhanced neutron fluxes and has considered that the combination of a high-current, pulsed proton accelerator and a heavy metal target, which produces neutrons by spallation and by fission, is especially attractive as the base for the next generation of

high flux neutron sources. The term spallation is used to describe those
nuclear reactions in which the target nucleus emits a rather large number of
nucleons or fragments. A spallation source, the SNS, is now being bulit at the
Science Research Council's Rutherford Laboratory and a similar project, the IPNS,
is planned for the Argonne Laboratory. Table 3 gives the main parameters of
the SNS. The facility makes use of existing building and plant released by the
closure of the 7 GeV proton accelerator, NIMROD, in June 1978. The injector
is a recently built 70 MeV linac designed for the NIMROD synchrotron but now
modified to increase its pulse repetition frequency to 50 Hz and to permit H^-
acceleration. The new synchrotron magnet ring is being constructed to replace
the NIMROD magnet. The H^- ions are stripped on injection to give a high

Proton energy	800 MeV
Proton injection energy	70 MeV (H^- ions)
Mean synchrotron radius	26 m
Proton intensity	2.5×10^{13} per pulse
Pulse repetition rate	50 Hz
Proton pulse length	0.4 μs (0.1 μs)
Target	^{238}U, water-cooled
Target heating	420 kW

Table 3 The main parameters of the Spallation Neutron Source now being built at the Rutherford Laboratory (Stirling 1978).

circulating proton intensity. The mean current will be ∼ 200 μA and extraction
will take place within one revolution, giving two 0.1 μs pulses within an
overall time envelope of 0.4 μs. (Stirling 1978).

The target material will be uranium, since spallation neutron yields increase
with target Z, plus an additional factor of about two for fissile targets
(Fraser et al 1965, Fullwood et al 1972). Each accelerator pulse will contain
some 2.5×10^{13} protons at 800 MeV. The uranium target gives 30 neutrons/
incident proton, so the overall neutron production rate will be 4×10^{16} ns^{-1}.
Four moderators will be placed in wing geometry above and below the water cooled
target and both ambient and cooled moderators will be provided.

3.4 Pulsed-Source Moderator Performance

We have seen that the resolution of pulsed source instruments will be adversely affected if the moderation process introduces an unnecessary broadening of the thermal neutron pulse width $\Delta t(\lambda)$. The design of suitable moderators therefore involves a compromise between maximum neutron flux and good time resolution (see for example Day and Sinclair (1969), Graham and Carpenter (1970) and Rief and Hartman, (1975)). Figure 8 illustrates the effect of moderator temperature on the spectral distribution. Figure 9 shows the neutron spectrum can be conveniently divided into two regimes:

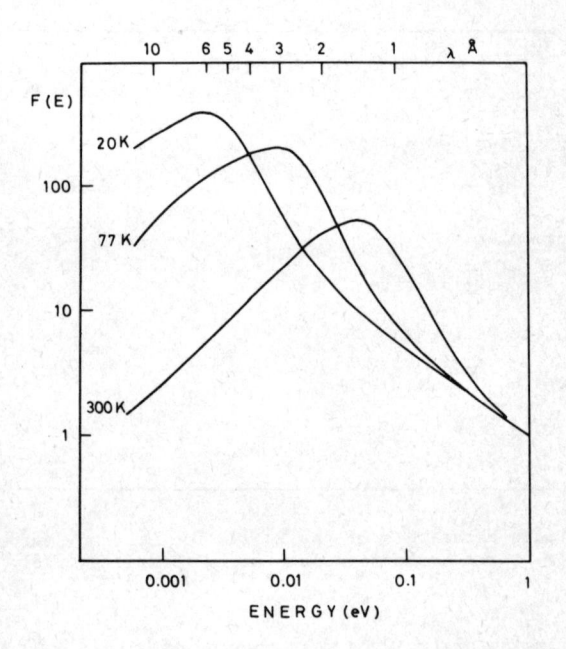

Figure 8 The dependence of pulsed thermal neutron flux $F(E)$ on energy as a function of moderator temperature (Stirling 1978).

(a) the epithermal or slowing down region in which $n_o(\lambda) = C/\lambda$ (3)
and

(b) the thermal region with its Maxwellian distribution. The Maxwellian amplitude is associated with a lengthened $\Delta t(\lambda)$ and is suppressed by poisoning the moderator. Alternatively, the Maxwellian distribution can be moved to lower energies by cooling the moderator.

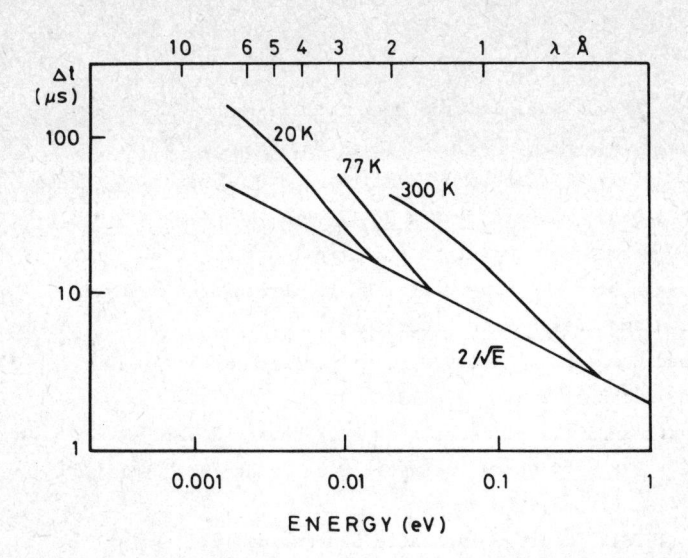

Figure 9   The dependence of neutron pulse width (Δt) on energy for different pulse source moderator temperatures (Stirling 1978).

In the epithermal region

$$\Delta t (\lambda) = B\lambda \text{ with } B \sim 7 \text{ μs if } \lambda \text{ is in Å}$$

The Harwell linac will have C (equation 3) $\approx 5.10^{10}$ ns^{-1} cm^{-2} Å^{-1}, whereas the SNS should have C $\approx 10^{13}$ ns^{-1} cm^{-2} Å^{-1}.

4   CONCLUDING REMARKS

High flux neutron sources are expensive to build and expensive to operate. Although it is clear that the move towards pulsed sources as a method of obtaining an enhanced effective neutron flux will be particularly effective for experiments which require epithermal or 'hot' neutrons, the gains over steady state source experiments at longer wavelengths will be less.

Detailed comparisons are by no means simple, as has recently been pointed out by Windsor (1978). They will inevitably depend on the ingenuity, innovation and effort which are brought to bear on the experimental apparatus associated with the two types of source.

REFERENCES

Carpenter J M (1977), Nucl Instrum Meth 145 91.

Day D H and Sinclair R N (1969), Nucl Instrum Meth 72, 237.

Fraser J S, Green R E, Hilborn J W, Milton J C D, Gibson W A, Gross E E and Zucker A (1965), Phys in Canada 21 17.

Fullwood R R, Cramer J D, Harman R A, Forrest R P Jr, and Schrandt R G (1972), "Neutron Production by Medium Energy Protons on Heavy Metal Targets", LA-4789 (Los Alamos Scientific Laboratory).

Graham K F and Carpenter J M (1970), Nucl Instrum Meth 85 163.

Kouts H (1963), J Nucl Energy 17 153.

Rief H and Hartman J (1975), Annals of Nucl Energy 2 521.

Stirling G C (1978), "Neutron Inelastic Scattering 1977" Vol 1,25 IAEA-SM-219/118, Vienna.

Windsor C G (1978), "Neutron Inelastic Scattering 1977" Vol 1,3 IAEA-SM-219/83 Vienna.

X-RAY SOURCES

R. Coisson

Istituto di Fisica, Università di Parma, Italy.

Introduction

We speak of "conventional" continuous sources (X-ray tubes) and synchrotron sources of "hard" X-rays (= that pass through Be window, $\lambda < 3$ or 4 Å) or "soft" X-rays.

"incoherent" source (phase of field emitted by different dS uncorrelated)

$d\Omega = \theta d\theta d\varphi$

Power emitted

$$P = \int b(x,y,\theta,\varphi,\nu)\,dS\,d\Omega\,d\nu \simeq b S \Omega \Delta\nu$$

We call b "spectral brilliance", $b\Delta\nu$ "brilliance", $bS\Delta\nu$ "brightness", etc... (remark: other authors call "brightness" what we call "brilliance", etc....)
ΩS area of (4-dim.) phase space, independent of distance and focusing (Liouville). Spectral and angular distribution might be linked: then in general S, Ω, ν not independent. Which are the most important parameters depends on the particular experiment (consider also time structure and polarization).

For an observer at distance R, irradiated area $\sim \Omega R^2 + S$

Practical unit conversions: $h\nu$ (keV) = $12.4/\lambda$ (Å) ; $\gamma = 1957\, E$ (GeV)
number of phot/sec = $5 \cdot 10^{14}$ power (Watts)·wavelength (Å).

X-ray tubes [8]

Characteristic X-rays: narrow-band spectrum ($\Delta\nu/\nu \sim 0.5 \cdot 10^{-3}$) due to rearrangement of atomic electrons after K-level ionization by the electron incident on the anode. K_α lines: $L \rightarrow K$ transitions (strongest); K_β lines: $M \rightarrow K$.
(energy a little bit lower than K absorption edge)
Empirical formula:
(E energy of electron,
E_k energy of K level; $E > E_k$
Z atomic number of anode)

$$\frac{\text{n.photons}}{\text{n.inc.electrons}} = 3 \cdot 10^{-4} \left(\frac{E-E_k}{1\text{ KeV}}\right)^{1.63} e^{-0.095 Z}$$

To find photons per unit $d\Omega$ at a given "take-off" angle α from anode surface, multiply by $1/4\pi$ and by a factor (≤ 1) higher for higher α and Z and for lower E (=0.5 for $\alpha = 6°$, 50 KeV electrons on Cu). (Lambert's law is not valid).

Bremsstrahlung: radiation due to short and intense accelerations of the incident electrons due to microscopic e.m. fields of atoms of anode: random shoot pulses give a white spectrum up to an energy E (electron K.E.). In a thick target, electrons gradually (on the average) slowed down: spectrum goes linearly to zero at $\nu = E/h$).

For E≪500 keV ($\gamma \ll 1$) isotropic radiation; for $\gamma \gg 1$ radiation peaked forward, within cone of aperture $\sim 1/\gamma$.

Semi-empirical formula for total Bremsstrahlung energy:

$$\frac{\text{X-ray energy}}{e^- \text{ beam energy}} = kZE \qquad \text{with } k \simeq 10^{-9} \text{ eV}^{-1}$$

<u>Power limitations</u> are due to the heat produced on the anode by the electron beam[6]. For stationary target: max.power that anode can withstand is proportional to thermal conductivity, to melting temperature minus room temp., and to linear dimensions of spot. (Then power per unit area (then brilliance) is inversely proportional to linear dimensions of anode spot). For 0.1 mm x 1 mm spot, $\sim$ 200 W. For rotating anode it is proportional to square root of velocity of anode surface. For 1 cm ϕ anode, 2500 r.p.m. it can be 6 to 12 times higher. (up to 10000 r.p.m. used).

Spot usually segment ax10a (1x10 mm; or 0.1x1 mm), and it is viewed at a "take-off" angle 6° (0.1 rad), so that it appears of dimensions axa. The Be windows (usually 2) at a distance of a few cm, transmit a solid angle of the order of 0.1 rad.

Anodes most commonly used (good thermomechanical characteristics):

Cu (K_α: λ=1.54 Å; a Ni absorber can filter out K_β and Bremsstrahlung)

Mo (K_α: λ=0.71 Å, a Zr absorber can filter out).

Al (K_α: λ=8.3 Å)

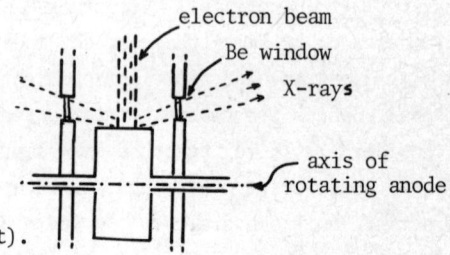

<u>Synchrotron radiation</u>

is radiation from ultrarelativistic electrons in magnetic fields (in practice, macroscopic electric fields are weaker than magnetic fields: for $\gamma \gg 1$, a 300 V/cm electric field is equivalent to 1 G magn. field, and static magn. fields of 50 KG can be produced).

For $\gamma \gg 1$, radiation sharply peaked forward ($\theta \lesssim 1/\gamma$ around direction of electron velocity). The "amplitude" $U(t) = \left(\frac{dP(t)}{d\Omega}\right)^{\frac{1}{2}}$ can be calculated by Liénard's formula (= Green's function of Maxwell's equations) and can be written $U(t) = C_0 \gamma^3 Bf$, where $f = (1 + \gamma^2 \theta^2)^{-3} [(1 - \gamma^2 \theta^2) + 4\gamma^2 \theta^2 \sin^2\varphi]^{\frac{1}{2}}$ angular distribution (=1 for θ=0), and f and magnetic field B must be evaluated at the time t' when the field has been emitted: t=t'+R(t')/c (R distance of observer), and $C_0 = 1.74 \cdot 10^{-7}$ Kg$^{-½}$ m sec^{½} Cb

To calculate spectrum:

$$\left(\frac{dP(t)}{d\Omega}\right)^{\frac{1}{2}} = U(t) \xrightarrow{\text{Fourier Trans.}} U(\nu) \xrightarrow{\text{mod.sq.}} |U(\nu)|^2 = \frac{dP}{d\Omega \, d\nu}$$

(in general it depends both on f(θ(t')) and on B(t')).

The total power emitted by an electron beam in a magnet is

$P = 4.22 \cdot IBE^3$ Watts per milliradian of orbit (B in Tesla, E in GeV, I in Ampères)

and most of it is within an angle $\pm 1/\gamma$ from the orbit plane.

Uniform magnet:

length of the pulse = $L_0(\frac{1}{V} - \frac{1}{C}) \simeq \frac{L_0}{2\gamma^2}$ = lag between field emitted at point A and at point B

beginning of the pulse (emitted when electron was in A)

end of the pulse

for $\gamma \gg 1$, $\frac{v}{c} \simeq 1 - \frac{1}{2\gamma^2}$

← direction of observer

$L_0 = \frac{mc}{eB}$

point at which the electron begins to be seen (after delay) by the observer

electron trajectory

Undulator: periodic field (period λ_0): same reasonment but observer continuously "illuminated"
e.m. field

obs. ────── λ_0 ────── electron trajectory

1) Uniform magnet. The observer is swept by the e.m. beam (of angular distribution $f(\theta)$) and sees a pulse of duration $\tau_c \sim \frac{m}{eB\gamma^2}$ and then a continuous spectrum extending up to a "critical frequency" $\nu_c \sim \frac{3}{4\pi \tau_c} = \frac{3eB\gamma^2}{4\pi m}$; while the electron moves on an arc of trajectory of length $L_0 \sim \frac{mc}{eB}$ (if B is uniform over the distance L_0). The number of photons emitted per second, per 0.1% bandwidth, per milliradian of orbit, integrated over the vertical (out of orbit) distribution (of width $\sim 2/\gamma$) can be written[2] (I in Ampères, E in GeV, B in Tesla):

$$\frac{dN}{dt}(0.1\%, \text{mrad}, \int \text{vert}) = 1.6 \cdot 10^{13} \, IE \, g(\lambda/\lambda_c), \qquad \lambda_c = \frac{18.64}{BE^2}$$

where $g(\lambda/\lambda_c)$ is a function plotted in fig.1, which has a maximum of 1 at $\lambda/\lambda_c \sim 3$ and decreases exponentially as $e^{-\lambda_c/\lambda}$ for $\lambda < \lambda_c$ ($\sim \frac{1}{25}$ at $\lambda_c/\lambda = 5$).
For example, if E = 1 GeV, B = 1 T, $L_0 \simeq 1.5$ mm, $\lambda_c = c/\nu_c$ = 18.6 Å.
In the plane of the orbit the radiation is linearly polarized in that plane, while out of that plane it is elliptically polarized.
The divergence of the photon beam will be $\sim (\sigma'^2 + 1/\gamma^2)^{\frac{1}{2}}$ where σ' is the divergence of the electron beam. If $\sigma' \ll 1/\gamma$, the spectrum depends on the angle from the plane of the orbit (the hardest components being near $\theta = 0$).

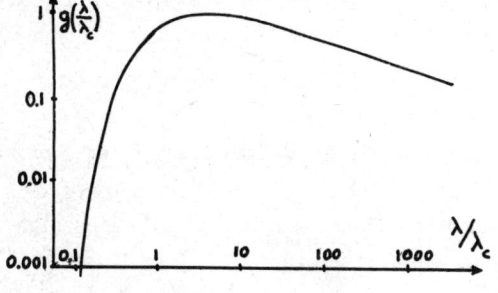

fig. 1: Normalized synchrotron radiation spectrum

2) Undulator. If the magnetic field B is "non uniform" (over a distance L_0) the spectrum depends also on the form at B(Z). With a sinusoidal B(Z) (period λ_0) the radiation emitted could be composed of a narrow ($\nu/\Delta\nu \sim$ number of periods) band at wavelength $\lambda_0/2\gamma^2$ (if $\lambda_0 \ll L_0$) or several narrow bands. But the frequency emitted is strongly dependent on the angle of emission, then the possibility of separating narrow bands depends on the angular spread of the electron beam: to get a bandwidth $\Delta\nu$, we must have $\sigma' < (\Delta\nu/\nu)^{\frac{1}{2}}/\gamma$ (then the angular spread must be $\ll 1/\gamma$).
While in a bending magnet the power is spread in the orbit plane, here it is all concentrated within a solid angle $\Omega_0 \sim \pi/\gamma^2$.
With suitable B(Z) any polarization could be produced: horizontal, vertical, or circular (with a helical field). In any case, in a given gap h (or diameter of tube), the shortest period λ_0 obtainable with appreciable magnetic field is $\sim$ h. For example, with a 2 cm gap, to reach $\lambda = 1$ Å the energy should be at least 5 GeV.
If a pinhole selects a band $\Delta = \Delta\nu/\nu$, spectral brilliance in case $\lambda_0 > L_0$ at a peak in spectrum is $\sim N/\Delta$ times higher than from a uniform magnet (same B, N=n. of periods). For strong field and large angular spread: spectrum $\sim$ "usual": this usually called "wiggler".
Another interesting case (but not for X-rays):
"Edge effect": shift to higher frequencies of spectrum from edge of magnet ($< L_0$). Can be used to see image in visible light of 300 GeV proton beam[7].
Storage rings (= synchrotrons where, particularly by use of a good vacuum (10^{-9} torr) the electron beam can be stored for a time $\sim$10 hours at constant energy, with r.f. cavities compensating at every turn the energy lost by synchr.radiation) are generally used as synchrotron sources because they work at constant energy, with high current and a stable beam.
At present most sources are designed for high energy physics, and not optimized for X-rays. Further developments:
1) "Wigglers": magnets, producing zero net deflection, inserted in straight sections of machine, which have locally a higher magnetic field (up to 6T if superconducting, to 2T if normal) than in bending magnets, possibly multiple-pole to add intensities: they allow a given machine to get lower λ_c (2 GeV machine can get $\lambda_c = 1$ Å).
2) Dedicated machines: with high currents (up to 1 A), low emittance (beam size x divergence), energy 2 GeV for X-rays, and 800 MeV for soft X-rays and vacuum UV.
3) (in the future): undulators, 50 or 100 periods; increase in spectral brilliance over small solid angles.

Pulsed sources

We cannot speak also of pulsed sources, but remark that:
1) Instantaneous power and brilliance from a pulsed X-ray tube can be 2 or 3 orders of magnitude greater than continuous ones.
2) In synchrotrons, instantaneous power and brilliance are 10^2 to 10^3 times higher than the average values (pulse duration $\sim$0.3 to 2 nsec, separation $\sim$0.1 to 1 μsec.)

3) Very interesting soft X-ray pulsed sources (particularly C K_α, $\lambda \simeq 40$ Å) (plasma in a capillary, bombarded by e^- beam (10^{-7} sec) or by pulsed laser (10^{-9} sec)) (see D.Sayre's talk in this workshop).
Can give 10^{15} phot/cm^2 at 20 cm in 10^{-7} sec, 1 p.p.s. with C K_α, then instantaneous spectral brilliance $\sim 7 \cdot 10^{18}$ phot/sec/0.1% band/mrad^2/mm^2 ($\sim 10^3$ times less for white spectrum).

Table

of orders of magnitude relevant to continuous-operation high power tubes and of some storage rings (examples taken from Europe: one low energy, soft X-rays, one designed for high energy physics, one 2 GeV dedicated, and a proposed high energy dedicated machine). For tubes, Cu K_α is only an example. For synchrotrons, we consider 1 mrad of orbit, but 10 mrad or more can be collected (and possibly focused, but brilliance does not change). The useful wavelength range extends down to $\lambda_c/4$. For intervals $\Delta\nu, \Omega$, S smaller than indicated, power scales down linearly; for greater interval, does not increase (or, for undulators, increase with $\Delta\nu$ or Ω but not both, up to $\Omega_0 \sim \pi/\gamma^2$). For undulators, the "integrated flux" is within a pinhole selecting a $\Delta\nu/\nu \sim 7\%$ band (and only $\sim 15\%$ of the power).

Table 1 - X-ray tubes and storage rings: some orders of magnitude of main parameters.

		wavelength range (Å)	$\Delta\nu/\nu$	apparent source (mm)	Ω (sterad)	integrated flux (W)	Spectral brilliance (phot/sec/ 0.1% band/ /mrad^2/mm^2)
sealed-off X-ray tube 2 kW, 50 kV	CuK Bremss.	1.54 0.25	0.05% white	1x1	0.1	0.01 0.02	$1.6 \cdot 10^8$ 10^5
V.H.power rot.-anode tube 50 kW 50 kV	CuK Bremss.	1.54 0.25	0.05% wh.	1x1	0.1	0.27 0.5	$4 \cdot 10^9$ $3 \cdot 10^6$
µfocus rot.-anode tube 3.5. kW 50 kV	CuK Bremss.	1.54 0.25	0.05% wh.	0.1x0.1	0.1	0.02 0.04	$2.8 \cdot 10^{10}$ $2 \cdot 10^7$
ACO (Orsay) (operating) 0.54 GeV, 150 mA, B=1.6T		λ_c=40 Å	wh.	0.5x0.6	$2 \cdot 10^{-6}$	0.16	$4.6 \cdot 10^{12}$
ADONE (Frascati) (operating) 1.5 GeV 100 mA, B=1T with 1.8T 5-pole wiggler		λ_c=8.3 λ_c=4.6	wh. "	1x0.4 1x0.1	$\frac{2}{3} \cdot 10^{-6}$	1.4 2.5	$3.5 \cdot 10^{13}$ $6.9 \cdot 10^{14}$
SRS (Daresbury) (under constr.) 2 GeV 300 mA, B=1.2T 4.5T 1 pole wiggler		λ_c=3.9 λ_c=1.1	" "	5x0.2	$\frac{1}{2} \cdot 10^{-6}$	12 45	$7.6 \cdot 10^{13}$ "
ESRF (European S.R.Facility, proposal) 5 GeV, 500 mA, B=0.7T 3T 1 pole wiggler undulator λ_0=5.6 cm, B=0.2T 5 m long (λ_5=5th harmonic)		λ_c=1 λ_c=0.023 λ_1=5 λ_5=1	" " 7% 7%	1x0.1 1x0.2	$\frac{1}{5} \cdot 10^{-6}$ $2 \cdot 10^{-9}$	185 790 78 26.4	$7.9 \cdot 10^{14}$ " $7 \cdot 10^{18}$ $4.7 \cdot 10^{17}$

References

1) E.E.Koch "Synchrotron radiation sources" in: "Interaction of radiation with condensed matter" vol.2, p.225-274 (Int.Centre for Theor.Phys., winter college 1976) IAEA, Vienna 1977.

2) A.Bienenstock, H.Winick "Synchrotron Radiation Research" SSRL Report, Stanford 1978.

3) Proceedings of course on Synchrotron Radiation Research (Alghero 12-24 Sept. 1976) (Ed. by A.N.Mancini and I.F.Quercia), vol.1.
   a) H.Winick: "Introductory lecture, wigglers, and considerations for the design of S.R.facilities", p. 3-61.
   b) D.W.Lynch "Comparison of S.R. with other sources" p. 298-300.
   c) P.Pianetta, I.Lindau "Phase-space analysis applied to X-ray optics" p. 372-387.

4) J.D.Jackson "Classical Electrodynamics" 2nd ed. J.Wiley 1975.
   R.Feynman, R.Leighton, M.Sands "Lectures on physics" Addison-Wesley.

5) R.Coïsson "Some remarks on radiation in non-uniform fields and undulators for X-rays. talk given at Wiggler Meeting, Frascati 29-30/6/78.

6) M.Yoshimatsu, S.Kozaki "High Brilliance X-ray Sources" in "X-ray optics; applications to solids" Ed. H.J.Queisser, p.9-33. Topics in Applied Physics vol.22 (Springer-Verlag 1977).

7) R.Coïsson, Opt.Comm. $\underline{22}$, 135 (1977); R.Bossart et al., submitted to Nucl.Instrum.& Meth. (1979).

8) A.Guinier, Théorie et technique de la radiocristallographie, Dunod,Paris.

ULTRASONIC WAVES AND OPTICS

A. Zarembowitch
Laboratoire de Recherches Physiques
Université Pierre et Marie Curie
Tour 22, 4 place Jussieu
75230 PARIS CEDEX 05

I - INTRODUCTION

The purpose of this paper is to stress some analogies between ultrasonics and different areas of physics (mainly optics).

Indeed, the use of analogy is not always a heuristic approach : for instance the concept of "caloric fluid" has delayed the development of thermodynamics ; it is likely that the importance of analogies drawn from mechanics has sometimes been a handicap to the development of other branches in physics ; in the case of FRESNEL's formulas on the reflection and refraction of light at the boundary of two dielectrics, mechanical analogies have been successful but started from a wrong basis.

However, the development of acoustics and that of optics have been so closely connected that it seems interesting to examine if the progress in one of the branches has been stimulating for the other.

The answer is not obvious ; for instance the laser revolution in optics has no equivalent in ultrasonics for a simple reason : from the early beginning ultrasonic waves have been obtained by using piezoelectric transducers which generate <u>coherent</u> waves and even now, piezo-electricity remains the main technique for generating ultrasounds from frequencies as low as 20.000 Hertz up to 20 Gigahertz; thus the requirement for spatial coherence is met automatically.

To analyse more in detail other analogies between ultrasonic and non-ultrasonic physics, this paper will be divided into three parts :
- the first part examplifies a successful analogy between optics and acoustics : the polarization, the reflection and the refraction of ultrasonic waves.
- the second part is devoted to non linear behaviour in acoustics and in other branches of physics.
- the third part shows some significant developments of ultrasonics non debitor to any

other field.

II - SOME SUCCESSFUL ANALOGIES BETWEEN OPTICS AND ULTRASONICS

A large number of optical effects can be transposed into the ultrasonic field. Some transpositions are trivial, for instance the realization of ultrasonics lenses ; some are sophisticated as the evidence for a "Goos-Hanschen" effects in acoustics (fig.1)

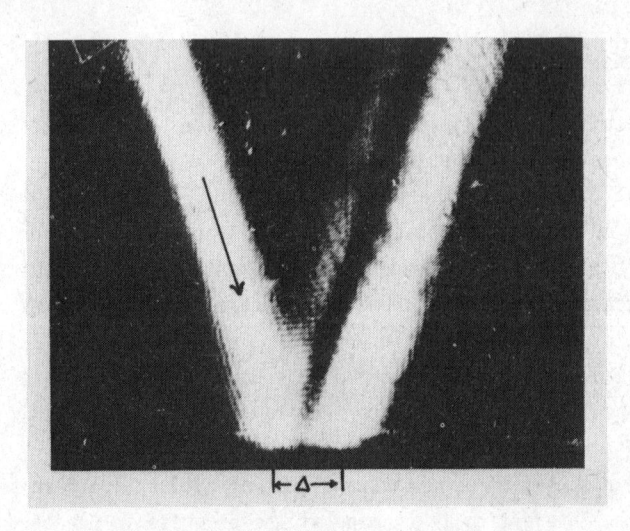

Figure 1 - Ultrasonic beam reflected from a water-glass interface - Δ is the lateral displacement of the reflected beam.

Obviously, we can expect fruitful analogies to be based on concepts rather than on devices ; a typical illustration is the analogy between optical and acoustical holography ; another is the close similarity between pure shear waves propagating in solids and electromagnetic waves, a similarity which allows us to build an ultrasonic polarimetry similar to the optic polarimetry.

Starting from this idea it is possible in acoustics :

a - to demonstrate the MALUS law

two Y-cut quartz transducers are used to generate and to receive transverse acoustic/waves in an isotropic medium ; they act as polarizers and analysers.

The amplitude of the electric signal received through the analyser is shown to be proportional to $\cos^2 \alpha$, where α is the angle between the directions of vibration of the two Y-cut quartz transducers (1).

b - <u>to generate circularly polarized waves</u> (FRESNEL's parallelepiped)

the phase difference between the reflected and the incident transverse elastic waves at the boundary between an isotropic solid and a gas can be chosen equal to $\frac{\pi}{2}$. As in optics, this property is used to generate circularly polarized waves (2) (3).

c - <u>to demonstrate a "POCKELS effect"</u> (acoustical birefringence)

a uniaxial stress is applied to an isotropic solid medium ; an acoustical birefringence occurs since the velocity of transverse waves polarized along or perpendicular to the stress is no longer the same (4).

d - <u>to give the evidence of a "FARADAY effect"</u>

the polarization plane of the acoustical transverse wave rotates under the influence of an external applied field in magnetic crystals (ferrimagnetic (5), paramagnetic (6), antiferromagnetic (7)).

e - <u>to find out a natural acoustical activity</u> (ARAGO effect)

using high frequency elastic waves (8) and Brillouin scattering (9) the polarization plane has been found to rotate without applied magnetic field in quartz crystals.

f - <u>to study a conical refraction effect</u>

in acoustics this effect is usual since the acoustical POYNTING's vector is no longer colinear with the wave vector for a transverse wave propagating along any (111) direction (10).

From these examples we can see that, in this field, analogies are very stimulating.

III - NONLINEAR BEHAVIOUR IN ACOUSTICS

Many branches of physics are debitors to acoustics from the point of view of nonlinearity. It is usual to find in acoustics striking examples to illustrate nonlinear physics : the aolien harp, the wind whistling through tall grass, the singing telephone wire...
So it seems interesting to examine more in detail some features of nonlinear behaviour in acoustics.

The starting point is that the acoustical behaviour is nonlinear for two reasons :

- the basic equations of acoustics are nonlinear ; for instance [1] is the wave equation for the adiabatic propagation of a plane wave in a non-viscous fluid.

$$\frac{\delta^2 \chi}{\delta t^2} = \frac{c^2}{1+(\frac{\delta \chi}{\delta a})^2} \frac{\delta^2 \chi}{\delta a^2} \qquad [1]$$

- the constitutive equation connecting stress to strain in a medium shows important departure from linearity even for small strain (that is very different from optics)

As a consequence, in ultrasonics we can observe specific behaviour that we will examine in connection with two phenomena, dispersion and absorption.

a. <u>influence of dispersion</u>

In optics, the refractive index depends on frequency ; as a result, in nonlinear optics, phase matching problems are very important for harmonic generation. On the contrary dispersion is very often negligible in ultrasonics ; for example the velocity of longitudinal waves propagating in NaCl is constant with an accuracy of 10^{-3} from 10^4 to 10^{10} Hertz. Consequently, when the wave propagates into the medium, the amplitude of generated harmonics can increase at the expense of the fundamental mode (cumulative effect) since phase matching conditions are satisfied. Because of harmonic growth the wave front becomes more straight and instabilities can occur if this growth is not limited by attenuation (fig. 2).

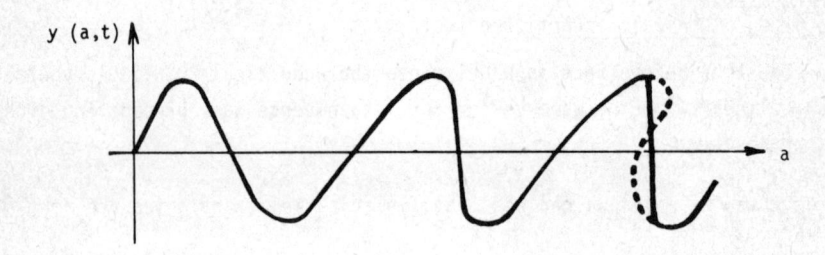

Figure 2

b. <u>influence of attenuation</u>

Many different processes are responsible for the attenuation of elastic waves in solids and fluids ; however, a large number of these processes leads to a regime in which the attenuation coefficient is proportional to the square of the frequency ; thus, high frequencies are strongly damped and the distorsion of the wave is usually limited by the attenuation. A very interesting, though particular situation, occurs in the case of an exact balance between harmonic generation and attenuation effects leading to the propagation of solitons (11).

Another interesting aspect of nonlinearity in ultrasonics is a parametric interaction used in the so-called "parametric arrays".

A basic problem in underwater acoustics is to find a compromise between directivity and attenuation. If high frequencies are used the ultrasonic beam will be directive but strongly damped ; at low frequencies directivity is lost. This severe limitation in the use of high frequencies can be overcome thanks to nonlinearity. Two high frequency ultrasonic beams whose frequencies are f and $f + \Delta f$ are generated by a transducer ; the nonlinear behaviour of water, though weak, is sufficient to allow the generation of a low frequency beam (frequency Δf) which is directive since it results from the beating of two directive beams. This technique (parametric arrays) is an important progress in ultrasonics within the last ten years (12) (13).

IV - ULTRASONIC SURFACE WAVES

If several areas of physics have been helpful to the development of ultrasonics, some parts of ultrasonics have been almost completely self operating. A good example is the generation and the use of ultrasonic surface waves.

The topic of surface waves is growing in importance on account of many possibilities offered by this technique. First of all, there is a large variety of elastic surface waves : RAYLEIGH waves, LOVE waves, BLEUSTEIN-GULAYEV waves, LAMB waves etc ... Next, techniques for generating elastic surface waves are easy and versatile. For example figure 3 shows how to generate ultrasonic Rayleigh waves : a thin metallic comb with interpenetrating fingers has been evaporated on the surface of a piezo-electric substrate ; the electrodes are connected to an electric generator whose frequency is chosen so that the elastic wavelength fit the distance between the fingers.

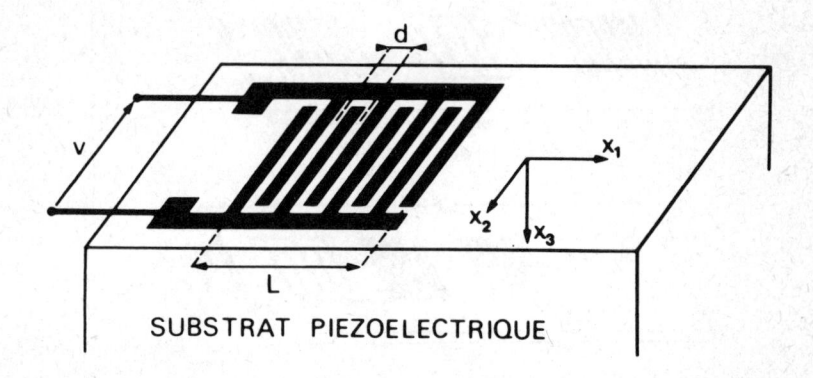

Figure 3 : Generation of ultrasonic Rayleigh waves.

Figures 4,5, show that fingers of different lengths or different spacement can be evaporated.

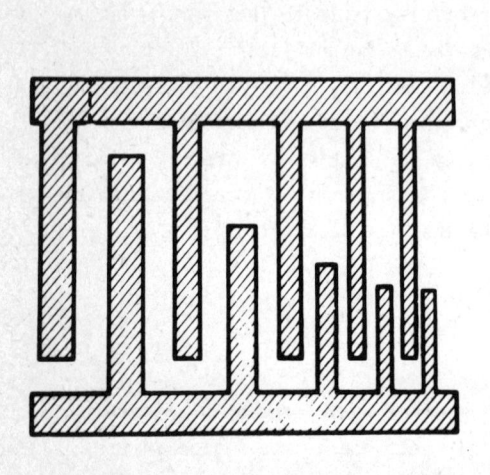

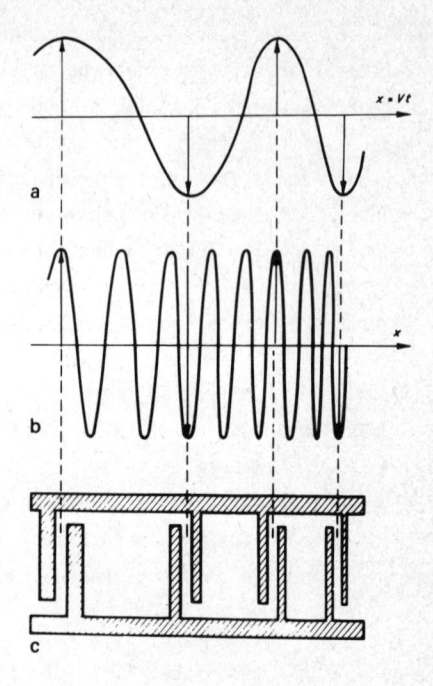

Figure 4

Figure 5 - Interdigital electrodes for pulse compression

Such combs are widely used for pulse compression (fig. 5), a powerful idea since it allows, in radar detection, to reconcile resolution power and long distance detection. Obviously, in this field also, analogies are not forbidden. Figure 6 shows a PEROT-FABRY resonator : the central comb generates a RAYLEIGH wave which is reflected by

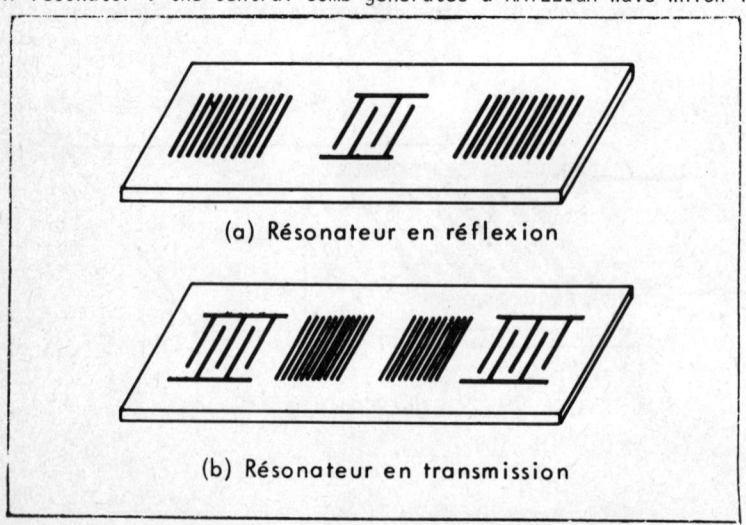

(a) Résonateur en réflexion

(b) Résonateur en transmission

Figure 6

two "mirrors" whose shape is that of a grating ; for BRAGG's condition the acoustical signal is completely reflected and a very good filter has been realised.

Many other devices are in progress : by ionic bombardment it is possible to modify locally the velocity of sound and to realize acoustical mirrors, beam splitters ....

In conclusion, this short review emphasizes the fact that ultrasonics has borrowed many of its concepts and techniques from optics and different branches of physics. Indeed ultrasonics arose in the early 20 th century and it is not unlikely that this young technique will be able in the future to stimulate the creativity in other domains of physics.

REFERENCES

(1)   PAUTHIER S., Ann. Phys. Paris 14, (1966) 195

(2)   PLICQUE F., FEUPIER J., ZAREMBOWITCH A., J. Acoust. Soc. Amer. 47, (1970) 168

(3)   FISCHER M., ZAREMBOWITCH A., Acustica 28, (1973) 259

(4)   FISCHER M., C.R. Acad. Sc. 274, (1972) 1115

(5)   MATTHEWS H., R.C. LECRAW Phys. Rev. Let. 8, (1962) 397

(6)   GUERMEUR R., JOFFRIN J., LEVELUT A., PENNE J., Sol. St. Com. 6, (1968) 519

(7)   BOITEUX M. and al., Phys. Rev. B (6) (1972) 2752

(8)   JOFFRIN J., LEVELUT A., Solid St. Com. 8, (1970) 1573

(9)   PINE A.S., Phys. Rev. B.2 (1970) 2049

(10) AULD B.A. - Acoustic fields and waves in solids - J. Wiley

(11) LIBCHABER J., TOULOUSE G., La Recherche n° 73 (1976)

(12) WESTERWELT P.J., J. Acoust. Soc. Amer. 32, (1960) 339

(13) BEYER R., Nonlinear Acoustics U.S. Navy p. 299

IMAGING PROCESSES AND COHERENCE IN OPTICS

A. MARECHAL
Institut d'Optique - Orsay
France

Optics is the domain of physics where the fundamental concepts of image formation developped : geometrical optics was used to interpret the behaviour of mirrors, lenses, of the eye etc... and allowed the realization of conventional optical instruments.

Physical optics opened new perspectives not only for the representation of the propagation of light but also for the formation of optical images and the existence of fundamental limitations imposed by diffraction. Phase contrast and interferential microscopy illustrate the role of physical optics in the development of potentialities of optical images ; they pointed out the importance of coherence and stimulated the study of the transition between coherence and incoherence i.e. the domain of partial coherence where the leaders have been Zernike, Hopkins, Wolf and Blanc-Lapierre in the period 1948-1960. Those basic concepts proved to be very important for understanding the formation of optical images in instruments like microscopes, optical projectors etc...

What is then the meaning of the word coherence ? According to tradition it expresses the ability to produce interferences when beams are superimposed ; we will use this term in this precise meaning, hoping that distorsions that could appear in other fields would not lead to appreciable "incoherences" with this original meaning.

We will first analyse the problem of coherence and then the problem of image formation where coherence is one of the basic factors.

I <u>Coherence, incoherence, partial coherence, space and time problems</u>

A) <u>Partial coherence</u>

Two examples will easily demonstrate the existence of intermediate situations between coherence and incoherence

1- Young fringes with an extended source : Let us imagine that we perform the classical experiment by Th. Young with a very small monochromatic source So illuminating two small holes T_1 T_2 (fig.1) :

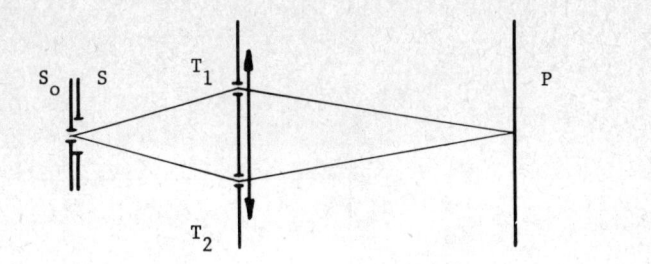

Fig. 1

fringes are observable in the plane P, and we conclude that vibrations from T_1 and T_2 are perfectly coherent. Let us now suppose that we use a larger source S : the fringe systems generated by the various elements of S (emitting incoherent vibrations) will be superimposed, the minima of illumination will no longer be zero, the contrast of fringes will decrease and it is very easy to increase the size of S and suppress interference phenomena : according to classical interpretation we can consider that this is the result of a blurring of interference phenomena having various positions in plane P. We also can declare that vibrations from holes T_1 and T_2 are no longer coherent : if the source S is large no interference phenomena can be observed and the vibrations from T_1 and T_2 behave like incoherent vibrations : the increase of dimensions of S opens the possibility of studying and intermediate domain between perfect coherence and perfect incoherence : the domain of partial coherence.

2-The Michelson interferometer with increasing OPD : Let us now consider a conventional Michelson interferometer illuminated with a very small monochromatic source (fig. 2) and let us vary the optical path difference (OPD) from zero to an importante value : it is well known that with conventional sources fringes will progressively vanish : their contrast will decrease from unity for a small OPD to zero when the OPD reaches values that depend on the spectral purity of the light : for conventional sources it is often expressed in fractions of a millimeter, reaching centimeters for good monochromatic sources and even one meter for sources used for the optical definition of standard lengths. Here the vibrations can be discomposed into spectral components contributing various interference fringe systems that are blurred if the OPD is important. We also can consider that the two vibrations coming from the two arms of the apparatus become incoherent if the OPD is important. In that case also a domain of partial coherence appears between perfect coherence and perfect incoherence but it is of completely different nature of the preceding case : the incoherence is closely related to the optical path difference Δ.

Fig. 2

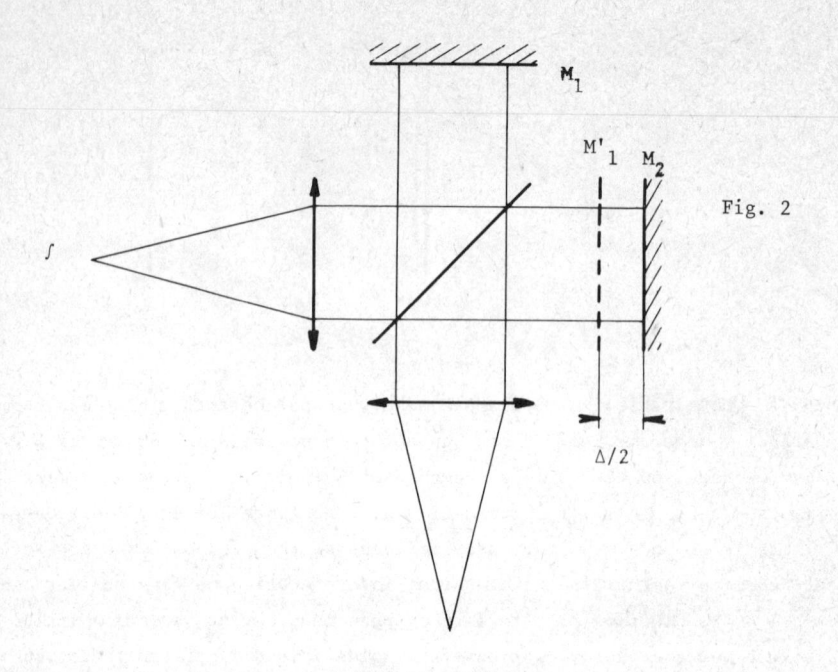

The interferometer combines vibrations emitted by the same small source S at instants t and t + Θ, if Θ is the time difference Δ/c between the times of propagation through the two arms : this experiment raises the question of coherence of the vibration emitted by a small source at instants separated by a time shift Θ. This could be called a problem of <u>temporal coherence</u>, in opposition with the preceding case where the relative position of holes T_1 T_2 was the dominant factor : the problem of coherence of T_1 T_2 in the Th Young's experiment is a problem of <u>spatial coherence</u>.

B) <u>The representation of vibrations by a fluctuating amplitude</u>

The existence of interference phenomena establishes the wave nature of light and justifies the representation of the vibration by the expression R (a exp jω t) where a is a complex amplitude. In fact this representation is very convenient for a limited interval of time : if we take the example of a good monochromatic source able to produce interference fringes with an OPD of 1cm this representation will be valid over a time domain of about $3 \cdot 10^{-11}$ sec. But if we increase the time delay Θ the experiment shows that fringes disappear : this representation is no longer valid for large OPDs or Θ s. In other words we have to consider that the amplitude a is no longer a constant but undergoes a relatively slow variation : during time interval of some periods T (T= $1.6 \cdot 10^{-15}$ s for λ =0,5μm) a can be considered as a constant but it varies appreciably if the time interval is increased to approximately $3 \cdot 10^{-11}$ s. In other words a becomes a function of time a (t) that does not vary in some periods but changes appreciably during a time interval defined by the OPD that produces blurring of the fringes. The light emitted by a source being in fact the sum of contributions of a large number of atoms (or molecules) it obeys statistical laws and it is convenient to consider a (t) as a fluctuating quantity whose mean value $<a(t)>$ is zero (otherwise there would be coherence for large values of θ) the mean square $<a(t) a^*(t)>$ representing the energy.

On the other hand we have to notice that the time of observation is generally large when compared with the time delay θ that affects coherence : the observable quantity is in fact $<a(t) a^*(t)>$.

C) The degree of partial coherence

Let us now combine two amplitudes $a_1(t)$ and $a_2(t)$ in such a way that we could obtain fringes : allowing the waves W_1 W_2 to present an angle α between them (of between the wave vectors k_1 and k_2) we shall obtain inference fringes in the observation plane with a fringe separation $i = \lambda / \alpha$ If ϕ is the phase difference $2 \pi \alpha y / \lambda$ at the point M (fig.3)

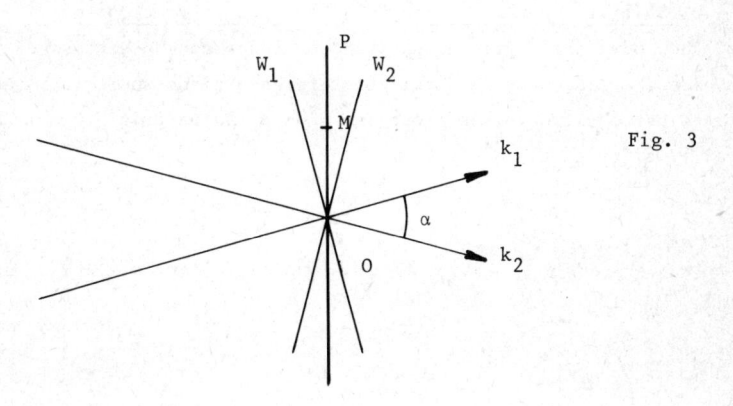

Fig. 3

the amplitude in M can be written, a_1 and a_2 being the amplitudes in O :

$$a_1(t) \exp(j \phi/2) + a_2(t) \exp(-j \phi /2)$$

The expression of the energy received in a point M where the phase différence is ϕ will be :

$$<a_1 a_1^*> + <a_2 a_2^*> + 2 R[<a_1 a_2^*> e^{j \phi}]$$

where the importance of the variation of illumination with ϕ depends upon the quantity $<a_1(t) a_2^*(t)>$

This quantity determines the quality of the interference fringes
- its modulus determines the contrast of fringes

- its argument determines the contrast of fringes

We will write :
$$\Gamma 1.2 = <a_1(t) \, a_2^*(t)>$$

and call this complex quantity the **degree of partial coherence**.
It is possible to normalise this quantity by writing

$$\Gamma_N = \frac{<a_1 \, a_2>}{(E_1 \, E_2)^{1/2}}$$

E_1 and E_2 being the energies corresponding of the two beams.
It is easy to check that
- if $\Gamma_N = 0$ (two independant beams) no fringes appear ; this is the case of incoherence.
- if $|\Gamma_N| = 1$ the beams are completely coherent.

In fact the coherence of two light beams is expressed by the correlation function of the two fluctuating amplitudes a_1, a_2 normalised if necessary.

D). Spatial coherence - The Zernike-Van Cittert theorem.

Let us now show how this degree of partial coherence can be expressed, in the schematic case where the source is far from the reference plane where we intend to express the coherence between two points : one is located in the origin O and the other has coordinates y. z. (fig. 4)

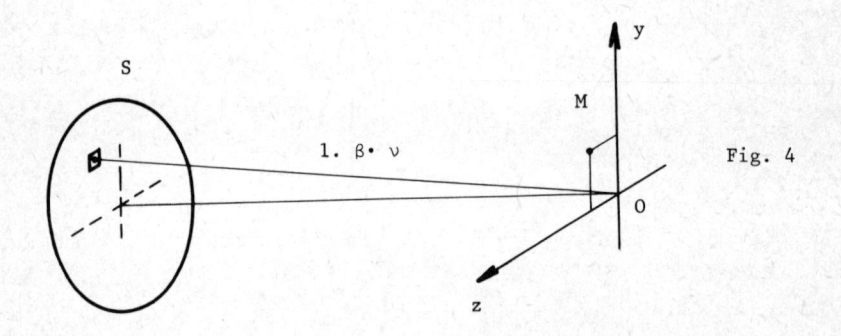

Fig. 4

If 1, β, γ are the components of a unit vector (β and γ being small) the density of energy on the source is a function $\varepsilon(\beta, \gamma)$ of the direction. Assuming that the various elements of the source are completely incoherent and supposing that the mth element produces an amplitude $a_m(t)$ in point O it produces an amplitude $a_m(t) \exp(jk(\beta_m y + \gamma_m z))$ in point M. If now we compute the degree of coherence between O and M

$$\Gamma \, O.M = <\Sigma \, a_m(t) \, \Sigma \, a_m^*(t) \exp(-j k (\beta_m y + \nu_m z))>$$

taking into account the incoherence of the various elements we obtain as a result expressed in terms of an integral :

$$\Gamma = \iint \varepsilon(\beta,\gamma) \exp(-j k (\beta y+\gamma z)) \, d\beta \, d\gamma$$

this is the Zernike-Van Cittert theorem :
The degree of coherence between two points receiving light from an extended source S is the Fourier transform of the energy luminance of the source.
This theorem applies very easily to the problem of the determination of stellar diameters by the measurement of fringe contrast as a function of the separation between two holes (or slits, according to the method of Michelson and Pease : assuming a uniform circular repartition for the star we obtain as degree of partial coherence $2J_1(Z)/Z$ with $Z = \frac{2\pi}{\lambda}\eta \, y$ where η is the angular radius of the star and y the separation between the holes. The first disparition of fringes will take place for $y = 1.2 \, \lambda/2\eta$
This theorem can be generalised in the following way : S be ing an element of the source, (S A) and (SB) two optical paths, $\int$ their difference the coherence between A and B can be expressed by the following relation

$$\Gamma = \iint \varepsilon(\beta,\gamma) \exp(-jk \, \delta(\beta,\gamma)) \, d\beta \, d\gamma$$

where A and B can be non longer be located in the same plane.

E) Temporal coherence and spectral finesse.

Let us now look at the relation between the degree of partial coherence $<a(t) \, a^*(t-\theta)>$ between vibrations emitted by the same source with a time delay θ and the spectral repartition of light ; it is well known that if we increase the "finesse" of the radiation i.e. if we decrease the width of the spectral band, we increase the possible O.P.D. in the interferometer and the possible delay θ. Let us now relate those various quantities :
If $f(t)$ is the vibration entering a spectrometer : in the spectral plane of the spectrometer we will receive a repartition of amplitudes representing the Fourier analysis of $f(t)$; if ν is the time frequency we will obtain a function $g(\nu)$ representing the spectral repartition of monochromatic amplitudes composing $f(t)$; this simply results from the additivity of amplitudes of various frequencies ; if only one frequency is present one exit slit only receives light in the spectral plane and if the light is complex its amplitude is "Fourier analysed" by the spectrometer. In fact the energy received in the spectral plane is represented by $g(\nu) \, g^*(\nu) = S(\nu)$ which is the spectral repartition of light.
On the other hand it is well known that the interferogram $I(\Delta)$ turns out to be related to the F.T. of S :

$$I(\Delta) = \int_0^\infty S(\nu)(1+\cos 2\pi \frac{\Delta \nu}{c}) \, d\nu$$

or more precisely the variable part of $I(\Delta)$ is :

$$i(\Delta) = I(\Delta) - I_o = \int S(\nu) \cos 2\pi\Delta\nu/c \, d\nu$$

from the comparison of those relations we deduce (fig.5) that the spectral

$$f(t) \xleftrightarrow{F.T.} g(\nu)$$

$$\left.\begin{array}{l} A(\theta) = \text{autocorr.} f(t) \\ = i(\Delta) \end{array}\right) \xleftrightarrow{F.T.} S(\nu) = gg^+$$

Fig. 5

repartition of energy is the F.T. of the auto-correlation function of $f(t)$: $A(\theta) = \langle f(t) f(t-\theta) \rangle$ what means that $A(\theta)$ should be identified with $i(\Delta)$. In fact this is not astonishing : the interferometer acts on the amplitude $f(t)$ as an autocorrelator, relating the amplitude at times t and $t - \theta$; a Michelson type interferometer could be called a time auto correlator for light amplitudes.

If now we develop the expression of $A(\theta)$ we obtain the following :

$$A(\theta) \propto R\left[\Gamma(\theta) \exp(j\omega\theta)\right]$$

which means that the two quantities $A(\theta)$ and $\Gamma(\theta)$ are very closely related : $A(\theta)$ is an oscillating quantity, the modulus being $|\Gamma(\theta)|$

Finally the energy spectrum $S(\nu)$ and the time partial coherence $\Gamma(\theta)$ are related as follows :

$$S(\nu) \propto \int_o^\infty A(\theta) \cos 2\pi\nu\theta \, d\theta = \int_o^\infty R|\Gamma(\theta) \exp j\omega\theta| \cos 2\pi\nu\theta \, d\theta$$

what shows that if $S(\nu)$ has a small extent, (very monochromatic light, high finesse) $\Gamma(\theta)$ has a large extent and the tolerable OPD on the interferometer is large.

F) The spatio temporal partial coherence.

In many cases the limitation of coherence results mainly from a predominant factor, (spatial or temporal) so that the relations established in D) or E) represent the situation. Nevertheless in some cases both geometry of the source and spectral finesse establish a limitation to coherence. In many of those cases it turns out that the O.P.D. does not vary appreciably when the point source moves on the source (otherwise the coherence would be very small) and if the source is homogeneous (the spectral repartition being everywhere the same) it is easy to dissociate spatial and temporal effects and the degree of partial coherence turns out to be the product of the two factors

$$\Gamma = \Gamma_S \Gamma_T$$

where Γ_S and Γ_T are spatial and temporal coherence expressed in sections D and E.

II Formation of optical images

We will mainly concentrate on the problem of the formation of optical images of extended objects : the basic mechanism is the superposition in the image plane of the images of the various elements of the objects : one little element of the object will contribute an element of the image represented by a diffusion function D (a diffraction image in the best cases, a more extended pattern in the presence of aberrations). In order to know how to combine the two images produced by two neighhouring elements of the object (fig. 6) we have to know whether vibrations coming from them are coherent, partially coherent or incoherent.

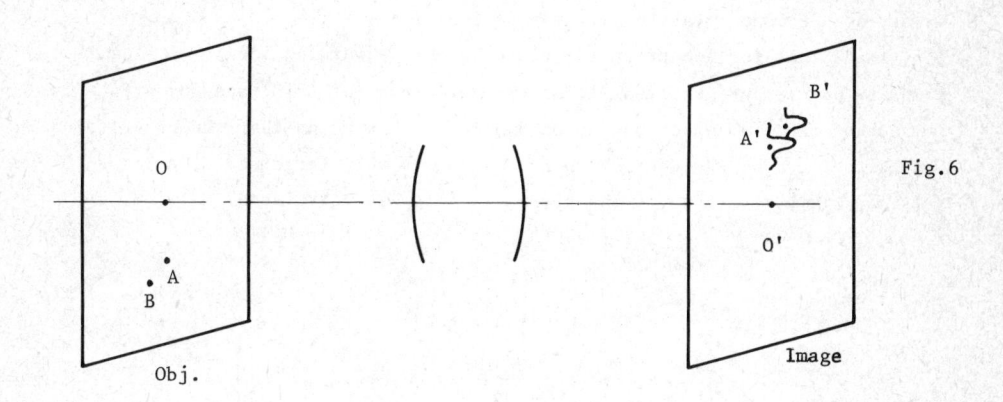

Fig.6

A) Incoherent illumination

The most common case is the incoherence : when taking an amateur photograph, an astronomical photograph, when separating various frequencies in a spectrograph as examples. In those conditions we have to add the energies coming from the various elements of the object, what leads to a convolution relation of the form $I = O \circledast D$: the image I is obtained by a convolution (in energy) between the object function (repartition of luminances) and the diffusion function D.

This can be expressed in the Fourier space : if i, o, and d are the Fourier transforms of I,O,D we can write $i = O \times d$ what means that the F.T. of the image is the product of transform of the object and of the transform of the diffusion pattern : the instrument can be characterised by an optical trnasfer function d that determines the way in which various spatial frequencies are transmitted by the instrument : this situation is very similar to that of the transmission of a temporal signal through an electronic or acoustic electronic device : the behaviour of the system can be known through a transfer function expressing the ability to transmit temporal frequencies. In optics we have to deal with spatial frequencies expressed in cycles per unit length and those components depend in fact on two parameters : the object or image are two

dimensional functions, as well as their transforms.
By application of the classical Fourier relations and taking into account the fact that
the diffracted amplitude is the F.T. of the repartition of amplitudes on the pupil
(pupil function) it turns out that the OTF is the autocorrelation of the pupil function:
in the case of a circular pupil the OTF is represented by the common surface to two
circles of radius α' (the angular radius of the pupil) shifted of a quantity proportionnal to the spatial frequency $1/p'$ if p' is the period of the component in the image
plane (fig. 7). In those conditions the OTF for a perfect instruments is represented
on fig.7 It has a strict cut off frequency $2\alpha'/\lambda$ what means that the instrument is
unable to transmit any information outside of that bandwidth. one of the consequences
is that it is possible to know completely the image by sampling the values at nodal
points of a crossed grating having a period $\lambda/4\alpha'$.
This means that for a perfect f/2 photo objective working for $\lambda = 0,5$ μm the number of
sampling points per square millemeter would be about $4 \cdot 10^6$ and for a 24×36^{mm} image
it reaches the tremendous number of $3.5 \cdot 10^9$: this means that the quantity of information that can be recorded by optical means is very large and vice versa the necessary bandwidth for transmitting good T.V. image is very large.

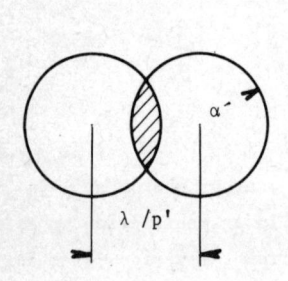

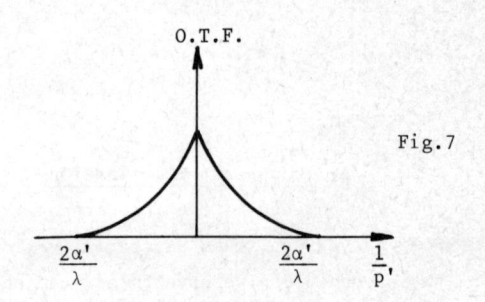

Fig.7

B) Coherent illumination

Fig.8 reprensents the typical mounting for performing completely coherent illumination:
a point source S illuminates an object having an amplitude transmission Ω and the various elements f_o the object receive perfectly coherent vibrations. It becomes obvious
that the image will be obtained by a convolution process applied to amplitudes :
The amplitude in the image is now the convolution of the amplitude in the object with
the repartition of amplitudes in the diffraction image i.e. the F.T. of the pupillar
repartition of amplitudes (the pupil function P). This process can also be established
by studying the way in which light undergoes two successive diffractions :
The pupil receives the F.T. of Ω (first diffraction), transmits the product $P \times FT(\Omega)$
and the second diffraction is responsible for producing in the image plane the

convolution F.T. (P) ⊛ Ω which is in complete agreement with the preceding conclusions. The optical transfer function turns out to be the pupil function P : according to the value of the spatial frequency the signal goes or not through the pupil : the OTF for amplitudes is a square function limited to the $\pm \alpha'/\lambda$; in coherent illumination the bandwidth is only one half of the value in incoherent illumination. Let us now point out one very interesting possibility of this coherent illumination : Abbe had already shown in his theory of the microscope that it is possible to transform the image of a periodic structure by eliminating various components for the diffraction phenomena in the pupil. We have shown how the attenuation of low spatial frequencies can help the perception of details and reinforce interesting informations : Fig. 9 is a typical example of those possibilities where a blurred text is treated in order to reinforce details and becomes readable by enhancement of details in the image. Another well known application is the technique of use of a matched filter suggested O'Neill and developped later by various authors : Fig. 10 is a typical example where the filter has been matched to the recognition of the letter e.

Fig. 8

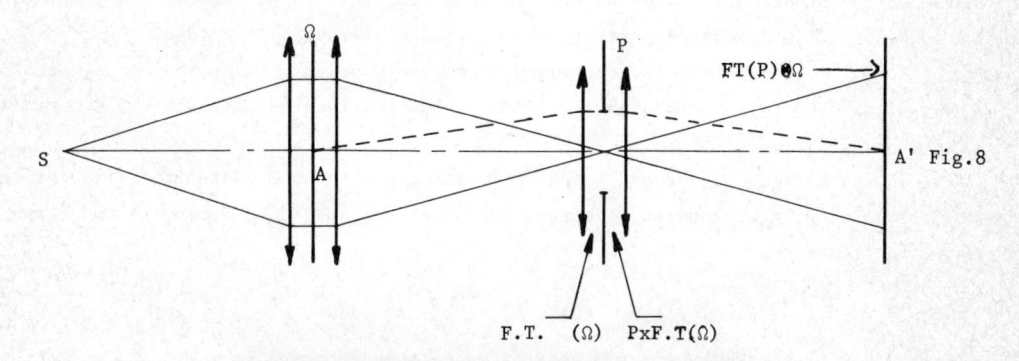

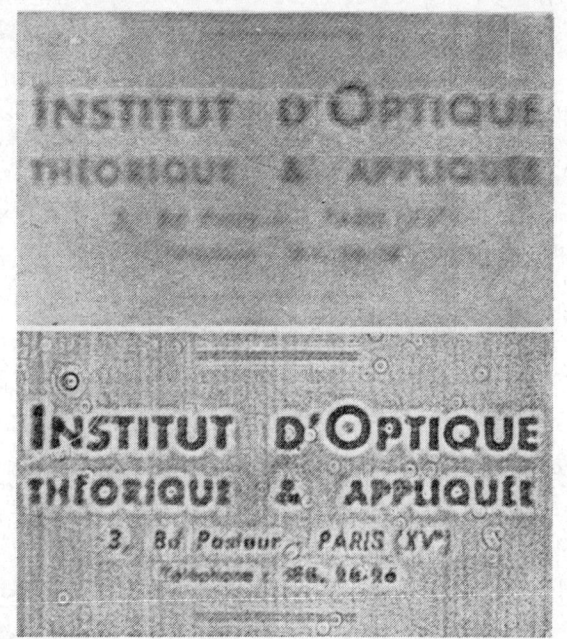

Fig. 9

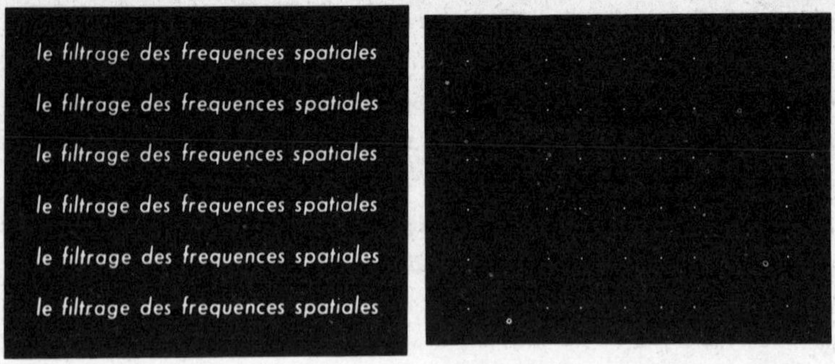

Fig. 10

Reconnaissance de la lettre e

C) <u>Partially coherent illumination</u>

The coherent illumination technique has the drawback that the illuminating point limits severely the total flux ; in many cases (microscopy, projection technique) it is necessary to use a source having an appreciable extension but this introduces a partially coherent illumination : if the source can be assimilated to a circular disc of angular radius α_c de degree of partial coherence is represented by a fonction $2 J_1 (Z)/Z$ with $Z = \frac{2\pi}{\lambda} \alpha_c y$; points that are very close are coherent ; then the coherence decreases and goes to zero for $y = 1.2 \lambda / 2 \alpha_c$ and oscillates and vanishes for distant points.

The relation between object and images becomes much more complicated : if Γ represents the degree of partial coherence between two points M_1 and M_2 according to the Zernike-Van Cittert theorem

$$\Gamma = \int \xi(S) e^{jkS(M_1 - M_2)} dS$$

where $\xi(S)$ represents the energy repartition on the condenser.

The repartition of illumination in the image is expressed as follows :

$$I(M') = \iint \Omega(M_1) E(M' - M_1) \Omega^*(M_2) E^*(M' - M_2) \Gamma(M_2 - M_1) dM_1 dM_2$$

In the general case of partial coherence the simple convolution relations are no longer valid and the relations are much more cumbersome. Nevertheless when the contrast of the object is small (what is often the case in optical microscopy) an approximation can be made and leads to a simple convolution relation

$$I(M') = I_o + 2R_e \left[\int \omega(M) E(M'-M) E_c^*(M'-M) dM \right]$$

where E and Ec are
- the repartition of the amplitude in the diffraction pattern produced by the full pupil
- the repartition in the pattern produced by the pupil limited to the aperture of the condenser.

Fig.11 represents the evolution of the "response function EEc for a perfect instrument where the pupil and the condenser are circular ; it is then possible to fill the gap between perfect coherence and perfect incoherence and study the evolution of the image or of the OTF (fig.12) between the two extreme cases.

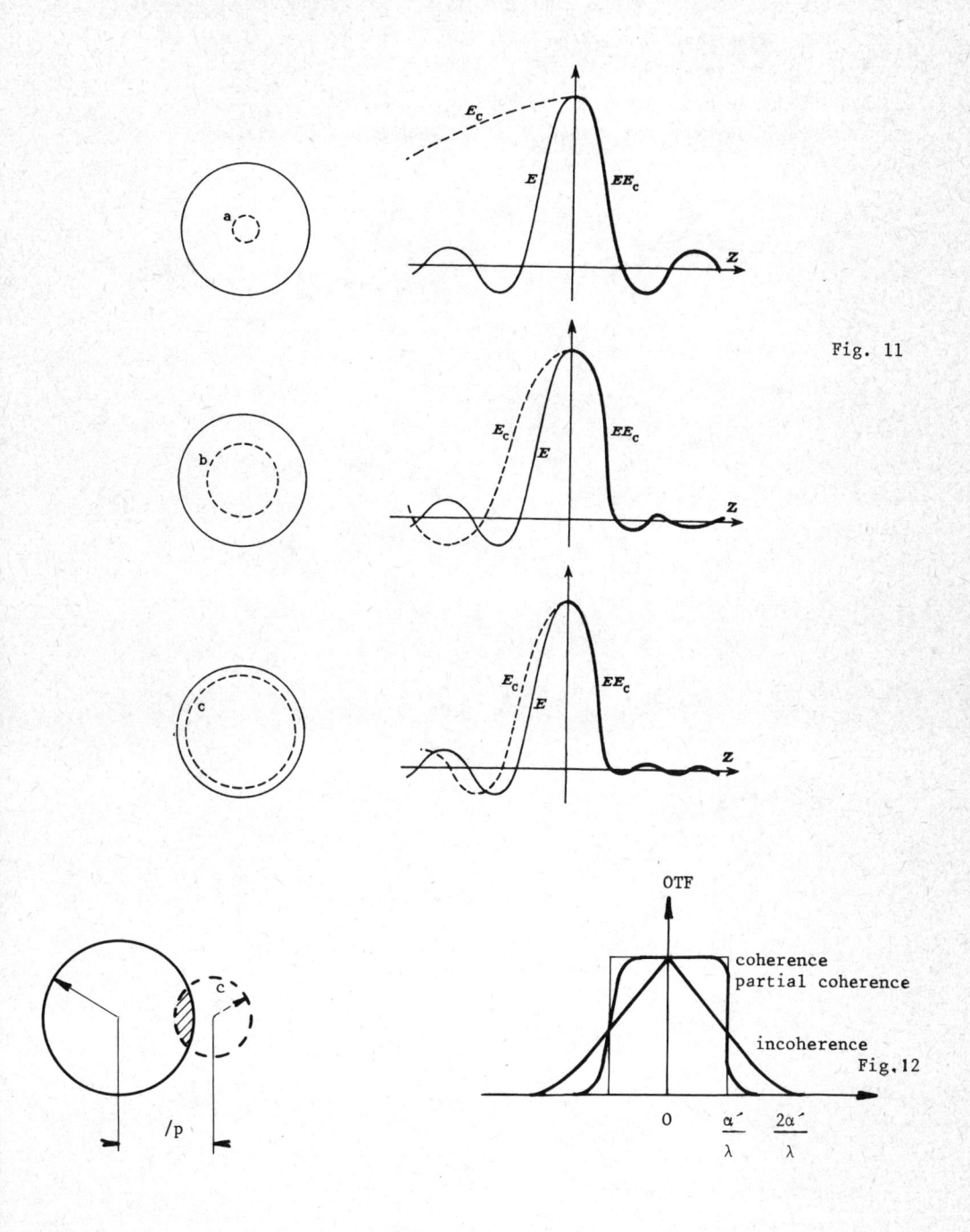

Fig. 11

Fig. 12

Bibliography

A.MARECHAL et M. FRANCON, Diffraction-Structure des Images Ed.Masson,Paris 1970
BORN M.& WOLF E., Principles of optics. Perg.Press.
BLANC-LAPIERRE & DUMONTET P., Rev.Opt.,34, 1955,1.
HOPKINS H.H., Proc. Roy. Soc., 217, 1953, 408
MENZEL E., Optik, 15, 1958, 460
SLANSKY S., Opt.Acta, 2, 1955, 119 ;J.Phys.,16,1955,13 S.
WOLF E., Proc. Roy. Soc.,225, 1954, 96 ; Proc. Roy. Soc.,230,1955,246
ZERNIKE F., Proc.Phys.Soc., 61,1948,158.

PROPAGATION OF NEUTRON BEAMS

A. Steyerl
Fachbereich Physik
Technische Universität München
8046 Garching, Germany

1. INTRODUCTION

In my view, the topic of this talk, "propagation of neutron beams", has two general aspects. One refers to the fundamental characteristics of propagation of the neutron as a matter wave, the other is connected with the specific problem of guiding slow neutrons, say, from a neutron source like a reactor, to the experimental site, or from one place in an apparatus to another. Evidently, the first and more fundamental of these aspects is very complicated, and it will be the subject of several other contributions at this meeting. It comprises not only phenomena like scattering or absorption, which occur when a beam of neutrons traverses matter, but it may even touch upon the very old puzzle of quantum mechanics as to the "reality" of a matter wave. To my knowledge, no one has ever given a completely satisfactory answer to that question.

In the present talk I shall concentrate mainly on the much easier problems of slow neutron guidance and confinement. The techniques used for these purposes are based on collective neutron interactions with matter, or with gravitational or magnetic fields. Within the restricted framework of the specific applications considered, these interactions may be described by the simple concept of an index of refraction.

2. INDEX OF REFRACTION AND NEUTRON MIRRORS

Refraction is well known as a common phenomenon for any radiation like sound, light, x-rays, electrons, or neutrons. From a microscopical point of view, refraction by matter may be described as a collective effect of coherent-elastic scattering in the forward direction. The elementary scattering centres may be atoms, nuclei, or other systems like small air bubbles in water as a source of sound wave scattering. The usual procedure of coherent superposition of the wavelets originating from all the scattering centres with the wave incident from outside the matter, then leads to a modification of the incident wavenumber k_o by an index of refraction n so that the wavenumber within the medium is $k = nk_o$. This result holds for any wave form and any shape of the region filled with matter. For nuclear scattering of slow neutrons the refractive index is given by

$$n = (1 - 4\pi N <b_{coh}>/k_o^2)^{1/2},$$

where b_{coh} is the coherent-scattering amplitude for the bound nucleus, and N is the number of nuclei per unit volume. The averaging refers to such cases where, on a microscopic scale, the medium consists of a mixture of atoms or isotopes with different scattering lengths, similarly as the "coherent"-scattering length refers to averaging the nuclear interaction over different relative orientations of neutron and nuclear spin, where applicable. In cases where the scattering distribution is non-isotropic, as in light or x-ray scattering, or in neutron scattering by paramagnetic atoms, the relevant scattering amplitude is that for forward scattering, i. e., b should be replaced by -f(0). Incoherent scattering effects including inelastic scattering and absorption may be incorporated in an imaginary part of the refractive index.

For neutrons or other massive particles the kinetic energy $E_{(o)} = \hbar^2 k_{(o)}^2 / 2M$ (M: mass) is proportional to the wavenumber squared, in the non-relativistic limit. Thus, refraction may be visualized as an effect of a "scattering potential" (or "optical potential") $U = 2\pi \hbar^2 N <b_{coh}>/M$, chosen in such a way that $E = E_o - U$.

In the framework of this picture it is very easy to extend the concept of neutron refraction also to situations where further interactions must be considered. Examples are the gravitational potential Mgz, where z is the height measured from an arbitrary origin, or the interaction energy $\pm \mu B$ of the neutron magnetic dipole moment $\vec{\mu}$ with a magnetic induction $\vec{B}$, where the sign depends on the mutual orientation of field and neutron spin. Numerically, $Mg = 10^{-7}$ eV/metre and $\mu = 6 \times 10^{-8}$ eV/tesla. Thus, the combined effect of these interactions may be expressed as a spatially variable index of refraction

$$n(r) = \{1 - [U(\vec{r}) + Mgz \pm \mu B(\vec{r})]/E_o\}^{1/2}.$$

Further interaction terms, for instance due to the weak interaction, could be included in the same way.

The slow neutron scattering potential is positive for most substances (which corresponds to a phase shift of $-\pi$ on scattering), thus $n < 1$. This is a necessary condition for the possibility of total reflection. Therefore, a slow neutron incident on a flat "mirror" of appropriate material will be totally reflected, provided that the angle of incidence, referred to the surface normal, lies in the range $\theta > \arcsin n$. This corresponds to the plausible condition that the kinetic energy of motion perpendicular to the surface should not exceed U, so that wave propagation within the medium is impossible. For the bulk of substances, U is of the order of 10^{-7} eV, therefore total reflection of thermal neutrons with typical energies $E_o \approx 25$ meV ($\lambda \approx 1.8$ Å) occurs only at very small glancing angles $\gamma = \pi/2 - \theta \leq (U/E_o)^{1/2} < 10^{-2}$. On the other hand, for neutrons with ultralow energies ("ultracold neutrons") E_o and U are comparable, and such neutrons may be totally reflected even at any angle of incidence. This is the basis for the possibility to store these neutrons in closed cavities ("neutron bottles") for fairly long times, presently up to 650 s /1/.

When slow neutrons interact with a magnetized ferromagnet, double refraction occurs because the index of refraction differs for the two possible neutron spin orientations relative to the internal induction $\vec{B}$. This phenomenon is widely used for slow neutron polarization and polarization analysis, choosing appropriate ferromagnetic mirrors for which $U + \mu B$ is large but $U - \mu B$ is nearly zero. Then only one spin state is totally reflected at glancing incidence while the other is not refracted and eventually absorbed in the mirror substance. Instead of reflection also the transmission through thin ferromagnetic films may be employed to obtain a polarized neutron beam, and this method is particularly useful for ultracold neutrons where a divergent beam may be polarized with 98 % efficiency by perpendicular transmission through thin monocrystalline iron films /2/.

The force experienced by a neutron in an inhomogeneous magnetic field, $F = \mp \mu \text{ grad } |B| = E_o \text{ grad } n^2$ may also be used for mirror reflection or beam focussing in appropriate magnetic field configurations /3, 4/. I shall mention one example later.

The beam divergences usable in all devices based on total mirror reflection are determined by the angular range of total reflection, which is quite small for thermal neutron energies. A possibility to overcome this limitation, the so-called "supermirror", has been developed and demonstrated by Schoenborn et al. /5/ and Mezei /6/. It consists in using

Bragg reflection from a synthetic, one-dimensional crystal made up of
thin evaporation layers with alternating scattering potentials, in much
the same way as dielectric multilayer systems used as high-reflectance
mirrors in light optics. In Mezei's version the lattice parameter varies
smoothly across the multilayer arrangement. In this way a wide wavelength band may be Bragg-reflected. Such devices are very efficient
polarizers and fairly good reflectors (with about 70 % reflectivity)
over an angular range exceeding the range of usual total reflection by
a factor of two.

3. NEUTRON GUIDE TUBES

The phenomenon of total reflection may be utilized for neutron beam
transport in "neutron guide tubes", in a similar way as light can be
conducted in light pipes. A neutron guide tube /7/ consists of an evac-
uated tube surrounded by mirror walls of high surface quality, usually
made of float glass with a thin nickel coating. Nickel is chosen as reflecting material because of its relatively high scattering potential
of about 2×10^{-7} eV.

The use of guide tubes offers the advantage over non-reflecting beam
tubes, that fairly large distances between neutron source and experiment can be bridged without significant loss in luminosity. This is
very important for time-of-flight measurements where long flight paths
are necessary for good resolution, but also for background considerations. The background due to γ-rays and epithermal neutrons from the
source can be practically eliminated by a guide tube because the glancing angle for these radiations would be extremely low, and direct sight
to the source may be avoided completely by appropriate curvature.

A great number of guide tubes have been installed until now at various
Institutes. For instance, guide tubes with a total length of over 500 m
provide thermal and cold neutron beams for a variety of instruments at
the High-Flux Reactor, Grenoble.

A further application of neutron guide tubes is the guide for polarized
neutrons /8/, which consists of magnetized ferromagnetic mirror walls.
The advantage of a polarizing guide over a simple mirror consists in
the possibility to provide larger beam divergences.

4. PECULIARITIES OF VERY-SLOW NEUTRON PROPAGATION

Now I want to draw your attention to some peculiarities of low-energy neutron propagation, connected with the effects of gravity and static magnetic fields. In applying the concept of refraction to inhomogeneous magnetic fields one must be aware of the limitation of this simple picture. It provides an adequate phenomenological description only as long as the neutron spin orientation relative to the field is a constant of motion, i. e., if the spin is able to follow adiabatically changes of field orientation along the neutron trajectory. The criterion for this is that the Larmor precession frequency in the local magnetic field should be significantly larger than the frequency of field rotation as "sensed" by the moving particle. This condition is usually satisfied for slow neutrons except possibly in critical regions where the magnitude of the field is small and its orientation changes rapidly. In such circumstances spin flip relative to the field may occur, and the simple formalism of an index of refraction breaks down.

The effect of gravity or magnetic fields is most pronounced for neutrons of very low energy, although the gravitational interaction has also been observed with thermal neutrons, using interferometric techniques /9/. Since the fields usually vary very slowly in space on a scale determined by the neutron wavelength, the WKBJ approximation as well as Ehrenfest's theorem of quantum theory provide a transition to classical mechanics. Thus, a neutron should describe the same trajectories under the influence of these potentials as a classical particle. This means, for instance, that a very slow neutron should describe a parabolic trajectory in the earth's gravitational field, but this result can also be obtained from the variation of the refractive index with height, similarly as light rays are curvilinear in media with a gradient of the refractive index.

The free fall of neutrons in the gravitational field has been utilized extensively for precise measurements of scattering lengths for slow neutrons /10/, making use of the fact that the energy of vertical motion accumulated by a particle upon falling through a given vertical distance may be determined very accurately. Thus, in the gravity refractometer built by Koester in Garching cold neutrons start from the reactor source horizontally at a height determined by a narrow entrance slit, and after having travelled for about 100 m they hit a horizontal mirror surface at a variable vertical distance H below the entrance slit. Then the limiting height H_{crit} for total reflection is simply related to the mean scattering length of the constituent atoms of the mirror. One of the results of such measurements is that the gravitational acceleration is the same for the neutron as for any other particle within an accuracy of less than 10^{-3}, which sets an upper bound to speculations about a possible spin dependence of gravity which would be the consequence of a postulated "gravitational dipole moment" /11/.

As another example of utilization of gravity, Fig. 1 shows the scheme of a gravity diffractometer for ultracold neutrons /12/. This instrument was used to study the diffraction of neutrons with wavelengths of about 1000 Å from a ruled grating. The neutrons again start horizontally from an entrance slit, fall in the gravitational field and are reflected from various mirrors in much the same way as in a billiard game. They first hit the vertical grating. After two further reflections from a horizontal and a vertical mirror they travel upwards and are collected by an exit slit at the heighest point of their flight parabola. The highest point is chosen because it is the focussing point for the beam which is much wider along other parts of the trajectory. In the diffraction process at the grating the neutrons gain vertical momentum, $2\pi\hbar m/d$ (d: groove spacing, m: order of diffraction), which may be measured by the change of the vertex height of the analyser flight parabola. One result of the experiment was a lower limit of 10^6 Å for an "intrinsic coherence length of the neutron wave train". This does not contradict the expectation of an infinite coherence length (apart from effects due to the finite neutron lifetime), derived from the simple consideration that the solutions of the Schroedinger equation in free space are plane waves without spatial limitation.

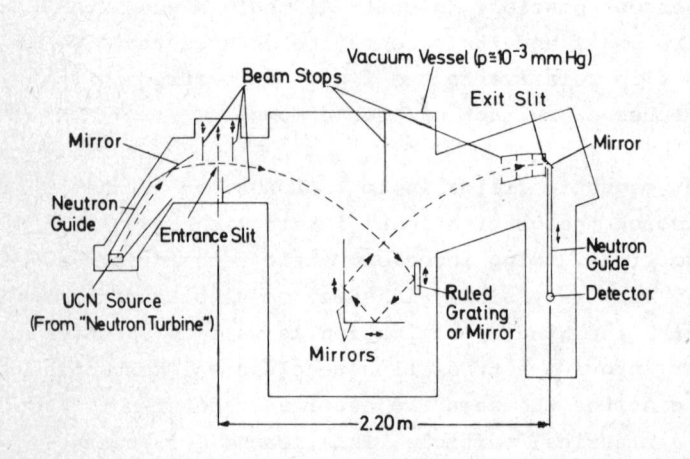

Fig. 1 Scheme of the neutron gravity diffractometer at Garching

Another example where the effect of gravity is significant is the problem of image formation with ultracold neutrons, say, with the futuristic aim of a neutron microscope. The variable refractive index of space in the presence of gravity, and its dependence on wavelength lead to chromatic aberration of space, which must be corrected for in an optical system. One possibility of compensation is provided by a "zone mirror" /13/, where diffraction and reflection at a concave reflecting zone plate are combined to achieve achromatism.

Let me conclude with an interesting example of the influence of magnetic fields on neutron propagation in the "neutron storage ring" developed and successfully tested by a group of the University of Bonn /14/. This device consists in a superconducting ring magnet as shown schematically in Fig. 2. The magnet generates a toroidal hexapole field with higher multipole components. Due to the magnetic dipole interaction very slow neutrons may be trapped in the toroidal region, and they were observed to circle around the ring until they disappear due to β-decay. It is hoped that magnetic storage in this or other forms may ultimately provide an improved value for the free neutron lifetime.

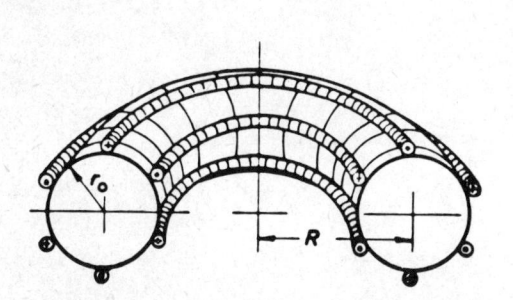

Fig. 2 Simplified scheme of a magnetic storage ring for neutrons

REFERENCES

/1/ Kosvintsev, Yu.Yu., Kushnir, Yu.A., Morosov, V.I., Stoika, A.D., Strelkov, A.V.: Preprint R3-11594, Joint Inst. Nucl. Research (Dubna) 1978.
/2/ Herdin, R., Steyerl, A., Taylor, A.R., Pendlebury, J.M., Golub, R.: Nucl. Instr. and Methods 148, 353 (1978).
/3/ Paul, W.: In Proc. Int. Conf. on Nucl. Phys. and Phys. of Fundamental Particles, Chicago, ed. by J. Orear, A.H. Rosenfeld, R.A. Schluter (Univ. of Chicago, 1951), p. 172.
/4/ Vladimirsky, V.V.: Sov. Phys. JETP 12, 740 (1961).
/5/ Schoenborn, B.P., Caspar, D.L.D., Kammerer, O.F.: J. Appl. Cryst. 7 508 (1974).
/6/ Mezei, F.: Comm. Phys. 1, 81 (1976); Mezei, F., Dagleish, P.A.: Comm. Phys. 2, 41 (1977).
/7/ Maier-Leibnitz, H., Springer, T.: J. Nucl. Energy 17, 217 (1963).

/8/ Berndorfer, K.: Z. Physik 243, 188 (1971).
/9/ Colella, R., Overhauser, A.W., Werner, S.A.: Phys. Rev. Lett. 34, 1472 (1975).
/10/ Koester, L.: Springer Tracts in Mod. Phys. 80, 1 - 55 (1977).
/11/ Golub, R.: Inst. Phys. Conf. Ser. No. 42, 104 (1978).
/12/ Scheckenhofer, H., Steyerl, A.: Phys. Rev. Lett. 39, 1310 (1977).
/13/ Steyerl, A., Schütz, G.: Appl. Phys. 17, 45 (1978).
/14/ Kügler, K.-J., Paul, W., Trinks, U.: Phys. Lett. 72B, 422 (1978).

COHERENCE OF ILLUMINATION IN ELECTRON MICROSCOPY

H.A. Ferwerda

Department of Applied Physics, State University at Groningen, The Netherlands.

1. Introduction

The aim of every kind of microscopy is to determine the structure of the object which has given rise to the observed image. In the case of quantitative microscopy one has to know for that purpose not only the optical properties of the imaging system but just as well the characteristics of the illumination of the object under investigation. We shall show that for conventional microscopy it is sufficient to know the second-order statistical properties of the illumination. In the case of microscopy with visible light the second-order statistical properties are described by the theory of partial coherence [1]. The knowledge which has been acquired in this field can directly be transferred to the case of electron microscopy. One might object that in electron microscopy one has to take into account the fact that electrons are fermions while photons are bosons. As spin effects are very small we can safely neglect the spins of the particles which take part in the imaging process. We may also ignore the difference in statistics, Fermi-Dirac for electrons and Bose-Einstein for photons, because the beams used in both types of microscopy have a low degeneracy parameter [2]. So there is no _essential_ difference between microscopy with light and electrons. Consequently it is to be expected, that the results obtained in this paper apply to many different types of imaging: light optical, electron optical, acoustical, etc.

The paper is organised as follows. As preparation for the study of the image formation we discuss in the next section the propagation of a wave function in free space. In Section 3 we shall study the coherence properties of the electron beam emitted by the cathode. Further we introduce the very useful concept of effective source. In Section 4 we treat the image formation in the conventional transmission electron microscope (CTEM) while in Section 5 the same problem is studied for the scanning transmission electron microscope (STEM). Finally the equivalence of the imaging by CTEM and STEM will be established.

2. Propagation of a wave function in free space

Let $\psi(\vec{r},t)$ denote a one-electron wave function. We will study the propagation of this wave function through free space. $\psi(\vec{r},t)$ satisfies the time dependent Schrödinger equation

$$-\frac{\hbar^2}{2m} \Delta \psi(\vec{r},t) = -\frac{\hbar}{i} \frac{\partial \psi(\vec{r},t)}{\partial t} \ . \qquad (2.1)$$

We shall almost exclusively have to deal with the corresponding stationary wave function $\psi(\vec{r})$ satisfying the time-independent Schrödinger equation:

$$-\frac{\hbar^2}{2m} \Delta\psi(\vec{r}) = E\psi(\vec{r}), \qquad (2.2)$$

where E is the energy. The relation between E and the wave number k is given by $E=\hbar^2 k^2/(2m)$. (2.2) then becomes

$$(\Delta+k^2)\psi(\vec{r}) = 0, \qquad (2.3)$$

which is the Helmholtz equation.

Let us suppose that $\psi(\vec{r})$ is known on the plane z=0 and that $\vec{r}_s$ is the two-dimensional vector describing the position of a point in this plane: $[\psi(\vec{r})]_{z=0} \stackrel{df}{=} \psi_o(\vec{r}_s)$. According to the Rayleigh-Sommerfeld formula [3] for a plane screen we find for the wave function in the space z>0:

$$\psi(\vec{r}) = -2 \int_\sigma \psi_o(\vec{r}_s) \frac{\partial}{\partial z} \frac{\exp[ik|\vec{r}-\vec{r}_s|]}{|\vec{r}-\vec{r}_s|} d^2 r_s, \qquad (2.4)$$

where it has been assumed that the wave function $\psi_o(\vec{r}_s)$ is only different from zero in the region σ of the plane z=0. (2.4) can be simplified further by noting that for 100keV electrons k is of the order $10^{12} m^{-1}$ so that the far field condition $k|\vec{r}-\vec{r}_s|\gg 1$ is very rapidly satisfied for vectors $\vec{r}$ lying to the right of the plane z=0. (2.4) then becomes approximately

$$\psi(\vec{r}) = -2ik \frac{e^{ikz}}{z} \int_\sigma \psi_o(\vec{r}_s) \exp\left[-2\pi i \frac{\vec{r}.\vec{r}_s}{\lambda z}\right] d^2 r_s. \qquad (2.5)$$

(2.5) gives the wave function in a plane z (z≠0). In order to introduce dimensionless coordinates we shall measure $\vec{r}_s$ in units z and $\vec{r}$ in units λ, the wavelength. With that convention (2.5) becomes

$$\psi(\vec{r}) = -2ikz\, e^{ikz} \int_\sigma \psi_o(\vec{r}_s) \exp(-2\pi i \vec{r}.\vec{r}_s) d^2 r_s. \qquad (2.6)$$

The proportionality constant z should not be disturbing because σ is measured in units z^2.

(2.6) is the fundamental formula describing the propagation of the wave function in free space. The wave function in the plane z≠0 is, apart from a multiplicative factor, the Fourier transform of the wave function $\psi_o(\vec{r}_s)$ in the plane z=0.

3. The electron gun. Effective source

The electrons are emitted by a heated filament or a cathode tip after which they are accelerated by a voltage of the order of 100kV. We shall characterize the electron beam when leaving the electron gun.

The electrons are emitted in a variety of quantum mechanical states: they have a range of different energies (approximately described by a Maxwell distribution), move in different directions, are emitted at different places on the cathode. The set of all electron wave functions compatible with the macroscopic properties of the source form an <u>ensemble</u>. The ensemble is completely specified when each member of the ensemble is assigned a statistical weight. We shall not endeavour to construct the ensemble explicitly because this is not essential for the development of the subsequent theory. We only have to use the existence of certain ensemble averages.

We assume that the electrons which are emitted at different places on the cathode at different times are uncorrelated. Strictly speaking this assumption is not rigorously true [11]: the electrons leaving the cathode have approximately a Maxwellian velocity distribution. If T is the cathode temperature the spread of the momentum of the electrons is of the order $(2mkT)^{\frac{1}{2}}$ (m is the electron mass and K Boltzmann's constant). From the Heisenberg uncertainty relation we find for the spread Δr of the wave packet, describing a single electron: $\Delta r \sim \hbar(2mkT)^{-\frac{1}{2}}$. For a cathode temperature of T=3000°K we find $\Delta r \sim 5$Å. When the wave packets of the electrons overlap we get interference between the electrons and they can no longer considered to be uncorrelated. Experience has shown [12] that in electron microscopy the transverse coherence length of the order of 5Å remains unobserved. So for practical reasons we make the subsequent calculations assuming that the emitted electrons are uncorrelated. This is the traditional picture of an <u>incoherent</u> source. We assume a planar source, where $\vec{r}_s$ describes the position of a point on the source. If $\psi(\vec{r}_s,t)$ is a wave function from the ensemble, the incoherence of the source is expressed by

$$<\psi(\vec{r}_s,t)\psi^*(\vec{r}_s',t')> = I(\vec{r}_s)\delta(\vec{r}_s-\vec{r}_s')\delta(t-t'). \qquad (3.1)$$

The angular brackets denote the ensemble average. $I(\vec{r}_s)$ will be referred to as the "intensity distribution across the source". A further discussion of this quantity can be found in Ref.2. We shall also see that the time coordinate becomes uninteresting because temporal coherence plays no important role in electron microscopy: taking an observation time $\Delta t=1$ sec., we can only observe (temporal) interference within an energy band of width

$$\Delta E \simeq \frac{\hbar}{\Delta t} \simeq 6 \times 10^{-16} \text{eV}. \qquad (3.2)$$

For this reason we can concentrate on <u>spatial</u> coherence. For $\psi(\vec{r},t)$ we take a stationary wave function $\psi(\vec{r})$ belonging to a fixed energy which will not be specified explicitly. (3.1) reduces to

$$<\psi(\vec{r}_s)\psi^*(\vec{r}_s')> = I(\vec{r}_s)\delta(\vec{r}-\vec{r}_s'). \qquad (3.3)$$

For the discussion of spatial coherence one introduces the <u>mutual coherence function</u>:

$$\Gamma(\vec{r},\vec{r}') = \langle\psi(\vec{r})\psi^*(\vec{r}')\rangle, \tag{3.4}$$

where $\vec{r}$ and $\vec{r}'$ can denote any two points. If $\vec{r}_s$ and $\vec{r}_s'$ both lie on an incoherent source we have because of (3.3):

$$\Gamma(\vec{r}_s,\vec{r}_s') = I(\vec{r}_s)\delta(\vec{r}_s-\vec{r}_s'). \tag{3.5}$$

The propagation of the mutual coherence can straightforwardly be deduced from (2.6). Taking the plane of the source to be z=0 the mutual coherence function in a plane z=ℓ_s with position vector $\vec{r}_o$ is given by

$$\Gamma_o(\vec{r}_o,\vec{r}_o') = \langle\psi_o(\vec{r}_o)\psi_o^*(\vec{r}_o')\rangle$$

$$= 4k^2 z^2 \int_{\sigma_s}\int_{\sigma_s} d^2r_s d^2r_s' \Gamma_s(\vec{r}_s,\vec{r}_s')\exp[-2\pi i\{\vec{r}_s\cdot\vec{r}_o - \vec{r}_s'\cdot\vec{r}_o'\}], \tag{3.6}$$

where σ_s is the region occupied by the source. For the incoherent source (3.5) we find

$$\Gamma_o(\vec{r}_o,\vec{r}_o') = 4k^2 z^2 \int_{\sigma_s} d^2r_s I(\vec{r}_s)\exp[-2\pi i\vec{r}_s\cdot(\vec{r}_o-\vec{r}_o')]. \tag{3.7}$$

It should be stressed that the result (3.6) only holds for propagation in <u>free</u> space. This is not true for a real electron gun where the electrons are accelerated after leaving the cathode. Formally that situation can easily be incorporated in our formalism: the motion of the electrons in the rotationally symmetric field of the gun is governed by the equation [4]:

$$\frac{-\hbar^2}{2m}\Delta\psi + \frac{e\hbar}{im}\vec{A}\cdot\nabla\psi + (-e\phi + \frac{e^2}{2m}A^2)\psi = E\psi, \tag{3.8}$$

where $\vec{A}$ is the vector potential and ϕ the scalar potential. As (3.8) is a linear equation the wave function in a plane $z=z_o>0$ can be expressed in the wave function $\psi_s(\vec{r}_s)$ in the plane z=0 using the appropriate Green function $L(\vec{r}_o,\vec{r}_s)$ [5]:

$$\psi_o(\vec{r}_o) = \int_{\sigma_s} L(\vec{r}_o,\vec{r}_s)\psi_s(\vec{r}_s)d^2r_s. \tag{3.9}$$

The mutual intensity in the plane $z=z_o$ is given by

$$\Gamma_o(\vec{r}_o,\vec{r}_o') = \int_{\sigma_s}\int_{\sigma_s} d^2r_s d^2r_s' L(\vec{r}_o,\vec{r}_s)L^*(\vec{r}_o',\vec{r}_s')\Gamma(\vec{r}_s,\vec{r}_s'). \tag{3.10}$$

In order to facilitate the discussion of the imaging process we introduce, following Hopkins [6], the concept of <u>effective source</u>: the effective source is that fictitious planar incoherent source which produces in its far field the same mutual coherence as the actual source, assuming that the radiation propagates through free space. Consulting the result (3.7) we see that $\Gamma_o(\vec{r}_o,\vec{r}_o')$ only depends on the difference $\vec{r}_o-\vec{r}_o'$. So only when the mutual coherence function depends on the difference of its coordinates an effective source can be introduced. In the remainder of this paper we assume this condition to be satisfied.

4. Image formation in CTEM

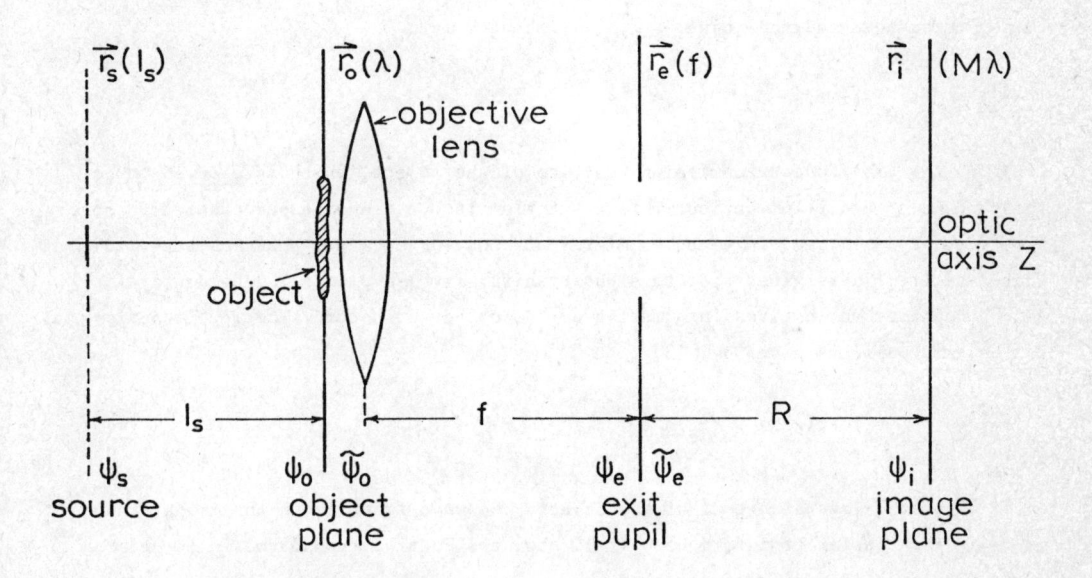

Fig.1: Image formation in CTEM

In Fig. 1 the schematic arrangement of the imaging process has been sketched. The source in this figure is the effective source. ℓ_s is the distance between the plane of the source and the object plane. The latter plane is a plane immediately behind the object. $\vec{r}_s$ is the coordinate in the source plane and is measured in units ℓ_s. $\vec{r}_o$ is the coordinate in the object plane and is measured in units of the wavelength λ. The exit pupil, also denoted as diffraction plane, is supposed to be situated in the back focal plane. $\vec{r}_e$ is the coordinate in the exit pupil and is measured in units f, the focal length. The image is recorded in the image plane, its coordinate $\vec{r}_i$ is measured in units $M\lambda$ where M is the lateral magnification. The distance R between the plane of the exit pupil and the image plane is approximately equal to the radius of the Gaussian reference sphere [7]. The lenses have schematically been shown in the figure. The magnification M=R/f. We shall calculate the intensity distribution in the image plane where we repeatedly use the Equations (2.6) and (3.6). We drop the factors in front of the integral by assuming them absorbed in the symbols of the wave functions in the different planes which have been marked in the figure. The wave function $\psi_s(\vec{r}_s)$ on the source gives rise to an illuminating wave function in the object plane which reads in the absence of the object:

$$\psi_o(\vec{r}_o) = \int_{\sigma_s} \psi_s(\vec{r}_s) \exp(-2\pi i \, \vec{r}_o \cdot \vec{r}_s) d^2 r_s. \qquad (4.1)$$

σ_s denotes the extent of the effective source. Because of the interaction between the electron beam and the object $\psi_o(\vec{r}_o)$ is changed into

$$\tilde{\psi}_o(\vec{r}_o) = \psi_o(\vec{r}_o) T(\vec{r}_o). \qquad (4.2)$$

$T(\vec{r}_o)$ is the amplitude transmission function of the object, which is <u>assumed</u> to be independent of the illumination. This assumption is questionable but crucial for the theory to be developed. In electron microscopy most objects are phase objects: in that case $T(\vec{r}_o)$ only gives rise to a phase shift: $T(\vec{r}_o) = \exp[i\alpha(\vec{r}_o)]$. Evidently $T(\vec{r}_o)$ is the quantity which gives information on the object structure. The wave function in the exit pupil is given by [13]

$$\psi_e(\vec{r}_e) = \int_{\sigma_o} \tilde{\psi}_o(\vec{r}_o) \exp(-2\pi i \, \vec{r}_e \cdot \vec{r}_o) d^2 r_o, \qquad (4.3)$$

where σ_o is the lateral extent of the object. The wave function in the image plane is again the Fourier transform of $\psi_e(\vec{r}_e)$. Now, two Fourier transforms in succession yield the inverted orginal function:

$$\psi_i(\vec{r}_i) = \tilde{\psi}_o(-\vec{r}_o). \qquad (4.4)$$

This situation applies to perfect optical imaging. In electron microscopy this is impossible as has been shown by Scherzer [8] for time independent, charge free rotationally symmetric fields. We have to take into account the aberrations of the optical system which modify $\psi_e(\vec{r}_e)$ by a phase factor into:

$$\tilde{\psi}_e(\vec{r}_e) = \psi_e(\vec{r}_e) \exp[ik\chi(\vec{r}_e)]. \qquad (4.5)$$

$\chi(\vec{r}_e)$ is called the wave-aberration function. Repeated application of (3.6) yields for the mutual coherence in the image plane

$$\Gamma_i(\vec{r}_i,\vec{r}_i') = \langle\psi_i(\vec{r}_i)\psi_i^*(\vec{r}_i')\rangle$$

$$= \int_{\sigma_e\sigma_e}\!\!\int d^2r_e d^2r_e' \int_{\sigma_o\sigma_o}\!\!\int d^2r_o d^2r_o' \int_{\sigma_s\sigma_s}\!\!\int d^2r_s d^2r_s'\; \Gamma_s(\vec{r}_s,\vec{r}_s')\; \times$$

$$\times \exp[-2\pi i\{\vec{r}_o\cdot\vec{r}_s-\vec{r}_o'\cdot\vec{r}_s'+\vec{r}_e\cdot\vec{r}_o-\vec{r}_e'\cdot\vec{r}_o'+\vec{r}_i\cdot\vec{r}_e-\vec{r}_i'\cdot\vec{r}_e'\}]\;\times$$

$$\times\; T(\vec{r}_o)T^*(\vec{r}_o')\exp[ik\{\chi(\vec{r}_e)-\chi(\vec{r}_e')\}]\;, \tag{4.6}$$

where we made use of the independence of $T(\vec{r}_o)$ and $\psi_o(\vec{r}_o)$ by pulling $T(\vec{r}_o)T^*(\vec{r}_o')$ outside the ensemble average. The intensity distribution in the image plane, $I_i(\vec{r}_i)$, is obtained by taking $\vec{r}_i=\vec{r}_i'$ in (4.6). Using (3.5), (3.7) and introducing the (amplitude) point spread function

$$D(\vec{r}) = \int_{\sigma_e} d^2r_e \exp[ik\chi(\vec{r}_e)]\exp[-2\pi i\,\vec{r}_e\cdot\vec{r}] \tag{4.7}$$

we find after a straightforward calculation:

$$I_i(\vec{r}_i) = \int_{\sigma_o\sigma_o}\!\!\int d^2r_o d^2r_o'\,\Gamma(\vec{r}_o-\vec{r}_o')T(\vec{r}_o)T^*(\vec{r}_o')D(\vec{r}_i+\vec{r}_o)D^*(\vec{r}_i+\vec{r}_o'). \tag{4.8}$$

The intensity distribution is recorded on a film which may be scanned subsequently by a densitometer. So the ultimately recorded quantity is

$$B(\vec{r}_i) = \int_{\sigma_i} d^2r_i'\, g(\vec{r}_i-\vec{r}_i')I_i(\vec{r}_i')$$

$$= \int_{\sigma_o\sigma_o}\!\!\int d^2r_o d^2r_o' \int_{\sigma_i} d^2r_i'\, g(\vec{r}_i')T(\vec{r}_o)T^*(\vec{r}_o')\Gamma(\vec{r}_o-\vec{r}_o')\;\times$$

$$\times\; D(\vec{r}_i+\vec{r}_i'+\vec{r}_o)D^*(\vec{r}_i+\vec{r}_i'+\vec{r}_o'). \tag{4.9}$$

$g(\vec{r}_i-\vec{r}_i')$ is the film/densitometer point spread function, σ_i is the extent of the image.

The fundamental problem to be solved in image evaluation is the extraction of $T(\vec{r}_o)$ from (4.9). As $T(\vec{r}_o)$ is a complex quantity one measurement is in general not sufficient for its determination. The problem can be solved by using two or more defocused images or the combination of the intensity distributions in the exit pupil and

image plane. For a survey of the litterature on this subject see [9].
Another remark has to be made in connection with the interaction between the electron beam and the object. In the theory developed above it has been assumed that the energy of the electrons participating in the imaging process is a well-defined quantity. This is necessary when the wavelength is used as a unit of length. So we have only considered the image formed by the elastically scattered electrons. It is an experimental fact that the greater part of the incident electrons are scattered inelastically and have a range of energy values. We shall show how these electrons manifest themselves in the image formation using an argument due to Huiser [9]. Let $\sigma(\theta)$ be the angular distribution of the inelastically scattered electrons where θ is the scattering angle in the laboratory system. This angular spread is superposed on the angular spread $f(\theta)$ of the incident electrons which is determined by the extent of the effective source. The total angular spread is given by the convolution

$$F(\theta) = \int f(\theta')\sigma(\theta-\theta')d\theta' \quad . \tag{4.10}$$

This increased angular spread can be regarded as an increase of the width of the effective source and thus leads to a decrease of the coherence of the illumination.

5. Image formation in STEM

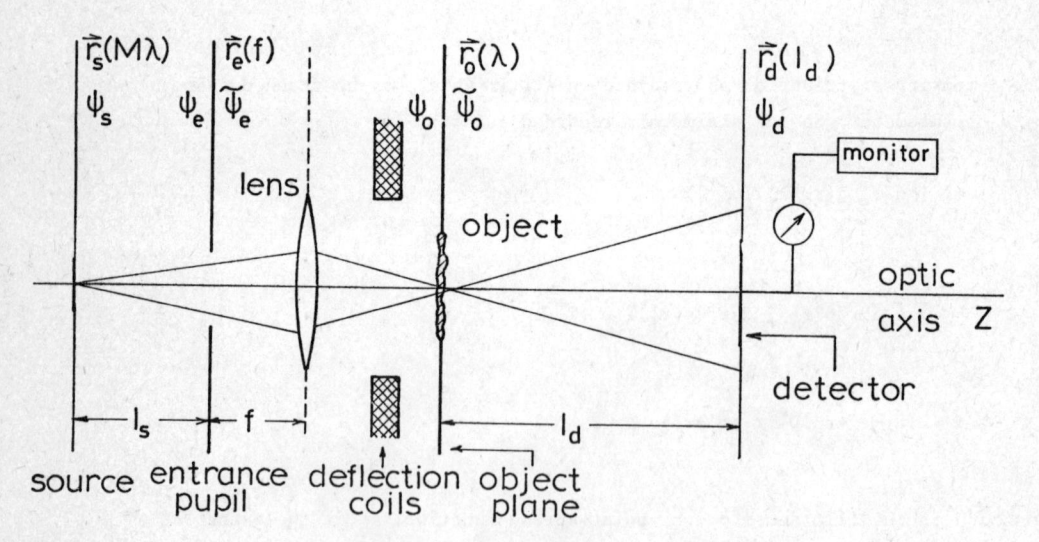

Fig. 2: Image formation in STEM

We first give a description of the imaging process in STEM. The arrangement has been sketched in Fig. 2. The (effective) source is imaged on the object with a large demagnification M^{-1}. In this way only a small portion of the object is illuminated. After interaction with the object the electrons are (partly) collected by a detector which gives rise to a current that is displayed on the monitor screen as a spot whose grey level is proportional to the value of the current. This spot contains only information on the illuminated area of the object. Next we move the illuminating spot across the object by changing the current through the deflection coils. By synchronizing the scanning spot on the object and the corresponding spot on the monitor screen we obtain an image on the monitor screen. We shall not dwell on the advantages or disadvantages of CTEM and STEM. We shall show that both methods of imaging are equivalent. We will see that STEM is in a sense a CTEM operated in reverse direction: when looking in the direction of the negative z-axis we see that there is a magnification $M=\ell_s/f$ between object plane and source plane; the entrance pupil plays the same role as the exit pupil in CTEM. Similarly as in CTEM we encounter the following succession of wave functions: $\psi_s(\vec{r}_s)$, the wave function on the source, gives rise to the wave function $\psi_e(\vec{r}_e)$ in the entrance pupil. Because of the reciprocity principle, which is only approximately satisfied [10], $\psi_e(\vec{r}_e)$ is modified by the same wave aberration function $\chi(\vec{r}_e)$ as in CTEM: the same aberration function applies to both directions of operation of the microscope. In this way $\psi_e(\vec{r}_e)$ becomes

$$\tilde{\psi}_e(\vec{r}_e) = \psi_e(\vec{r}_e)\exp[ik\chi(\vec{r}_e)]. \qquad (5.1)$$

$\tilde{\psi}_e(\vec{r}_o)$ gives rise to the wave function $\psi_o(\vec{r}_o)$ in the object plane, which is modified by the interaction with the object into

$$\tilde{\psi}_o(\vec{r}_o) = \psi_o(\vec{r}_o)T(\vec{r}_o), \qquad (5.2)$$

where it is assumed that the amplitude transmission function $T(\vec{r}_o)$ does not depend on $\psi_o(\vec{r}_o)$. $\tilde{\psi}_o(\vec{r}_o)$ propagates in free space to the detector plane where it gives rise to the wave function $\psi_d(\vec{r}_d)$. In Fig. 2 the units are given for the coordinates in the various planes. Also in the present case we study the image created by the elastically scattered electrons.

The current leaving the detector is given by

$$j_d(\vec{r}_i) = \int_{\sigma_d} h(\vec{r}_d)I_d(\vec{r}_d,\vec{r}_i)d^2r_d, \qquad (5.3)$$

where $h(\vec{r}_d)$ is the detector sensitivity function and σ_d is the area occupied by the detector. $I_d(\vec{r}_d,\vec{r}_i)$ is the intensity in the detector plane when the effective source has been displaced over a distance $\vec{r}_i$ by exciting the deflection coils. Introducing

$$\gamma(\vec{r}_o-\vec{r}_o') = \int_{\sigma_d} d^2r_d h(\vec{r}_d)\exp[-2\pi i\, \vec{r}_d\cdot(\vec{r}_o-\vec{r}_o')] \qquad (5.4)$$

and the amplitude point spread function

$$D(\vec{r}) = \int_{\sigma_e} d^2 r_e \exp[ik\chi(\vec{r}_e)] \exp(-2\pi i \, \vec{r}_e \cdot \vec{r}) \tag{5.5}$$

we find, using the same technique as in the previous section:

$$j_d(\vec{r}_i) = \int_{\sigma_s} d^2 r_s \int\int_{\sigma_o \sigma_o} d^2 r_o d^2 r'_o \, I(\vec{r}_s) T(\vec{r}_o) T^*(\vec{r}'_o) \, \times$$

$$\times \, \gamma(\vec{r}_o - \vec{r}'_o) D(\vec{r}_s + \vec{r}_i + \vec{r}_o) D^*(\vec{r}_s + \vec{r}_i + \vec{r}'_o). \tag{5.6}$$

The results of the imaging in CTEM and STEM can now easily be compared. Comparing (4.9) and (5.6) and also (3.7) and (5.4) we see that the corresponding quantities are as listed in the table below

CTEM	STEM
$I(\vec{r}_s)$	$h(\vec{r}_d)$
$D(\vec{r}_i + \vec{r}_o)$	$D(\vec{r}_i + \vec{r}_o)$
$\Gamma(\vec{r}_o - \vec{r}'_o)$	$\gamma(\vec{r}_o - \vec{r}'_o)$
$g(\vec{r}_i - \vec{r}'_i)$	$I(\vec{r}_s - \vec{r}_i)$

From (5.4) we see that the coherence conditions in STEM are determined by the width of the detector. In CTEM the coherence is governed by the width of the source. The detector sensitivity in STEM corresponds to the intensity distribution across the source in CTEM. The film response in CTEM corresponds to the source intensity distribution in STEM.

References

1. M. Born and E. Wolf: <u>Principles of Optics</u>, 4th ed. (Pergamon Press, Oxford 1970), Chap. 10. See also the lectures by Maréchal, this volume
2. H.A. Ferwerda: Optik <u>45</u>, 411 (1976)
3. J.W. Goodman: <u>Introduction to Fourier Optics</u> (McGraw-Hill, New York 1968) pp. 44,45, where we neglected the obliquity factor which is certainly allowed in electron microscopy because of the small angular apertures
4. W. Glaser: <u>Grundlagen der Elektronenoptik</u> (Springer, Vienna 1952), Eqn. (159-1)
5. See Ref.4, formula (180, 32)
6. H.H. Hopkins: Proc. Roy. Soc. <u>A208</u>, 263 (1951), ibid. <u>A217</u>, 408 (1953)
7. Ref.1, pp. 204, 205
8. O. Scherzer: Z. f. Physik <u>101</u>, 602 (1936)
9. H.A. Ferwerda: <u>Inverse Source Problems in Optics</u>, H.P. Baltes, Ed. (Springer, Berlin, 1978), Chap.2. See also Groningen Univ. Theses Huiser and van Toorn (1979)
10. H.J. Butterweck: A.E.Ü. <u>31</u>, 335 (1977)
11. H.A. Ferwerda and M.G. van Heel: Optik <u>47</u>, 357 (1977)
12. J. Frank, S.C. McFarlane, K.H. Downing: Optik <u>52</u>, 49 (1978) and cited references
13. Ref.3, Section 5.2

PROPAGATION OF SOUND AND ULTRASOUND

IN NON-HOMOGENEOUS MEDIA*

Sébastien M. CANDEL

Office National d'Etudes et de Recherches Aerospatiales

92320 Châtillon, France

and Ecole Centrale des Arts et Manufactures

92290 Chatenay-Malabry

<u>Summary</u> This paper describes numerical techniques which may be used to analyse (acoustic) wave motion in non-homogeneous media. The methods specifically considered are based on (1) one dimensional direct and inverse scattering theory (2) the parabolic approximation (3) the geometrical approximation (4) direct Fourier synthesis of the wave field.

1. INTRODUCTION

Imaging processes in physics generally involve an interaction mechanism between waves and matter. It is clearly important to be able to analyse this interaction in order to conceive, develop or improve imaging techniques. Such studies may now be carried in many complicated situations by direct numerical simulation of the wave propagation process. The aim of the present paper is to describe some typical numerical techniques and the underlying approximations which are currently used to deal with acoustic wave motion. It should be stressed that the same (or very similar) techniques are applied in other fields of physics like optics and electromagnetics. In that sense the methods described have a fair degree of generality. We shall not attempt to present, in this limited space, the standard material which may be found in classical books dealing with wave motion like those of WHITHAM[1], LIGHTHILL[2], MORSE & INGARD[3], MIKLOWITZ[4], ACHENBACH[5], OFFICER[6], EWING, JARDETSKY & PRESS[7], BREKHOVSKIKH [8], KLINE & KAY[9], FELSEN & MARCUVITZ[10], MARCUSE[11]. These texts cover in great detail analytical methods and important results. However, only CLAERBOUT'S monography[12] on geophysical data precessing seems to include a detailed discussions of numerical methods for wave propagation. Thus our aim will be to illustrate such numerical treatments and to this purpose we shall briefly consider four problems : (1) wave scattering in one dimension (direct and inverse), (2) multidimensional wave scattering by weak inhomogeneities (3) three dimensional wave refraction in inhomogeneous media, (4) source radiation in a layered medium. The methods applied to these problems will be respectively (1) one dimensional scattering theory and the forward scattering approximation (FSA), (2) the parabolic equation method (PEM), (3) the geometrical approximation (4) direct Fourier synthesis (DFS) of the wave field. In selecting these topics and techniques we do not intend to cover the whole field of numerical wave propagation but only aim to discuss some powerful tools.

Parts of the work presented were performed at Universite de Technologie de Compiègne and at the California Institute of Technology. The author gratefully aknowledges The support and encouragement given to this research by President Guy Denielou, Professor Frank Marble and Dr. J. Taillet (ONERA).

The choice is also a reflection of the author's own research involvment in the subject for the past few years [13] [14] [15] [16] [17]. No attempt has been made to cite the relevant litterature dealing with the problems treated. Numerous references may be found in the classical textbooks [1] to [12] or in the author's publications.

2. WAVE SCATTERING IN ONE DIMENSION

Plane wave propagation in one dimensional inhomogeneous media has been studied extensively. The problem arises in numerous practical situations, it is a good benchmark for approximation techniques and a prototype for more complicated multidimensional inhomogeneous wave propagation. Scattering theory in one dimension is well described by LAX & PHILIPPS [18] or more concisely by ROSEAU [19] and in numerous textbooks on quantum mechanics. Our objectives will be here (1) to describe numerical techniques for solving the <u>direct</u> problem (2) to show that the forward scattering approximation is well suited to the calculation of the reflection response of the medium (3) to show that the inverse scattering problem (i.e. the problem of deducing the structure of the medium from its reflection response) may be handled with a simple algorithm only involving FFT operations and numerical integration of ordinary differential equations.

<u>Formulation and analysis</u>

We consider a one dimensional inhomogeneous medium in which the sound speed c and density ρ_0 only depend on the z coordinate (figure 1). A monochromatic plane wave propagating in the homogeneous region 1 ($z \leq 0$) penetrates in the inhomogeneous region 3 ($0 < z < \delta$) where it is reflected and transmitted. At $z = \delta$ the medium becomes homogeneous and the wave reaching $z = \delta$ propagates in region 2 ($z \geq \delta$) with no further changes of its amplitude and direction. If the xOz plane is chosen to contain the initial wave vector then all wave vectors belong to that plane because the medium is stratified. Furthermore the wavenumber on the transverse x direction is conserved ($k_\perp = k_0 \sin\theta_0$, where θ_0 designates the initial incidence angle).

We assume that the acoustic wave motion is described by the following set of linear equations

$$\frac{\partial \rho}{\partial t} + \rho_0 \nabla \cdot \underline{v} = 0$$
$$\rho_0 \frac{\partial \underline{v}}{\partial t} + \nabla p = 0 \qquad (1)$$
$$p = c^2 \rho$$

where $p, \underline{v}(u, v, w), \rho$ represent the acoustic pressure, velocity and density perturbations and where ρ_0 and c designate the local density and sound speed. In accord with the above discussion and assumptions the wave field may be cast in the form

$$\begin{bmatrix} p \\ u \\ v \\ w \end{bmatrix} = \begin{bmatrix} p(z) \\ u(z) \\ 0 \\ w(z) \end{bmatrix} e^{ik_\perp x - i\omega t} \qquad (2)$$

which when substituted in (1) yields

$$\frac{d}{dz}\begin{bmatrix} p \\ w \end{bmatrix} = \begin{bmatrix} 0 & i\rho_0\omega \\ ik_\shortparallel Y_\shortparallel & 0 \end{bmatrix}\begin{bmatrix} p \\ w \end{bmatrix} \qquad (3)$$

where $k = \omega/c$, $k_\shortparallel = (k^2 - k_\perp^2)^{1/2}$ is the local longitudinal wave number and $Y_\shortparallel = \frac{1}{\rho_0 c}\frac{k_\shortparallel}{k}$ is a longitudinal admittance (for a plane wave in a homogeneous medium $Y_\shortparallel$ would be equal to the ratio between w and p). The set of equations (3) may be integrated numerically from $z = \delta$ by providing suitable initial conditions in that section. This is accomplished by giving an arbitrary value to $p(\delta)$ for instance $p(\delta) = 1$ and noting that the only wave to exist in that section is plane and propagates outwards so that $w(\delta) = Y_{\shortparallel}(\delta) \, p(\delta)$. A similar procedure is applied in the calculation of the reflection and transmission coefficients : the field in region 1 may be written as a superposition of an incoming and on outgoing wave

$$p(z) = A \exp ik_{\shortparallel 1}z + B \exp -ik_{\shortparallel 1}z$$
$$w(z) = Y_{\shortparallel 1}(A \exp ik_{\shortparallel 1}z - B \exp -ik_{\shortparallel 1}z) \quad , \quad z \leq 0 \quad (4)$$

Combining these two expressions at $z=0$ yields A and B and the transmission and reflection coefficients :

$$R(0) = B/A = [p(0) - w(0)/Y_{\shortparallel}(0)]/[p(0) + w(0)/Y_{\shortparallel}(0)] \quad (5)$$

$$T(0) = 1/A = 2/[p(0) + w(0)/Y_{\shortparallel}(0)] \quad (6)$$

As an illustration let us consider the case of plane waves at normal incidence ($k_{\perp}=0$) impinging on a medium of constant density exhibiting a linear variation of refraction index N (here $N = Y = \rho_0 c^o / \rho_0 c$) figure 1). The reflection coefficient modulus descreases with oscillations when the reduced frequency $\omega_* = \omega \delta /c^o$ increases (figure 2a). At $\omega_* = 0 (\lambda = \infty) \; |R(0)| = |1 - Y_2/Y_1|/|1 + Y_2/Y_1|$, which is the value it would take for a medium with a steplike refraction index. The phase of R increases monotonically with ω_* (figure 2b) but its variation is not linear and this behavior corresponds to the dispersion in time delay of the reflected waves. Other examples may be found in [16].

<u>Local wave decomposition</u>

A very useful concept in the analysis of wave motion is that of the local wave representation. Instead of describing the global evolution of p and w, the field is written as a sum of two local waves. In accord with the usual conventions in geophysics the axis points downwards and we designate by U and D the up-and downgoing local waves. These two waves may be defined <u>a priori</u> (other definitions are possible) by the following expressions and their reciprocal

$$U = \tfrac{1}{2}(p - w/Y_{\shortparallel}) \qquad p = U + D$$
$$D = \tfrac{1}{2}(p + w/Y_{\shortparallel}) \quad (7) \qquad w = Y_{\shortparallel}(D - U) \quad (8)$$

The definitions are such that for a homogeneous medium U and D coincide with up and downgoing plane waves having $k_{\shortparallel}$ as longitudinal wavenumber.

Substitution of (8) into (3) yields a coupled set of differential equations for U and D :

$$\frac{d}{dz}\begin{bmatrix}U\\D\end{bmatrix} = \begin{bmatrix}-ik_{\shortparallel} & 0\\0 & ik_{\shortparallel}\end{bmatrix}\begin{bmatrix}U\\D\end{bmatrix} - \frac{1}{2Y_{\shortparallel}}\frac{dY_{\shortparallel}}{dz}\begin{bmatrix}1 & -1\\-1 & 1\end{bmatrix}\begin{bmatrix}U\\D\end{bmatrix} \quad (9)$$

Note that in a homogeneous medium the two equations decouple and that $U = U(0)\exp -ik_{\shortparallel}z$ $D = D(0)\exp ik_{\shortparallel}z$. Now the form of system (9) suggests a forward scattering approximation which consists of (1) neglecting U in the calculation of D (2) using D obtained in the first step to calculate U. In practice this reduces to replacing system (9) by

$$\frac{d}{dz}\begin{bmatrix}\tilde{U}\\\tilde{D}\end{bmatrix} = \begin{bmatrix}-ik_{\shortparallel} & 0\\0 & ik_{\shortparallel}\end{bmatrix}\begin{bmatrix}\tilde{U}\\\tilde{D}\end{bmatrix} - \frac{1}{2Y_{\shortparallel}}\frac{dY_{\shortparallel}}{dz}\begin{bmatrix}1 & -1\\0 & 1\end{bmatrix}\begin{bmatrix}\tilde{U}\\\tilde{D}\end{bmatrix} \quad (10)$$

This set differs from (9) by a single coefficient (the third coefficient in the second square matrix). A direct numerical check of the FSA is thus straight-forward. Figures 3a & b show the reflection coefficient modulus calculated in this approximation for the medium of figure 1. They only slightly differ from the exact values of figures 2a & b. Many other cases treated in this way lead to the same conclusion that the FSA is an excellent tool for solving reflection problems. What makes this result important is that in the FSA it is also possible to give an analytic expression for the reflection coefficient :

$$\tilde{R}(\omega) = \tilde{U}(0)/\tilde{D}(0) = -\int_0^\delta \frac{1}{2Y_{\shortparallel}} \frac{dY_{\shortparallel}}{dz'} \exp(i\int_0^{z'} 2k_{\shortparallel} dy) \, dz' \quad (11)$$

The reflection coefficient appears as a nonlinear Fourier transform of the relative variation of longitudinal admittance.

Inverse scattering in one dimension

There is an extensive literature on this problem, a detailed presentation is given by CHADAN & SABATIER [20]. Our approach [16] is to devise an approximate solution from expression (11). The method consists of inverting the nonlinear Fourier transform and integrating the result as well as an additional equation for the position to obtain $Y_{\shortparallel}(z)$. The sound speed structure may then be determined if the density is known or similarly the density may be computed if the sound speed is known. If both density and sound speed structures are not known they may be determined from reflection responses obtained in two directions [16]. Figures 4a, b, c show reconstructed and original admittance profiles for three typical media.

For these calculations the waves were normally incident ($k_\perp = 0$, $k_{\shortparallel} = k$) and the density was assumed constant. The reconstructed profiles closely follow the exact profiles except for oscillations associated to the limited bandwidth of the incident waves.

3. THE PARABOLIC EQUATION METHOD (PEM)

The parabolic approximation may be introduced like the FSA of the previous section by performing a local wave decomposition of the wave field. Let us assume for simplicity that the pressure field is governed by the nonhomogeneous Helmholtz equation

$$\nabla^2 p + k^2 p = 0 \quad (12)$$

where $k = \omega/c(\underline{x}) = k_0 N(\underline{x})$

The refraction index $N = c_0/c(\underline{x})$ varies weakly in the domain of propagation. Now, consider waves travelling around a particular axis (for instance the z axis), then up- and downgoing waves may be defined by the following expressions

$$U = \frac{1}{2}(p + \frac{i}{k_0}\frac{\partial p}{\partial z}) \qquad p = U + D$$
$$D = \frac{1}{2}(p - \frac{i}{k_0}\frac{\partial p}{\partial z}) \quad (13) \qquad \frac{\partial p}{\partial z} = ik_0(D-U) \quad (14)$$

Other definitions are possible, those selected are such that in a homogeneous medium, U and D exactly coincide with up- and downgoing plane waves propagating along the z-axis. Equations for U and D may be conveniently derived by first writing (12) in the form :

$$\frac{\partial}{\partial z}\begin{bmatrix} p \\ \frac{\partial p}{\partial z} \end{bmatrix} = \begin{bmatrix} 0 & 1 \\ -\nabla_\perp^2 - k^2 & 0 \end{bmatrix}\begin{bmatrix} p \\ \frac{\partial p}{\partial z} \end{bmatrix} \quad (15)$$

After some algebra one obtains

$$\frac{\partial D}{\partial z} = \frac{i}{2k_0}(k^2 + k_0^2 + \nabla_\perp^2)D + \frac{i}{2k_0}(k^2 - k_0^2 + \nabla_\perp^2)U \quad (16)$$

$$\frac{\partial U}{\partial z} = \frac{i}{2k_0}(k_0^2 - k^2 - \nabla_\perp^2)D - \frac{i}{2k_0}(k^2 + k_0^2 + \nabla_\perp^2)U \quad (17)$$

To decouple this set of equations and in analogy with the treatment of the previous section, we may neglect U in the first equation describing D and then use D obtained in this first step to calculate U (the process may be iterated). Now, in this approximation the equation for D takes the form

$$\frac{\partial D}{\partial z} = \frac{i}{2k_o}(k^2+k_o^2+\nabla_\perp^2)D \qquad (18)$$

It is more meaningful to cast D in the form of a nonhomogeneous plane wave $D = \Psi(x,y,z)e^{ik_o z}$. The amplitude function Ψ is then solution of the following parabolic equation

$$2ik_o\frac{\partial \Psi}{\partial z} + \nabla_\perp^2 \Psi + k_o^2(N^2-1)\Psi = 0 \qquad (19)$$

The parabolic approximation consists of solving this equation instead of (12).* The initially elliptic problem is thus replaced by a Cauchy problem which may be solved with little difficulty by stable accurate and fast finite difference techniques. Further mathematical details are available in references [15] [21] and the literature cited there in (see also BUXTON [22] in the present volume).

Typical example of calculation

We shall here describe one single calculation based on the PEM : the scattering of plane waves by a cylindrical inhomogeneity of circular cross section (figure 5a). The modulus and phase of D in three axial sections are represented on figures 5b & c. Their general appearance is that of Fresnel gratings which arise as a result of the interference between a plane wave (the incident wave) and a cylindrical wave (the scattered wave). The interference fringes are correspondingly located on $kx^2/2z$ = Cste parabolas (figure 5d). The fringe pattern described may be observed experimentally (figure 5e). The image, kindly supplied by B. BAERD was obtained with the Bragg diffraction apparatus developed at ONERA . For this classical problem initially solved by RAYLEIGH the PEM and the Born approximation give comparable results. However for more complicated situations in particular for inhomogeneous domains of large extent the PEM is superior to the classical techniques (like the Born approximation, the Method of Smooth Perturbation, the Geometrical Approximation).

4. THE GEOMETRICAL APPROXIMATION

Despite the conclusion of the last section the geometrical approximation is still quite useful for describing the propagation of waves in media whose characteristic scale l and characteristic time τ greatly exceed the wavelength λ and period $T = 2\pi/\omega$. The geometrical solution is constructed in the following three steps : (1) a dispersion relation between the angular frequency ω and the wavevector $\underline{k}$ is obtained (2) the characteristic lines or "rays" of this relation are determined (3) the field amplitude is calculated along the characteristic lines. Descriptions of the geometrical approximation may be found in WHITHAM [1], LIGHTHILL [2] FELSEN & MARCUVITZ [10], KLINE & KAY [9] and details on its numerical implementation are given in [13], [14]. Briefly outlined, the problem is to construct the field $\underline{u}(\underline{x},t)$ solution of a set of equations of the general form

$$L(\frac{\partial}{\partial t},\nabla;\underline{x},t)\underline{u}(\underline{x},t) = 0 \qquad (20)$$

---

* Other derivations of this equation are available, for instance TAPPERT [21].

The solution is sought in the form of a local plane wave
$$\underline{u}(\underline{x},t) = \underline{u}^{o}(\underline{x},t) \exp i \Psi(\underline{x},t) \quad (21)$$
with $\underline{k} = \nabla \Psi$, $\omega = -\partial \Psi/\partial t$

$\underline{u}^o$ is on amplitude vector and Ψ a phase function. Expession (13) is substituted in (12) and the amplitude vector is then expanded in an asymptotic series $\underline{u}^o = \underline{u}_o + \underline{u}_1 + \underline{u}_2 \ldots$
This procedure yields to the zeroth order a dispersion relation $D(\omega,\underline{k};\underline{x},t) = 0$
and to the next orders equations for the field amplitudes. In the case of acoustic wave propagation. The dispersion relation has the form
$$\omega^2 = k^2 c^2 \quad (22)$$
and a conservation equation for the wave action $A_o = \frac{(\pi_o)^2}{\rho_o c^2} \frac{1}{\omega}$ is obtained at the first order :
$$\frac{\partial}{\partial t} A_o + \nabla \cdot \underline{c}_g A_o = 0 \quad ; \quad \underline{c}_g = \frac{\partial \omega}{\partial \underline{k}} \quad (23)$$
This equation determines the pressure field amplitude π_o. If the propagation medium is time independant and characterized by the refraction index $N = c_o/c$ the rays are given by the following set of equations
$$\frac{d\underline{x}}{ds} = \frac{\underline{p}}{N^2} \quad , \quad \frac{d\underline{p}}{ds} = \frac{\nabla N}{N} \quad (24)$$
where $\underline{x}$ designates the current position on the ray and $\underline{p} = \underline{k}/k_o$ is a reduced wavevector. The phase function is then of the form
$$\Psi(\underline{x},t) = k_o S(\underline{x}) - \omega t \quad (25)$$
and equation (23) becomes
$$\nabla \cdot (\frac{\pi_o^2}{\rho_o c} \underline{\nu}) = 0 \quad (26)$$
where $\underline{\nu} = \underline{k}/k$ designates the unit vector in the wavevector direction.
The rays may be obtained by integrating the set of ODE (24). To obtain the field amplitude, one consistent way is to integrate equation (26) on a "wave volume". One obtains as a result that
$$\frac{\pi_o^2}{\rho_o c} \delta \Sigma = cste \quad (27)$$
along each ray. In this expression $\delta \Sigma$ designates the elementary wavefront section and may be obtained by integrating twelve additional ordinary differential equations [14]. Thus the calculation of the geometrical field reduces to an integration of 18 ODE's. The initial source position $\underline{x}_o$, initial wavevector angles θ_o and α_o are to be specified together with 12 other initial conditions. To illustrate this process we only show some typical ray tracings and refer the reader to [14] for amplitude calculations. Figure 6 gives the ray diagram for a source radiating in an ocean 6000 m deep, characterized by a "Munk canonical sound speed profil". The rays are refracted by the medium and reflected by the surface and bottom. The pattern formed is typical of the SOFAR channel. Figures 7a, b, c display three-dimensional ray tracings for a source placed in a subsonic jet. In this case the medium is inhomogeneous in its mean velocity and sound speed. Downstream rays are refracted away from the jet axis while upstream, they curve towards the axis. Further comments on this problem may be found in [13] or [14].

Ray tracings are particularly useful in delineating wave paths and they provide an intuitive picture of the propagation phenomenon. However the geometrical approximation does not account for wave diffraction and in many instances the rays form envelopes (caustic surfaces) and the calculation of the field amplitude requires special techniques which are not easy to implement numerically.

5. DIRECT SOLUTIONS

Up to now emphasis was placed on approximation methods. However in some instances direct numerical solutions may be constructed. One possible and rather general idea is to solve the problem by Fourier transform technique and then synthesize the solution in real space by making use of the fast Fourier transform. This idea is well suited to source radiation problems in bayered media. This kind of problem has been extensively studied in geophysics [6][7][8] and electromagnetics [10] but it is difficult fo find a complete graphical representation of the wave field even in the simplest situation. We consider a line source situated at a height $z=-h$ in the presence of a plane interface ($z = 0$) separating a region ($z<0$) of density ρ_1 and sound speed c_1 from a region ($z>0$) characterized by ρ_2 and c_2. The source radiates monochromatic waves of angular frequency ω and its amplitude is 2π. The Fourier transform solution of this problem has the form

$$p_1(x,z) = \int \frac{e^{-\gamma_1 |z+h| + i\alpha x}}{\gamma_1} d\alpha - \int \frac{e^{\gamma_1(z-h) + i\alpha x}}{\gamma_1} d\alpha$$
$$+ \int e^{-\gamma_1 h + \gamma_2 z + i\alpha x} \frac{2\rho_2}{\rho_1 \gamma_2 + \rho_2 \gamma_1} d\alpha \quad , \quad z<0 \quad (27)$$

$$p_2(x,z) = \int e^{-\gamma_1 h - \gamma_2 z + i\alpha x} \frac{2\rho_2}{\rho_1 \gamma_2 + \rho_2 \gamma_1} d\alpha \quad , \quad z>0 \quad (28)$$

where $\gamma_1 = (\alpha^2 - k_1^2)^{1/2}$, $\gamma_2 = (\alpha^2 - k_2^2)^{1/2}$

Replacing the continuous Fourier transforms by discrete Fourier transforms and taking some precautions in their evaluation, one obtains a direct solution of the problem without much effort and only little computation time. Figures 8a, b show one such solution for $\rho_1 = \rho_2$ and $c_2 = 2c_1$. The total number of points of the FFT used in this calculation was N = 2048 but only the first 128 points are represented. In region 1, the direct, reflected and refracted waves combine to form an interference pattern while a transmitted wave propagates outwards in region 2. Other cases as well as a comparison with approximate solutions obtained by the stationary phase method will appear in a forthcoming paper [17].

REFERENCES

1. G.B. WHITHAM - Linear and Nonlinear Waves. John Wiley, New York (1974).
2. M.J. LIGHTHILL - Waves in fluids. Cambridge University Press, Cambridge (1978).
3. P.M. MORSE & K.U. INGARD - Theoretical Acoustics. Mc Graw Hill, N.Y. (1968).
4. J. MIKLOWITZ - The theory of Elastic Waves and Waveguides. North Holland, Amsterdam (1978).
5. J.D. ACHENBACH - Wave Propagation in Elastic Solids. North Holland, Amsterdam (1975).
6. C.B. OFFICER - Introduction to Sound Transmission. Mc Graw Hill, N.Y. (1958).
7. W.N. EWING, W.S. JARDETZKY & F. PRESS. Elastic Waves in Layered Media. Mc Graw Hill N.Y. (1957).
8. L.M. BREKHOVSKIKH - Waves in layered media. Academic Press N.Y. (1960).
9. M. KLINE & I.W. KAY - Electromagnetic theory and geometrical optics. John Wiley, N.Y. (1965).

10. L.B. FELSEN & N. MARCUVITZ - Radiation and scattering of waves. Prentice Hall, Englewood Cliffs (1973).

11. D. MARCUSE - Light transmission optics. Van Nostrand N.Y. (1972).

12. J.F. CLAERBOUT - Fundamentals of geophysical data processing. Mc Graw N.Y.(1976).

13. S.M. CANDEL - Analyse theorique et experimentale de la propagation acoustique en milieu inhomogène et en mouvement. Thèse de Doctorat es Sciences. Univ. de Paris VI (1977). ONERA Publication 1/1977.

14. S.M. CANDEL - Numerical solution of conservation equations arising in linear wave theory : application to aeroacoustics. J. Fluid Mech. 83 (1977), 465-493.

15. S.M. CANDEL - Numerical solution of wave scattering problems in the parabolic approximation. J. Fluid Mech. 90 (1979) 465-507.

16. S.M. CANDEL - F. DEFILLIPI & A. LAUNAY - Determination of the inhomogeneous structure of a medium from its plane wave reflection response. Part 1 and 2. Submitted to J. Sound Vib. (1979).

17. S.M. CANDEL & C. CRANCE. Direct Fourier synthesis of wave fields in layered media. To be published (1979).

18. P.D. LAX & R.S. PHILIPPS - Scattering theory. Academic Press N.Y. (1967).

19. M. ROSEAU - Asymptotic wave theory. North Holland, Amsterdam (1976).

20. K. CHADAN & P.C. SABATIER - Inverse problems in quantum scattering theory. Springer N.Y (1977).

21. F. TAPPERT - The parabolic approximation method in wave propagation and Underwater Acoustics, J.B. KELLER & J.S. PAPADAKIS ed. Springer, Berlin (1977).

22. B.F. BUXTON - The elastic scattering of fast electrons. This volume.

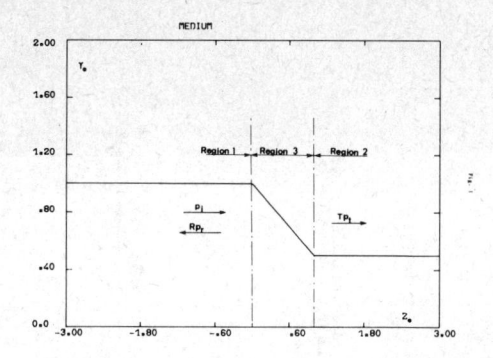

Fig. 1 Geometry of the one dimensional scattering problem.

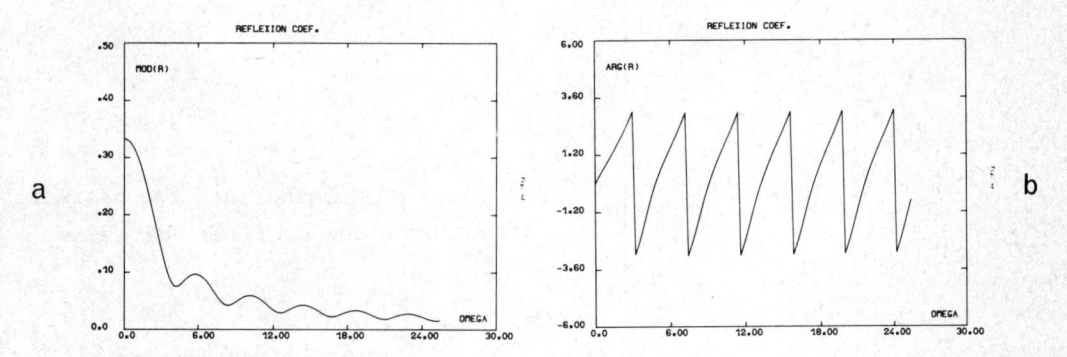

Fig. 2 Propagation of plane waves in a medium characterized by a linear distribution of refraction index. (a) reflection coefficient modulus. (b) Reflection coefficient phase.

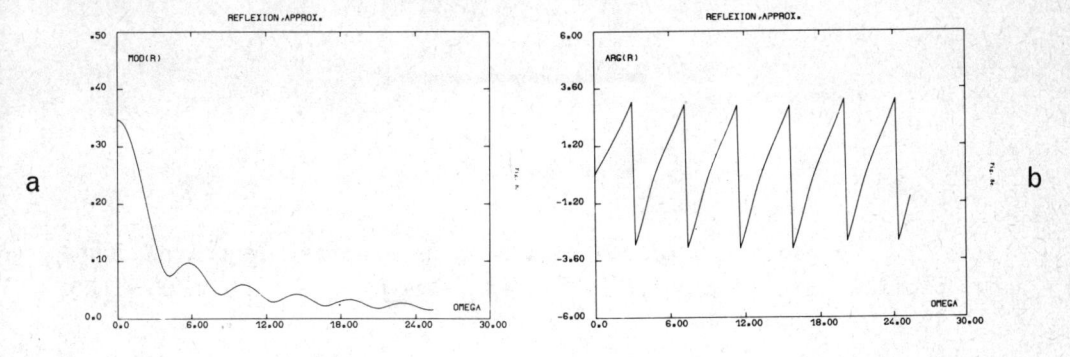

Fig. 3 Reflection coefficient calculated in the forward scattering approximation for the medium of fig. 1. (a) Reflection coefficient modulus. (b) Reflection coefficient phase.

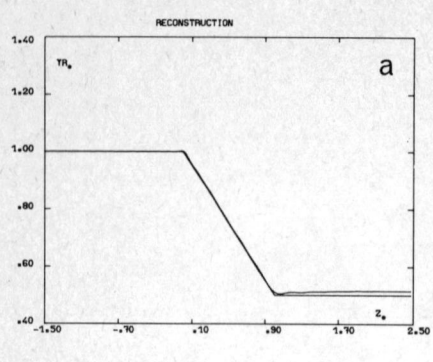

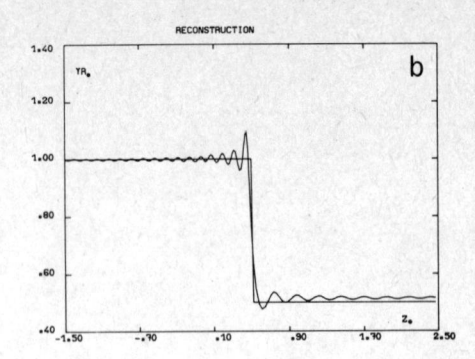

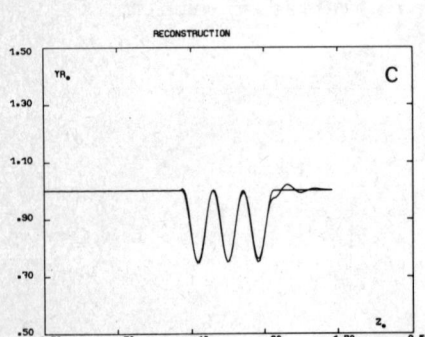

Fig. 4a,b,c Original and reconstructed refraction index profiles for three typical media

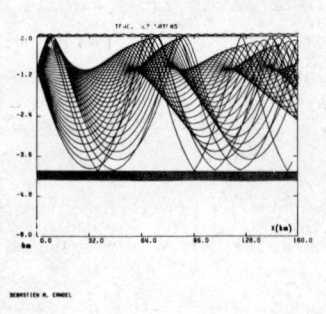

Fig. 6 Ray diagram for a source radiating underwater. Canonical sound speed profile $c(z) = c_0 + c_0 \epsilon (e^{-\eta} - 1 + \eta), \eta = \frac{2(z-z_A)}{B}$ $z_A = 1000$ m, $B = 1000$ m, $\epsilon = 0,57 \times 10^{-2}$, $c_0 = 1500$ m/s.

Note: Figure 5 appears on page 105

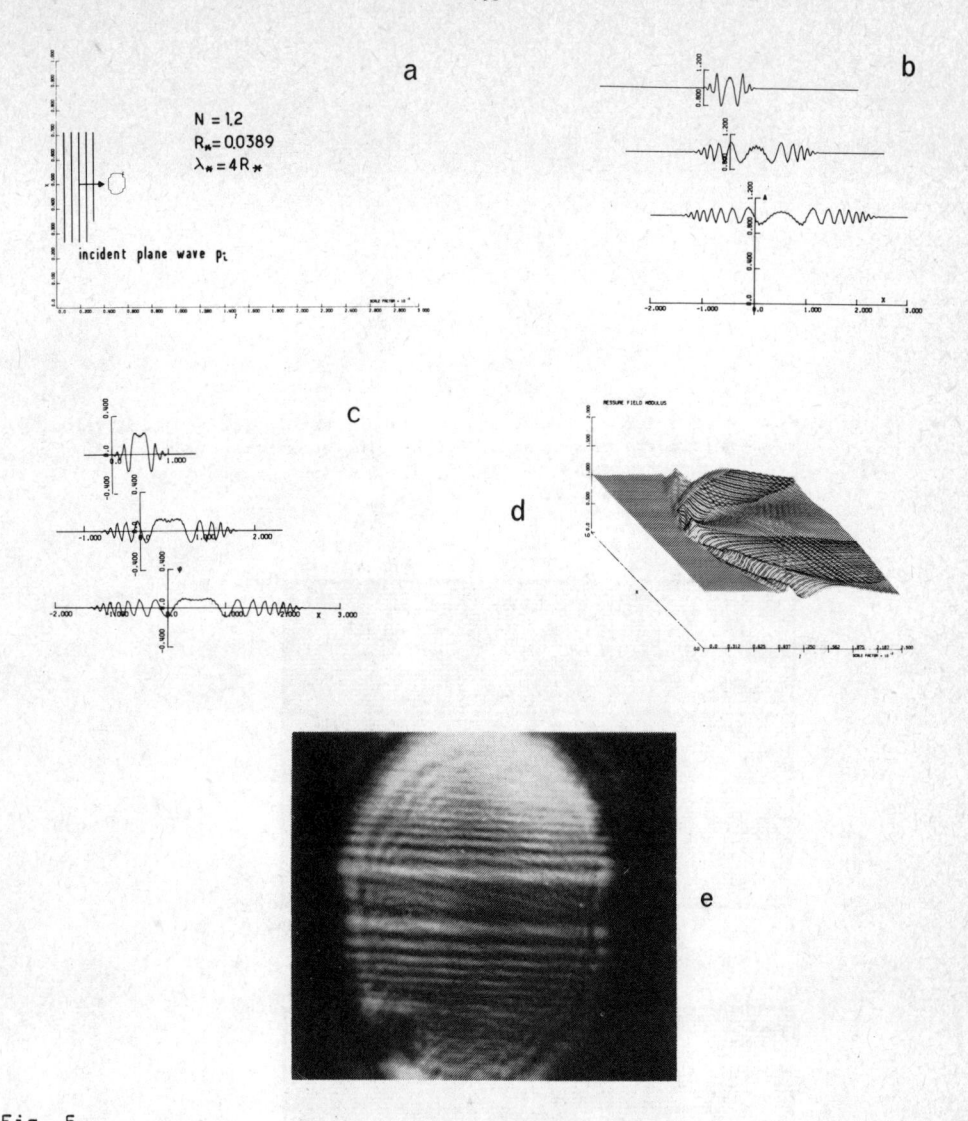

Fig. 5

Scattering of a plane wave by a cylindrical inhomogeneity of uniform refraction index N=1.2. The incident wave propagates in the positive z direction with a wavelength λ_0= 4R. (a) Geometry of the problem. (b) Field modulus in three axial sections. (c) Phase calculated with respect to the incident wave and represented in three axial sections. (d) Field modulus in perspective (fig. extracted from ref. [15]). (e) Fringe pattern observed with a Bragg diffraction apparatus. Ultrasound of wavelength λ_0 = 1 mm impinges on a ϕ 0.4 mm copper wire. The fringes are observed at a 63.8 mm from the wire.

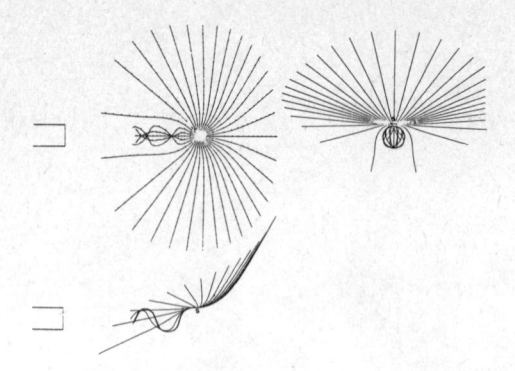

Fig. 7a, b, c Ray tracing for a source situated in a subsonic jet (initial velocity U_0 = 390 m/s, initial temperature T_0 = 870 K). The source is at x_1^0 = 6D, x_2^0 = 0, x_3^0 = 0.6D (fig. extracted from ref. [15]).

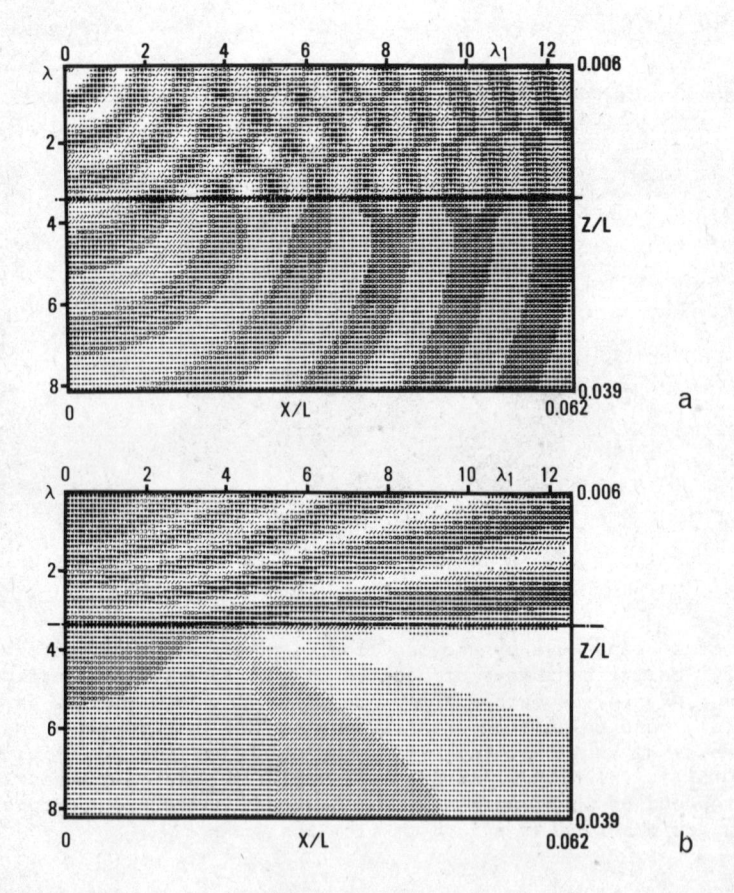

Fig. 8 Source radiation in a layered medium. The solution is obtained by direct Fourier synthesis of the wavefield. (a) Real part of the pressure field. The 10 level grey scale covers the range -1 to +1. (b) Modulus of the pressure field. The 10 level grey scale covers the range 0 to 2. (Figure extracted from ref. [17]).

CLASSICAL APPLES AND QUANTUM POTATOES

Jean-Marc Lévy-Leblond
Laboratoire de Physique Théorique
et Hautes Energies
Université Paris VII
75221 Paris Cedex 05

My subject indeed is about coherence, in our thinking, that is, and imaging processes, namely, mental ones.

What *are* these "things" we are theorizing upon and experimenting with : photons (visible, X-rays, etc.), neutrons, electrons, phonons, etc. ? They show a strange behaviour ; at least, "they are all crazy in the same way"[1]. It has become customary to describe these "things" (let us adopt for the moment this non-committing terminology) in terms of waves and particles, since these were familiar ideas at the beginning of the quantum era.

Stripped down to its essentials, the *particle* concept is that of a *localized* and *discrete* object (where ? how many ? are meaningful questions), while the *wave* concept is that of an *extended* and *continuous* object (a "field") ; both the extension in space-time and the continuous nature accounts for the occurence of new patterns in a region of space-time where two previously separated waves recombine (superpose, if the theory is linear), patterns which we describe as interference or diffraction effects.

Now, the "things" of the quantum level can be counted, that is, they have a *discrete* nature, as they transport quanta of energy and momentum, and they are *extended* as well, since they give rise to diffraction phenomena. It may well be understood why they were first described in terms of particles and waves, since these were the familiar, *classical*, ideas. Depending on the prominent features of the specific situation, a particle-like or wave-like picture suggested itself. Of course, there was a serious logical difficulty in using two contradictory ideas for describing the same "thing". It was the role of the so-called "complementarity" principle to give a sort of philosophical insurance against the risks of conceptual conflicts.

It must be realized to-day that this view of the quantum world, adapted as it was to its first explorations, is totally outdated. In the past fifty years, we have accumulated sufficient familiarity, theoretical as well as experimental, with the quantum world to no longer look at it through classical glasses[2]. Indeed, the very development of quantum physics has shown that our "things" are not waves or particles, not even waves and particles ; in fact, they are neither waves, nor particles. As such, they deserve a new name, "quantons", for instance[3]. Of course, the question is not one of pure terminology, and we may well use old words - waves or particles - provided we know and let know that they have a new meaning, in order to prevent dangerous confusions. Similar situations prevail in daily life : if, on a French menu, you read "bifteck, pommes frites", it means "beefsteak, French fries" and the "pommes" are not "apples" : an old world, "pommes (de terre)", was used in the 17th century for a new thing, "potatoes". However, and this shows the possibility of misunderstanding, on the same menu, "boudin aux pommes" means "black pudding with mashed apples" and the word here keeps its old meaning ; the dish is a very different one !

Various examples of such confusions in quantum theory could be given. Let me mention one : the Landé pseudo-paradox. Landé, as several physicists of his generation, never fully accepted quantum mechanics, and tried to explain it away in terms of an underlying classical theory. Landé stressed what he thought was a serious inconsistency at the very basis of quantum theory, namely, the De Broglie relationship $p = h/\lambda$ between the momentum of a quanton (as "particle") and its wavelength (as "wave"). Indeed, he argued[4], this relationship is incompatible with our views on space-time and the equivalence of inertial frames. In Galilean relativity theory, the momentum of course depends on the frame, while the wavelength is an invariant ; De Broglie's relationship thus cannot be true in all equivalent frames, and must be dismissed as a foundation of quantum mechanics (Landé then showed by an elaborate scheme while, nevertheless, it seemed to work). The amusing point about this argument is that it has nowadays become rather difficult to grasp for quantum experimentalists. Indeed, neutron physicists, especially, know quite well that λ _is_ related to p by $\lambda = h/p$ and _does_ change from one inertial frame to the other. This effect has been experimentally checked and is even put to use in neutron technology[5]. How then can Landé state that λ is an invariant ? This is precisely the crux of the matter ! In classical wave theory, λ indeed is an invariant : the crest-to-crest distance of sea waves is the same to an aircraft pilot and to a lighthouse keeper. However, a quantum wavefunction is essentially different from a classical wave and its wavelength indeed changes, according to quantum theory, so that it matches perfectly the momentum in De Broglie's relationship[6]. In other words, Landé failed to recognize the specific nature of quantum theory and the change in meaning it brings about for "wave functions", "wavelengths", etc. He mistook "pommes (de terre)" for "pommes (de l'air)", and ordered apples with his steak ; no wonder he did not like it. The irony is that to-day we have become so familiar with quantum waves, as shown by the neutron physicists reaction, that we might forgot about classical waves ; it would be a pity, though, not to appreciate the subtleties of "boudin aux pommes".

In this one example, the quantum features of the wavelength are readily understood. But there are many instances, in recent times, of situations where naive interpretations, by not recognizing the abuses of language implied by the "wave" or "particle" terminologies within quantum physics, either led to lengthy arguments about the observability of a predicted physical phenomenon, or to disagreements concerning its explanation once observed, or still to the overlooking of potentially useful effects. As examples, let me give for the first case, the recent controversy about the Aharonov-Bohm effect which is now claimed to be nonexistent[7] ! The Hanbury Brown and Twiss experiment and intensity interferometry in general, offer an illustration of the second case, with the dispute over its quantal or otherwise nature. Finally, the elegant "spin-echo" technique in neutron spectroscopy invented by F. Mesei a few years ago[8], clearly shows the power of a full quantum-mechanical picture beyond simple-minded images.

It still remains to understand the rather large domain of applicability of classical "wave" or "particle" ideas. Let me propose here an elementary metaphor. Consider a cylinder. It presents both circular and rectangular features :

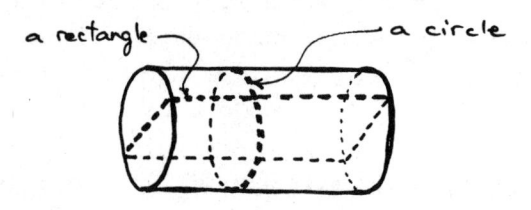

However, it _is_ not a circle or a rectangle, and the ideas of "complementarity" or "duality" of circle and rectangle would be of little use in understanding and describing it. The situation is the same for quantons with respect to waves and particles. A general quantum situation cannot be analyzed in purely classical wave or particle terms, exactly as a cylinder in general transcends the circle or rectangle descriptions. It is true, yet, that very good approximate descriptions of a cylinder can be given with circle or rectangle pictures, exactly as wave or particle pictures may be used for quantons. The conditions for the validity of such approximations nevertheless deserve some attention. One should distinguish, it seems to me, two very different cases where such classical descriptions may be valid, namely 1) a specific description of a general object, 2) a general description of a specific object.

Consider case 1). Using our metaphor, take a general cylinder. Now it is true that there are two specific points of view under which the cylinder is seen either as a circle (along its axis), or as a rectangle (on its side).

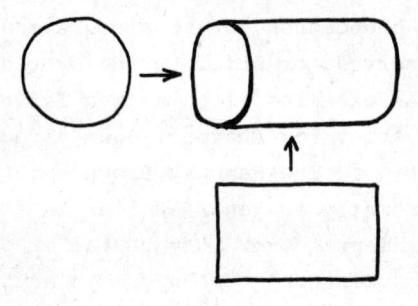

If one sticks close to one of these views, the cylinder will appear little different from a circle or a rectangle. Of course, these points of view do not overlap, so that there is no contradiction in these descriptions ("complementarity"), and they suffice in principle to reconstruct the entire object ("duality"). It is clear, however, that, from a general point of view, the cylinder will not reduce to one or the other of these pictures. The situation is much the same for a quantum object. It is only under very specific circumstances that either a particle, or a wave description will exhaust its appearance. Only in definite experimental setups where the superposition principle is crucial, such as for diffraction effects, will the wave aspect prevail, while the particle aspect will impose itself for instance in collision experiments where the discrete nature of quantons is essential. Any non-specific situation, such as the one of the electron in the hydrogen atom for instance, can only be analysed by using the full conceptual apparatus of quantum theory, as I have already stressed.

Case 2), conversely, concerns the general description of specific objects. Consider, for instance, a very flat cylinder or a very long one :

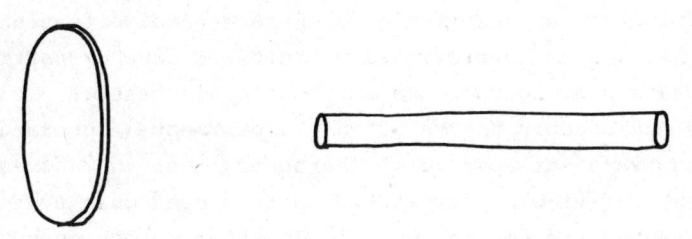

In the first case, a description of the circular faces almost exhausts the appearance of the object, while in the second the side surface only can easily be seen, from practically any point of view. The situation is reversed : one has to adopt a very specific point of view for the rectangular side of the flat cylinder or the circular end of the long one to appear prominently. Similarly, most macroscopic objects, will usually appear either as discrete, localized, chunks of matter ("particles" : pieces of solid material), or as continuous, extended fields (waves", as for the electromagnetic field). If one considers the above flat and long cylinders to be special arrangements of "elementary" cylinders :

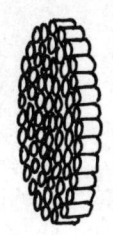

exactly as solids or fields are built out of quantons, the analogy is complete.

Of course, it is the very prevalence in our macroscopic world of these rather specific objects which led us to build the naive ideas sufficient to describe them. As long as there seems to be only very flat or very long cylinders, the ideas of circles and rectangles are sufficient. As long as quantons only show up in large amounts with either wave or particle properties, it is not possible to build concepts pertaining to the description of more general situations ! Also, this explanation of the approximate validity of wave or particle ideas in case 2) has a bearing upon the explanation in case 1). It is because of the macroscopic nature of most experimental apparatus and their ensuing "wave" or "particle" character, that the corresponding specific points of view appear as the most natural ones even when looking at quantum systems in general. In other words, most experimental situations, up to rather recent times, automatically selected particular "windows" through which quantons usually only showed either their wave, or their particle sides.

An important question remains, though : how comes that most macroscopic objects look either as waves, or as particles ? How comes that arrangements of the small cylinders which are not either very long or very flat

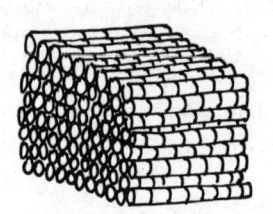

are so uncommon ? Of course, we now know how to build macroscopic systems which exhibit major quantal properties and do not reduce to a classical wave or particle description : a vessel full of superfluid helium or a laser beam are but two examples. Still, all the more so, the question is certainly not a trivial one and much work yet remains to be done before a satisfying answer exists, showing why classical theories (particle mechanics and field theory) give such a good approximation to quantum theory (see a discussion of some recent work in ref. 2).

In other words, while we now have a good grasp, practical and conceptual, of quantum physics, it is classical physics which requires a better understanding...

REFERENCES

(1) R.P. Feynman in The Character of the Physical Law (M.I.T. Press, Boston).

(2) For an overall picture of quantum theory in a modern perspective, see J.-M. Lévy-Leblond in Quantum Mechanics, A Half a Century Later, J. Leite Lopes and M. Paty eds. (Reidel, Dordrecht 1977), pp. 171-206 ; also J.-M. Lévy-Leblond, Int. J. Qu. Chem. 12, suppl. 1, 415(1977).

(3) See for instance M. Bunge, Philosophy of Physics (Reidel, Dordrecht 1973).

(4) A. Landé, Am. J. Phys. 43, 701(1975).

(5) C.G. Shull and N.S. Gingrich, J. Appl. Phys. 35, 678(1964). See also the High Energy Resolution Backscattering Spectrometer with "Doppler drive" (IN 10) at the I.L.L. in Grenoble.

(6) J.-M. Lévy-Leblond, Am. J. Phys. 44, 1130(1976).

(7) P. Bocchieri and A. Loinger, Nuovo Cimento 47A, 475(1978) ; P. Bocchieri, A. Loinger and G. Siragusa, Nuovo Cimento (to be published). G. Casati and I. Guarneri, Phys. Rev. Let. 42, 1579 (1978).

(8) F. Mesei, Z. Phys. 255, 146(1972) ; Physica 86-88B, 1049(1977) ; Neutron Inelastic Scattering 1977, (IAEA), vol. 1, 125(1978), and contribution at this Workshop.

It is a pleasure to thank Françoise Balibar for many discussions and F. Tasset and F. Mesei for supplying experimental informations.

OUTLINE OF A COHERENT THEORY OF SKIING

F. Mezei

Institut Laue-Langevin, 156X, 38042 Grenoble, France

and

G. Joubert

U. E. R. E. P. S., University of Grenoble
38400 Saint-Martin d'Hères, France

I. Introduction

The word "theory" has to be used with caution in connection with something as practical and, consequently, as complex as skiing. As in many (if not most) cases of practical interest, what theory can do here is a sound analysis of the empirical findings, in an attempt to single out the promising ones, or at least to point out the worst errors.

Skiing has changed enormously in the last 10 years. Besides the internal logics of the strive for perfection and performance (especially in a competition sport) this evolution is largely due to the introduction of sophisticated design, new materials and technology into the fabrication of skiing equipment. The properties of today's, even middle priced skis and boots are mostly determined by the purpose for which they were made and not by a restricted choice of materials, as was the case earlier. For example, ski boots made in the last 2-3 years can be at the same time rigid in one direction and elastic in another.

The evolution of the technique of skiing, in interaction with the improvement of the equipment, seems to happen in a very logical direction : towards a simplification. As compared to his predecessors, today's skiers have to do less to achieve the same result or they can go further with the same effort and risk. The body movements become more and more reduced, sober and esthetic at the same. Essentially, modern skiing tends towards an extreme economy of gestures with the main emphasis on the fine movements of maintaining the equilibrium of the body on the skis ; in this respect some of the movements are analogous to the spontaneous balance keeping in everyday walking.

In what follows we will try to outline a coherent image of the key physical aspects of modern skiing. We will consider the basic mechanical properties of skis and boots, the main effects of the ski-snow interaction forces, the mechanisms for piloting the skis and for maintaining the balance.

As far as references are concerned in the literature of skiing it is not customary, and practically not possible, to give a proper referencing of ideas and achievements as it would be ideally meant in science. In any case to our knowledge, the considerations as presented in this paper are original and they are based on a concept of modern skiing developed by one of the authors (G.J.) and described in his latest book (1).

II. The skis and boots

The exact behaviour of a pair of skis on the snow depends on a large number of mechanical parameters. Some dozen of these (shape, resistance to bending and twisting, frequency and attenuation of vibrational modes, etc.) are actually measured and checked by the manufacturers of top quality skis. However, even these data turn out to be insufficient to characterize how skis "feel" in action. At least for the moment, the mechanical analysis is way behind the practical experience. In any case, the most pertinent features are the followings.

The shape of modern skis, narrower at the center than at both the tip and the tail (as exageratedly shown in Fig. 1)

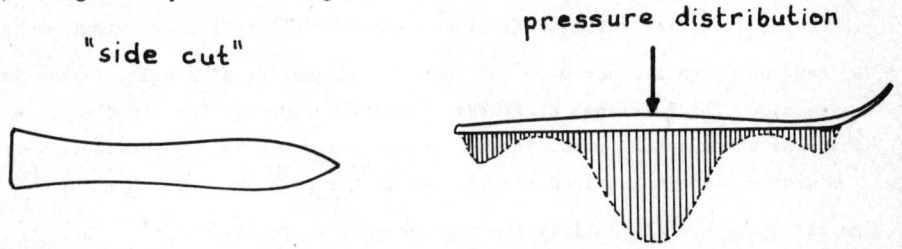

Fig. 1

serves two main purposes :
- the ski put on its edge has a natural tendency to turn following the curved line of the edge, the "side cut".
- the widening towards the tail provokes an extra friction on the snow at this level, which has been found to be vital for the directional stability of the skis at higher speeds.

The typical distribution of pressure under the ski on a flat snow surface is also shown in Fig. 1. It has been found empirically that this is the type of pressure distribution which is best suited for downhill skiing. (For cross country skis the distribution has to be more uniform). The pressure maxima at the extremities and the exact shape (width) and position of the main pressure peak at the ski center are determined by the camber and the stiffness distribution of the ski, respectively. More pressure, for example, towards the tip produces more friction and snow packing resistance at this point, which results in lesser directional stability, i.e. in a sense the ski is easier to turn. Therefore downhill racing skis have relatively soft tips, which not only improve their stability but also make them faster because the deformation of the snow under the ski tips happens more gradually.

The notion of "stiff" and and "soft" here is meant to be with respect to an empirically established "neutral" average. In much of what follows we have to assume that we are dealing with such "neutral", well balanced skis. In particular skis, which have lost their original elasticity due to a long wear and tear, will most often respond quite differently. In general, our analysis leads to the demonstration of tendencies which are clearly perceptible and effective close to an equilibrium situation, but can be obscured by serious deviations. E.g. a bad directional stability of the skis in straight gliding can be due to a habitual forward lean of the skier, and not to the skis. (Furthermore in such a case the compensation of this effect by the use of soft tipped skis might only give very temporary results, since the skier is likely to give freer reign to his, now less dangerous, bad habit). More generally, the modification of the above ski-snow contact pressure distribution by a displacement of the point of application of the body weight, ideally between the centre of the boot sole and a point a few centimeters behind it, is a fine but crucial element in controlling the behaviour of the skis.

The most delicate and not very well understood quality of a pair of skis is its hold on icy snow. The torsional stiffness, vibrational properties and the "side cut" certainly play an essential role. For an average skier the main perceptible difference between a medium and a top priced ski is just in the grip on ice.

The ski boots have to satisfy two main mechanical conditions :
- They should be completely rigid laterally to permit a precise control of the ski edge.
- They should be elastic within certain limits in the forward-backward direction. This latter property has been given much emphasis in the last few years. It is essential for the fine control of the skis and for a smooth maintenance of equilibrium, because it permits delicate displacements of the body centre of gravity with respect to the feet without parasitic torques. Let us consider the example of an off-balance forward lean of the skier. If the boots block any movement of the shins, immediately the torque puts an extra weight on the ski tips producing an increased snow resistance, which in turn tends to add to the forward lean. If, on the other hand, the shin has a certain freedom, the skier will have a chance to reestablish a perfect balance by pushing his feet forward without putting much torque on the boots. However, if he should fail to do so, finally the boots have to step in by gradually blocking the shin at a limit of forward bend and providing the skier with a support for a "passive" recovery of the equilibrium. The same argument applies to a backward lean too but in this case the amplitude of the movement is about 3 times less. This relative freedom of the shin (of about 15° total amplitude) in the forward-backward direction is also necessary for a soft, balanced up-down motion of the skier, e.g. to absorb the shock of bumps.

II. Ski-snow contact forces and the steering of the skis.

From the mechanical point of view, downhill skiing can be divided into three main components : control of the skis, the maintenance of balance and the trajectory of the body centre of gravity.

The particularity of the control of the skis is due to their large moment of inertia, in comparison with the moment of inertia of the bare feet or even that of the whole straightened out body with respect to the vertical axis through the centre of gravity. Thus it is understandable that in most cases the sheer muscular force of the skier(the "active pivoting")proves to be insufficient in itself to change the direction of the long axis of the skis rapidly enough.

In actual fact, the most frequently used mechanism for rotating the skis into a desired direction procedes by the utilization of the friction and snow resistance forces acting on the skis which we call "ski-snow forces". We have already mentioned some of their effects : namely those which assure a good directional stability to the skis when they are flat on the snow and sliding in a straight line. We now come to the point where we show how the ski-snow forces can also make the skis turn ("passive pivoting").

One of these cases has also been mentioned above : the ski carving the snow on an edge tends to turn following the curved line of the edge. The curvature of the ski-snow contact line is even increased by the bending of the edged ski under the weight of the body.This "carved turning effect" plays a major role in performance skiing (high speeds, icy slopes) since it produces little friction and allows the skier to avoid undue slide-slips, but it cannot result in sharp turns. In addition, it often requires an extraordinary muscular "fixation" effort to keep the ski on the edge, and the roughness of the snow surfaces and/or the vibrations of the ski can make continuous carving impossible. Thus the "carved turns" more often than not represent but an objective for the top level skier to be approached by a constant effort in both holding the ski and readjusting its position in order to minimize the sideslip.

The predominantly utilized mechanism of "passing pivoting" of the skis is what we call the "sideslip turning effect". This is effective when the skis are put more or less on the edge, but not sufficiently to produce a carving effect, thus they sideslip rather freely. The friction and snow resistance forces acting on the sideslipping ski (Fig. 2)

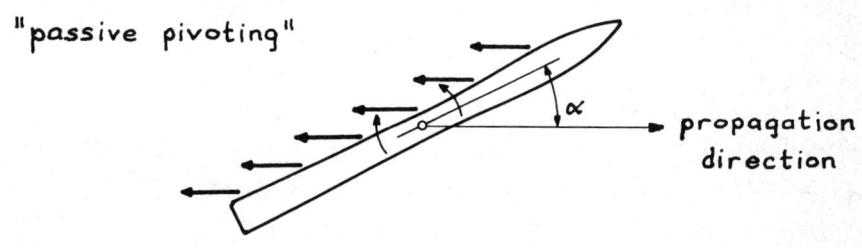

Fig. 2

will in general produce a resultant torque with respect to the ski centre of gravity which can be in either sense, depending on the distribution of these forces along the ski.Thus, putting more weight towards the tip increases the ski-snow forces in front of the ski centre, which results in a net torque turning the ski even further from the propagation direction, i.e. increasing the sideslip angle α. To the contrary, more pressure on the tails tends to turn the skis into the direction of sliding, i.e. to reduce α.

In practice, consciously or not, skiers use to different degrees all of the three pivoting mechanisms we have considered. Up to a good medium level one can steer one's skis with the help of, nearly exclusively, the "sideslip turning effect" and a very little of "active pivoting". In competition skiing the strive for a minimum friction requires a preferential use of "active pivoting" and "carved turning effects".

In addition, there is a dynamical aspect of these turning effects, which plays an essential role in wedeln, i.e. in a fast succession of left and right turns. At the final phase of a turn the skis are subject to a maximum of pressure,vertically and sideways, which makes them bend, since they are more or less on the edge. At the end of e.g. a left turn there is a release of this pressure and the unbending ski rebounds from the snow upwards and to the left. If the tail of the ski was under bigger pressure, i.e. bent more strongly, then the rebound will also turn the ski to the right, similarly to what has been discussed in connection with Fig. 2. At the same time the muscles in the body compressed and stretched to the maximum by the twist of the feet to the left will react like springs at the release producing an upward rebound and a strong turning effect on the skis to the right. Thus at the end of a turn the pressure release creates both an "active" (muscular) and a "passive" rebound pivoting, which both act in the sense of starting an opposite turn if the pressure was predominantly on the ski tails.

III. Maintenance of equilibrium

The particularity of keeping balance on skis as compared with our everyday experience is the low friction in the direction of the ski axis. In the "forward-backward" plane S defined by the direction of the ski and the vertical the condition of equilibrium is that the application line of the resultant force $\vec{F}$, exerted by the snow surface on the skier's feet via the skis and the boots, passes through the body centre of gravity C. (Fig. 3 a). The sum of $\vec{F}$, of the gravity $\vec{G}$ and of the air resistance $\vec{R}$ produces the acceleration of the skier. Due to the small friction the force $\vec{F}$ is close to perpendicular to the ski-snow line in the plane S. The principal method of equilibration is then to displace the feet, i.e. the action line of $\vec{F}$ with respect to the centre of gravity C, which we call a "lean". E.g. by pushing the feet forward, $\vec{F}$ will pass in front of C, and this creates a backward torque. Notice that this "backward lean" does not mean necessarily a backward inclination of the body.

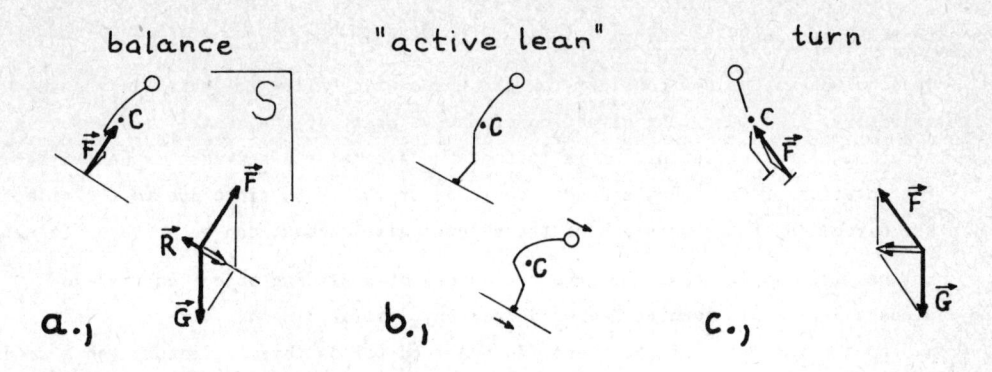

Fig. 3

There are basically two ways of producing a lean. The "active lean" consists of pushing the feet and the upper part of the trunk together with the head and the arms simultaneously in the same direction (Fig. 3 b). The basic feature of this manoeuvre is that it does not involve a change of the angular momentum of the skier : the movement of the feet is compensated by that of the upper body i.e. the skier-ski system behaves as a closed system. This is why it can be executed very rapidly. This is one of the fundamental element of modern skiing.

We call "passive" those leans which involve a change of the body angular momentum. These are obtained either by the torque produced by a pressure against the boots and/or by just letting the body lean under the effect of an off-balance. The maintenance of balance is ideally achieved by a judicious combination of reaction and anticipation. The unconscious reactions of continuous balance readjustment are in fact active leans. Pressure against the uppers of the boot can only serve as an emergency solution.

Anticipation is particularly essential at higher speeds when the skier has to adjust his position by "active" or "passive" leans in view of the obstacles to come. E.g. a slowing down due to a patch of deep snow has to be anticipated by a backward lean, an acceleration due to a steeper slope by a forward lean.

We have seen above that the fine piloting of the skis requires a precise positioning of the weight along the skis which means relative displacements of the body centre of gravity with respect to the feet. Thus the balance keeping "leans" have at the same time an effect on the turning of the skis. E.g. a forward lean during a (sideslipped) turn will tend to pivot the skis more into the turn, while a slight backward lean will tend to straighten out the turn. This is precisely the key tendency in modern skiing: minute adjustments of the centre of gravity serve at the same time the maintenance of equilibrium and the control of the skis. The virtuosity in the use of these fine controls is one of the main qualities of the very good skiers, which makes them slide faster and with less effort even on very easy slopes.

IV. The trajectory and the turns.

To conclude we can now consider the ski turns which allow the skier to choose his trajectory. In a turn, the direction of the velocity of the centre of gravity is modified. This innocent definition is in fact fundamental : a ski turn is not primarily the rotation of the body, and not even that of the skis, since due to the sideslip the direction of the skis and of the skiers' displacement can be quite different.

Let us first notice that the free trajectory of a sliding object on a slope is obviously a parabola turning downward from any initial diagonal direction and approaching the fall-line of the slope. In order to follow this trajectory (on a smooth snow) the skier only has to let his skis be as loose and free as possible, i.e. to keep them flat while in a perfectly balanced stance. On the other hand, the skier has to make an effort to leave the free trajectory, e.g. to make the turn sharper right from the beginning or to turn beyond the fall-line.

In a turn, the skier has to exert a laterally oblique pressure on the snow, the reaction of which produces the centripetal acceleration of the centre of gravity. The lateral equilibrium is maintained by an inward lean of the centre of gravity C with respect to the ski which carries most of the pressure. This lean is achieved more naturally and with less displacement of C if most of the weight is on the foot external to the turn (Fig. 3 c). The internal foot serves as security for the case when the external ski would not hold laterally on the snow.

The main problem the skier has to solve is to steer his skis with respect to the desired trajectory of the centre of gravity, on a trajectory which produces the necessary oblique pressures to provide the centripetal force and corresponds at the same time to a perfect lateral and forward-backward equilibrium. The delicacy of the problem is that, as we have seen, balance and pivoting of the skis are strongly related. Let us consider a very important practical aspect of this. The outside ski, which carries most of the skier's weight and thus meets more snow resistance, moves along a longer trajectory in the turn compared with the rest of the body. If the skier does not make a specific effort of pushing his outside ski forward during the turn, it will drag behind i.e. a forward lean will occur. The situation is aggravated by the fact that in the second half of the turn, where the skier's path deviates most from the free trajectory, there will be maximum pressure on the skis, quite similar to that experienced when going through a depression. This pressure tends to increase any initial (forward) off-balance, and makes the balance correction very difficult. As we have seen above, the forward lean tends to increase the sideslip angle, and the skier ends his turn in a position badly adapted for starting the next turn, especially since he has to recover balance first. On the other hand, a slight extra pressure on the ski tail creates a rebound which literally projects the skis into the next turn.

Finally we are left to consider the beginning of a turn. The passive pivoting of the skis assumes some sideslip with the ski tips toward the inside of the turn.

To initiate this sideslip on a smooth snow surface a minimum of active pivoting is
sufficient in the first part of the turn which corresponds to the free trajectory.
To do this in a heavy, catchy snow, one has to unweight, i.e. to accomplish the
initial pivoting while the skis are more or less in the air during a rebound or even
a jump. In this case the conservation of angular momentum requires an angular acceleration of the body in the opposite sense, i.e. opposite to the turn being commenced.
In effect the moment of inertia of the straight body is fairly small, but it becomes
appreciable in a somewhat bent position with the seat, the shoulders the head and
the widely opened arms moved away from the vertical axis. Thus instead of an apparent
twist of the body, the angular momentum change rather corresponds to taking the so
called "position of angulation", i.e. moving the slightly bent upper body towards outside to the turn. In addition, particularly, in deep snow where there is no possibility
of freeing the skis, an extra momentum can be given to the turn starting pivoting by a
"hip projection". This means an initial rotation of the slightly bent body in the sense
of the turn to come and a pivoting of the skis during the subsequent de-acceleration
of the angular angular momentum acquired in the body rotation.

V. Conclusion

The key points of modern skiing can be summarized as follows. Most of the piloting
of the skis can be obtained by utilizing the "free-trajectory" tendency of the skis to
approach the fall line and by "passive pivoting" based on a fine control of the ski-snow
contact forces. This latter is achieved by delicate adjustment of the equilibrium,
which is predominantly maintained by "active leans" i.e. closed system displacements
of the feet with respect to the body centre of gravity. The moment of inertia of the
body does not play an essential role in the turning of the skis, except under extreme
conditions, like heavy snow, certain competition turns and - when the skier, out of
balance, has no other means left at his disposal.

Reference

(1) G. Joubert, Le Ski: un art.....une technique.
   (Arthaud, Grenoble 1978)

WAVE-MATTER INTERACTIONS : A GENERAL SURVEY

Jean Sivardière
Centre d'Etudes Nucléaires de Grenoble
DRF/Laboratoire d'Interactions Hyperfines
85 X - 38041 Grenoble Cédex, France

1. The formalism of collision theory

The angular distribution of particles scattered by a fixed potential is conveniently described in term of a differential cross section $d\sigma/d\Omega$ (θ,ϕ). If the potential is spherical as we assume now, the total cross section σ is given by :

$$\sigma = 2\pi \int_0^\pi \frac{d\sigma}{d\Omega}(\theta) \sin\theta \, d\theta \qquad (1)$$

σ is the section of a rigid sphere which classically would give the same scattering effects.

For an elastic process, the wave function is a solution of the Schrödinger equation :

$$\left[\frac{-\hbar^2}{2m}\nabla^2 + V(r)\right]\psi(\vec{r}) = E\,\psi(\vec{r}) \qquad (2)$$

As usual in a collision problem, we look for the behavior of $\psi(\vec{r})$ at great distances from the scattering region. An asymptotic form of $\psi(\vec{r})$ to the order $1/r$ is :

$$\psi(\vec{r}) = e^{ikz} + f(\theta)\frac{e^{ikr}}{r} \qquad (3)$$

The first term is a plane wave of wave vector $\vec{k}_o = k\vec{z}$ and represents a particle of mass m, linear momentum $\vec{p}_o$ and energy E : $\vec{p}_o = \hbar\vec{k}_o$, $E = \hbar^2 k^2/2m$. The second term represents a particle moving radially outward. $f(\theta)$ is called the scattering amplitude or length, and is related to the cross section by :

$$\frac{d\sigma}{d\Omega}(\theta) = |f(\theta)|^2 \qquad (4)$$

2. The Born approximation

The equation (2) is equivalent to:

$$(\nabla^2 + k^2)\psi(\vec{r}) = U(r)\psi(\vec{r}) \qquad (5)$$

with $U(r) = 2mV(r)/\hbar^2$. Introducing the Green's function:

$$G(\vec{r}, \vec{r'}) = -\frac{1}{4\pi}\frac{e^{ik|\vec{r}-\vec{r'}|}}{|\vec{r}-\vec{r'}|} \qquad (6)$$

which is a solution of:

$$(\nabla^2 + k^2) G(\vec{r},\vec{r'}) = \delta(\vec{r}-\vec{r'}) \qquad (7)$$

we get the general solution of (3):

$$\psi(\vec{r}) = \psi_0(\vec{r}) + \int U(r')\psi(\vec{r'}) G(\vec{r},\vec{r'}) d^3\vec{r'} \qquad (8)$$

The integral form (8) of the Schrödinger equation (5) is well adapted to the calculation of $\psi(\vec{r})$ by successive approximations. If $U(r) = 0$, $\psi(\vec{r}) = \psi_0(\vec{r})$. If $U(r)$ is a perturbation ($V \ll E$), $\psi(\vec{r})$ can be approximated by:

$$\psi_i(\vec{r}) = \psi_0(\vec{r}) + \int U(r')\psi_{i-1}(\vec{r'}) G(\vec{r},\vec{r'}) d^3\vec{r'} \qquad (9)$$

We may write also:

$$\psi(\vec{r}) = \psi_0(\vec{r}) + \int U(r')\psi_0(\vec{r'}) G(\vec{r},\vec{r'}) d^3\vec{r'}$$
$$+ \iint U(r')U(r'')\psi_0(\vec{r''}) G(\vec{r'},\vec{r''}) G(\vec{r},\vec{r'}) d^3\vec{r'}\, d^3\vec{r''} + \ldots \qquad (9a)$$

The scattered wave is a sum of partial terms $\psi^{(k)}$, each of them representing the incident wave scattered k times. This multiple scattering can be neglected if $V \ll E$.

In the first order Born approximation, we choose: $\psi_i(\vec{r}) = \psi_1(\vec{r})$ and use the Fraunhofer approximation:

$$k|\vec{r}-\vec{r'}| \simeq kr - \vec{k}\cdot\vec{r'} \qquad (10)$$

with $\vec{k}$ parallel to $\vec{r}$ and $|\vec{k}| = k$, whence:

$$\psi(\vec{r}) = e^{ikz} - \frac{e^{ikr}}{4\pi r}\int U(r') e^{-i\vec{k}\cdot\vec{r'}} \psi_0(\vec{r'}) d^3\vec{r'} \qquad (11)$$

of the form (3) with :

$$f(\theta) = -\frac{1}{4\pi} \frac{2m}{\hbar^2} \int V(r') e^{i\vec{q}\cdot\vec{r'}} d^3\vec{r'} \qquad (12)$$

θ is the scattering angle and $\vec{q} = \vec{k}_o - \vec{k}$, $|\vec{q}| = q = 2k \sin\frac{\theta}{2}$. Choosing $\vec{k}$ as the polar axis, we get as well :

$$f(\theta) = -\frac{2m}{\hbar^2 q} \int_0^\infty V(r') r' \sin qr' \, dr' \qquad (13)$$

$f(\theta)$ can be written as a matrix element :

$$f(\theta) = -\frac{1}{4\pi} \frac{2m}{\hbar^2} \langle e^{i\vec{k}\cdot\vec{r}} | V(r) | e^{i\vec{k}_o\cdot\vec{r}} \rangle \qquad (12a)$$

whence :

$$\frac{d\sigma}{d\Omega} = -\frac{1}{4\pi} \frac{2m}{\hbar^2} |\langle k|V|k_o\rangle|^2$$

which is just the Fermi golden rule.

$f(\theta)$ is maximum for $\theta = 0$ and independant of E : $f(0) \sim \langle V(r)\rangle$. If θ increases from zero, the waves reemitted by the different points of the scattering region begin to interfere, so that $f(\theta) \simeq 0$ for $\theta > 1/qa$ where a is the range of the potential.

However if $ka \ll 1$, whence $qa \ll 1$, these interference effects do not manifest and the scattering is isotropic. This is the case if $V(r) \sim \delta(r)$ or $a \simeq 0$, or for small incident energies (but still larger than V for the Born approximation to be valid) : in either case, the wavelength λ of the incident wave is much larger than a, and the wave is not sensitive to the details of the scattering region.

3. The method of partial waves and phase shifts

This method is more general and mostly useful for low incident energies. Since the potential is spherical, we may develop $\psi(\vec{r})$ in spherical harmonics :

$$\psi(\vec{r}) = \sum_\ell R_\ell(r) P_\ell(\cos\theta) \qquad (14)$$

$R_\ell(r)$ satisfies :

$$\frac{1}{r^2} \frac{d}{dr}\left(r^2 \frac{dR}{dr}\right) + \left[\frac{2m}{\hbar^2}(E-V) - \frac{\ell(\ell+1)}{r^2}\right] R = 0 \qquad (15)$$

For $r \gg a$, $R_\ell(r)$ has the general form:

$$R_\ell(r) = A_\ell \left[\cos \delta_\ell \cdot j_\ell(kr) + \sin \delta_\ell \cdot n_\ell(kr) \right] \quad (16)$$

($\delta_\ell = 0$ if $V(r) = 0$) and for $r \gg 1/k$:

$$R_\ell(r) = A_\ell \frac{\sin\left(kr - \ell\frac{\pi}{2} + \delta_\ell\right)}{kr} \quad (17)$$

δ_ℓ is called the phase shift. We must now identify the two asymptotic forms (2) and (17). Using the Bauer formula:

$$e^{ikr \cos \theta} = \sum_{\ell=0}^{\infty} (2\ell+1) i^\ell j_\ell(kr) P_\ell(\cos \theta) \quad (18)$$

We get:

$$f(\theta) = \frac{1}{2ik} \sum_\ell (2\ell+1) \left(e^{2i\delta_\ell} - 1 \right) P_\ell(\cos \theta) \quad (19)$$

whence, using the orthogonality of the Legendre polynomials, the total cross section:

$$\sigma = \frac{4\pi}{k^2} \sum_\ell (2\ell+1) \sin^2 \delta_\ell \quad (20)$$

The partial wave method is interesting for the following reasons:
- the expression (19) is valid whatever the value of V compared to E (case of slow electrons);
- the function $j_\ell(kr)$ is proportionnal to r^ℓ for small values of r and presents a maximum for $r \sim \ell/k$. Suppose that $a \ll \ell/k$: the partial wave ℓ is almost zero within the potential region and $\delta_\ell \simeq 0$, so that the summation in (19) can be limited to $\ell = ka = a/\lambda$. The physics is very simple: if $ka < \ell$ the particle does not enter the potential region since $p = \hbar k$ and the angular momentum is: $\hbar\ell = pr_0 = \hbar k r_0$, $r_0 \gg a$; r_0 is the impact parameter.

3.1. Born approximation for the phase shifts

If $V \ll E$, $\hbar^2/ma^2$, we get from a perturbation treatment the value of δ_ℓ for fast particles:

$$\tanh \delta_\ell = -\frac{2m}{\hbar^2} k \int_0^\infty V(r) j_\ell^2(kr) r^2 dr \quad (21)$$

If $ka \gg 1$ or $|V| \ll \hbar^2/ma^2 \ll E$, all the δ_ℓ are small ($\delta_\ell \sim 1/ka$ or $1/\sqrt{E}$) so that :

$$e^{2i\delta_\ell} - 1 \simeq 2i\delta_\ell \qquad (22)$$

whence :

$$f(\theta) = \frac{1}{k} \sum_\ell (2\ell+1)\, \delta_\ell\, P_\ell(\cos\theta)$$

$$= -\int_0^\infty r^2 V(r) \left[\sum_\ell (2\ell+1)\, j_\ell^2(kr)\, P_\ell(\cos\theta) \right] dr \qquad (23)$$

which is just the Born scattering length, since :

$$\sum_\ell (2\ell+1)\, j_\ell^2(kr)\, P_\ell(\cos\theta) = \frac{\sin qr}{qr} \qquad (24)$$

If $ka \ll 1$ or $|V| \ll E \ll \hbar^2/ma^2$ (short range potential) it is found that $\delta_\ell \sim (ka)^{2\ell+1} = E^{\ell+1/2}$, only the s-part ($\ell = 0$) of the incident wave is diffracted, and the scattering is isotropic and independant of E : $\delta_o = \alpha k$, $f(\theta) = \delta_o/k = \alpha$ (α and V have opposite signs). More generally the scattering of the ℓ-wave is negligible if $\ell/k \geqslant a$, $ka \leqslant \ell$.

3.2. Scattering of slow particles

If $E \ll V$, the Born approximation is no longer valid. δ is calculated by fitting the function $R_\ell(r)$ for $r < a$, which may have an analytical form, to the asymptotic solution.

For example in the case of a perfectly rigid sphere of radius a, we have :

$$\tanh \delta_\ell = \frac{j_\ell(ka)}{n_\ell(ka)} \qquad (25)$$

$$\simeq (ka)^{2\ell+1}$$

In the low energy limit $ka \ll 1$, only $\delta_o = ka$ has to be considered ; the scattering is isotropic and independant of E : $d\sigma/d\Omega = a^2$, $\sigma = 4\pi a^2$. In the high energy limit, $ka \gg 1$, $\sigma = 2\pi a^2$ which is twice the classical result, and the scattering is strongly anisotropic whereas $d\sigma/d\Omega$ classical $= a^2/4$.

4. Scattering of X-rays by electrons

The classical Thomson theory for free electrons is valid for the bound electrons of atoms : the electronic Rayleigh scattering is isotropic (except for a factor 1 or $\cos\theta$ depending on the polarization of the incident wave), and the scattering length $f(\theta)$ is equal to the classical radius of the electron $r_e = e^2/me^2 = 2,82.10^{-13}$ cm $= 2,8.10^{-5}$ Å. This result agrees with the general considerations above : the electron behaves as a rigid sphere of radius $a = r_e \ll \lambda \sim 1$ Å, only the s-partial wave is scattered in the low-energy limit.

Of course the scattering by the electronic distribution of an atom is no longer isotropic, since λ is of the same order of magnitude as the atomic radius R_{at}. The scattering length is then equal to $r_e f_x(\theta)$, where the form factor $f_x(\theta)$ is the Fourier transform of the electronic density. $f_x(0) = Z$, the number of electrons of the atom.

Rayleigh scattering by nuclei of mass M is negligible since the corresponding scattering length is equal to $e^2/Mc^2 = r_e/1840$. Finally the magnetic scattering length is equal to $r_e \lambda_c/a_o = r_e/137$ (the first Bohr radius, $a_o = \hbar^2/me^2 = 0.529$ Å and the Compton wavelength $\lambda = \hbar/me = 3.86 \times 10^{-3}$ Å).

5. Scattering of electrons by a Coulomb potential

We consider the screened Coulomb potential

$$V(r) = q_1 q_2 \frac{e^{-\alpha r}}{r} = V_o \frac{e^{-\alpha r}}{r} \qquad (26)$$

In the Born approximation we get from (13) :

$$f(\theta) = -\frac{2mV_o}{\hbar^2 q} \int_0^\infty e^{-\alpha r'} \sin qr' \, dr' \qquad (27)$$

The integral is equal to $q/(q^2+\alpha^2)$ whence for $\alpha \to 0$:

$$f(\theta) = -\frac{2mV_o}{\hbar^2 q^2} = \frac{q_1 q_2}{E} \frac{1}{\sin^2\frac{\theta}{2}} \qquad (28)$$

By accident, this is just the classical Rutherford formula. The total cross section σ is infinite, but this singularity can be discarded since unscreened Coulomb potentials are never found in nature.

We calculate now the scattering of an electron by an atom of atomic number Z. $V(\vec{r}) = e\,\phi(\vec{r})$. From the Poisson equation, we obtain the Fourier components of $\phi(\vec{r})$ from the Fourier components of the charge density $\rho(\vec{r})$:

$$\phi_q = \frac{4\pi}{q^2}\,\rho_q \tag{29}$$

whence

$$\phi_q = \frac{4\pi}{q^2} \int \rho(\vec{r})\,e^{i\vec{q}\cdot\vec{r}}\,d^3\vec{r}$$

$$= Ze - e\int n(\vec{r})\,e^{i\vec{q}\cdot\vec{r}}\,d^3\vec{r} \tag{30}$$

$n(\vec{r})$ is the electronic density :

$$\rho(\vec{r}) = Ze\,\delta(\vec{r}) - e\,n(\vec{r}) \tag{31}$$

We get finally :

$$f(\theta) = -\frac{2me^2}{\hbar^2}\,\frac{Z - f_x(\theta)}{q^2} \tag{32}$$

$$f(0) = -\frac{1}{a_0} \int n(\vec{r})\,r^2\,d\vec{r} \tag{33}$$

which is finite and of the order of the atomic radius.

6. Scattering of slow neutrons

Consider the partial incident p-wave ($\ell = 1$). The corresponding impact parameter r_o is given by : $r_o = 1/k \sim \lambda$ so that $r_o \sim 1\,\text{Å} \gg R$, which is the radius of the nucleus : only the s-wave is scattered. Let us put : $f(\theta) = -b$. We may then use the so-called Fermi potential :

$$V(\vec{r}) = \frac{2\pi\hbar^2}{m}\,b\,\delta(\vec{r}) \tag{34}$$

b is the Fermi length, which neglecting spin, is given by the Breit-Wigner formula :

$$b = R - \frac{\Gamma_n}{2k\,E_r} + i\,b'' \tag{35}$$

E_r is the resonance energy of the compound nucleus, Γ_n the width of the level ; b" describes absorption. In general $b \simeq R > 0$, and $\delta_o = -\pi$.

If we now consider spin, we must consider the two possible compound nuclei with spins $I \pm \frac{1}{2}$ and introduce two Fermi lengths b^+ and b^-. We get then :

$$\frac{d\sigma}{d\Omega} = |\bar{b}|^2 + B^2 \, I(I+1) \qquad (36)$$

with :

$$\bar{b} = b^+ \frac{I+1}{2I+1} + b^- \frac{I}{2I+1}$$

$$B = \frac{b^+ - b^-}{2I+1} \qquad (37)$$

The first term of $d\sigma/d\Omega$ leads to coherent scattering, the second term to incoherent scattering due to the disorder in the spin state of the compound nucleus. Since in this case scattering may involve a change in the spin state of the nucleus, $f(\theta)$ is an operator ($\vec{\sigma}$ is the spin operator of the neutron) :

$$f(\theta) = \bar{b} + B \, \vec{\sigma} \cdot \vec{I} \qquad (38)$$

Considering now electromagnetic scattering by the magnetic moment $\vec{M} = 2\mu_B \vec{S}$ of the electronic target, we find :

$$f(\theta) = b + p \, \vec{Q} \cdot \vec{\sigma} \qquad (39)$$

where $\vec{Q}$ is the projection of $\vec{S}$ on the diffusion plane and :

$$p = \gamma \, \frac{r_e}{2} \, f_m \qquad (40)$$

$\gamma = 1,91.f_m$ is the magnetic form factor, that is the Fourier transform of the density of magnetic electrons. If the target is paramagnetic, there is only incoherent magnetic scattering. If the target is magnetic and if moreover the incident neutron beam is polarized, nuclear and magnetic scattering interfere.

7. Refractive index

Consider forward scattering : the phase shift in the scattering process induces a change of the phase velocity of the beam, and the refractive index is given by :

$$n = 1 - \frac{\lambda^2}{2\pi} F(0)$$

$$\simeq 1 - \frac{V_0}{2E} \qquad (41)$$

where $F(\theta)$ is the structure factor of the unit volume and :

$$F(0) = \frac{2m}{4\pi \hbar^2} V_0 \qquad (42)$$

For X-rays, we know from Maxwell theory that : $n^2 = \varepsilon$. ε is the dielectric constant, which is related to the uniform electric susceptibility χ_0 by : $\varepsilon = 1 + 4\pi \chi_0$. From Debye's theory :

$$\chi_0 = - \frac{\mathcal{N} e^2}{m \omega^2} \qquad (43)$$

where $\mathcal{N}$ = NZ is the number of electrons per unit volume and N the number of atoms per unit volume. Since $n \simeq 1$, we get :

$$n = 1 - \frac{2\pi \mathcal{N} e^2}{m \omega^2}$$
$$= 1 - \frac{\lambda^2}{2\pi} r_e \mathcal{N} f_x(0) \qquad (44)$$

in agreement with formula (41).

Similarly for neutrons (μ_n is the neutron moment and B the induction)

$$n = 1 - \frac{\lambda^2}{2\pi} b \mathcal{N} \pm \frac{\mu_n B}{2E} \qquad (43)$$

Finally for electrons, the index is larger than 1 and given by (41) with V_0 negative.

8. Applications

We compare the use of X-ray, electron and neutron diffraction in structure determination. <u>X-ray diffraction</u> is widely used. However it has some limits :
- light atoms are not easily seen, since $f_x(\theta) \sim Z$,
- atoms of neighbouring Z are difficult to distinguish,
- X-rays are not sensitive to magnetic structure.

<u>Fast electron diffraction</u> ($\lambda \sim 0.05$ eV) is a good complementary method, which is more sensitive to light atoms but not to superstructures involving neighbouring Z atoms. <u>Slow electrons</u> ($\lambda \sim$ Å) do not penetrate crystals : they are used to study surface structures, but dynamical effects are important.

<u>Neutron diffraction</u> is necessary to solve magnetic structures. It is useful also for solving crystallographic structures, since b does not vary regularly with Z : the position of light atoms can be determined (e.g. metallic hydrides) and superstructures resolved (e.g. FeCo).

Complex structures are determined by the X-N and X-X methods : neutron diffraction (or large angle X-ray diffraction) gives the position of the nuclei (or core electrons) and the temperature factor, then the electronic distribution (or the external electronic one) is determined from the above results and the X-ray (or small angle X-ray) diffraction diagram.

References

G.E. BACON,
Neutron diffraction, Oxford, 1962.

A. GUINIER,
X-ray diffraction, Freeman, 1963.

N.F. MOTT and H.S.W. MASSEY,
The theory of atomic collisions, Oxford, 1949.

R. OMNES,
Introduction à l'étude des particules élémentaires, Ediscience, 1970.

L.I. SCHIFF,
Quantum Mechanics, Mc Graw Hill, 1968 (see chapters 5 and 9 for collision theory)
See also textbooks by Messiah, and Landau and Lifschitz.

B.K. VAINSHTEIN,
Structure analysis by electron diffraction, Pergamon Press, 1964.

BASIC X-RAY INTERACTIONS WITH MATTER

L. Gerward
Laboratory of Applied Physics III
Technical University of Denmark
DK-2800 Lyngby, Denmark

1. Interaction processes

In the photon energy range below 1 MeV the electromagnetic interaction between x-rays and matter leads to the following processes:

1) Photoelectric absorption (true absorption). In this process a photon disappears and an electron is ejected from an atom. The electron carries away all the energy of the absorbed photon minus the energy binding the electron to the atom.

2) Coherent scattering (Rayleigh scattering). This is a process by which photons are scattered by bound atomic electrons and in which the atom is neither ionized nor excited. The scattering from different parts of the atomic charge distribution is then coherent, i.e. there are interference effects. For an assemblage of atoms the scattering from the different atoms may add up coherently or incoherently depending on the atomic arrangement.

It is often assumed that the Rayleigh scattering is elastic. However, the scattering from a free atom is never strictly elastic because of the recoil energy. In a crystal lattice the recoil is negligible because it is absorbed by the crystal as a whole. However, the interaction with the lattice vibrations (phonons) may give rise to inelastic thermal diffuse scattering. This scattering is at least partially coherent. In conclusion, the Rayleigh scattering from an assemblage of atoms may be coherent or incoherent and elastic or inelastic.

3) Incoherent scattering (Compton scattering). This process can be visualized as a collision between the photon and one particular electron. The photon loses some of its energy and its wavelength is accordingly modified. Thus the scattering is inelastic. No interference takes place between radiation scattered by different electrons of the material system.

The interaction processes also produce fluorescent x-ray and a number of emitted electrons, namely photo-electrons, Auger electrons and Compton recoil electrons. These x-rays and electrons can be analysed by spectroscopic methods and give information about the element composition and the electronic structure of the sample.

The total photon-atom interaction cross section can therefore be written

$$\sigma_{tot} = \tau + \sigma_{coh} + \sigma_{incoh} \qquad (1)$$

where τ is the photoeffect cross section. The magnitudes of the cross sections are shown in Fig. 1 for germanium.

2. Rayleigh scattering

The high-energy limit of the atomic scattering factor, f_0, for an atom of atomic number Z is defined as the matrix element

$$f_0(\vec{q},Z) = \sum_{n=1}^{Z} <\psi_0|\exp(i\,\vec{q}\cdot\vec{r}_n)|\psi_0> \qquad (2)$$

where ψ_0 is the ground-state wave function of the atom and $\vec{q}$ is the scattering vector. The atomic scattering factor may also be expressed as the Fourier transform of the electron density. In the forward direction one has $f_0(0,Z) = Z$ (Fig. 2).

Only in the case of the hydrogenic atoms (single electron) can the Schrödinger equation for the ground-state wavefunctions, ψ_0, be solved in a simple analytic form. For many-electron atoms the Coulomb repulsion between the electrons prevents exact solutions, and so a variety of approximations have been used.

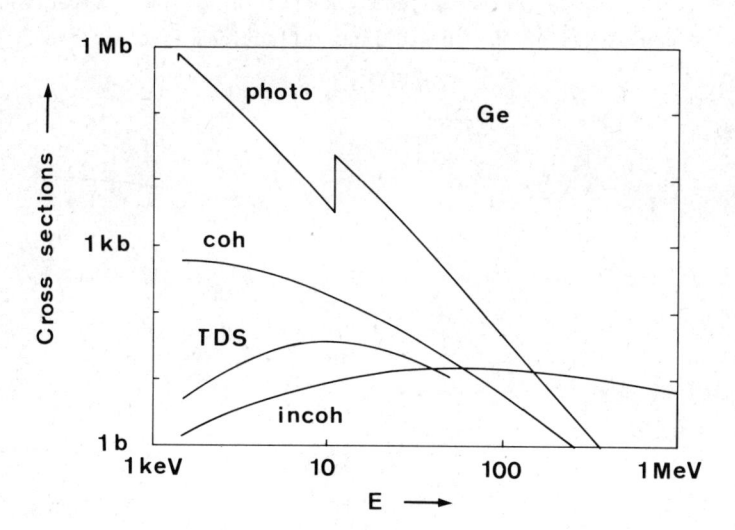

Fig. 1. Cross sections for the photoeffect, coherent and incoherent scattering, and thermal diffuse scattering in germanium. Data from Ref. 3.

Hydrogenic solutions can be obtained by assuming each electron of the atom to move in a hydrogen-like field reduced from the nuclear field by a screening constant, with the screening constant different for each electron group. In the self-consistent-field method each electron is assumed to move in the field of the nucleus and in an average field due to the other electrons. The most successful such one-electron scheme is the Hartree-Fock method in which the total wave function is written as a determinant of one-electron wavefunctions.

The differential scattering cross section is given by

$$d\sigma_{coh}/d\Omega = r_e^2 \, f_0^2 \, \sin^2 \alpha = f_0^2 (d\sigma^{Th}/d\Omega) \tag{3a}$$

where r_e is the classical electron radius, α the angle between the observed direction and the electric field of the incident wave, and σ^{Th} the cross section for a free electron according to the classical Thomson formula. For unpolarized radiation one obtains

$$d\sigma_{coh}/d\Omega = r_e^2 \, f_0^2 \cdot \tfrac{1}{2}(1 + \cos^2 \theta) \tag{3b}$$

where θ is the scattering angle.

In a condensed matter the coherence can extend to electrons of different atoms and give rise to more striking interference effects, such as Bragg-law diffraction by crystal lattices. The amplitudes of the crystalline reflections is described by the structure factor F:

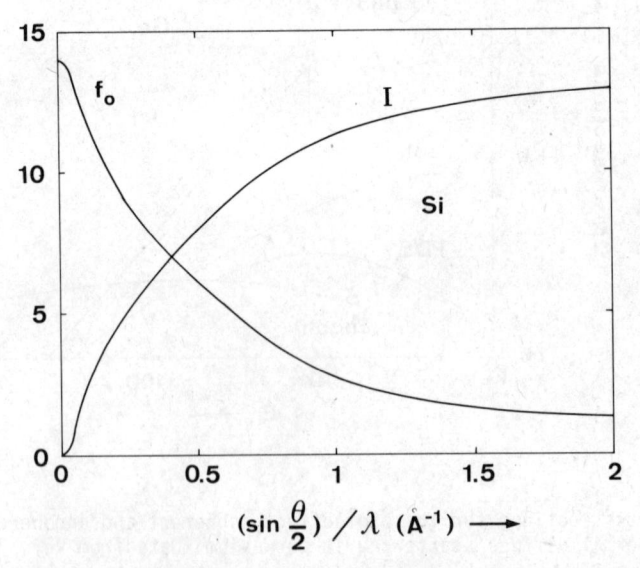

Fig. 2. The atomic scattering factor, f_0, and the incoherent scattering function, I, for silicon.

$$F = \sum_n f_n \exp(i \vec{q} \cdot \vec{r}_n) \qquad (4)$$

where the summation over n involves the positions $\vec{r}_n$ of the different atoms in the unit cell. The Bragg condition implies that the scattering vector $\vec{q}$ equals a reciprocal lattice vector. In a position in which no Bragg reflection occurs, the total Rayleigh scattering from a crystal is in general much less than the sum of intensities scattered by the individual atoms. The observed intensity is in this case due to all deviations from crystal periodicity (thermal vibrations, impurities etc.).

3. Anomalous scattering and absorption

Dispersion corrections to the atomic scattering factor have to be taken into account because of the interaction between the perturbing electromagnetic field and the excited states of the electrons. In non-relativistic quantum mechanics one has

$$f = f_0 + f' + i f'' \qquad (5)$$

where f' and f'' are related through a Kronig integral:

$$f'(\omega) = \frac{2}{\pi} \int_0^\infty \frac{\omega' f''(\omega')}{\omega^2 - \omega'^2} d\omega \qquad (6)$$

According to relativistic quantum mechanics there will be an additional, real term on the right-hand side of equation (5) [1,2].

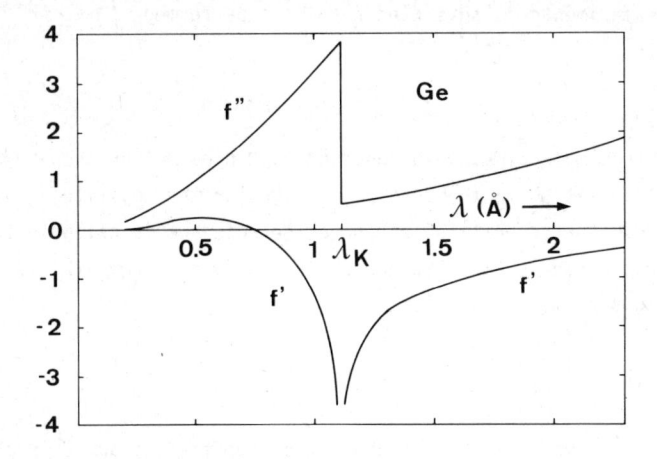

Fig. 3. Anomalous dispersion corrections of germanium, Data from Ref. 1.

The imaginary part f" is directly related to the photoeffect cross section:

$$f''(\omega) = (\omega/4\pi r_e c)\tau(\omega) \qquad (7)$$

Theoretical cross sections have been calculated rigorously using relativistic wavefunctions by Cromer and Liberman[1], Storm and Israel[3] and others. On the other hand relatively simple formulae for the calculation of hydrogen-like photoeffect cross sections have been communicated by Wagenfeld[4]. These formulae provide not only the possibility of a fast and easy calculation of the normal photoeffect cross section but also of the angular dependent term, $F''_{hk\ell}/F''_{000}$, governing the anomalous absorption of wavefields in perfect crystals (Borrmann effect).

Hydrogen-like photoeffect cross sections in a medium energy range (5 to 25 keV) have been published by Hildebrandt, Stephenson and Wagenfeld[5] for silicon and germanium and then expanded[6] to all elements in the range Z = 6 to Z = 54. Later Stephenson[7] has added further data for weaker energies. The agreement between theoretical and experimental cross sections is very satisfying in the medium energy range. For higher energies the hydrogen-like cross sections turn out to be somewhat too large, whereas the agreement between the measured cross sections and those of Storm and Israel is remarkably good as shown by Gerward and Thuesen[8].

Experimental values of τ over a large enough frequency range will then allow the determination of f' by integrating equation (6). The additional term, which should appear at the right-hand side of equation (5) according to relativistic quantum mechanics, cannot be determined from photoelectric absorption measurements[2]. It can only be determined by an experiment where the total real part of the atomic scattering factor is measured directly, for example by measuring the x-ray refractive index. However, very few accurate experiments of this kind have been performed[9-15].

3.1. EXAFS

Absorption spectra from molecules, solutions and condensed matter show a modulation of the absorption coefficient above the absorption edges of the constituent atoms. This modulation is observed for several 100 eV above the edge and is called extended x-ray absorption fine-structure or EXAFS. The normalized EXAFS spectrum $\chi(k)$ is defined in terms of the x-ray absorption coefficient μ by

$$\chi(k) = (\mu-\mu_0)/\mu_0 \quad ; \quad k = (2m\,E/\hbar^2)^{\frac{1}{2}} \qquad (8)$$

where μ_0 is the smoothly varying average absorption coefficient and E is the energy of the photoelectron, measured relative to the absorption edge.

The basic mechanism of the EXAFS is the interference between the outgoing photoelectron wave from the x-ray absorbing atom and the backscattered waves from the surrounding atoms. Fourier analysis of the EXAFS data can locate the positions of the atoms surrounding the absorbing atom [16-18]. Since EXAFS measures the immediate environment of a given type of atom it does not require that the sample be single crystal or even crystalline. With the recent availability of x-ray synchrotron radiation, there has been a renewed interest in the use of EXAFS studies.

3.2. The Borrmann effect

According to the dynamical theory of diffraction a number of wavefields are produced in a perfect crystal set for Bragg diffraction. The effective absorption coefficient of a particular wavefield depends on the polarization state and the deviation from the exact Bragg condition. The absorption coefficient of the wavefield having minimum absorption is given by

$$\mu_{min} = \mu(1 - \varepsilon)/\cos \theta_B \tag{9}$$

where $\varepsilon = |F''_{hk\ell}/F''_{000}|$ (centrosymmetric structure assumed) and θ_B is the Bragg angle. This phenomenon, known as the Borrmann effect, can be used for the imaging of lattice defects, such as dislocations.

The factor ε has been calculated by Wagenfeld[4] using hydrogen-like photoeffect cross sections. In the case of perpendicular polarization one has

$$\varepsilon^\perp = a\left[1 - 2(\tau^Q/\tau)\cdot\sin^2 \theta_B\right]\cdot\exp(-M) \tag{10}$$

where a is a geometrical factor, which equals unity when all atoms scatter in phase, and τ^Q the quadropole component of the photoeffect cross section.

4. Compton scattering

The basic theory of this effect, assuming the electron to be initially free and at rest, is that of Klein and Nishina[19]. Over most of the region in which Compton scattering is a major part of the total cross section, the Klein-Nishina theory is directly applicable. The electron binding effects are taken into account by writing the differential scattering cross section as

$$d\sigma_{incoh}/d\Omega = I(\vec{q},Z)d\sigma^{KN}/d\Omega \tag{11}$$

where σ^{KN} is the Klein-Nishina cross section and $I(\vec{q},Z)$ the incoherent scattering function. For large scattering vector $I(\vec{q},Z)$ approaches Z (Fig. 2).

References to text-books dealing with one or more topics of the present review.

R.W. James, The Optical Principles of the Diffraction of X-rays, Bell, London 1950, Chap. I, III and IV.
B.E. Warren, X-Ray Diffraction, Addison-Wesley, Reading 1962, Chap. 1 and 11.
L.V. Azároff (Editor), X-Ray Spectroscopy, McGraw-Hill, New York 1974, Chap. 6.
L.V. Azároff (Editor), X-Ray Diffraction, McGraw-Hill, New York 1974, Chap. 1.
L.H. Schwartz and J.B. Cohen, Diffraction from Materials, Academic Press, New York 1977, Chap. 4.

References to recent tabulations of scattering factors, cross sections and attenuation coefficients.

International Tables for X-ray Crystallography Vol. III (1962); Vol. IV (1974), Kynoch Press, Birmingham.
J.H. Hubbell, Vm. J. Veigele, E.A. Briggs, R.T. Brown, D.T. Cromer and R.J. Howerton, J. Phys. Chem. Ref. Data 4, 471-538 (1975); erratum in 6, 615-616 (1977).
J.H. Hubbell and I. Øverbø, J. Phys. Chem. Ref. Data 8, 69-105 (1979).

References in the text.

1. D.T. Cromer and D. Liberman, J. Chem. Phys. 53, 1891 (1970).
2. L. Gerward, G. Thuesen, M. Stibius Jensen and I. Alstrup, Acta Cryst.A (in press).
3. E. Storm and H.I. Israel, Nuclear Data Tables A7, 565 (1970).
4. H. Wagenfeld, Phys. Rev. 144, 216 (1966).
5. G. Hildebrandt, J.D. Stephenson and H. Wagenfeld, Z. Naturforsch. 28a, 588 (1973).
6. G. Hildebrandt, J.D. Stephenson and H. Wagenfeld, Z. Naturforsch. 30a, 697 (1975).
7. J.D. Stephenson, Z. Naturforsch. 30a, 1133 (1975).
8. L. Gerward and G. Thuesen, Z. Naturforsch. 32a, 588 (1977).
9. C. Malgrange, E. Velu and A. Authier, J. Appl. Cryst. 1, 181 (1968).
10. U. Bonse and H. Hellkötter, Z. Physik 223, 345 (1969).
11. D.C. Creagh and M. Hart, phys. stat. sol. 37, 753 (1970).
12. U. Bonse and G. Materlik, Z. Phys. B 24, 189 (1976).
13. C. Cusatis and M. Hart, Proc. R. Soc. Lond. A 354, 291 (1977).
14. T. Takeda and N. Kato, Acta Cryst. A 34, 43 (1978).
15. T. Fukamachi and S. Hosoya, Acta Cryst. A 31, 215 (1975).
16. E.A. Stern, Phys. Rev. B 10, 3027 (1974).
17. F.W. Lytle, D.E. Sayers and E.A. Stern, Phys. Rev. B 11, 4825 (1975).
18. E.A. Stern, D.E. Sayers and F.W. Lytle, Phys. Rev. B 11, 4836 (1975).
19. O. Klein and Y. Nishina, Z. Physik 52, 853 (1929).

SCATTERING EXPERIMENTS IN ULTRASONIC SPECTROSCOPY

F. Cohen-Tenoudji

Groupe de Physique des Solides de l'Ecole Normale Supérieure, Université PARIS VII
2, Place Jussieu, 75221 Paris Cedex 05, France
Laboratoire associé au C N R S.

INTRODUCTION

In the field of object characterization by ultrasounds, the aim of Spectroscopy is to get information on the object by Fourier analysis of scattered echoes when the sample is irradiated by short ultrasonic pulses. This technique has been proposed in 1960 by O.R. Gericke [1] in material evaluation. Since then, it has been used in all fields of investigation by ultrasounds mainly non destructive testing of materials and biological tissue characterization.

PRINCIPLE OF ULTRASONIC SPECTRAL ANALYSIS

In evaluating the shape of an object by ultrasounds, one is concerned with the diffraction problem. Indeed, ultrasonic sources are spatially coherent sources. For a given frequency, there is a unique phase relationship between the waves received by two points of the objects. The shape and the intensity of the signal given by a receiver will then depend on the time of coherence of the emitter. If the coherence time is great, echoes received from two scattering centres will not be separated; It is the diffraction regime ; And, for a given geometry, the amplitude of the sum of two echoes is changing according to the frequency of the emitter. For a correct evaluation of the scatterers, it will be necessary to vary frequency. The difficulty of great coherence time is that it exists a possibility to include spurious signals coming from scatterers located in the neighbourhood of those of interest.

On the contrary, a short coherence time will allow a better time separation of signals. In ultrasonic spectroscopy, the sample is insonified with short ultrasonic pulses with very broad band Fourier spectrum. One can Fourier analyse some selected parts of the signal, creating anew the situation of long coherence time without spurious echoes. Most of the existing scattering theories are concerned with monochromatic waves and their results are given as functions of that frequency. So, with ultrasonic spectroscopy using signals with wide frequency band, one can easily compare experimental Fourier amplitude as a function of frequency with theories describing the scattering of monochromatic waves.

Besides, one can use short pulses technique to make signal processing operations such as deconvolution, improvement of signal over noise, etc...

EXPERIMENTAL EQUIPMENT

In order to be able to transmit very short pulses, the emitter and receiver of ultrasounds have to be non resonant. One uses now ferroelectric ceramics , Titanate Zirconate of lead PZT, Barium Titanate Ba Ti O_3; These ceramics are cut as discs whose thickness is half the wavelength in the ceramic of the sound of the highest frequency that has to be emitted. The back faces of discs are damped by contact with an absorbing material of the same acoustical impedance of the ceramic, and so the resonance of the ceramic is highly damped. When the plated faces of the disc are submitted to a short electric pulse, the ceramic sends out a short ultrasonic pulse with broad band frequency content. To improve the frequency bandwidth, it is useful to be able to adjust the rising time and the width of the electric pulse. Typically bandwidth at - 20 dB is one and a half the centre frequency (Fig.1).

For ultrasounds propagation, the transducers are in contact with the sample to analyse or immersed in water for coupling with the sample. Ultrasonic echoes may be detected by the emitter or by another transducer.

In the electronic part , an analogic gate selects the part to be analysed in the received signal. (Fig.2). The selected part can be Fourier analysed by an analogic spectrum analyser or digitized for further numerical treatment. The digitizing process mostly used for frequencies up to 30 MHz is a sweeping gate sampler.

CHARACTERIZATION EXPERIMENTS BY ULTRASONIC SPECTROSCOPY

We will restrict the examples given here to the propagation of one type of waves in a linear medium. The problem of the echo created by a given surface is then the resolution of the propagation equation $\Delta p - 1/c^2 \ (\partial^2 p/\partial t^2) = 0$
where p is the acoustic pressure, taken in account the boundary conditions on the surface.

The exact solution cannot generally be obtained and several approximations are made for peculiar cases; The mostly used |2-8| is the Kirchhoff approximation which assumes that the radiation scattered by a surface element is distributed evenly over a solid angle of 2π neglecting secondary diffraction of one surface element over another one. This approximation should be applied only when the radii of curvature of the surface are large compared with wavelength and if the surface slopes of the scattering surfaces are small, but it gives good results in less restrictive cases.

The formalisms which are usually used are, the Helmholtz integral, Green's functions, potential method. They lead, in Kirchhoff's approximation, to results very similar as Huygens' optical principle do.

We will now show three types of applications :

1/ <u>Scattered pressure by an object wholly contained in the field created by the emitter</u>

This concerns the important problem of characterizing a defect in non destructive testing of materials.

Using Kirchhoff approximations, A.Freedman has shown that the echo to a short incident pulse was made of separate pulses, each one being created at a discontinuity of the solid angle w(r) subtended by the surface as seen by the transducer at the distance r. E.Lloyd |3,4| developped the method in back scattering geometry to practical cases and showed that the electrical signal given by the transducer was proportionnal to the second time derivative of the solid angle where $t = \frac{r}{2c}$ (c is the sound velocity).

Figure 3 gives, after Lloyd the formation of the two pulses in opposite phase given by a sphere and the resultant amplitude spectrum.

L.Adler and H.L. Whaley |5| have shown that Ultrasonic Spectroscopy is a powerful method to get evaluation of defects size in materials. They use a method based on extracting the defect size from the minima in the backscattered spectrum. Their experiences are well explained by D.M.Johnson |6| who, following an argument based on Huygens' principle, showed that the backscattered pressure at the frequency f was given by

$$P_S(f) = - \frac{2i\pi fB}{2\pi r^2 c} \iint_A \cos\theta \exp\left(-\frac{i4\pi rf}{c}\right) dA$$

where dA is a scattering surface element at the distance r of the emitter inclined of θ on the direction of wave incidence. B is an amplitude factor.|Fig.4|.

We can see that this formulation is very practical for numerical calculus

Figure 5, after Johnson, shows the good agreement of this formulation with the experience of Adler and Whaley on the diffraction by a disc for different inclinations θ.

The method of using the minima in the backscattered spectrum to characterize the size of a defect may be also used for evaluating surface defects by surface waves |7| where one can say in first approximation that the two edges of the defect are two points scatterers out of phase. But it has to be pointed out that the inverse problem, i.e. the determination of the shape of an object is not solved by one spectroscopic experiment. In order to get more information, some other parameters, such as angles of incidence, have to be varied.

2/ <u>Surface roughness characterization by Ultrasonic Spectroscopy</u>

The scattered ultrasonic pressure by a rough surface can give parameters of the surface roughness |8-10|. Ultrasonic Spectroscopy has been applied for both periodic and random roughness.

a) One dimensional periodic surface |11-12|

In the Kirchhoff approximation, following Beckmann |9|, the backscattered pressure amplitude at angle θ of incidence in the far field is given by :

$$A(f,\theta) = B \int_{-L}^{+L} \exp - \frac{4i\pi f}{c} (x \sin \theta - \zeta(x) \cos \theta) \, dx$$

where B is a coefficient, 2 L is the insonified part of the grating, $\zeta(x)$ is the equation of the surface profile.

In the case of small roughness, when $\left|\frac{4\pi f \zeta(x)}{c} \cos \theta\right| \ll 1$ one can expand the exponential in series and get

$$A(f,\theta) = B \int_{-L}^{+L} \left(1 + \frac{4i\pi f}{c} \zeta(x) \cos \theta\right) \left(\exp - \left(\frac{4i\pi f}{c} x \sin \theta\right)\right) dx$$

which is the sum of two integrals, the first one is negligible for $\theta \neq 0$, the second one appears to be proportional to Fourier transform of the surface profile.

In function of frequency, the backscattered intensity will be made of peaks whose amplitudes are proportional to the Fourier coefficient of the decomposition of the surface profile in series. The two first coefficient of gratings have been shown to be well estimated by ultrasonic measurements |11|.

On Fig.6, are shown both ultrasonic and computed amplitude spectra after mechanical measurement of the profile and correction by the frequency response of the transducer for two samples. They appear in good agreement.

On Fig.7, are shown the spectra obtained for a sample having defects in periodicity. In both experimental and computed spectra and in the same order of magnitude, it appears ghosts peaks at half harmonics characteristic of defects in periodicity.

b) Randomly rough surfaces

As shown by C.S. Clay and H. Medwin |13|, the intensity scattered in the specular direction allows the evaluation of statistical parameters of the surface.

Following Beckmann |9|, one can write the mean scattered intensity at a point B as a sum of two terms. The first one called coherent intensity is dominant at low frequency, the second one called incoherent intensity is dominant at high frequency. $<I> = <P_S^*> <P_S> + \text{Variance} |P_S|$, $<P_S>$ being the mean value of the acoustical pressure.

We will restrict here to the case of normal incidence.

In low frequency regime

$$<P_S> = - \frac{ik}{2\pi} \iint_A D_o \frac{e^{2ikR_o}}{R_o^2} <e^{-2ik\zeta}> \, dx \, dy$$

k is the wave number, D_o the directivity function of the emitter, R_o the distance from the emitter to the plane $\zeta = 0$, $\zeta(x,y)$ is the height of the surface.
When $\zeta = 0$, P_S is the scattered pressure, P_R, by a plane surface so that one can write
$$<P_S> = P_R <e^{-2ik\zeta}>$$
This expression is valid even in the Fresnel approximation.
If we note $W(\zeta)$ the probality density function of the height one has
$$<e^{-2ik\zeta}> = \int_{-\infty}^{+\infty} W(\zeta) e^{-2ik\zeta} d\zeta$$

which is the characteristic function of the surface.

In the general case $<P_S>/P_R$ is a complex quantity and one can see that it is theoretically possible to measure the parameters of the heights distribution function. When $W(\zeta)$ is gaussian with h root mean square deviation from zero height, one has in low frequency $<I> = I_R e^{-4k^2h^2}$
So, the departure of the intensity from the plane surface case gives a measurement of h.
At high frequency, for a gaussian autocorrelation function, with $\sqrt{2}\,h/L$ being the mean slope, one has
$$I = I_R \frac{\pi}{4Ak^2} \frac{L^2}{h^2}$$
with I_R intensity scattered by a plane surface, A being the insonified area.

Ultrasonic spectroscopic experiments have been done |14| and it has been shown that one can evaluate the roughness h from the low frequency part of one backscattered spectrum and get the value of the autocorrelation length L from the intensity at high frequency.

Comparisons are made with mechanical measurements of the surface for samples with 8 μm < h < 60 μm ; 80 μm < L < 200 μm
The agreement for h is good 10% for L the method is less accurate (error in range of 30%).
On Figure 8 (a et b) are plotted the intensities for two different samples.
We see the quickest decrease of the roughest sample at low frequency and the smaller high frequency value of the sample with the greatest h/L ratio, i.e. the greatest slopes. The method can be applied to characterize roughness even for inside surfaces after crossing a wall.

3/ Characterization of scatterers distributed in volume

N.F. Haines et al |15| have studied layered media by spectroscopy, both in amplitude and in phase. They obtain very good agreements between the experimental and theoretical values.

F.Lizzi and M.A. Laviola |16| used the concepts of layered media in ophtamology to study the detachment of the retina and the detection of tumors behind the retina; From typical minima in the amplitude spectrum, they get the size of the different layers. Several authors use spectroscopy for tissue characterization in biology. D.Nicholas and C.R.Hill |17| use the concept of Bragg diffraction to get correlation of the backscattered intensity versus the angle of incidence with several tissues structures. E.Holasek et al |18| and P.P. Lele |19| using transmission or backscattering experiments obtained attenuation measurements characterizing many biological tissues.

To conclude, Ultrasonic Spectroscopy is a technique which can give quickly some parameters of the scattering objects specially in non destructive testing of materials and biological tissue characterization.

One can think that the possibility of signal processing will enlarge the results already obtained in the inverse problem.

BIBLIOGRAPHY

1. O.R. Gericke , J. Acoust. Soc. Am. , 35 ,364-368 (1963)
2. A. Freedman , Acustica , 12 , 247-258 (1962)
3. E. Lloyd , Ultrasonics International Conference Proceedings , IPC London 54-57 (1975)
4. E. Lloyd , Eighth World Conference on non destructive testing , Cannes , 2B10 (1976)
5. L. Adler and H. L. Whaley , J.Acoust. Soc. Am. , 51 , 881-887 (1972)
6. D. M. Johnson , J. Acoust.Soc. Am. , 59 , 1319-1323 (1976)
7. A. Jungman,F. Cohen-Tenoudji and B.R. Tittmann , FASE 78,Warsaw , Published by the Polish Academy of Sciences , 1 , 191-193 (1978)
8. C. Eckart , J. Acoust. Soc. Am. , 25 , 566-570 (1953)
9. P. Beckmann and A. Spizzichino , The scattering of Electromagnetic Waves from rough surfaces , (Mac Millan , New York) (1966)
10. P. J. Welton , J. Acoust. Soc. Am. , 54 , 66 , (1973)
11. A. Jungman , F. Cohen-Tenoudji and G. Quentin , Ultrasonics International Conference Proceedings , IPC London , 385-396 (1977)
12. F. Cohen-Tenoudji, M. Joveniaux, A. Jungman and G. Quentin , Eighth World Conference on non destructive testing , Cannes , 3F4 (1976)
13. C. S. Clay and H. Medwin , J. Acoust. Soc. Am. 47 ,1412 (1970)
14. F. Cohen-Tenoudji and G. Quentin , Colloquium on Scattering of Ultrasounds , GPS Paris (1979) . Published in Revue du Cethedec (In press)
15. N. F. Haines , J.C. Bell and P.J. McIntyre , J. Acoust. Soc. Am. , 64 , 1645-1651 (1978)
16. F. Lizzi and M.A. Laviola , IEEE Ultrasonics Symposium , 29-32 (1975)
17. D. Nicholas and C.R. Hill , Ultrasonics International Conference Proceedings 269-272 (1975)
18. E. Holasek, W.D.Jennings, A. Sokollu and E.W. Purnell , IEEE Ultrasonics Symposium, 73-76 (1973)
19. P.P. Lele, A.B. Mansfield, A.I. Murphy, J. Namery and N. Senepati , Ultrasonic Tissue Characterization , NBS Special Publication , 453 , 167-196 (1975)

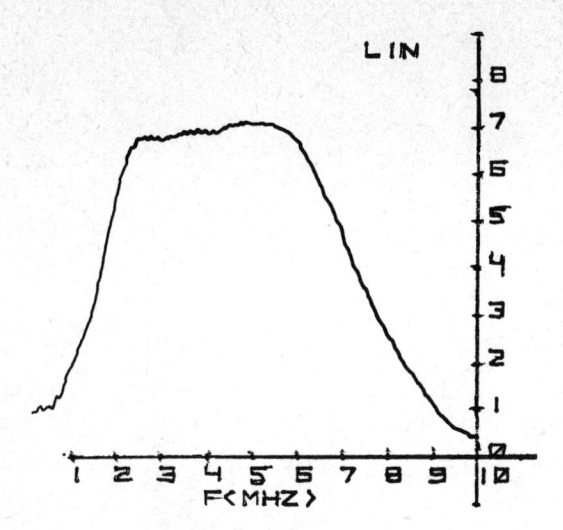

Fig.1 Amplitude spectrum of a typical echo in the range 1.5-10 MHz

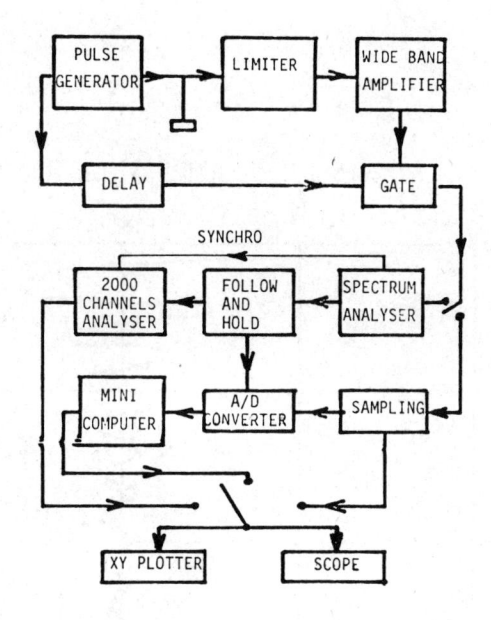

Fig.2 Experimental equipment used in an experience of ultrasonic spectroscopy.

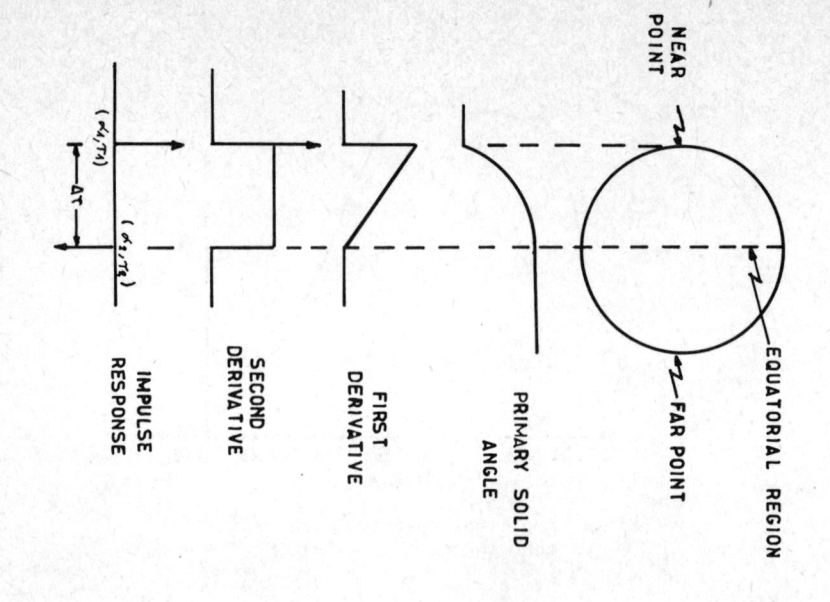

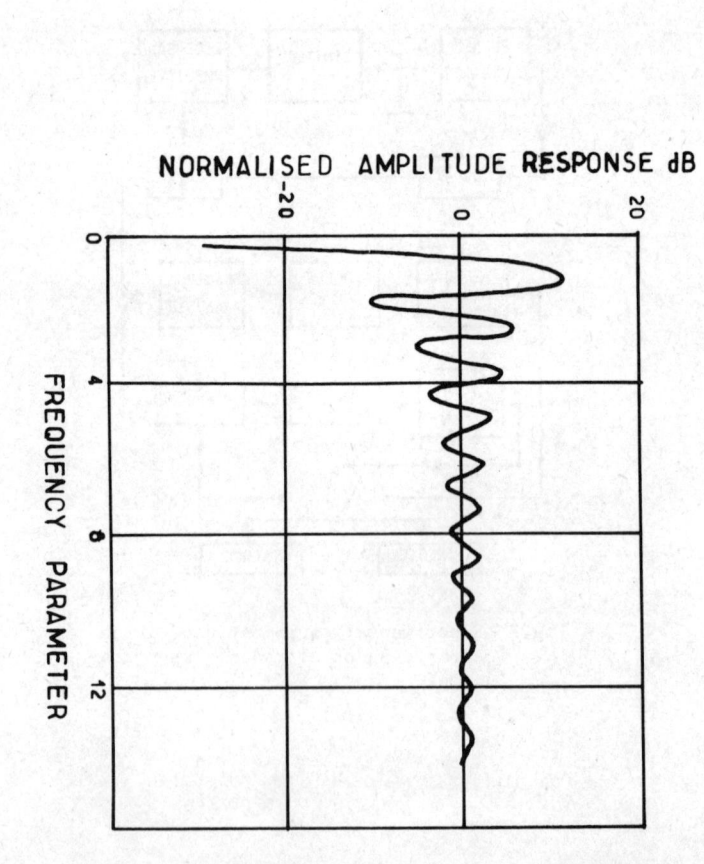

Fig.3 Formation of impulse response and amplitude spectrum for a sphere (After Lloyd)

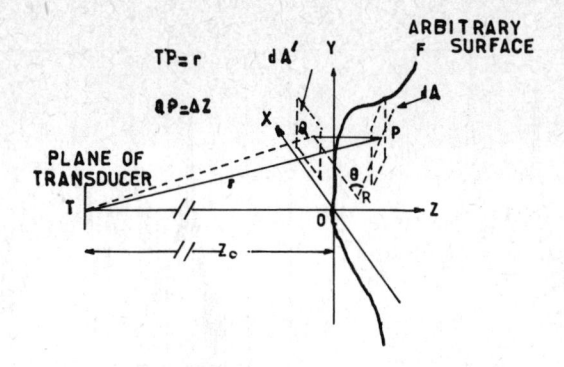

Fig.4 Geometry in the backscattering experiment (After Johnson)

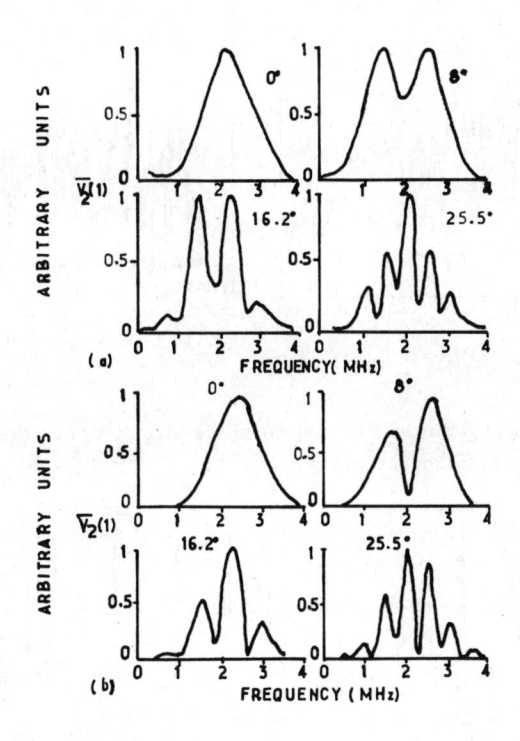

Fig.5 Experimental (a) and theoretical(b) frequency spectra for reflection of an inclined disc (After Johnson)

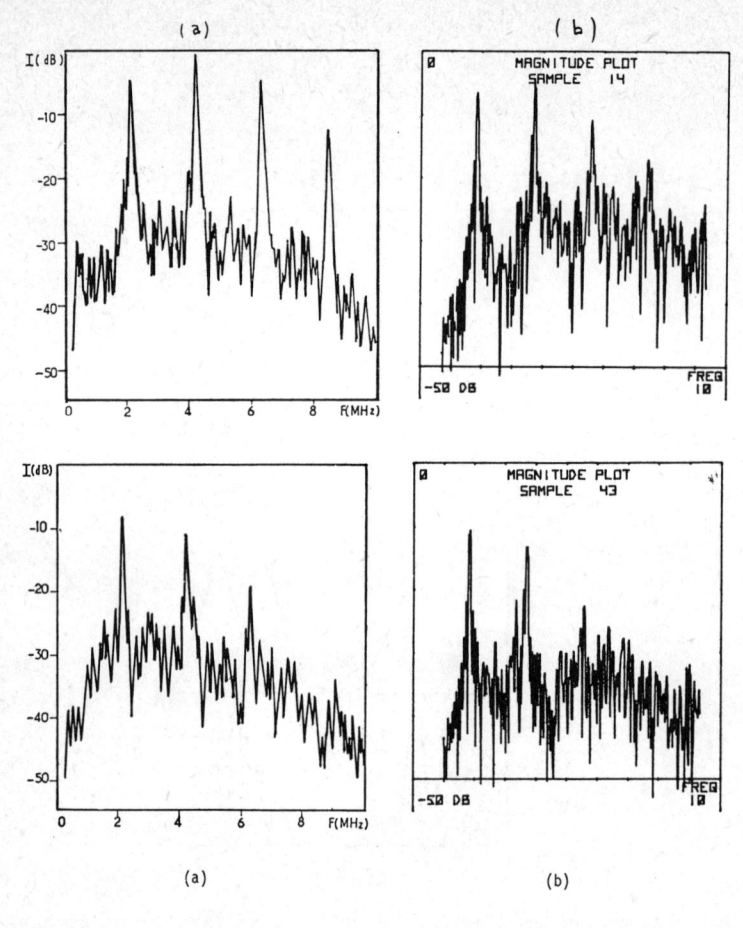

Fig.6 Experimental (a) and computed (b) spectra obtained for two different gratings |11|

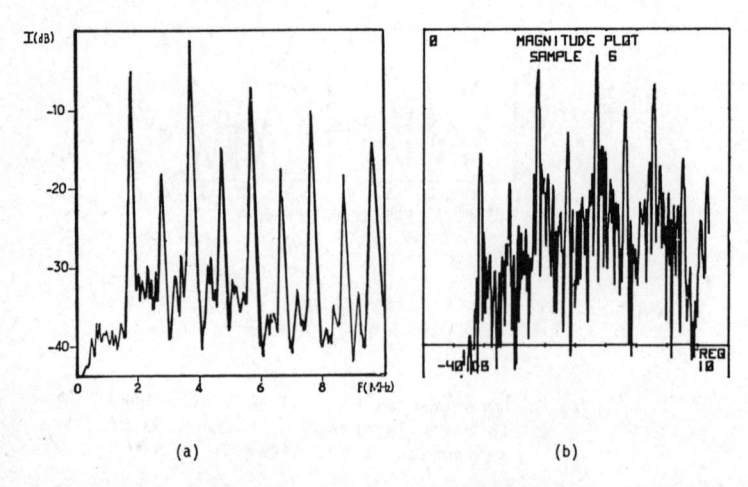

Fig.7 Experimental (a) and computed (b) for a grating having defects in periodicity |11|

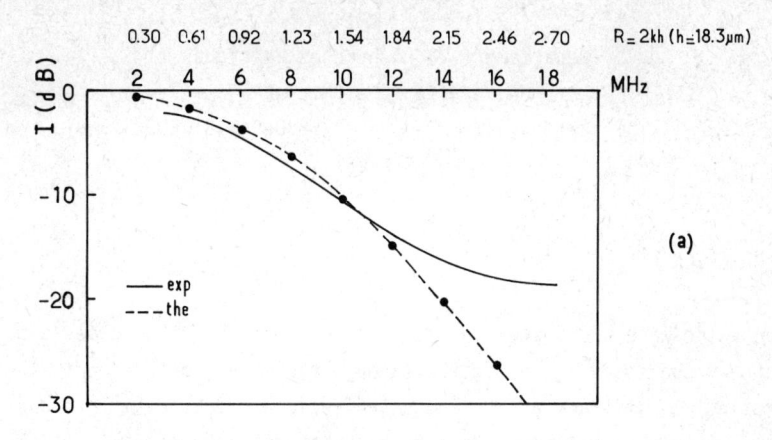

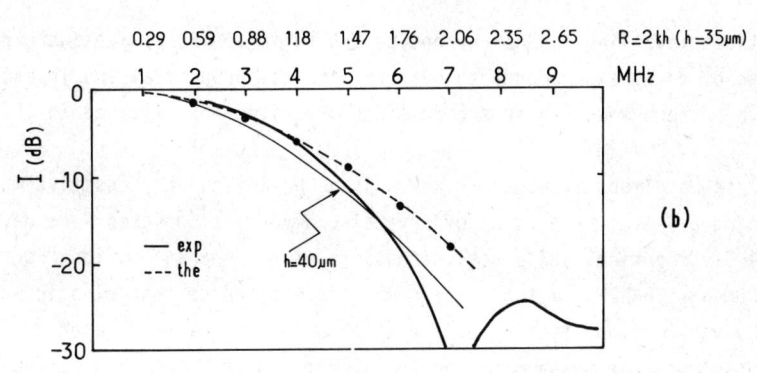

Fig.8 Backscattered intensity at normal incidence for two samples
a) h = 18,3 μm , L = 180 μm
b) h = 35 μm , L = 110 μm
Full curve: Experimental ; Pecked curve: exp (− $4k^2h^2$)

ACOUSTO-OPTICAL INTERACTIONS
F. MICHARD
Departement de Recherches Physiques
Université P. et M. Curie
Tour 22 - 4, pl. Jussieu - 75230 PARIS CEDEX 05
FRANCE

I - INTRODUCTION

When an elastic wave propagates in a transparent medium, it produces a periodic modulation of the index of refraction through the elastooptic effect. The moving phase grating may diffract a fraction of an incident light beam into one or more directions. This phenomenon is known as acoustooptic interaction.

After the original suggestion of BRILLOUIN in 1922 |1| that elastic waves would diffract light beam, experimental evidence of this effect involving thermal waves was first pointed out by GROSS in 1930 |2| then independently by LUCAS-BIQUARD |3| and DEBYE-SEARS |4| in 1932 using ultrasonic waves.

Investigations involving *incoherent* optics and sonic *coherent* waves were developed and reviews of early experiments to measure material properties and investigate the diffraction pattern under various conditions are given for instance in |5| |6|.

The scattering of light by acoustic waves is a convenient and sensitive method of investigating physical properties of crystals. Several techniques were developed to measure elastooptical and elastic constants of transparent solids. For example, in the BERGMANN-SCHAEFFER method, the sample is excited so that elastic waves propagate in many directions, setting up a grating of elastic strain. When an incident light beam illuminates the sample, the *image* of the diffracted light exhibits the elastic anisotropy of the crystal (cf. figure 1) |7|. Scattering of light by ultrasonic waves is also a very accurate method to probe the phase velocity of elastic waves |8|: In the case of figure 2 scattering of light is achieved by stationnary ultrasonic waves excited between two parallel faces of the sample. The frequencies which correspond to the resonance of the sample are detected by observing the diffraction of light between crossed polarizers

The recent development of techniques for the generation of high frequency *coherent* elastic waves, combined with the advent of laser as source of *coherent* light has given rise to a fruitful new field of research and application based on the interaction of these two coherent waves.

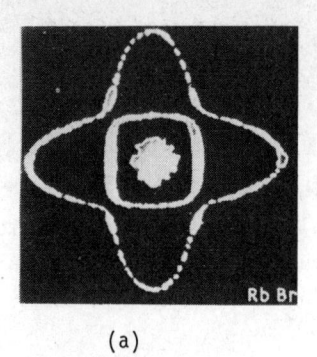

Figure 1

BERGMANN -SCHAEFER diagram for Rb Br :
light propagating along
a) cubic axes,   b) cube diagonals
(L. BERGMANN, Z. NATURFORSCH 12a, 229 (1957))

Figure 2

Stationnary ultrasonic waves in a glass sample |8|

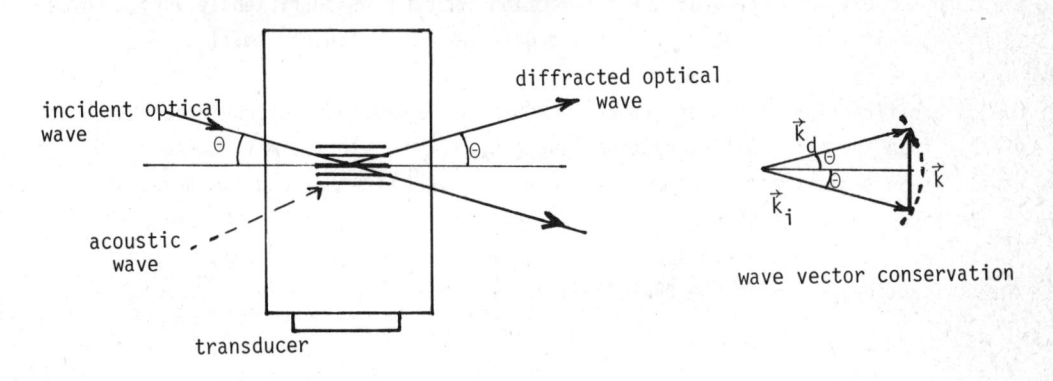

Bragg scattering geometry in case of an isotropic material

Figure 3

More recently progress in acoustic surface waves and optical guided waves have prompted theoretical and experimental work in acoustooptic interaction in guided wave structures |9|.

After a short recapitulation of the basis principle of acoustooptic interaction we give a few elementary comments about bulk wave device applications of acoustooptic interaction and then we confine ourselves to a discussion of some problems in solid state physics which are stimulated by the scattering of light by hypersonic waves.

II - ACOUSTOOPTICAL INTERACTION

Phenomenological theory of the elastooptic effect is based on POCKEL's theory which states that the change of the inverse dielectric tensor ΔB_{ij} set up by the acoustic wave is proportional to the acoustic strain S_{kl} :

$$\Delta B_{ij} = p_{ijkl} S_{kl} \qquad (1)$$

The elastooptical coefficients p_{ijkl} are generally assumed to be symmetrical with respect to ij and kl. NELSON and LAX (1970)|10| have shown this, in general, not necessarily true. They pointed out that POCKEL's formulation is incomplete and that rotation effects, arising from shear waves, should be also included in strongly birefringent media. In place of the acoustic strain, one must use the displacement gradient as the pertinent variable.

Acoustooptic diffraction can be represented as a parametric interaction |11|. Via the elastooptic effect, the incident optical wave mixes with the acoustic waves to generate a number of polarization waves at the combination frequencies. The polarization waves in turn will generate optical radiation at these new frequencies.-provide that phase matching between these two waves is achieved-. In general multiple diffraction to higher orders may occur, if the interaction length L is sufficiently large however light intensity in the high diffraction orders become negligibly small.

Multiple diffraction to higher orders (RAMAN-NATH regime) occurs providing that L is smaller than the critical interaction length $L_c = \frac{n \Lambda^2}{\lambda_0}$. This limit is increasingly difficult to satisfy according to simultaneous increase in the acoustic frequency. For instance in fused silica for λ_0 = 0,633 µm, L_c = 8,4 mm at 100 MHz and L_c=0,93mm at 300 MHz. Thus L_c is approximately the minimum interaction length to insure that the diffraction occurs in the BRAGG regime.

Typical arrangement to perform acoustooptic interaction is shown figure 3. Ultrasonic waves are generated into a material medium (usually a solid) by means of a piezoelectric transducer bonded to one face of the sample. With well enough collimated optical and ultrasonic beams it is possible to set out the geometry so that the amplitude of the light scattered by adjacent volumes adds constructively and scattering is obtained for a *single* angular phase matched direction.

In case of infinite plane wave the interaction process conserves the momentum and the energy :

$$\vec{k}_{diffracted} = \vec{k}_{incident} \pm \vec{k}_{acoustic} \qquad (2)$$

$$\omega_{diffracted} = \omega_{incident} \pm \Omega_{acoustic} \qquad (3)$$

With regard to the small difference in frequency between the two optical waves (the acoustic frequency beeing small compared to the optical)- the locus of the scattering interaction in momentum space is a circle with radius k. So in case of isotropic medium we obtain the condition :

$$\sin \Theta_B = \frac{1}{2} \frac{|\vec{K}|}{|\vec{k}|} = \frac{\lambda_o}{2 n \Lambda} \qquad (4)$$

Λ acoustic wavelength
λ_o free space optical wavelength
n index of refraction of the material

This is named "BRAGG condition" by formal analogy with XR diffraction by lattice planes. Furthermore it can be pointed out that the intensity of the diffracted light beam depends on the acoustic intensity. In particular for low acoustic power the intensity of the diffracted beam is proportional to the acoustic power. Furthermore the ability of a material to diffract lightbeam depends on its figure of merit :

$$M = \frac{p^2 n^6}{\rho V^3}$$

p : effective photoelastic constant
ρ : density
V : acoustic velocity

III - DEVICE APPLICATIONS

We can expect from the previous considerations that acoustooptic interaction is very convenient for the design of devices intended to perform various optical beam control functions |11| |12| |13|. These basic acoustooptic devices are characterized by three distinct regimes of interaction geometry depending on the parameter $a = (\frac{\delta \Theta_o}{\delta \Theta_a})$, the ratio of divergence angles of the optical beam and the acoustics beam.

In the limit a << 1, the device acts as a deflector, for a ~ 1 as a modulator and in the limit a >> 1 it can be used as an optical filter.

We discuss here deflector and modulator functions

- Deflector :

From (4) we can expect that the direction of the diffracted beam can be varied by changing the driving frequency. In a deflection system the main parameters are resolution and speed. Resolution is defined as the range of deflection angles divided by the angular spread of the diffracted beam :

$$N = \frac{\Delta \Theta_B}{\delta \Theta_B} \qquad (5)$$

Where from (4) the total angle of deflection for a frequency change Δf is :

$$\Delta \Theta_B = \frac{\lambda_o \Delta f}{n V \cos \Theta_o} \qquad (7)$$

(V : acoustic wave velocity)

In the limit a << 1, the divergence of the diffracted beam is equal to that of the incident beam. Resolution N is proportional to $\tau \Delta f$, where τ is the acoustic transit time across the optical beam. τ is also a measure of the speed of the deflection system. So the deflector bandwith Δf which is proportional to the resolution and speed product is the main parameter to characterize the efficiency of the deflector.

The first application of an acoustooptic deflector was the horizontal deflection in a laser TV display |14|. The discovery of new efficient acoustooptic materials such as $PbMoO_4$ and TeO_2 allowed to increase the deflector bandwith significantly (150 to 300 MHz) with a resolution of about 1000 spots. A good review of main applications of acoustooptic deflectors is given for instance in ref |11|.

- Modulators :

The acoustooptic interaction may also be used to modulate light. Both amplitude and frequency modulators can be achieved. For proper modulator operation, the divergence of the optical beam must be about equal to that of the acoustic beam : $a = \frac{\delta \Theta_o}{\delta \Theta_a} \sim 1$

Scattering with amplitude modulated acoustic frequency removes energy from the incident light beam and produces an amplitude modulated scattered beam.

The requirement of equal diffraction angles is necessary to assure the full use of all the acoustic and optical energy.

The success of acoustooptic modulations increases in recent years due to their many advantages such as low drive power, high extinction ratio (ratio between the maximum intensity of the diffracted laser beam) all types of modulation available...

Design and construction of a acoustooptic modulator for communication in the infrared is reported for instance in |15|, |16|.

Another fruitful field of applications is the use of acoustooptic modulator inside a laser cavity. These intra cavity application involve Q-Switching, mode locking and cavity dumping.

In some TV experiment deflector and modulator are associated |17|.

IV - ACOUSTOOPTIC INTERACTION AS LOCAL PROBE TO STUDY THE PHYSICAL PROPERTIES OF TRANSPARENT MEDIA.

From the previous considerations (cf § 2), we can expect that the scattering of light by elastic waves is a convenient and sensitive investigation method of some physical properties of transparent media.

a - The BRAGG angle is proportional to the ratio of elastic wave frequency to velocity. This proportionality can be used to measure the velocity.

b - The intensity of the diffracted light is proportional to the acoustic intensity which provides a probe of the acoustic intensity within the sample and consequently of the attenuation of hypersonic waves...

c - Measurements of the diffraction efficiency can be used to determine elastooptical coefficients of materials.

d - The phase shift of the diffracted light can be used by mixing with the undiffracted light to probe the phase of the elastic wave.

In an other way scattering of light by coherent elastic waves (BRAGG scattering) is very complementary of scattering of light by thermal waves (BRILLOUIN scattering).

These thermal waves grow spontaneously, oscillate at the sound frequency Ω (k) and decay away at a charateristic rate Γ (k) only to be replaced by another wave with a different starting phase and a similar temporal history |18|, |19|. The spectrum of the scattered light, deriving from correlations in the fluctuations in the thermal phonons, consists of a pair of doublets, whose shape is Lorentzian if the correlation function for the sound wave amplitude dies off exponentially. The width of the spectral lines gives the sound waves lifetime $1/\Gamma$ (k) and the splitting of the doublets around the unshifted frequency is equal to the frequency of the scattering sound wave :

$$\frac{\Omega}{2\pi} = \frac{V}{\Lambda} = \frac{2 V n}{\lambda_o} \sin \frac{\Theta}{2}$$

(Θ is the angle between the incident and the diffracted light beam) Thus the scattering angle Θ determines the wavelength of the scattering sound wave. By scattering through larger and larger angles sound waves of shorter and shorter wavelength are probed. The shortest sound wave responsible for light scattering being equal to one half the wavelength of the light in the medium.

Until the advent of laser, the breadth of the spectrum of the incident light made it difficult to resolve the BRILLOUIN doublets and almost impossible to determine their natural linewidth. The use of laser as monochromatic, intense and unidirectional light source gives a review of interest for this technique and in conjunction with high resolution spectroscopy its allows to determine the velocity and attenuation of sound waves whose frequency may be as high as 20-30 GHz (in case of back-scattering)

We want now exemplify by a choice of some problems in the domain of solid state physics the importance of acoustooptic interaction as local probe to study the physical properties of transparent media.

1) Phonon-phonon interaction

Experimental studies of the hypersonic attenuation in crystals and its dependence on the frequency are important from the standpoint of the theory of phonon-phonon interaction. In case of BRAGG scattering, the measurement of the diffracted light intensity as a function of the distance to the transducer allows to obtain the *intrinsic* attenuation in the crystal getting rid of spurious effects such that due to the bonding of the sample or the non parallelism of its faces that can occur in classical pulse-echo techniques that provides a *transit time* in the sample.

From the previous considerations BRAGG scattering used in conjunction with BRILLOUIN scattering is a very convenient method to probe in a wide range of frequency (typically 100 MHz to 30 GHz) the attenuation of elastic waves |19|.

2) Phase transitions

Another especially promising application of light scattering by elastic waves is the study of crystals which experience phase transition |20|. It can be used for instance to investigate the critical phenomena occuring in the vicinity of phase transitions such as domain structure, interaction of acoustic waves with domain walls. Damping and velocity of elastic waves may be simultaneously measured near the phase transition. The possibility of carrying out measurements over displacements of the order of 1 mm inside a domain of the crystal allows first to determine the high value of the attenuation and, second practically to eliminate the effect on the results of temperature gradients.

3) Acoustic activity

Another facinating problem is the phenomenon of acoustic activity which is the mechanical analogy of optical activity. Acoustic activity is related to the spatial dispersion of the elastic moduli :

$$C_{ijkl}(\Omega, \vec{K}) = C_{ijkl}(\Omega) + i \gamma_{ijklmn}(\Omega) K_m + \ldots$$

The acoustic gyrotropy tensor γ lifts the degeneracy for transverse elastic waves at finite wave vector. In case of quartz crystal the effect was previously exhibited in neutron scattering results |22| and by BRILLOUIN scattering |23|.

The acoustic activity was also measured directly by classical ultrasonic experiments |24| |25|. Experiments of this kind yield results that are averaged over the entire length of the crystal. Acoustic activity in quartz crystal was investigated by means of BRAGG scattering |26| along the Z axis of a quartz crystal. The intensity of the scattered light oscillates with a period which depends of the specific rotation ability of the crystal. So it is possible to probe the effects of gyrotropy at arbitrary points of the sample.

4) Characterization of the local homogeneity of sample.

We refer now to the use of light scattering by acoustooptic interaction to obtain both the amplitude and the phase of the ultrasonic wave as is propagates in the sample.

The phase of the elastic wave may be detected by beating the diffracted field with a reference (a fraction of the incident undiffracted light beam). By translating the sample continuously in the same direction as the acoustic wave vector $\vec{K}$, the beating signal changes sinusoidally to give both the amplitude and the phase of the elastic wave. The device described in reference|27| allows to measure the acoustical wavelength in the frequency range 200 MHz - 1 GHz with a spatial resolution of the order of 10 µm. By means of this device, local ultrasonic velocity fluctuation of the order of 10^{-4} can be detected. Thus, applications such as control of material doping (homogeneity, impurity gradients etc...) and fundamental studies of phonon propagation in thermal or strain gradients may be anticipated.

REFERENCE

1. L. BRILLOUIN "Diffusion de la lumière et des rayons X", Ann. Phys. (Paris) 17, 88-122, (1922)
2. E. GROSS "The splitting of spectral lines at scattering of light by liquids" Nature 126, 400, (1930)
3. R. LUCAS, P. BIQUARD, "Optical properties of solids and liquids under ultrasonic vibrations" J. Phys. Radium, 3, 464-477, (1932)
4. P. DEBYE, F. SEARS, "On the scattering of light by supersonic waves" Proc. Nat. Acad. Sci. 18, 409-414 (1932)
5. M. BORN, E. WOLF, Principle of optics, third Ed, Pergamon Press, New York, ch. 12, (1965)
6. C.F. QUATE, C.D.W. WILKINSON, D.K. WINSLOW, "Interaction of light and microwave sound", Proc. IEEE 53, 1604-1623, (1965)
7. C.S. SCHAEFER, L. BERGMANN Naturwissenchaften, 23, 799, (1935).
8. A. ZAREMBOWITCH, "Etude théorique et détermination optique des constantes élastiques de monocristaux". Bull. Soc. Franç. Minér. Crist. 88, 17-49, (1965)
9. G.I. STEGEMAN, "Optical probing of surface waves and surface wave device", IEEE Trans. Sonics and Ultrasonics, S.U. 23, 33-63, (1976).
10. D.F. NELSON, M. LAX, "New Symmetry for acousto-optic scattering", Phys. Rev. Lett. 24, 379-380, (1970)
11. I.G. CHANG, "Acoustooptic devices and applications", I.E.E.E. Trans. Sonics and ultrasonics, 1, 2-22, (1976)
12. E.I. GORDON "A review of acoustooptical deflection and modulation devices" Proc. I.E.E.E. 54, 1931-1401, (1966)
13. R.W. DIXON "Acoustooptic interactions and devices" IEEE Trans. on electron devices ED-17 , 229-235, (1970)

14 A. KORPEL, R. ADLER, P. DESMARES, W. WATSON, "A television display using acoustic deflection and modulation of coherent light. Proc. IEEE 54 1429-1437, (1966)

15 A.W. WARENER, D.A. PINNOW, "Miniature acoustooptic modulator for optical communications" J. Quantum Electron. , QE9, 1155-1157, (1973)

16 I.C. CHANG, G. MORADIAN, "Frequency modulated acoustooptic modulates for 10.6 µm laser communications" Electrooptics system desing" Conf. San Francisco (1974)

17 R. TORGUET "Etude théorique et expérimentale de l'interaction de la lumière avec des ondes acoustiques de forte puissance" thèse d'Etat Université Paris VI (1973)

18 G.B. BENEDEK, K. FRITSCH "Brillouin scattering in cubic crystals", Phys. Rev. 149, 647-662 (1966)

19 G.E. DURAND, A.S. PINE, "High-resolution low level Brillouin spectroscopy in solids" IEEE Journal of quantum Electronics QE4, 523-528, (1968)

20 F. MICHARD, F. SIMONDET, L. BOYER, R. VACHER "A first step towards zero sound", Phonon scattering in solids Ed Challis Plenum Publ. Corp. 87-89 (1976)

21 S. KH. ESAYAN, B.D. LAIKHTMAN, V.V. LEMANOV, "Elastic and photoelastic properties of gadolinium molybdate crystals near a phase transition", Sov. Phy. JETP 41 342-345 (1975)

22 M.M. ELCOMBE, "Some aspects of the lattice dynamics of quartz" Proc. Phys. Soc (London) 91, 947-958 (1967).

23 A.S. PINE -Linear wave-vector dispersion of the shear-wave phase velocity in α quartz" J. Acoust. Soc. Am. 47, 73 (1970)

24 J. JOFFRIN, A. LEVELUT, "Mise en évidence et mesure du pouvoir rotatoire acoustique naturel du quartz-α Solid state commun. 8, 1573-1575 (1970

25 A.S. PINE 'Direct observation of acoustical activity in α quartz" Phys. Rev. B2, 2049-2054, (1970)

26 M.F. BRYZHINA, S. Kh ESAYAN, V.V. LEMANOV. "Investigation of acoustic activity in crystals by the method of BRAGG reflection of light". JETP Letters, 25 483-485, (1977)

27 F. MICHARD, B. PERRIN "New optical probe for local elastic study in transparent media" J. Acoust. Soc. Am. 64, 1447-1456, (1978).

H.F. PHONON TRANSMISSION AS A PROBE
OF CONDENSED MATTER

J.P. Maneval
Groupe de Physique des Solides de l'E.N.S.
24 rue Lhomond, 75231 Paris 5, France

1. INTRODUCTION

Phonons are quanta of mechanical vibrations in a solid, with a spectrum extending from audio-waves up to frequencies in the terahertz range (1 THz = 10^{12} Hz), in the infrared lattice absorption bands. Those high frequencies which are not generated by coherent ultrasonic techniques are named H.F. by convention. To characterize the quantized energy, the following equivalences are useful : $h\nu \leftrightarrow k_B T \leftrightarrow eV \leftrightarrow hc(1/\lambda_{opt})$. Here, h, k_B, e, c are respectively the Planck constant, the Boltzmann constant, the electronic charge and the velocity of light. In numbers, we have :

$$1 \text{ THz} \leftrightarrow 50 \text{ K} \leftrightarrow 4 \text{ meV} \leftrightarrow 30 \text{ cm}^{-1} \qquad (1)$$

By nature, the phonons are virtually coupled to any departure (defect or excitation) from the ideal solid. In the range specified by Eq. 1, they are in energetic resonance with superconducting gaps, Zeeman splittings, vibronic levels, etc... Besides, the typical phonon wave-length at 1 THz being $\sim$ 30 Å, H.F. phonons may be put in spatial interference with, for instance, Bohr orbits of shallow impurities in semiconductors, or free electrons, or semi-macroscopic objects such as evaporated thin films. We shall show in this article how the transport of H.F. acoustical phonons can bring information about (a) crystal acoustics, (b) phonon interactions and selection rules and (c) represents a possibility for imaging, restricted to low temperatures unfortunately.

2. BALLISTIC PROPAGATION

H.F. phonons achieve sufficiently long free paths at helium temperatures to exhibit ballistic (non-diffusive) propagation over macroscopic distances. This is seen in heat pulse experiments where a packet of thermal phonons produced by Joule effect in a metal film are allowed into a single crystal ; the temperature signal is then detected at the opposite end by a thin-film bolometer (Fig. 1a). Three separate signals appear in the course of time corresponding to the three acoustic branches of the ν versus q (wavevector) dispersion relation (Fig. 1b). L is for longitudinal, T_1 and T_2 are respectively for the fast and slow transverse polarizations. The velocities of sound are thus measured to within 0,5 % accuracy.
Additional information is obtained by using quantum detectors instead of wide band bolometers. Suppose (Fig. 1b) the detector threshold is 2Δ. In the case of normal dispersion, the fastest incoherent phonons will travel with the group velocities

$V_g = \partial\omega/\partial q$ taken at this particular energy, which may be significantly different (Fig. 1c) from the velocity of sound as determined by the bolometer trace. It appears indeed under magnification that the leading edges are very clearly delayed respective to one another.

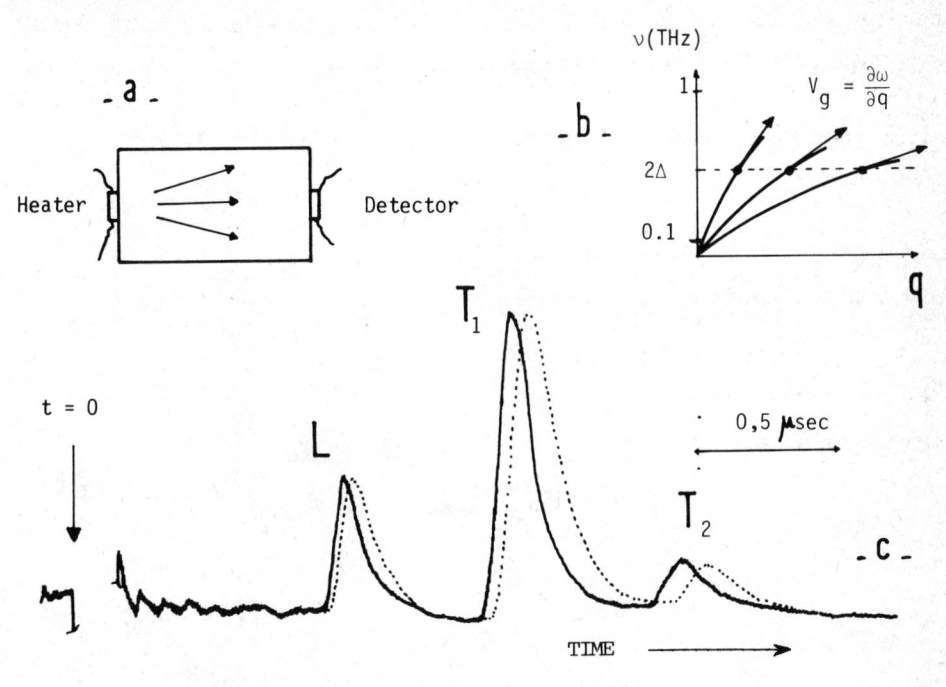

Fig. 1 : (a) InSb sample, oriented along [110], provided with thin film phonon transducers. (b) Dispersion relations for acoustic modes. (c) Bolometer (full line) and tin superconducting junction (dotted line) traces versus time as responses to a Dirac excitation. T_o = 1.3 K. Gap of tin : $2\Delta(Sn)$ = 1.145 meV.

Chromatic dispersion is in turn a useful tool for characterizing the threshold of a given detector, such as a superconducting junction in a magnetic field, or for measuring H.F. phonon frequencies when no other spectroscopy is available.

On the other hand, ballistic heat-pulse experiments provide also direct evidence of acoustic anisotropy and self-focussing effects (Ref. 1).

3. SOURCES AND RECEIVERS

- Thermal transducers. Evaporated metal films (Au, Cu, Constantan ...) are very efficient pulsed-heat radiators, with a damage limit at about 1 kW/mm^2. They can be driven by an electrical current or, alternatively, by a laser pulse. They are essentially broadband phonon generators, whose emission, in a first approximation, can be

assimilated to a blackbody spectrum (Ref. 1). These sources are adequate to study
the diffusion of heat taking place at $\sim$ 10 K and above. Second sound propagation has
also been observed in exceptionaly pure NaF single crystals (Ref. 2).
Superconducting bolometers, made of Al, Sn, Pb, Pb-Bi, etc..., are the most widely
used wide-band detectors, featuring sensitivity and speed of response ($\sim 10^{-8}$ sec).
The main drawback is their sensitivity to the magnetic field.

- Quantum transducers. Consider the scheme of electron energies in a symmetric superconducting tunnel diode (Fig. 2).

Fig. 2 : Superconducting tunnel junction as a quantum detector. Pairs are broken by the incident phonons ; a current of quasiparticles passes through the oxide barrier.

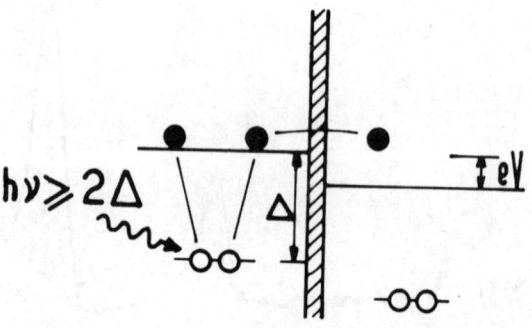

The interelectrode oxide barrier is so formed that it is transparent to quasiparticles (normal electron tunnelling), while pair-wavefunctions on both sides remain un-correlated. The mechanism by which a H.F. phonon (or a microwave photon) is quantally detected is well established (Ref. 3). If the incident energy $h\nu$ is larger than the superconducting gap 2Δ, a Cooper is broken, and a non-thermal quasiparticle current is then recorded as a signal. Fig. 1 is an example of such a process. Penetration of a smaller than critical magnetic field allows tunability to a certain extent. Fig. 3 shows how the frequency range 10 to 1000 GHz can be scanned by superconducting junctions.

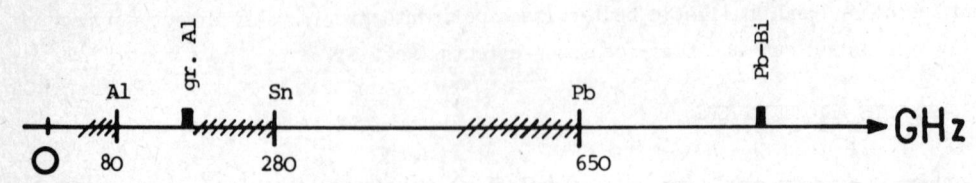

Fig. 3
Detection threshold of superconducting tunnel junctions. The hatched sections indicate the range of magnetic tunability. Granular Al and Pb-Bi are non-tunable.

As quasimonochromatic generators (Ref. 3), the superconducting junctions emit either recombination phonons ($h\nu = 2\Delta$) at a fixed frequency, or "bremstrahlung" phonons ($h\nu = eV - 2\Delta$), at a frequency depending upon the bias voltage V. This is a very attractive device since tunability can be achieved by voltage modulation. Although the quasimonochromatic power is very limited ($\sim$ 10 μwatts for typically 5 GHz bandwidth), very fine spectroscopy could be achieved by this means. Examples are impurity levels in Al_2O_3, phonon-magnon coupling in MnF_2 (Ref. 4).

4. STIMULATED PHONON EMISSION (Ref. 5)

The three-level electronic system of V^{4+} ions in Al_2O_3 is well known (Fig. 4) : above the $E_{3/2}$ ground-state lies a first $E_{1/2}$ excited state 28.1 cm^{-1} apart, while the second excited state, which also transforms as $E_{1/2}$, is itself 24.7 cm^{-1} above. All three transitions, labelled A_{31}, A_{32} and A_{21}, are possible via photon or phonon emission ; only 3 $\rightarrow$ 2, however, is associated with L-polarized phonons along the c-axis of Al_2O_3.

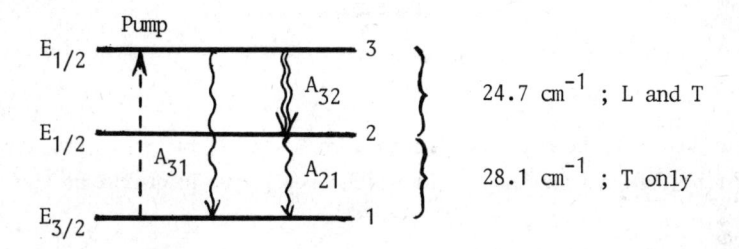

Fig. 4 : Energy diagram for the lowest three levels of the d^1 configuration of V^{4+} in Al_2O_3. Only levels 2 and 3 are coupled via an L phonon pointing along the c-axis.

A transient population inversion was produced by pumping the ground-state level by the 52.6 cm^{-1} radiation of a CH_3F laser. Stimulated emission manifests itself, at high pump powers, by an L-peak in the ballistic phonon signal. At pump intensities below threshold, the L component disappears, indicating that A_{31} is the main decay process.
To date, no volume amplification could be detected.

5. COHERENT FAR I.R. GENERATION

Limitation of coherent H.F. ultrasound stems for two main causes : the requirement of surface flatness which becomes an exceedingly difficult problem below 1000 Å, and the lack of powerful enough sources in the microwave region. On the other hand, far infrared radiation from molecular lasers overlap with the R.F. submillimeter band. The idea of exciting hypersound thanks to the electric field of an I.R. beam originated some years ago (Ref. 6).

A molecular laser beam is focussed onto the surface of a piezoelectric material (quartz), where it is partially reflected and partially transmitted (Fig. 5).

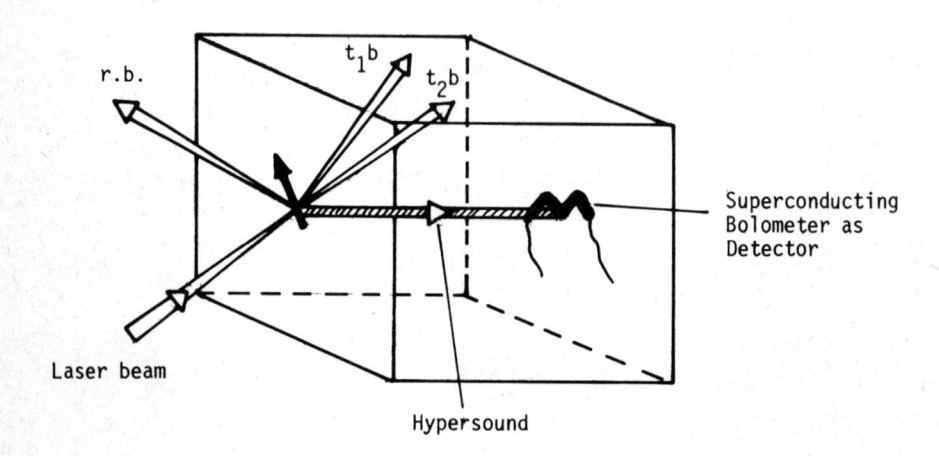

Fig. 5 : The incident I.R. beam gives rise to a reflected beam (r.b.), two transmitted beams (t_1b and t_2b), and a hypersound beam which propagates in quartz at the velocity of lattice waves.

However, the I.R. field is also coupled to mechanical motion of the quartz surface and this results in an additional beam of H.F. ultrasound, at the <u>frequency of the incident photons</u>, a generation process which must be distinguished from bulk absorption.

Unlike the incoherent mechanisms (thermal generation ; relaxation in a superconductor) described earlier, piezoelectric conversion is a coherent process provided the surface smoothness is perfect enough. For conditions of coherent I.R. excitation, this should produce a collimated beam of H.F. phonons. This property was unambiguously checked by moving the laser spot in front of the detector ; it was observed that no spreading at all occurs over several millimeters of propagation.

Assuming continuum mechanics, the rate of power (P) conversion reads :

$$\frac{P \text{ (acoustical)}}{P \text{ (infrared)}} = 4 \rho d^2 \frac{v_s^3}{\varepsilon_0 c (1 + \sqrt{\varepsilon})^2} \tag{2}$$

where ρ is the density of mass, v_s the velocity of sound, and d is the appropriate piezoelectric modulus. ε_o and ε are respectively the dielectric permittivity of vacuum and the relative permittivity of the medium. For quartz, the ratio is as low as 10^{-6} or less. In spite of this, coherent H.F. ultrasound could be produced at 0.89 THz with an HCN laser, and more recently at 2.53 and 3.4 THz.

This fascinating new technique, which boosts the domain of coherent ultrasonics by a factor of 100 (no coherent detection has been reported though), raises also some new problems such as the unexpectedly long propagation length (several millimeters) of THz waves.

6. IMAGING ?

Due to their limited mean free path, their sensitivity to imperfections and poor transmission characteristics at the interfaces, one can question the applicability of H.F. phonons in imaging. In fact, they constitute an unexcelled probe in some instances :

- image of the Fermi surface : in a collision between an acoustical phonon and a free electron, the maximum momentum imparted to the electron is $2k_F$, where k_F is the Fermi radius. This interference phenomenon between the lattice wave and the de Broglie wave of the electron is related to the famous Kohn anomaly. Spectroscopy by H.F. incoherent phonons using quantum detectors led to a value of the Fermi wavelength. $\lambda_F \simeq 280$ Å for a semiconductor containing about 4×10^{17} electrons/cm^3 (Ref. 7).

- Measurement of Bohr orbits in solids : relatively large orbits are associated with donor impurities in most semiconductors, due to the reduced effective mass and large dielectric constant. If a phonon mode is coupled to the impurity levels, the coupling constant is modulated by the structure factor $(1 + a^2 q^2)^{-2}$, where a is the Bohr radius, which will cause a very rapid decrease of the phonon absorption at $q \gtrsim a^{-1}$. Absorption spectroscopy thanks to chromatic dispersion (see § 2) led to a = (39 ± 1) Å for the Bohr radius of the Sb donor in Germanium (Ref. 8).

- Thin-film resonances : although measurement of evaporated thin films is possible, it cannot compete with other standard methods. On the contrary, H.F. phonon interferences are well adapted to the thicknesses of superfluid helium films (~ 100 Å or less). Helium film thicknesses were thus measured with good resolution thanks to the spin-phonon spectrometer (Ref. 9) made of magnetically tunable T_m^{++} ions in SrF_2.

In conclusion, spectroscopy with H.F. phonons, rather than imaging, has been exploited as yet. The latter area is likely to develop, especially if it becomes possible to handle conveniently coherent phonon beams.

REFERENCES

1. R.J. Von Gutfeld, Physical Acoustics, edited by W.P. Mason, (Academic, New York, 1968), Vol. V.

2. S.J. Rogers, Phys. Rev. B $\underline{3}$, 1440, (1971) ; and Dieter W. Pohl and V. Irniger, Phys. Rev. Letters, $\underline{36}$, 480, (1976).

3. W. Eisenmenger, in Physical Acoustics, edited by W.P. Mason, (Academic Press, New York, 1976), vol; XII, p. 80.

4. J. Mattes, P. Berberich and H. Kinder, J.de Physique (Paris), $\underline{39}$, Suppl. C.6-988 (1978).

5. W.E. Bron and W. Grill, Phys. Rev. Letters $\underline{40}$, 1459, (1978).

6. W. Grill and O. Weis, Satellite Symposium of the 8th International Congress of Acoustics on "Microwave Acoustics", Lancaster 1974, Editor : E.R. Dobbs and J.W. Wigmore, p. 179. Published by Institute of Physics, London.

7. D. Huet, B. Pannetier, F.R. Ladan and J.P. Maneval, J. de Physique, Paris, $\underline{37}$, 521 (1976).

8. D. Huet, Ph.D. thesis, Paris, 1978, unpublished.

9. Anderson C.H. and Sabisky E.S., in Physical Acoustics, ed. by W.P. Mason (Academic Press, N.Y.) 1971, Vol. VIII.

INFRARED DETECTORS

INFRARED IMAGING SYSTEMS

A. HADNI

University of Nancy I
54037 NANCY cedex - FRANCE

The problem of infrared detectors is difficult to introduce in a short time. It is a very interesting topics because it involves the problem of thermal fluctuations, and because of the improvements recently obtained with major applications to spectroscopy, astrophysics and imaging systems.

INTRODUCTION.

In any infrared detector the photon energy is transformed into some kind of excitation. There are two types of detectors depending on the use of this excitation [1].

1 - Quantum detectors :

The excitation of the detector leads to an instantaneous change of an easily measurable physical property (i.e. electrical conductivity) which is detected before thermalization, i.e. before thermal equilibrium occurs, i.e. before any change of temperature.

This explains why a germanium photo conductor can be immersed into liquid helium and give a photoconductive signal. The temperature has not to change.

Let us look at two examples of quantum detectors to see how thermal equilibrium is destroyed by absorption of a photon:

Ex. 1 - Photoconductivity (fig.1).

Ex. 2 - Ruby Quantum Counter (fig.2).

The energy levels of Cr^{3+} in Al_2O_3 are given in fig.2.

At 2 K the population of the 2 A level is negligible and nearly no R_2 light is absorbed.
Far infrared quanta at 29 cm^{-1} are absorbed, thermal equilibrium is destroyed and the R_2 wavelength is emitted at 6922 Å. The sensitivity is 1 µw.

Besides photo-conductors and quantum counters we can cite a number of infrared quantum detectors : photo-voltaïc cells, photographic plates, electronic bolometers, Josephson junctions, photon-drag detectors etc...

2 - Thermal detectors.

There is some kind of excitation. Electrons or phonons are excited either inside the detector itself or in a thin black layer deposited on the detector. This excitation is not directly detectable. We have to insulate the detector from the heat sink to get, after some time, an increase ΔT of temperature. Now many physical properties, may be all physical properties, are temperature sensitive and we have to choose one of the most sensitive to detect the infrared absorbed energy.

The response needs some time and a thermal time constant $\tau = \frac{\mathcal{C}}{\mathcal{G}}$ is introduced where $\mathcal{C}$ is the heat capacity and $\mathcal{G}$ the thermal losses for a unit temperature difference between detector and thermal sink. The thermal losses are made by radiation, convection and conduction. When they are limited to radiation the order of magnitude for τ is 1 or 2 seconds, at room temperature: the result is that a thermal detector cannot be immersed into the heat sink which should prevent any increase of temperature.

I - NOISE, NOISE EQUIVALENT POWER [3] [4].

When a shutter is placed in front of the infrared detector to stop any infrared radiation from the source, some noise is still observed.

When the detector is opened the noise N is added to the signal S. We define the responsivity as $R = \frac{S}{\emptyset}$. The result is that Responsivity can be normalized by taking noise as a unit. This gives the detectivity $D = \frac{R}{N}$ (w^{-1}).
The inverse of detectivity is the NEP $= \frac{N}{R}$ (w).
The time constant $\tau \simeq \frac{2\pi}{\Delta f}$ and in most cases :

$$D \propto \frac{1}{\sqrt{A}} \quad \frac{1}{\sqrt{\Delta f}} \quad ; \text{ hence a normalized detectivity :}$$

$$D^* = D \sqrt{A} . \sqrt{\Delta f} \quad (w^{-1} \text{ cm. } Hz^{1/2}).$$

The problem of noise which is now encountered in every field of physics has been concerning the infrared physicists since the earliest time.

I think it is because the only broad band infrared source since the pioneer work of Rubens until now has been the blackbody, the brilliancy L_ν of it decreases dramatically towards low frequencies : $L_\nu \propto \nu^2 T$.

I - 1 - Noise sources internal to the detector.

a - Johnson noise.

Johnson's noise is due to the random motions of the charge carriers. It is given by Nyquist's formula :

$$\overline{e^2} = 4 R k T \Delta f$$

R is the real part of the detector impedance, Δf is the band width. Johnson's noise is a white noise.

b - Thermal noise.

It occurs from temperature variations in the detector. In the case of thermal detectors the signal being sensitive to temperature, fluctuations give a noise. These fluctuation come on one hand by conduction and convection processes. These ones can be avoided. On the other hand they occur by random emission of photons. These ones cannot be avoided and lead to a NEP

(detector photon noise) = $2\sqrt{2\alpha\sigma k T_1^5}$. For $T_1 = 300$ K, we get NEP (detector photon noise, 300 K) = $5.5 \cdot 10^{-11}$ w.
At 3 K, we should have NEP (3 K) = $5.5 \cdot 10^{-16}$ w.

I - 2 - Background noise.

The photon noise due to fluctuations in background emission is equal to $2(2 k T_2^5 \alpha\sigma)^{1/2}$ and leads to a maximum detectivity D^* (Background limited) = $\dfrac{1}{(2 k T_2^5 \alpha \sigma)^{1/2}}$ which could be observed with quantum detectors since the thermal noise does not affect them. This noise cannot be avoided with quantum detectors but it is limited to the photons which have the right wavelength to be detected ($\lambda < \lambda_o$) or to photons transmitted by a suitable filter. This noise can also be reduced with thermal detectors when they have to look at a reduced wavelength interval. Cold filters are introduced.

In conclusion the ideal detectivity of a thermal detector is limited by photon noise both in the detector at temperature T_1 and in the background at temperature T_2 and

$$D^{*ideal}_{thermal\ detector} = \frac{1}{2(2 k T_1^5 \alpha \sigma)^{1/2} + 2(2 k T_2^5 \alpha \sigma)^{1/2}}$$

α being the absorption coefficient, σ and k the Stefan and Boltzmann constants respectively.

Fig.3 gives D^* vs background temperature T_2 for two chosen detector temperature $T_1 = 290$ K and $T_1 = 77$ K. It is seen that for both detector and background at 290 K, D^*ideal = $2 \cdot 10^{10}$ w^{-1}cm Hz$^{1/2}$.

Lowering the background temperature (or the detector temperature) down to zero gives only an improvement factor of $\sqrt{2}$.
Fig.4 gives D^*_{ideal} vs λ for a thermal detector : it is a constant equal to $2 \cdot 10^{10}$ w^{-1} cm Hz$^{1/2}$ as we have seen it.
For a photoconductive detector where λ is the limit of sensibility, the background noise is limited to photons with wave-length shorter than λ and detectivity increases as λ is reduced. As far as detectivity is concerned, quantum detectors operated at room temperature are better than thermal

detectors for $\lambda < 10$ μm. It is seen however that, they are far from the background limit (i.e. PbS).

II - COMPARISON OF THERMAL AND QUANTUM DETECTORS.

We have seen that with cold filters the detectivities of both detectors are comparable. Of course the sensitivity range is reduced in both cases.

The time constant $\tau = \dfrac{\mathcal{C}}{\mathcal{G}}$ of thermal detectors is higher. It can be decreased in two ways :
- increase $\mathcal{G}$: but detectivity is reduced
- decrease $\mathcal{C}$: either reduce thickness d ($\mathcal{C} \propto d$) or decrease temperature ($\mathcal{C} \propto T^3$).

III - INDIRECT DETECTIVITY (HETERODYNING).

It is easy to show that assuming same noise in both detections we have :

$$\frac{D^*_{\text{heterodyne}}}{D^*_{\text{direct}}} = \sqrt{\frac{\emptyset_L}{\emptyset_S}}$$

($\emptyset_L$ = local power; $\emptyset_S$ = signal power).

However the best advantage is to obtain a high spectral resolution. Heterodyne detection translates the problem of spectral resolution into a lower frequency domain where it is easier to build very narrow filters. With direct detection, high resolution must be obtained by the use of very long pathlength differences before detection and this becomes difficult with band-width less than 0.01 cm^{-1} [7].

IV - APPLICATIONS TO IMAGING.

IV - 1 - The pyroelectric vidicon.

The pyroelectric detector has a good detectivity at room temperature, uncommon high speed for a thermal detector and a great variety of possible configuration. For instance a very thin plate of a pyroelectric crystal can be cut perpendicular to the pyroelectric axis and receive an infrared

image. It gives a relief of temperature and thus a relief of polarization, (i.e. bound charges). These charges can be read with an electron beam as in a classical vidicon. We made the first proposal in 1963 [6]. It has taken 10 years to get useful images [8] [9].
The best results up to now are obtained with triglycine sulfate (TGS) single crystal plates. This is the Pyricon. The images show a 200 × 200 spatial resolution (5 lines pair per mm), a 0.5 K thermal resolution on the object with 10 images per second.
Improvements are still expected and TV in the dark is close to be competitive with visible TV.

IV - 2 - Infrared surface detection.

We shall cite :

1) the evaporography which gives an image every 10 s by specific evaporation of a liquid on the hot spots of the infrared image.

2) the Marangoni effect. In the "panicon" [10] a thin oil film deposited on a solid base has its thickness modulated by the unfalling infrared radiation. It is not a problem of evaporation but surface deformation due to local variation of surface tension which is very temperature sensitive. The Marangoni effect is faster than evaporography.
The Marangoni effect has been made more sensitive by Mr Loulergue and Mr Levy from the "Institut d'Optique" [9] recently by using liquid-liquid interface.
They claimed to get 5 images per second with 5 lines/mm on the bolometer and a thermal resolution of 0.5 K on the object.

BIBLIOGRAPHY

[1] "Essentials of Modern Physics applied to the Study of the Infrared", A. HADNI, Pergamon Press 1967.

[2] H. LENGFELLNER and K.F. RENK - I. EEE Journal of Quantum Electronic vol. Q.E. 13 (1977) 421.

[3] "L'Infrarouge Lointain", A. HADNI, Presses Universitaires de France, 1969.

[4] "Elements of Infrared Technology", P.W. KRUSE, L.D. Mc GLANCHLIN and R.B. Mc QUISTER, Edited by John Wiley 1962.

[5] A.M. BRADSHAW and F.M. HOFFMAN - Surf. Sc. 72 (1978) 573.

[6] A. HADNI - Possibilités actuelles de détection du rayonnement infrarouge - J. Phys. 24 (1963) 694.

[7] E.C. SATTON and C.H. TOWNES - Appl. J. 208 (1976) L 145.

[8] D. CHARLES, F. Le CARVENEC - Adv. Electr. Elect. Phys. 33A (1972) p. 73.

[9] P. FELIX, X. GERBAUX, A. HADNI, J. MANGIN, G. MOIROUD, G. MORLOT, R. THOMAS, S. VERON - Far Infrared applications of pyroelectric retina TV tubes - Optics and Laser Technology 8 (1976) 75.

[10] J.C. LOULERGUE, Communication à la 2e réunion européenne de "Infrared-Physics" ; Zurich (Mars 1979).

[11] J.C. LOULERGUE : Brevet Anvar n° 78.28.041

174

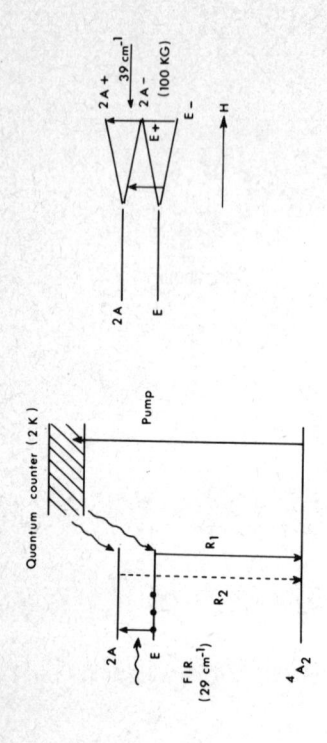

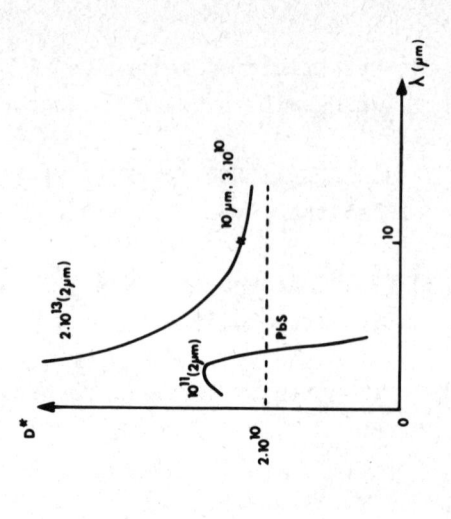

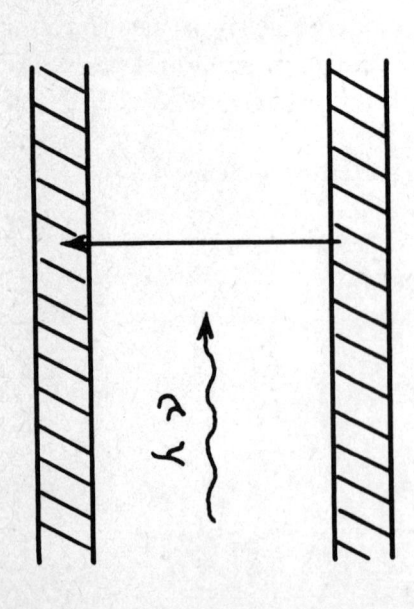

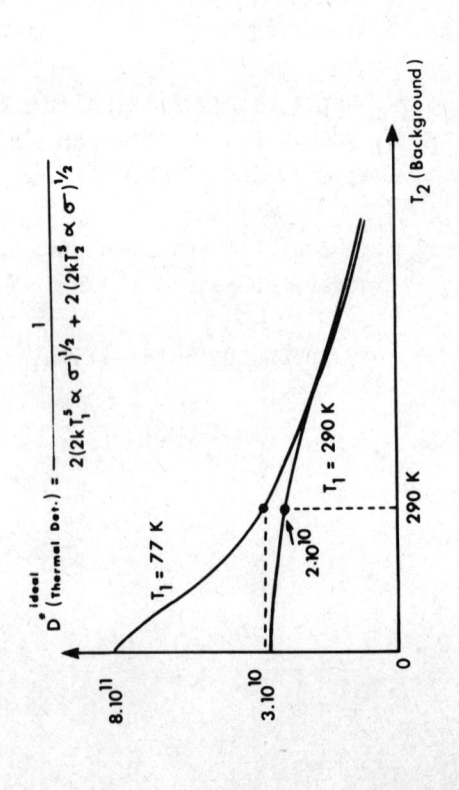

THE ELASTIC SCATTERING OF FAST ELECTRONS

B.F. Buxton

The Cavendish Laboratory,
Cambridge, CB3 OHE/UK

1. Elastic scattering in High Energy Electron Diffraction (HEED)

Since electrons interact strongly with matter in order to interpret the diffraction patterns and images observed in an electron microscope we nearly always have to solve the Schrödinger equation (Sivardiere: this volume)

$$\left[\nabla^2 + k^2 - U(\underline{r})\right] \psi(\underline{r}) = 0 \tag{1}$$

without invoking the Born approximation or other weak scattering models. In fact, by writing down a single particle equation such as (1) we have already considerably restricted the scope of our theory because this simple Schrödinger equation with a potential $U(\underline{r})$ having no internal degrees of freedom cannot describe the inelastic scattering of the electron beam by a real solid.

At best, (1), can only describe the elastic scattering if an optical potential is used, i.e.

(i) $U(\underline{r})$ is complex to allow for absorption of the elastically scattered electrons due to the inelastic processes (Dederichs 1972). It is called an optical potential by analogy with the use of a complex dielectric constant in optics or X-ray diffraction.

(ii) The effect of the thermal vibration of the atoms is included not only in $\mathrm{Im}(U(\underline{r}))$ where the possibility of inelastic phonon scattering is catered for, but also by using for the real part of the potential an average $\langle U(\underline{r}) \rangle$ in which each atomic potential is smeared or averaged by convolution with the probability density function for thermal vibrations. In a crystal this is equivalent to multiplying each Fourier coefficient of U by a Debye-Waller factor.

(iii) Some relativistic effects are included because the kinetic energy E of the electron beams usually used in HEED ($E \gtrsim 100$ keV) is comparable to the rest mass energy $m_o c^2$ of an electron (511 keV). Thus, the wave number k ($= 2\pi/\lambda$) and the mass m of the incident electrons are calculated relativistically:

$$k = (c/\hbar) \sqrt{m^2 - m_o^2}$$

and
$$m = m_o(1 + E/m_o c^2) \quad . \tag{2}$$

Since the effective potential in the Schrödinger equation (1) is given by

$$U(\underline{r}) = 2m\, V(\underline{r})/\hbar^2 \qquad (3)$$

in terms of the potential energy $V(\underline{r})$, the strength of U increases as the accelerating voltage of the incident electron beam is raised.

2. Forward scattering approximation

In HEED we want solutions of the Schrödinger equation describing the scattering of a beam of fast electrons which is fired through a thin sample (fig. 1).

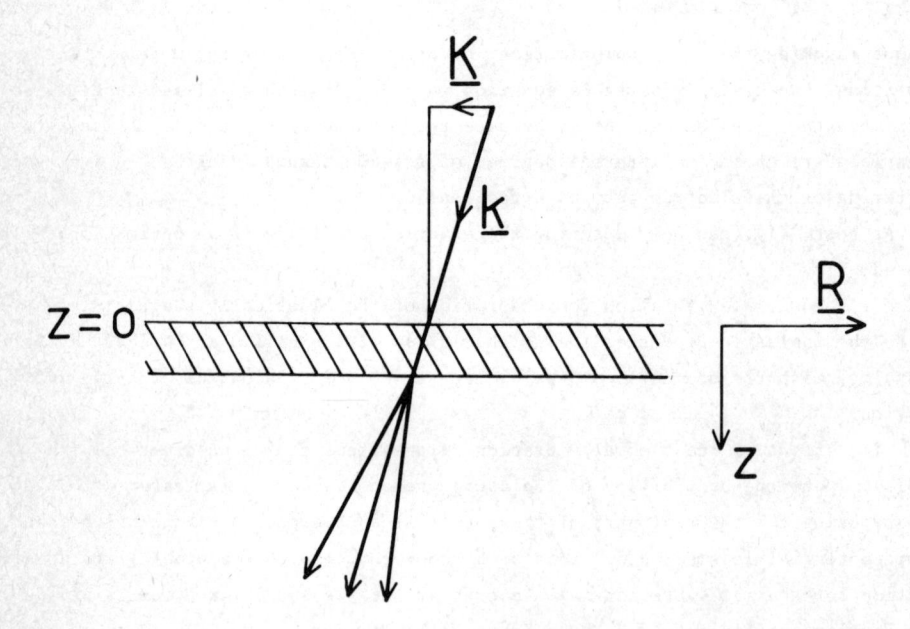

Fig. 1.

As the electron wavelength is very small compared to atomic dimensions ($\lambda < 0.05\text{Å}$ if $E \gtrsim 100$ keV), the scattering is strongly peaked in the forward direction in HEED and we may simplify (1) if the electrons are fired into the sample at near normal incidence. We let

$$\psi(\underline{r}) = e^{ikz}\, \tau(\underline{R}, z) \qquad (4)$$

and neglect the second derivative $\frac{\partial^2 \tau}{\partial z^2}$ on the assumption that $\tau(\underline{R}, z)$ is slowly varying. Thus:

$$2ik \frac{\partial \tau}{\partial z} = \left[-\nabla_{\underline{k}}^2 + U(\underline{R}, z) \right] \tau(\underline{R}, z) , \qquad (5)$$

which is like a 'time' dependent Schrödinger equation. Back scattering is neglected in this approximation so τ satisfies the initial condition at the entrance surface $z = 0$ that it matches onto the incident plane wave, i.e.

$$\tau(\underline{R}, 0) = e^{i \cdot \underline{K} \cdot \underline{R}} \qquad (6)$$

where $\underline{K}$, the transverse component of the incident wave vector $\underline{k}$, defines the orientation of the incident beam (fig. 1).

3. The phase object approximation

A wide variety of techniques may now be used to solve (5). In the simplest approximation we neglect the $-\nabla_{\underline{R}}^2$ term which represents the transverse kinetic energy of the waves. (5) can then be integrated i.e.

$$\tau(\underline{R}, z) = \exp\left[\frac{-i}{2k} \int_0^z dz' \, U(\underline{R}, z') \right] \tau(\underline{R}, 0) \qquad (7)$$

which is just the eikonal approximation (Glauber 1959). Although it is only valid for moderately thin samples, (7) is a kind of dynamical approximation as the potential U enters non-linearly in contrast to the kinematic (or Born) approximation which is linear in U and may be obtained by expanding the exponential to first order when

$$\frac{1}{2k} \int_0^z dz' \, U(\underline{R}, z') \ll 1 \qquad (8)$$

(7) is then equivalent to the first Born approximation provided the forward scattering assumption is incorporated in the latter and it is further assumed that $\underline{K}^2 z / 2k \ll 1$.

Note that since both U and k increase as m increases, the Born approximation is not a high energy approximation in HEED, but is valid for very thin samples when $\int_0^z dz' U(\underline{R}, z')$ is small. The phase object approximation is therefore often used to indicate the qualitative effects of dynamic diffraction e.g. Cowley (1978) uses it to discuss high resolution imaging in electron microscopy. It should also be noted that (7) can be applied just as easily to disordered or amorphous materials as to crystals and that it is invertable, i.e. given the wave function $\tau(\underline{R}, z)$ we can infer the projected potential $\int_0^z dz' \, U(\underline{R}, z')$. Attempts have therefore been made to find explicit formulae similar to (8) but of wider validity. Jap and Glaeser (1978) adopt an approach based on the Feynman path integral which has been discussed by Van Dyck (1975).

4. The multi-slice method

However, the phase object approximation neglects the transverse kinetic energy $-\nabla_R^2$ in (5) which imparts the wave nature to the propagation of the electrons through the solid and must be included in more sophisticated approximations. One way is to convolute (7) with the _Fresnel_ approximation to the wave propagator as we know that the important secondary wavelets travel in the forward direction almost parallel to z:

$$\tau(\underline{R}, z) = \int d^2 \underline{R}' \left(\frac{k}{2\pi i z} \right) \exp\left[i \frac{k(\underline{R}-\underline{R}')^2}{2z} - \frac{i}{2k} \int_0^z dz' U(\underline{R}', z') \right] \tau(\underline{R}', 0) \; . \quad (9)$$

This formula, which is easily shown to be a solution of the Schrödinger equation (5) to first order in z (which is assumed to be small) is the basis of the multi-slice method when it is used to describe the evolution of the wavefunction from a plane at depth z_n to $z_{n+1} = z_n + \Delta z$ rather than from 0 to z as above. Propagation through the whole sample is accomplished by dividing it into many such thin slices and repeatedly applying (9).

Although (9) is the form of the multi-slice employed by Ishizuka and Uyeda (1977), a slightly different equation is usually used in which the phase-grating $\exp\left[-i/2k \int_0^z dz' \, U(\underline{R}, z') \right]$ is evaluated at $\underline{R}$ after the initial wave has propagated to the plane z. This is an equally good solution of the Schrödinger equation (5) to first order in z. Van Dyck (1978) has recently exploited this ambiguity in the order of propagation to obtain a more accurate solution.

The multi-slice has most frequently been used to calculate the amplitudes of the waves diffracted by a perfect crystalline sample when (9) is Fourier transformed and applied in reciprocal space. However, in recent work it has been used to calculate high resolution images of defects in crystals (e.g. : Fields and Cowley 1978, Cowley and Fields 1979).

5. The Howie-Whelan equations

We do not however have to use the multi-slice to calculate the diffracted wave amplitudes: we can follow the spirit of Darwin's original treatment of the dynamical theory of X-ray diffraction and represent the wavefunction inside a crystalline sample as a sum of diffracted waves:

$$\tau(\underline{R},z) = e^{-ikz} \sum_{\underline{g}} a_{\underline{g}}(z) e^{i(\underline{k} + \underline{g}) \cdot \underline{r}} \qquad (10)$$

If we similarly expand the periodic potential within a crystalline sample in a Fourier series,

$$U(\underline{r}) = \sum_{\underline{g}} U_{\underline{g}} e^{i\underline{g} \cdot \underline{r}} \qquad (11)$$

the Schrödinger equation (5) will be satisfied if the amplitudes $a_{\underline{g}}$ change with depth z according to:

$$\frac{da_{\underline{g}}}{dz} = -is_{\underline{g}} a_{\underline{g}} - i/2k \sum_{\underline{g}'} U_{\underline{g}-\underline{g}'} a_{\underline{g}'} \qquad (12)$$

where, the excitation error

$$s_{\underline{g}} = \left[(\underline{K} + \underline{G})^2 - K^2 \right] /2k + g_z \qquad (13)$$

has been introduced. ($\underline{G}$ is the component of $\underline{g} \perp$ to z, g_z its component $\parallel$ to z). $s_{\underline{g}}$ is the distance in the z direction from the Ewald sphere to the reciprocal lattice point $\underline{g}$ (fig. 2).

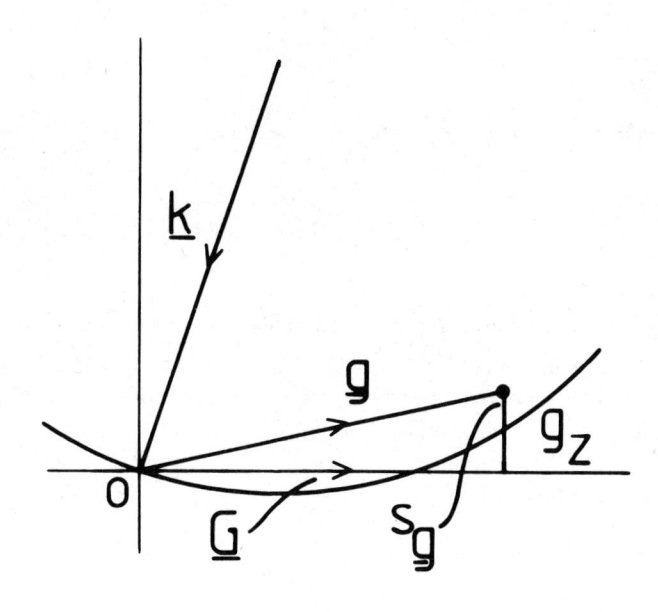

Fig. 2.
The excitation error $s_{\underline{g}}$. For clarity the curvature of the Ewald sphere has been much exaggerated.

It should be noted that (12) cannot be derived from the Schrödinger equation (5) because (11) is only valid for z within the sample (U vanishes outside the crystal). In general, there are too many diffracted wave amplitudes a_g to be uniquely determined by (5), but (12) is a convenient choice which ensures that $\tau(\underline{R},z)$ satisfies the Schrödinger equation. Similarly, we choose

$$a_g(0) = \delta_{g,0} \qquad (14)$$

which automatically ensures that $\tau(\underline{R},0)$ is equal to the incident plane wave at z = 0.

It is well known that the Howie-Whelan equations (12) can be integrated analytically if only two diffracted wave amplitudes are included (the incident beam a_0 and one other strong reflexion) to obtain the 'pendulum' solution.

$$a_0 = \exp\left[\frac{-iz}{2}(s_g + U_0/k)\right]\left\{\cos\left(\frac{\pi z}{\xi_g}\right) + \frac{is_g\xi_g}{2\pi}\sin\left(\frac{\pi z}{\xi_g}\right)\right\} \qquad (15)$$

$$a_g = \frac{-iU_g\xi_g}{2\pi k}\exp\left[\frac{-iz}{2}(s_g + U_0/k)\right]\sin\left(\frac{\pi z}{\xi_g}\right), \qquad (15)$$

in which intensity is 'periodically' transferred from the incident beam to the diffracted beam and back again in the extinction distance,

$$\xi_g = 2\pi/\sqrt{s_g^2 + U_g U_{-g}/k^2} \qquad (16)$$

Of more interest here is the fact that (12) can be formally integrated

$$\underline{a}(z) = \exp\left[\frac{-izM}{2k}\right]\underline{a}(0) \qquad (17)$$

(see for example Hirsch et al 1965), by writing the amplitudes a_g as a column vector $\underline{a}$ and utilizing the scattering matrix

$$S = \exp\left[\frac{-izM}{2k}\right] \qquad (18)$$

where M is the matrix of the coefficients in the Howie-Whelan equations (12). The matrix exponential in (18) can be evaluated from its series expansion, or recursively, or by using the Bloch wave method described in the next section. However, we should first note that the scattering matrix itself has been used in the calculation of electron microscope images of defects (Hirsch et al 1965).

6. Bloch waves and the dispersion surface

As mentioned above, we could introduce the Bloch waves as a formal device for evaluating the scattering matrix, but let us instead return to the physics. It is known from solid state physics that electrons in a crystal do not propagate as plane waves (which are scattered by the periodic potential as (12) shows) but as Bloch waves $\psi_j(\underline{r})$ each of which can be written as a sum of plane waves

$$\psi_j(\underline{r}) = \sum_{\underline{g}} c_{\underline{g}}^{(j)} e^{i(\underline{k}^{(j)} + \underline{g}) \cdot \underline{r}} \qquad . \qquad (19)$$

Since we are considering elastic scattering the Bloch waves excited in the crystal all have the same energy k^2 and, to match onto the incident plane wave at $z = 0$ they all have the same transverse component of the Bloch wavevector $\underline{k}_\perp^{(j)} = \underline{K}$, but the z components of the wavevectors $k_z^{(j)}$ differ in general. The loci of the wavevectors $k_z^{(j)}$ as functions of $\underline{K}$ or the incident beam orientation form the dispersion surface. We could calculate the dispersion surface and the Bloch waves by substituting (19) in the original Schrödinger equation (1), but we really only want those solutions which correspond to waves propagating almost parallel to the z axis so it is easier to use (5) which neglects back-scattering even though from (5) the curvature of the dispersion surface is only approximately correct. To calculate the Bloch waves we assume that (11) is valid both inside and outside the sample. Thus, if we let

$$\gamma_j = k_z^{(j)} - k_z \simeq k_z^{(j)} - k + \underline{K}^2/2k \qquad , \qquad (20)$$

we find that

$$\sum_{\underline{g}'} \left[(\gamma_j + s_{\underline{g}}) \delta_{\underline{g}\,\underline{g}'} + \frac{1}{2k} U_{\underline{g}-\underline{g}'} \right] c_{\underline{g}}^{(j)} = 0 \qquad . \qquad (21)$$

These are the many-beam equations which, being of standard eigenvalue-eigenvector form can be solved easily on a computer. The only problem is that it may be necessary to use a rather large matrix ($\sim$ 100 x 100) to obtain accurate results, especially for heavy materials at high voltages or when the incident beam is nearly parallel to a low index zone axis of the crystal. In the latter case there are many strong reflexions in the zero Laue zone with small excitation errors $s_{\underline{g}}$ which must be included in the many-beam matrix. However, the major asset of the Bloch wave method is not that it is a particularly efficient computational scheme (for large calculations the multi-slice is usually faster and can handle many more reflexions) but, the insight it provides via the intermediate concept of the Bloch waves which propagate independently through the crystal. The electron wavefunction is thus given by a sum of Bloch waves

$$\psi(\underline{r}) = \sum_j \varepsilon_j \psi_j(\underline{r}) \qquad (22)$$

with constant excitation amplitudes ε_j determined by matching (22) to the incident wave at $z = 0$. The diffracted wave amplitudes may be obtained by substituting (19) into (22) and comparing with (10) from which we see that the diffracted wave intensities are largely determined by interference of the Bloch waves excited in the crystal.

Finally, let us note that the $\text{Im}(U(\underline{r}))$ is small so that absorption effects can usually be included by perturbation theory. The eigenvectors obtained from the many-beam equations (21) are then orthogonal and we can evaluate the excitation amplitudes:

$$\varepsilon_j = C_0^{(j)*} \tag{23}$$

However, the eigenvalues γ_j and the Bloch wavevectors $k_z^{(j)}$ become complex when $\text{Im}(U(r))$ is included showing that, due to the inelastic scattering, the elastically scattered intensity decreases as the waves propagate through the crystal.

7. Conclusion

We can connect the approaches described here by noting that in the Bloch wave method we are effectively calculating the scattering matrix S by diagonalizing the matrix M. It is not surprising therefore that the many-beam equations (21) can be solved when only two plane waves (C_0 and C_g are included in the expansion of the Bloch wave (19):

$$\gamma_{\pm} = -1/2 \left[s_{\underline{g}} + U_0/k \mp \sqrt{s_{\underline{g}}^2 + U_{\underline{g}} U_{-\underline{g}}/k^2} \right] \tag{24}$$

and, if we ignore $\text{Im}(U(r))$, the eigenvectors are given by:

$$\left. \begin{array}{l} C_0^{(\pm)} = \dfrac{1}{\sqrt{2}} \sqrt{1 \mp s_{\underline{g}} \xi_{\underline{g}}/2\pi} \\[1em] C_{\underline{g}}^{(\pm)} = \pm \dfrac{e^{i\arg(U_{\underline{g}})}}{\sqrt{2}} \sqrt{1 \pm s_{\underline{g}} \xi_{\underline{g}}/2\pi} \end{array} \right\} \tag{25}$$

The dispersion surface is thus obtained in the familiar two beam approximation as sketched in fig. 3 near the Brillouin zone boundary perpendicular to $\underline{g}$. Like the phase object approximation this simple approximation has also been extensively used to discuss qualitatively the effects produced by dynamical diffraction in electron microscopy. However, it should be emphasized that the usefulness of the Bloch wave concept goes far beyond the two beam theory, for example in the explanation of channelling effects, the critical voltage effect, higher order Laue zone diffraction effects or the multiple refraction effects observed when steep wedge shaped crystal specimens are used.

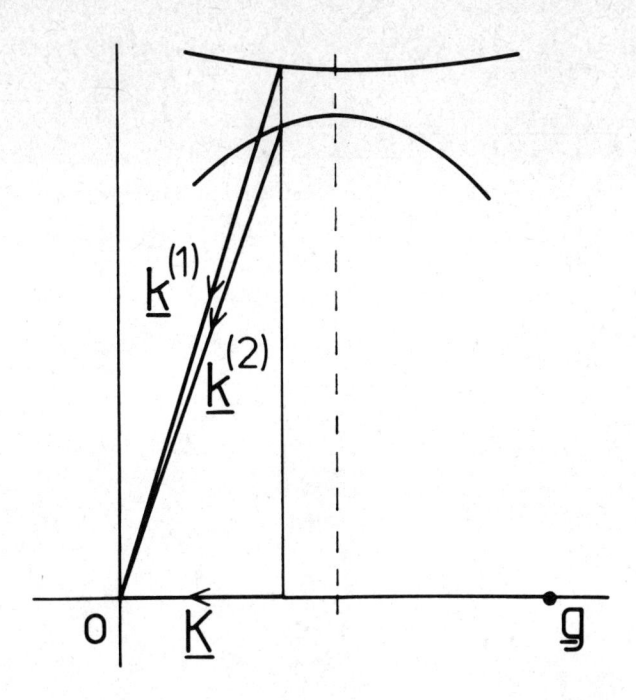

Fig. 3. The two beam dispersion surface (the curvature is exaggerated for clarity).

Finally, it should be noted that low energy electron diffraction has not been discussed at all here. A different kind of dynamical theory is usually used in LEED in which the scattering of the electron waves by the atoms in a solid is described by using the phase shifts as discussed by Sivardiere and these scattered waves are in turn themselves scattered by other atoms in the sample. The details of this multiple scattering cannot be discussed here (Pendry 1974), but let us note that such an approach would not be fruitful in HEED because of the large number of phase shifts that would be required to describe the scattering of the 100 keV electrons by an atom.

References

Cowley J.M. 1978 in 'Electron Microscopy 1978' III, pp. 207-271, ed. by J.M. Sturgess (Ontario: Microscopical Society of Canada).
Cowley J.M. and Fields, P.M. 1979, Acta.Cryst. A35, 28-37.
Dederichs P.H. 1972, Sol.St. Phys. 27, 134-236.
Fields P.M. and Cowley J.M., 1978, Acta.Cryst. A34, 103-112.

Glauber, R.J., 1959, in 'Lectures in Theoretical Physics' I, 315-414. ed. by W.E. Brittin and L.G. Dunham (New York: Interscience).
Hirsch, P.B., Howie, A., Nicholson, R.B., Pashley, D.W. & Whelan, M.J., 1965, Electron Microscopy of Thin Crystals (London : Butterworths).
Ishizuka, J, and Uyeda, N. 1977, Acta. Cryst. A33, 740-749.
Jap, B.K. and Glaeser, R.M., 1978, Acta.Cryst. A34, 94-102.
Pendry, J.B., 1974, Low Energy Electron Diffraction (London : Academic Press)
Van Dyck, D., 1975, Phys.Stat.Sol. (b), 72, 321-336.
Van Dyck, D., 1978, in 'Electron Microscopy 1978' Vol.I. pp. 196-197, ed. by J.M. Sturgess. (Ontario : Microscopical Society of Canada).

INELASTIC ELECTRON SCATTERING

C. Colliex, C. Mory and P. Trebbia
Laboratoire de Physique des Solides associé au CNRS
Bâtiment 510, Université Paris-Sud
91405 Orsay   France

1. Elastic and inelastic processes.

Various types of collision processes between complex particles are generally classified, see for instance Messiah[1], as follows :
- <u>elastic</u> collisions of type $A + B \rightarrow A + B$ with only a change in relative momentum between the initial and final states which correspond to the same channel,
- <u>inelastic</u> collisions of type $A + B \rightarrow A^* + B^*$ which involve a change of internal quantum state and consequently different input and output channels but the same interaction potential in these channels,
- <u>rearrangement</u> collisions of type $A + B \rightarrow C + D$ which involve exchange of particles between the two partners of the collision.

In the specific case of scattering of high energy electrons by solids, exchange effects are neglected and one only considers elastic and inelastic collisions. In the transmission electron microscope the beam of quasimonochromatic electrons of primary energy E_o interacts with the specimen so that most of the outgoing electrons have suffered changes in direction and in energy. From the experimental point of view, the term inelastic process with an energy loss ΔE is used when the measured change in energy (ΔE) is larger than the smallest detectable one (δE) with the spectrometer which is used. In most cases this lower limit is set by the width of the primary beam at about 1 eV.

A complete knowledge of the distribution in momentum ($\vec{q}$) and energy (ΔE) of the transmitted electrons is therefore necessary to characterize the scattering properties of the target and consequently to reveal its static and dynamical behaviour . Rather few measurements of this type have been achieved ; this is mainly due to the large dynamic range which is necessary to compare the intensities at q and $\Delta E = 0$ and at large q and ΔE : it easily reaches ratios of the order of 10^6 to 10^8 : 1. To obtain satisfactory data over a large number of experimental points, computer aided systems combining an automatic recording of spectra such as $I_{\Delta E}(q)$ and (or) $I_q(\Delta E)$ and electron counting capabilities are required.

The results must be displayed as a three-dimensional chart I(ΔE,q) for an amorphous or polycrystalline specimen with a cylindrically symmetric angular distribution of scattered electrons. The first systematic work of this type is due to Batson (2) for the valence electron excitation spectrum in aluminium. His results are displayed as intensity contour maps as a function of ΔE in eV and q in Å^{-1}, similar to the example shown in figure 1. We are now extending such measurements to other elements (carbon, gold ...) ; some partial data are shown in figure 2, as a set of intensity traces $I_q(\Delta E)$ for a fixed scattering angle (Trebbia and Colliex(3)).

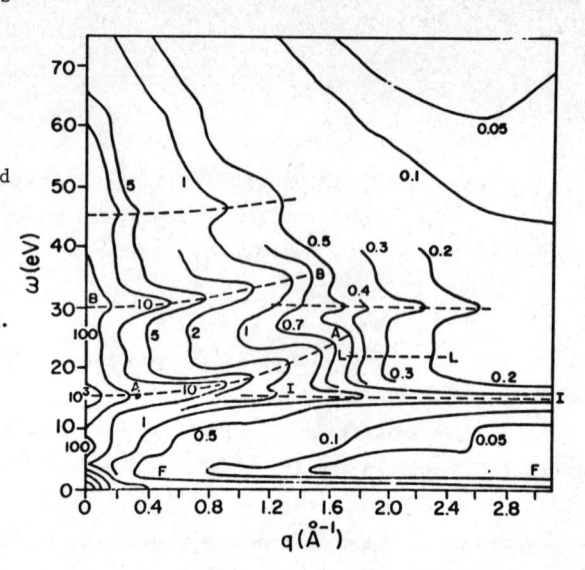

Fig. 1 : Intensity contour map I(q,ΔE) for a 500 Å thick aluminium sample with primary electrons of 75 kV (from Batson (2)).

In this simple description in terms of elastic and inelastic collisions, one omits the very important contribution of multiple scattering, that is the probability for an incident electron to be scattered m times elastically (without change in energy) and n times inelastically (with changes in energy ΔE_1, ΔE_2 ... ΔE_n) so that the total change in energy is $\Delta E = \sum_i \Delta E_i$.

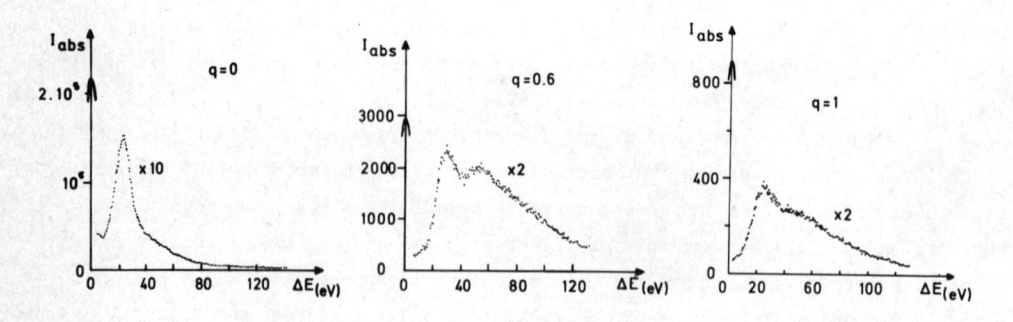

Fig.2 : Energy distribution of 50 kV electrons transmitted through a 250 Å thick carbon film for various momentum transfers (from Trebbia and Colliex (3)). q-values are given in Å^{-1}.

Misell (4) has proposed a classification of these various processes within a probabilistic approach using a statistical Poisson law for multiple events. A single process is characterized by the distribution of the scattered electrons as function of the angle of scattering and the transferred energy, that is for a single elastic collision :

$$I^1_{el}(\theta,\Delta E) = I^1_{el}(\theta) \cdot \delta(\Delta E)$$

and for a single inelastic collision :

$$I^1_{in}(\theta,\Delta E) = F^1_{in}(\theta) \cdot G^1_{in}(\Delta E)$$

if one assumes that it is possible to separate the behaviour in energy and in angle (the validity of this hypothesis will be discussed later on). By performing some simple integrations, one finds that the total probability for single elastic scattering is : $I^1_{el} = \int_\theta \int_{\Delta E} I^1_{el}(\theta,\Delta E) d\theta d\Delta E = \frac{t}{\Lambda_e}$ where t is the specimen thickness and Λ_e the elastic mean free path. Similarly for single inelastic scattering : $I^1_{el} = \int_\theta \int_{\Delta E} I^1_{in}(\theta,\Delta E) d\theta d\Delta E = \frac{t}{\Lambda_i}$ where Λ_i is the inelastic mean free path. For 100 kV primary electrons orders of magnitude of mean free paths are :

Å	Carbon	Aluminium	Gold
Elastic Λ_e	1300	700	60
Inelastic Λ_i	600	500	400
Z	6	13	79

These values have been shown to obey an approximate law : $\frac{\Lambda_e}{\Lambda_i} \simeq \frac{20}{Z}$; it results that for moderate specimen thicknesses the probability of single scattering becomes greater than 1. Multiple events must be taken into account and the above expressions are to be modified as follows :

−) multiple elastic scattering :
Angular distribution : $I^m_{el}(\theta) = I^{m-1}_{el}(\theta) * I^1_{el}(\theta)$

$$I_{el}(\theta) = e^{-t/\Lambda_e} \sum_{m=1}^{\infty} (\frac{t}{\Lambda_e})^m \cdot \frac{I^m_{el}(\theta)}{m!}$$

Total probability : $I_{el} = \int_\theta I_{el}(\theta) d\theta = 1 - e^{-t/\Lambda_e}$

−) multiple inelastic scattering :

$$I^n_{in}(\theta,\Delta E) = I^{n-1}_{in}(\theta,\Delta E) * I^1_{in}(\theta,\Delta E) = F^n_{in}(\theta) \cdot G^n_{in}(\Delta E)$$

$$I_{in}(\theta,\Delta E) = e^{-t/\Lambda_i} \sum_{n=1}^{\infty} (\frac{t}{\Lambda_i})^n \frac{I^n_{in}(\theta,\Delta E)}{n!} = F_{in}(\theta) \cdot G_{in}(\Delta E)$$

Total probability :

$$I_{in} = \int_\theta \int_{\Delta E} F_{in}(\theta) G_{in}(\Delta E) d\theta d\Delta E = 1 - e^{-t/\Lambda_i}$$

-) combined elastic-inelastic scattering :

The probability of m elastic interactions and n inelastic ones is given by the Poisson distribution :

$$P_{mn} = \frac{e^{-t/\Lambda_e}}{m!}(\frac{t}{\Lambda_e})^m \cdot \frac{e^{-t/\Lambda_i}}{n!}(\frac{t}{\Lambda_i})^n$$

Three main contributions can then be defined, the unscattered one corresponding to m = 0 and n = 0, the elastic one with m varying from 1 to ∞ and n = 0, the inelastic one with n varying from 1 to ∞.

A summation rule can then be written :

$$1 = I_{un} + I_{el} + I_{in}$$

where :

*) $I_{un} = e^{-t/\Lambda_t}$ for m = 0 and n = 0

The total mean free path is defined by $1/\Lambda_t = 1/\Lambda_e + 1/\Lambda_i$

*) $I_{el} = e^{-t/\Lambda_i}(1 - e^{-t/\Lambda_e})$ for $\sum_{m=1}^{\infty}$ and n = 0

The correction term expresses the fact there is no inelastic scattering.

*) $I_{in} = (1 - e^{-t/\Lambda_i})$ for $\sum_{m=0}^{\infty}$ and $\sum_{n=1}^{\infty}$

In this classification the inelastic component contains the single inelastic process and all multiple elastic-inelastic processes with at least one inelastic collision. It can be refined when one considers more specifically the angular dependence :

$$I_{in}(\theta) = \sum_{n=1}^{\infty} (\frac{t}{\Lambda_i})^n \frac{1}{n!} \left[e^{-t/\Lambda_i} \cdot F_{in}^n(\theta) + I_{el}(\theta) * F_{in}^n(\theta) \right]$$

In the bracket the first term represents the convolution of the true inelastic component with the unscattered one and the second term with the elastically scattered one. As it will be shown later that the angular spread of the inelastically scattered electrons is narrower than the elastic one, four main classes of scattered particles can be roughly distinguished within a graph in energy and momentum transfers (figure 3).

Fig.3 :

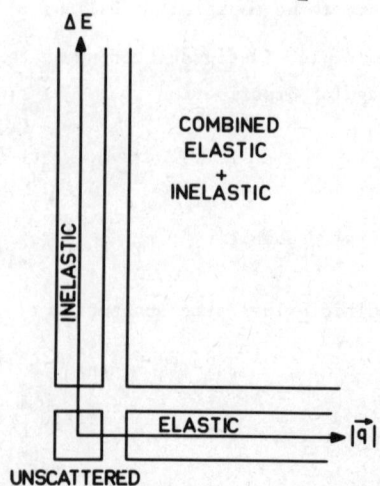

2. Single inelastic process.

2.1. Extraction of the single loss term :

As the elastic scattering is considered in Buxton's paper (this volume) the following part of this paper deals with the inelastic term, as defined in the previous section. This inelastic contribution is generally revealed as an energy loss spectrum, that is :

$$I_\alpha(\Delta E) = \int_{\theta=0}^{\alpha} I(\Delta E,\theta) \cdot 2\pi \sin\theta\, d\theta$$

It represents the energy distribution of the electrons transmitted within an aperture of semi angle α. There does not yet exist a completely satisfactory deconvolution procedure to extract the single inelastic process from such an experimental spectrum and the generally used solutions implicitly assume a separation of the angular and energy dependence which is most satisfied when the angular acceptance α remains small.

In this small angle limit, various procedures have been developed to extract the single loss profile from an experimental one. Misell and Jones (5) have shown that the probability distribution of multiple scattering can be handled so that

$$I_1(\Delta E) = I_{exp}(\Delta E) - \frac{1}{2I_o}\left[I_{exp} * I_{exp}\right] + \frac{1}{3I_o^2}\left[I_{exp} * I_{exp} * I_{exp}\right] + \ldots$$

where I_o denotes the unscattered peak intensity. To solve this equation it can be convenient to work in the Fourier spectrum. It has been used by Johnson and Spence (6), and this last author (7) has recently established how the Fourier coefficient $\widetilde{I_1}(\tau)$ can be deduced from $\widetilde{I_{exp}}(\tau)$ with the help of the simple algebraic relation :

$$\widetilde{I_1}(\tau) = \widetilde{M}(\tau) \cdot \text{Log}\left[\frac{1}{I_o} \cdot \frac{\widetilde{I_{exp}}(\tau)}{\widetilde{G_o}(\tau)}\right]$$

This expression actually incorporates the instrumental impulse response with $\widetilde{G_o}(\tau)$ and a gaussian filter $\widetilde{M}(\tau)$ to level out the noise which is introduced through these mathematical manipulations.

Another technique to solve these deconvolution equations is to calculate by iteration the spectrum for double, triple ... losses using step by step increments of the energy loss from the zero one. This procedure has been first described by Daniels et al. (8) in a general frame, then simplified by Wehenkel (9) and by ourselves (10) to be easily applied to the deconvolution of an energy loss spectrum between 0 and 150 eV.

Figure 4, concerning a 600 Å thick titanium foil, clearly displays the distribution of multiple events. The double plasmon appears as a satellite at about 35 eV on the onset of the peak which corresponds to the excitation of the 3p electrons. Similarly a faint structure on the experimental curve at 70 eV is due to double processes of type plasmon + 3p excitation. Both contributions are no longer visible on the deconvoluted curve.

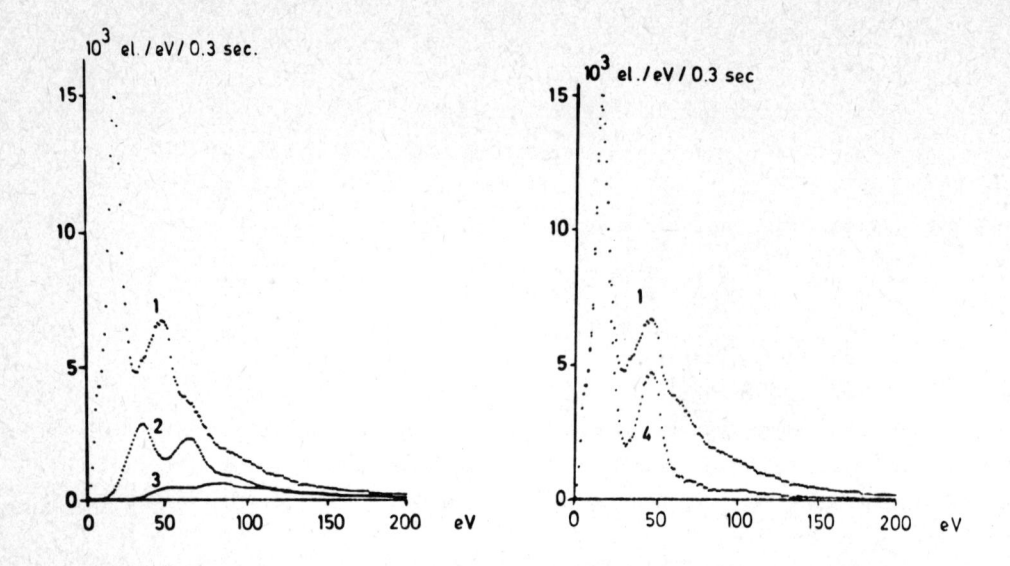

Fig.4 : Deconvolution of multiple energy losses by an iterative technique :
  a) Curve 1 : experimental spectrum for a 600 Å titanium foil. Plasmon peak is at 17 eV and the maximum after 45 eV corresponds to the M_{23} edge.
  Curve 2 : calculated double loss spectrum.
  Curve 3 : calculated triple loss spectrum.
  b) Curve 4 : result of the deconvolution, equal to Curve 1 - (2+3).

The validity of these data handling procedures has been checked for gold specimens of different thicknesses. From experimental spectra exhibiting quite noticeable differences it is possible to obtain a unique solution for the single loss profile.

2.2. <u>Content of the single loss term</u> :

It is beyond the scope of this paper to describe all the features of an energy loss spectrum. From two typical examples shown in figure 5, concerning a metal foil and a biological specimen, it is however possible to extract the general rules which govern the energy distribution of the electron excitation spectrum.

The first curve deals with a chromium specimen. Apart from the zero loss peak, it contains various structures corresponding to the excitation of the different electron populations in the target. From lower to higher energy losses they successively concern the conduction and the core electrons (atomic 3p at 50 eV and 2p at 575 eV). The first term appears as a peak at 24 eV generally labelled "plasmon" peak and the other ones as structures or edges superposed over a continuously decreasing background. Moreover, on this background, one detects stray

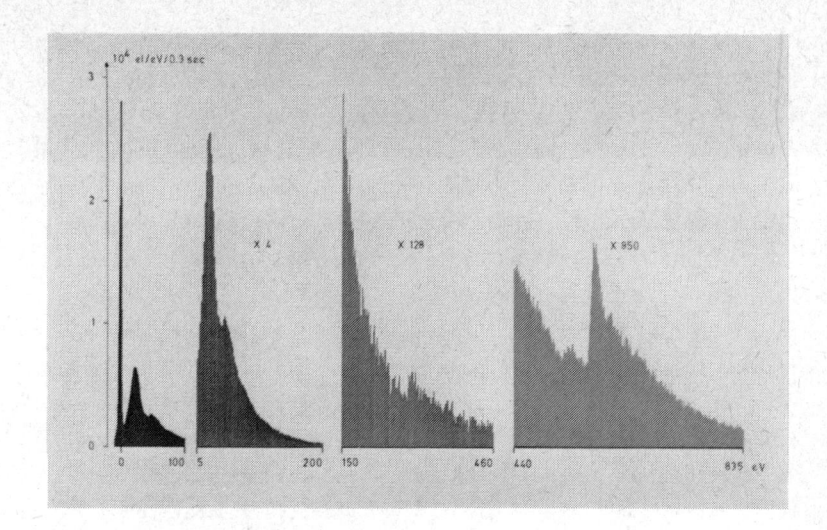

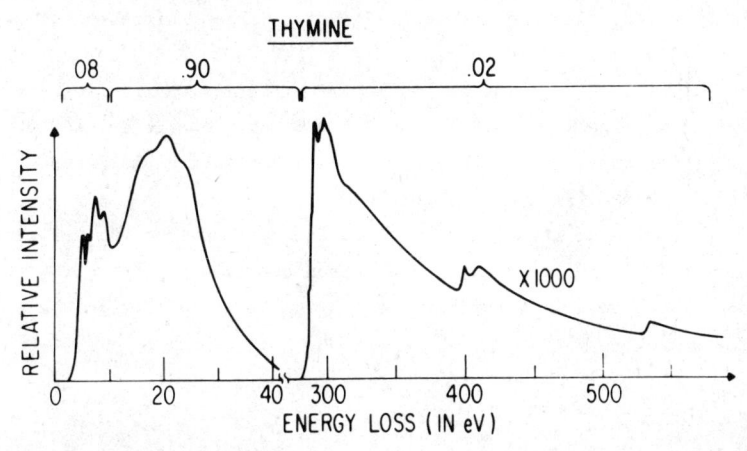

Figure 5 : a) Energy loss spectrum of 50 kV electrons transmitted through a 250 Å chromium foil. Because of the thinness of the specimen the multiple processes have not been subtracted. Semi angle of collection is 10 mrad. From left to right one sees the unscattered peak, the plasmon peak at 24 eV and the M_{23} edge at 50 eV. Contamination by carbon and by oxygen is revealed through their K signals respectively at 285 eV and 530 eV. They are followed by the chromium L_{23} edge at 580 eV. Notice the different intensity scales on the various curves.

b) Characteristic electron energy loss spectrum for a $\simeq$ 500 Å thick film of the nucleic acid base thymine ($C_5N_2O_2H_6$) supported on a $\simeq$ 20 Å thick carbon substrate. Primary energy of electrons is 25 keV. Structure due to both the valence shell and the inner shell excitations are shown. Above the spectrum, the relative fraction of the total inelastic scattering cross section is indicated for various energy regions (from Isaacson (11)).

signals due to the K edges of contaminant carbon at 285 eV and oxygen at 530 eV. There cannot be any confusion between the processes involving conduction and 3p and 2p electrons which require very distinct amounts of energy. On the other hand, it is not possible to distinguish between the nearly free 4s and tightly bound 3d electrons occupying the conduction band, which consequently differs from the standard jellium model. However, there still exists a typical plasmon energy, the value of which deviates by a few eV from that calculated in the free electron model.

An energy loss spectrum for an important biological component, the thymine molecule, is also shown in figure 5. It has been recorded by Isaacson (11). The interpretation of the various structures involves successively the electrons in the molecular orbitals for the fine structures between 5 and 10 eV, all the valence electrons from the various atoms of the molecule which give rise to a non characteristic "collective" term at about 25 eV which is visible for any carbonaceous material, and finally the atomic K edges for carbon, nitrogen and oxygen superposed on the background at respectively 285 eV, 405 eV and 545 eV. Such a spectrum is therefore interesting at two levels : the low energy fine structures, for instance the weak peaks at 4.8 eV and 7.4 eV, which characterize the molecular bonds with π orbitals, and the high energy core losses which can be used to determine the chemical composition of the molecule.

As a conclusion it is clear that an energy loss spectrum contains quite complete information concerning all the electron orbitals in the solid, whether they are rather localized in an atomic shell or delocalized through the conduction states of the solid.

3. Quantum mechanics formalism for an inelastic collision. Application to the study of the angular and energy distributions of the inelastic electrons

The analysis of the content of an energy loss spectrum shows that the solid behaves either as a collection of noninteracting atoms (for the characteristic core edges) or as a gas of interacting electrons (for the excitation of conduction electrons). It is consequently useful to treat the inelastic scattering of electrons either by an isolated atom, or at the other extreme, by an assembly of free electrons, a jellium, which consitutes the simplest approach for the description of the conduction electrons.

3.1. General formalism for an inelastic electron scattering :

For an inelastic collision the initial state is defined by the bracket $|\vec{k}_o 0\rangle$ that is a primary wave of wave vector $\vec{k}_o$ and the target in its fundamental state $|0\rangle$. The final state is described by $|\vec{k}_n n\rangle$, corresponding to an emerging electron with wave vector $\vec{k}_n$ and the target in the excited state $|n\rangle$. The total hamiltonian of the system is : $\mathcal{H} = \mathcal{H}_o + p^2/2m + \mathcal{H}_i$

where $\mathcal{H}_o|0\rangle = \varepsilon_o|0\rangle$ and $\mathcal{H}_o|n\rangle = \varepsilon_n|n\rangle$ define the eigenstates and the eigenvalues of the target.

The conservation of energy involves :

$$(\hbar^2/2m)(k_o^2 - k_n^2) = \varepsilon_n - \varepsilon_o$$

and the conservation of momentum :

$$\vec{q} = \vec{k}_n - \vec{k}_o$$

which gives, for the inelastic process depicted in the following sketch :

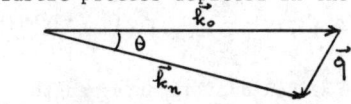

$$q^2 = (k_o - k_n)^2 + 4 k_o k_n \sin^2\theta/2$$

The transition probability between the initial and final states is determined by the transition amplitude which, in the first Born approximation, reduces to :

$$T^{(B)}_{o \to n} = \langle \vec{k}_n n | \mathcal{H}_i | \vec{k}_o 0 \rangle$$

The interaction hamiltonian represents the Coulomb interaction between the incoming fast electrons and the electric charges in the solid target. Following the procedures which are developed in any quantum mechanics textbook, such as Messiah (1), one can calculate :

- the transition probability per unit time

$$W_{on} = \frac{2\pi}{\hbar} |T^{(B)}_{o \to n}|^2 \delta\left[\varepsilon_n - \varepsilon_o - \frac{\hbar^2}{2m}(k_o^2 - k_n^2)\right]$$

- the differential cross section between states 0 and n with ejection of the scattered particle of wave vector $\vec{k}_n$

$$\frac{d\sigma_{o \to n}}{d\Omega(\vec{k}_n)} = \left(\frac{m}{2\pi\hbar^2}\right)^2 \cdot \frac{k_n}{k_o} |T^{(B)}_{o \to n}|^2 \delta\left[\varepsilon_n - \varepsilon_o - \frac{\hbar^2}{2m}(k_o^2 - k_n^2)\right]$$

The last term can be shortened into $\delta(E_{tot})$ to express in simple terms the conservation of energy. The next step is to evaluate the matrix element for the transition in both models (isolated atom or jellium). As there exists strong analogies between both hamiltonians, one expects similar expressions. In the case of the single atom it has been established by Bethe (12) and a complete review is due to Inokuti (13). The jellium model has been thoroughly studied in a series of papers by Pines and Nozières and a good survey can be found in the textbook by Pines (14).

3.2. Bethe theory for the inelastic scattering on a single atom :

Our purpose is to extract the useful formulae to evaluate the differential cross section in energy and in solid angle from which it is possible to deduce the angular and energy dependence for each type of inelastic process.

3.2.a. Cross section :

The initial and final wave functions are atomic ones, $|o\rangle$ describing the ground state one and $|n\rangle$ an excited one which can lie either in the discrete spectrum or in the continuum, in which case it is labelled as $|\varepsilon\rangle$ where ε represents the energy above the vacuum level.

$$\frac{d\sigma_{on}}{d\Omega(\vec{k_n})} = \left(\frac{m}{2\pi\hbar^2}\right)^2 \cdot \frac{k_n}{k_o} \left|\langle n|\sum_{j=1}^{z}\frac{e^2}{|r-r_j|}e^{i\vec{q}\cdot\vec{r}}|o\rangle\right|^2 \cdot \delta(E_{tot})$$

can be transformed into :

$$\frac{d\sigma_{on}}{d\Omega(\vec{k_n})} = \left(\frac{4\pi e^2}{q^2}\right)^2 \cdot \left(\frac{m}{2\pi\hbar^2}\right)^2 \cdot \frac{k_n}{k_o} \cdot \left|\langle n|\sum_{j=1}^{z} e^{i\vec{q}\cdot\vec{r}_j}|o\rangle\right|^2 \cdot \delta(E_{tot})$$

Introducing the concept of generalized oscillator strength (GOS) defined as :

$$f_{on}(\vec{q}) = \frac{2m(\varepsilon_n-\varepsilon_o)}{\hbar^2 q^2} \left|\langle n|\sum_{j=1}^{z} e^{i\vec{q}\cdot\vec{r}_j}|o\rangle\right|^2$$

for transitions towards discrete states, and by its equivalent form :

$$\frac{df(\vec{q},\varepsilon)}{d\varepsilon} = \frac{2m(\varepsilon-\varepsilon_o)}{\hbar^2 q^2} \left|\sum_{\Omega}\langle\varepsilon\Omega|\sum_{j=1}^{z} e^{i\vec{q}\cdot\vec{r}_j}|o\rangle\right|^2$$

for transitions towards a final state in the continuum, defined by its energy ε above the vacuum level and by a set Ω of all other quantum numbers such as the angular momentum, the cross section can be expressed as :

$$\frac{d^2\sigma(\vec{q},\Delta E)}{d(\Delta E)\cdot d\Omega} = \frac{2me^4}{\hbar^2\Delta E} \cdot \frac{1}{q^2} \cdot \frac{df(\vec{q},\Delta E)}{d(\Delta E)}$$

in the solid angle

$$d\Omega = \frac{2\pi}{k^2}\cdot q\,dq$$

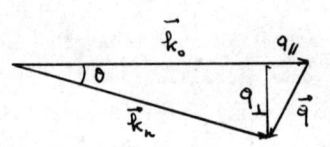

Using the simple relation extracted from the above scheme :

$$q^2 = q_{//}^2 + q_{\perp}^2 = k^2(\theta^2 + \theta_E^2)$$

where $\theta_E = \Delta E/2E_o$ is a characteristic inelastic angle for an energy loss ΔE suffered by a primary electron of energy E_o :

$$\frac{d^2\sigma(\theta,\Delta E)}{d(\Delta E)d\Omega} = \frac{2e^4}{mv^2\cdot\Delta E} \cdot \frac{1}{\theta^2+\theta_E^2} \cdot \frac{df(\vec{q},\Delta E)}{d(\Delta E)}$$

3.2.b. Angular dependence :

In the small angle limit, one can develop : $e^{i\vec{q}\cdot\vec{r}_j} = 1 + i\vec{q}\cdot\vec{r}_j + \ldots$ and the GOS can be extrapolated into its small q limit, that is the optical oscillator strength (OOS).

$$\frac{d^2\sigma(\theta,\Delta E)}{d(\Delta E)d\Omega} = \frac{4}{a_o^2 k^2} \cdot |x_{on}|^2 \cdot \frac{1}{\theta^2+\theta_E^2}$$

where the dipole matrix element is q-independent. As a consequence, the angular distribution of the scattered electrons behaves as $1/(\theta^2+\theta_E^2)$ at small angles, for any inelastic transition in an atomic collision. At large angle, deviations from this Lorentzian profile are introduced by the GOS. A three dimensional plot of $df(\vec{q},\Delta E)/d(\Delta E)$ in the coordinate system $(\vec{q},\Delta E)$ is known as the Bethe surface. As this surface embodies all the information concerning the inelastic scattering of charged particles by atoms, many experimental investigations as well as theoretical calculations of the GOS have been performed in atomic and plasma physics.

In the electron microscope it governs the behaviour of the electrons which are detected in the background or in an atomic core loss in the energy loss spectrum.

α) In the background case this atomic model can be used when most of the inelastic electrons have excited conduction electrons towards high energy states in the continuum. These electrons are reasonably considered as free because their binding energy is small when compared to the recoil energy which they get in the collision. Egerton (15), Leapman and Cosslett (16) have used this approach to interprete the angular distribution of the background electrons in an energy window located just below the K edge in carbon. The intensity is then peaked at an angle $\theta = (\Delta E/E_o)^{1/2}$; this maximum is known as the Bethe ridge, and corresponds in classical terms to close collisions with individual free electrons.

β) For core level excitations the GOS has recently been calculated for various inner shell electrons by using atomic wave functions $|n\ell>$ and $|\epsilon\ell'>$ for the initial and final states (Leapman (17)). The GOS remains forward peaked at the edge, but at higher energy losses, it peaks at non zero momentum transfer corresponding to the Bethe ridge. Since the total angular distribution is weighted by the term $1/(\theta^2+\theta_E^2)$, it remains forward peaked but its width increases.

3.2.c. Energy dependence :

The energy loss spectrum, that is the energy dependent term, is estimated by the integral :

$$\frac{d\sigma_\alpha}{d(\Delta E)} = \int_{q_{min}}^{q_{max}} \frac{d^2\sigma}{d(\Delta E).d\Omega} d\Omega$$

where $q_{min} = k\theta_E$, and $q_{max} = k\alpha$ is set by the collection aperture. The energy differential cross section thus defined represents the gross shape of the energy distribution of a core loss signal. Profiles concerning the boron K edge, the magnesium L_{23} edge and the molybdenum M_{45} edge have been calculated by Rez and Leapman (18). Quite clear differences appear following the symmetry of the atomic orbitals which are involved in the transition.

3.3. Collective response for the jellium model :

3.3.a. Cross section :

The mathematical support is very similar to the one which has been used in the Bethe theory. Some minor changes are introduced in the terminology but they all express the close similarity between the matrix elements of the electron density (written as $\sum_{j=1}^{z} e^{i\vec{q}\cdot\vec{r}_j}$ in the atom and $\rho_{-\vec{q}} = \int \rho_{\vec{r}} \, e^{i\vec{q}\cdot\vec{r}} \, d\vec{r}$ in the jellium), taken between the initial and final states of the transition. Starting from this fundamental analogy which has been established formally by Fano (19), the useful formulae are deduced by simple manipulations from those established in the Bethe theory. But instead of involving the GOS, they are generally expressed in terms of the energy loss function $-\,\mathrm{Jm}\, 1/\varepsilon(\vec{q},\omega)$, using the dielectric description of the electron gas introduced by Nozières and Pines. The cross section is then currently written as :

$$\frac{d^2\sigma(\vec{q},\Delta E)}{d(\Delta E).d\Omega} = \frac{1}{2\pi^2 E_o a_o} \cdot \frac{1}{\theta^2 + \theta_E^2} \cdot \mathrm{Jm}\, -\frac{1}{\varepsilon(\vec{q},\omega)}$$

The energy loss function contains all the useful information concerning the response of the electron gas to the perturbation involved by the incident electron and is equivalent to the Bethe surface introduced in the atomic case.

3.3.b. Angular dependence

It is governed by the $(1/q^2).\mathrm{Jm}\, 1/\varepsilon(\vec{q},\omega)$ factor. In the small angle scattering limit which corresponds to $|\vec{q}| \to 0$, one generally assumes that the momentum dependence of the energy loss function can be neglected so that :

$$\mathrm{Jm}\, \frac{1}{\varepsilon(\vec{q},\omega)} \to \mathrm{Jm}\, \frac{1}{\varepsilon(o,\omega)} = \mathrm{Jm}\, \frac{1}{\varepsilon(\omega)}$$

Similarly to the atomic case the angular distribution of the electrons which have been inlastically scattered by the conduction electron gas is mainly governed by the simple Lorentzian profile $1/(\theta^2+\theta_E^2)$ and this effect has been checked by Kunz (20).

However in this small angle approximation one misses the information concerning the electron excitation spectrum which is contained in the $\vec{q}$-dependence of the energy loss function. For instance the behaviour of the pole of this function $\omega_p(\vec{q})$ defines the dispersion law of the plasmon mode and, more generally, inelastic electron scattering experiments constitute one of the few efficient tools to investigate non vertical transitions for the electron gas of the solid. Raether (21) has recently reviewed some of the problems of interest which are presently investigated in this field : the anisotropy of the dispersion curves reflecting band structure features, the behaviour of the plasmon mode at large q values and its strong decay into electron hole pairs ... Such studies are still rather tricky because of the great difficulties encountered when one wants to extract a single inelastic profile from a large number of multiple elastic-inelastic processes which constitute the majority of the detected signal at large q. Though most of the experimental work has been devoted to the plasmon mode in aluminium there still remains uncertainties about its behaviour beyond the cut-off wave vector (Batson, Chen and Silcox (22)).

3.3.c. Energy dependence

In the small angle limit the energy loss function $- \text{Jm } 1/\varepsilon(\omega)$ has been measured over an energy range extending from zero to a few tens of eV for many substances. For simple metals (alkali, Be, Mg, Al ...) and semiconductors such as Ge, Si ... it mainly consists of a single peak located at a frequency $\omega_p = (n_o e^2/m\varepsilon_o)^{1/2}$ which corresponds to the well known plasmon frequency. In such cases the jellium model is a reasonable approach and the dielectric coefficient can be calculated in the random phase approximation (RPA). For the plasmon energy $\hbar\omega_p$, the value of $-\text{Jm } 1/\varepsilon(\omega_p)$ provides an estimate of the screening of the external electric perturbation by the dynamic collective response of the electron gas. When one fits the energy loss function with a Lorentzian profile centered at ω_p, such as

$$- \text{Jm} \frac{1}{\varepsilon(\omega)} = \frac{\omega_p \cdot \omega_p^2 \cdot 1/\tau}{(\omega^2 - \omega_p^2)^2 + \omega^2/\tau^2}$$

the parameter τ is related to the energy half width of the plasmon peak $\Delta E_{1/2}$ by $\tau = \hbar/\Delta E_{1/2}$ and represents the lifetime of the plasmon. Values of about 10^{-16} s are quite typical.

For other materials such as compounds, insulators, transition and noble metals, the conduction electron gas differs from the jellium model because of the strong influence of more tightly bound electrons, such as the d-band ones. Standard procedures have then been established to extract the dielectric coefficients $\varepsilon_1(\omega)$ and $\varepsilon_2(\omega)$ from an energy loss measurement (see Daniels et al (8)).

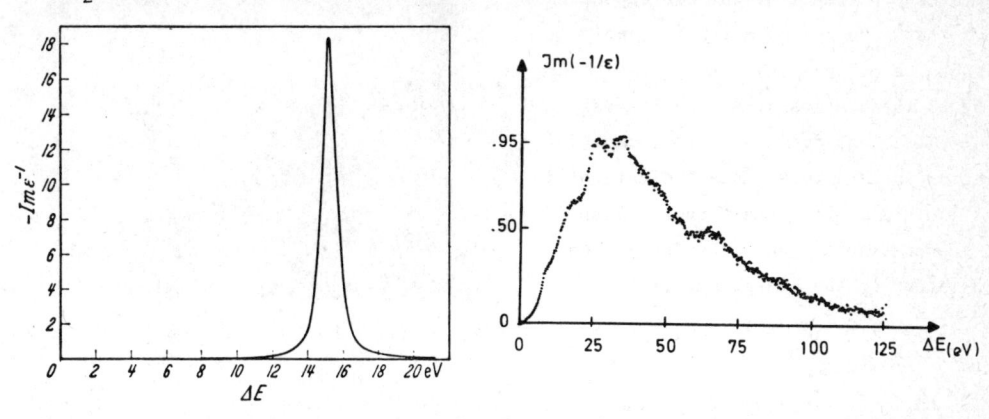

Figure 6 : Measured energy loss function in the limit $q \to 0$ for aluminium with a well defined plasmon peak, and for gold in which case the observed structures correspond to a complex mixture of interband and collective processes.

4. Microanalytical applications

When combined with the high resolution capabilities of the electron microscope, the use of electron energy loss spectroscopy (EELS) as an essential component for chemical microanalysis is a rapidly developing field, the limits of which remain to be firmly established. The general idea is to select a small volume of a sample and to detect useful signals in the energy loss spectrum of the transmitted electron beam. Although the project of obtaining elemental information from EELS was suggested by Hillier and Baker more than thirty years ago (23), it is only during these last few years that some obvious results could be obtained with this technique (Isaacson and Johnson (24), Colliex, Cosslett, Leapman and Trebbia (25), Colliex and Trebbia (10), Jouffrey et al (26)).

In the energy loss spectrum the chemical information is displayed over the whole distribution of inelastic electrons. One can use either the strong and poorly characteristic signals associated to the excitation of valence electrons (mainly plasmon lines) or the weak and truly characteristic core loss signals due to the ionization of atomic levels. Both types of information have actually been taken into account in various examples concerning material and biological problems. It is however generally admitted that the characteristic core level signals which provide an unambiguous tool for elemental recognition, offer greater promise for microanalytical purposes.

In a first operating procedure the energy loss spectrum is recorded for a selected specimen area, the size of which can be as small as 10 Å with a high resolution scanning transmission electron microscope (STEM). The detection of several excitation edges identifies the presence of the corresponding elements. Moreover quantitative analysis can be achieved by some simple handling of these signals. A core loss signal $S_c(\alpha,\Delta)$ is defined as the number of counts detected in a given conical solid angle of collection (half angle α) and corresponds to all characteristic events superposed on the background :

$$S_c(\alpha,\Delta) = \sum_{\Delta E=E_c}^{E_c+\Delta} (N_T(\Delta E) - N_B(\Delta E))$$

N_T and N_B represent the total and background number of counts in the channel located at ΔE in the loss spectrum.

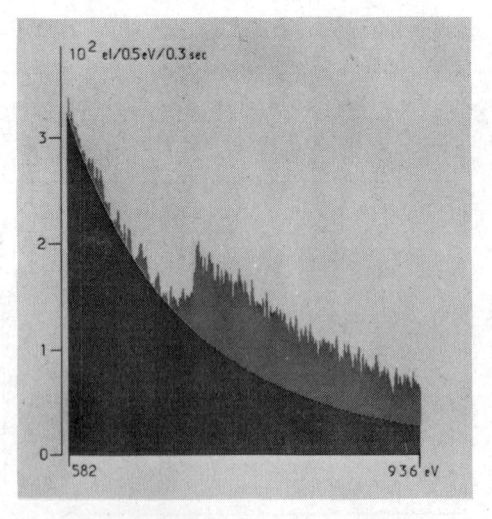

Fig.7 : Definition of the characteristic L_{23} signal for iron at 720 eV by extrapolation of the background following a power law.

Figure 7 shows a L_{23} edge for iron. The background is fitted with a power law a ΔE^{-r} by minimizing the standard deviation between the experimental and the model curves over a large energy domain, approximatively 150 eV broad, before the edge. One can then estimate the total number of atoms responsible for this signal, by using the formula proposed by Egerton (27) :

$$N = \frac{S_c(\alpha,\Delta)}{I_0(\alpha,\Delta)\sigma_c(\alpha,\Delta)}$$

where : - $\sigma_c(\alpha,\Delta)$ is the core loss cross section measured in a collection angle of half-width α and in an energy window Δ above the edge at E_c,
- $I_0(\alpha,\Delta)$ is the low lying energy loss contribution measured in the same conditions, that is involving the unscattered peak and the valence electron inelastic component from 0 to Δ.

In a binary alloy estimations of the concentration can be achieved by a ratio method with a rather good accuracy (Trebbia and Colliex (28)). The limits of the technique have been estimated as follows :
- the minimum detectable number of atoms is set by stability conditions at about one hundred atoms in present STEMS for a specimen which is not beam sensitive ; but radiation damage, which is particularly important for biological specimens, can raise this limit at about 10^6 to 10^8 atoms ;
- the low limit concentration depends on the extension of the analysed volume of material and it is evidently not possible to detect minority elements in a reduced assembly of atoms.

The resulting optimal domain of application for EELS is for the analysis of major elements in small volumes of typical size lying between 20 and 100 Å.

For many applications (for instance in the biological and mineralogical fields) the mapping of given elements over larger areas is most useful, and another working procedure, the filtered image energy selecting mode, provides an interesting solution. It consists of displaying an image of the sample with the electrons transmitted in a given energy window $\Delta E \pm \delta E$. By selecting plasmon images, El Hili (29) could thus characterize some well defined precipitates in aluminium based alloys. The systematic extension of this method however suffers severe limitations when one wants to use core loss signals. Besides the increased recording time associated with the weakness of the signal, the contrast in a filtered image us not only representative of the element associated with the edge, but also contains contributions from background variations due to thickness, density or crystal orientation changes. Jeanguillaume, Trebbia and Colliex (30) have discussed the limit of the method and suggested a few solutions to overcome this difficulty at the expense of more elaborate detection systems.

1. A. MESSIAH (1960), Mécanique quantique II, 687, Ed. Dunod
2. P.E. BATSON (1976), Ph.D Thesis, Cornell University
3. P. TREBBIA, C. COLLIEX (1978), Jap. J. Appl. Phys., Suppt. 17-2, 234
4. D.L. MISELL (1973), Advances in Electronics and Electron Physics, 32, 63
5. D.L. MISELL, A.F. JONES (1969), J. Phys. A, 3, 540
6. D.W. JOHNSON, J.C. SPENCE (1974), Proc. 8th Int. Conf. Electron. Microscopy, I, 386
7. J.C. SPENCE (1977), Proc. 35th EMSA Meeting, Boston, 234
8. J. DANIELS, C.V. FESTENBERG, H. RAETHER, K. ZEPPENFELD (1970) Springer Tracts in Modern Physics, 54, 77
9. C. WEHENKEL (1975), Ph.D Thesis, Université d'Orsay
10. C. COLLIEX, P. TREBBIA (1978), Electron Microscopy, State of the Art III, 268 Ed. J.M. Sturgess, Microscopical Society of Canada
11. M. ISAACSON (1975), Techniques in Electron Microscopy and Microprobe Analysis, 247, Ed. B. Siegel and D. Beaman, J. Wiley N.Y.
12. H. BETHE (1930), Ann. Physik, 5, 325
13. M. INOKUTI (1971), Rev. Mod.Physics, 43, 297
14. D. PINES (1963), Elementary excitations in solids, Ed. Benjamin N.Y.
15. R.F. EGERTON (1975), Phil. Mag.31, 199
16. R.D. LEAPMAN, V.E. COSSLETT (1977) Vacuum 26, 423
17. R.D. LEAPMAN (1979) Ultramicroscopy, 3, 413
18. P. REZ, R.D. LEAPMAN (1979) to be published
19. U. FANO (1956), Phys. Rev. 103, 1202
20. C. KUNZ (1961), Phys. Stat. Sol. 1, 441
21. H. RAETHER (1978), Jap. J. Appl. Phys., Suppt. 17-2, 227
22. P.E. BATSON, C.H. CHEN, J. SILCOX (1976) Phys.Rev.Letters, 57, 937
23. J. HILLIER, R.F. BAKER (1944), J. Appl. Phys. 15, 663
24. M. ISAACSON, D. JOHNSON (1975), Ultramicroscopy 1, 33
25. C. COLLIEX, V.E. COSSLETT, R.D. LEAPMAN, P. TREBBIA (1976), Ultramicroscopy 1, 301
26. B. JOUFFREY, Y. KIHN, J.P. PEREZ, J. SEVELY, G. ZANCHI (1978), Electron Microscopy, State of the Art III, 292, Ed. J.M. Sturgess
27. R.F. EGERTON (1978), Ultramicroscopy 3, 243
28. P. TREBBIA, C. COLLIEX (1979), J. Microsc. Spectrosc.Electron, to be published.
29. A. EL HILI (1966), J. Microscopie 5, 669
30. C. JEANGUILLAUME, P. TREBBIA, C. COLLIEX (1978), Ultramicroscopy 3, 237

INTERACTION OF THERMAL NEUTRONS WITH MATTER

G. L. Squires
Cavendish Laboratory, Madingley Road,
Cambridge CB3 OHE, England

Introduction

Thermal neutrons are a powerful tool for investigating a variety of properties of condensed matter. Their usefulness arises from the lack of charge on the neutron, the value of its mass, and the fact that it has a magnetic moment. Neutrons at room temperature have a wavelength of about 2 Å, which is of the order of interatomic distances in solids and liquids, and an energy of about 25 meV, which is of the same order as the energy of many excitations in condensed matter. Thus the scattering of thermal neutrons yields information on the positions and motions of the particles in the scattering system.

Neutrons interact with matter in two ways. The first is via the nuclear force between the neutrons and the nuclei in the scattering system. The second is via the interaction between the magnetic moment of the neutron and the magnetic field of unpaired electrons in the atom. Thus the nuclear interaction is always present. The magnetic interaction occurs additionally when the scattering atoms are magnetic, and provides information on the magnetic properties - both static and dynamic - of the scattering system.

Nuclear scattering

We consider first scattering by a single nucleus fixed at the origin. The wavelength of thermal neutrons is very much larger than the range of nuclear forces. The scattering is therefore spherically symmetric, and the wavefunction for the incident and scattered waves may be written in the form

$$\psi(\vec{r}) = e^{ikz} - \frac{b}{r} e^{ikr}. \qquad (1)$$

The quantity b is known as the <u>scattering length</u> and represents the strength of the scattering from the particular nucleus.

We may distinguish two types of nuclei. In the first type the scattering length is complex and varies rapidly with the energy of the neutron. The scattering is associated with the formation of a compound nucleus with energy close to an excited state. Examples of such nuclei are ^{113}Cd and ^{157}Gd. The imaginary part of the scattering length corresponds to absorption, and such nuclei are highly absorbing.

In the second type of nuclei, which comprises the great majority, the compound nucleus is not formed near an excited state. The scattering length is almost entirely real and is independent of neutron energy. We shall confine the subsequent discussion to such nuclei.

The value of the scattering length depends on the nucleus (i.e. its Z, N values) and also on the spin value of the nucleus-neutron system. If the nucleus has spin $I(\neq 0)$ it can form two states with the nucleus with spins $I \pm \frac{1}{2}$. Each state has its own scattering length denoted by b^+ and b^-. For a nucleus with $I = 0$ there is only one scattering length. As we do not have a proper theory of nuclear forces for calculating the scattering lengths, we have to treat them as parameters to be determined experimentally. Several methods are available, e.g. the gravity refractometer, giving values accurate to about 1 in 10^3. Unlike their counterparts in X-ray scattering which vary smoothly with the atomic number Z, neutron scattering lengths vary erratically from one nucleus to its neighbour. A selection of values is given in Table 1.

Table 1   Values of scattering length b (in units of 10^{-15} m)

Nuclide:	^{1}H		^{2}H		^{23}Na		^{59}Co	
T:	1	0	3/2	1/2	2	1	4	3
b:	10.8	−47.5	9.5	1.0	6.3	−0.9	−2.8	9.9

T is the spin of the nucleus-neutron system.

If we calculate $\psi(\vec{r})$ using the Born approximation, we obtain the result

$$\psi(\vec{r}) = e^{ikz} + \frac{f(\theta)}{r} e^{ikr}, \qquad (2)$$

where

$$f(\theta) = -\frac{m}{2\pi\hbar^2} \int V(\vec{r}) e^{i\vec{q}\cdot\vec{r}} d^3r . \qquad (3)$$

$V(\vec{r})$ is the potential of the nucleus-neutron force, and m is the mass of the neutron. This result is consistent with (1) provided

$$V(\vec{r}) = \frac{2\pi\hbar^2}{m} b \, \delta(\vec{r}) . \qquad (4)$$

Only this δ-function form of $V(\vec{r})$, known as the Fermi pseudopotential, gives the required result of isotropic scattering in the Born approximation.

For a general system of particles the scattering is obtained by adding the waves scattered by each nucleus with due regard to phase. If the nuclei were fixed in position the cross-section would be

$$\frac{d\sigma}{d\Omega} = \left| \sum_j b_j \exp(i\vec{q}\cdot\vec{R}_j) \right|^2 = \sum_{jj'} b_j b_{j'} \exp\{i\vec{q}\cdot(\vec{R}_j - \vec{R}_{j'})\} , \qquad (5)$$

where b_j is the scattering length, and $\vec{R}_j$ the position of the j th nucleus. The scattering vector $\vec{q}$ is defined by $\vec{q} = \vec{k}_o - \vec{k}$, where $\vec{k}_o$ and $\vec{k}$ are the wavevectors of the incident and scattered neutrons. For this scattering system, only elastic scattering can occur, i.e. $k_o = k$.

In practice the nuclei are not fixed in position. In liquids they move around, and in solids they oscillate about their equilibrium positions. Inelastic scattering is therefore possible. We put $\hbar\omega = E_o - E$, where $E_o (= \hbar^2 k_o^2/2m)$ and $E (= \hbar^2 k^2/2m)$ are the initial and final energies of the neutron. The cross-section has the form

$$\frac{d^2\sigma}{d\Omega dE'} = \frac{k}{k_o} \frac{1}{2\pi\hbar} \sum_{jj'} b_j b_{j'} \int_{-\infty}^{\infty} dt \, \exp(-i\omega t)$$
$$\times \left\langle \exp\{-i\vec{q}\cdot\vec{R}_{j'}(0)\} \exp\{i\vec{q}\cdot\vec{R}_j(t)\} \right\rangle. \qquad (6)$$

In this expression $\vec{R}_j(t)$ is the Heisenberg operator for the position of the j th nucleus, i.e.

$$\vec{R}_j(t) = \exp(iHt/\hbar) \, \vec{R}_j \, \exp(-iHt/\hbar), \qquad (7)$$

where H is the Hamiltonian of the scattering system. The symbol $\langle \ \rangle$ denotes the thermal average of the operator enclosed, at the temperature of the scattering system. Classically the quantity $\vec{R}_j(t)$ may be regarded as the position of the j th nucleus at time t.

Coherent and incoherent scattering

Consider a scattering system consisting of a single element where the scattering length b varies from one nucleus to another owing to nuclear spin or the presence of isotopes or both. We rewrite (6) as

$$\frac{d^2\sigma}{d\Omega dE'} = \sum_{jj'} b_j b_{j'} \langle j, j' \rangle, \qquad (8)$$

where $\langle j, j' \rangle$ stands for the entire j, j' term on the right-hand side of (6) apart from the two scattering lengths. We assume there is no correlation between the value of the scattering length and nuclear site, and average (8) over all distributions of the scattering lengths among the sites. This gives

$$\frac{d^2\sigma}{d\Omega dE'} = (\bar{b})^2 \sum_{\substack{jj' \\ j \neq j'}} \langle j, j' \rangle + \overline{b^2} \sum_j \langle j, j \rangle$$
$$= (\bar{b})^2 \sum_{jj'} \langle j, j' \rangle + \{\overline{b^2} - (\bar{b})^2\} \sum_j \langle j, j \rangle. \qquad (9)$$

The first term in this expression is the <u>coherent</u> cross-section, and the second is the <u>incoherent</u> cross-section. Coherent scattering depends on the relative positions of all the nuclei and therefore contains interference terms. It corresponds physically to the scattering that would occur from a hypothetical system containing the same atomic positions and motions, but with all the scattering lengths equal to the mean of the actual scattering lengths. Incoherent scattering does not show interference and is the scattering due to the random variation of the scattering lengths about their mean value. If all the scattering lengths are equal, then $\overline{b^2} = (\overline{b})^2$, and the incoherent scattering is zero.

For a single isotope with spin I, the scattering lengths b^+ and b^- occur with relative weights $I + 1$ and I. For several isotopes, each scattering length is further weighted by the relative abundance of the isotope. These factors determine the quantities $(\overline{b})^2$ and $\overline{b^2} - (\overline{b})^2$. Numerical values are usually quoted in terms of

$$\sigma_{coh} = 4\pi (\overline{b})^2 \quad \text{and} \quad \sigma_{inc} = 4\pi \left\{ \overline{b^2} - (\overline{b})^2 \right\}. \tag{10}$$

Some representative values are given in Table 2. The extension of the theory to systems containing more than one element is readily made.

Table 2   Values of σ_{coh} and σ_{inc} (in units of 10^{-28} m^2)

Element or nuclide:	^{1}H	^{2}H	C	O	Mg	Al	V	Fe	Cu	Zn
σ_{coh}:	1.8	5.6	5.6	4.2	3.6	1.5	0.02	11.5	7.5	4.1
σ_{inc}:	80.2	2.0	0.0	0.0	0.1	0.0	5.0	0.4	0.5	0.1

<u>Correlation functions</u>

From (6) and (9) the coherent cross-section may be written as

$$\frac{d^2\sigma}{d\Omega dE'} = \frac{\sigma_{coh}}{4\pi} \frac{k}{k_0} N S(\vec{q},\omega), \tag{11}$$

where

$$S(\vec{q},\omega) = \frac{1}{2\pi\hbar} \int G(\vec{r},t) \exp\{i(\vec{q}\cdot\vec{r} - \omega t)\} d^3r \, dt, \tag{12}$$

$$G(\vec{r},t) = \frac{1}{(2\pi)^3 N} \int \sum_{jj'} \left\langle \exp\{-i\vec{q}\cdot\vec{R}_{j'}(0)\} \exp\{i\vec{q}\cdot\vec{R}_j(t)\} \right\rangle \exp(-i\vec{q}\cdot\vec{r}) d^3q. \tag{13}$$

N is the number of particles in the scattering system. Thus the cross-section is proportional to the scattering function $S(\vec{q},\omega)$, which is the Fourier transform in space and time of the time-dependent pair-correlation function $G(\vec{r},t)$. The quantity $G(\vec{r},t) d^3r$ may be interpreted classically as the probability that, given a particle at the origin at time $t = 0$, any particle (including the origin particle) is in the element of volume d^3r at position $\vec{r}$ at time t.

The relations (11) to (13) are an extension of the result for Fraunhofer diffraction of light that the amplitude of the scattered wave is the Fourier transform in space of the density function of the scattering system. In the case of light (and X-rays) the radiation travels so fast that the particles in the scattering system do not have time to move in the interval between the incident radiation reaching neighbouring atoms in the system. The radiation 'sees' the particles frozen in some configuration, and the scattering depends only on the value of $G(\vec{r},t)$ at $t = 0$. Thermal neutrons, however, have velocities of the order of 10^3 m s^{-1}. The time for the neutron to travel between particles separated by distances of about 10^{-10}m is $\sim 10^{-13}$ s. This is of the order of the period of oscillation for atoms in solids and liquids. Thermal neutron scattering therefore gives information on $G(\vec{r},t)$ at general values of t, which is one of the reasons neutrons are so useful for investigating the properties of condensed matter.

Magnetic scattering

We confine the discussion to the scattering that arises from the interaction between the magnetic moment of the neutron and the magnetic field due to the spins of the electrons, i.e. we ignore orbital effects. For a single electron at the origin the field $\vec{B}$ at position $\vec{R}$ is

$$\vec{B} = \text{curl }\vec{A}, \qquad \vec{A} = \frac{\mu_0}{4\pi}\frac{\vec{\mu}_e \times \vec{R}}{R^3}, \qquad \vec{\mu}_e = -2\mu_B \vec{s}, \tag{14}$$

where $\vec{\mu}_e$ is the magnetic moment of the electron, $\vec{s}$ its spin angular momentum in units of $\hbar$, and μ_B is the Bohr magneton.

The magnetic moment of the neutron may be expressed as

$$\vec{\mu}_n = -\gamma \mu_N \vec{\sigma}, \tag{15}$$

where μ_N is the nuclear magneton, $\vec{\sigma}$ the Pauli spin operator of the neutron, and $\gamma = 1.91$. The magnetic potential is

$$V_{mag} = -\vec{\mu}_n \cdot \vec{B}. \tag{16}$$

Inserting this in the Born approximation expression (3) we obtain for the magnetic counterpart of the scattering length

$$b_{mag} = -\gamma r_e \vec{\sigma} \cdot \vec{Q}_\perp, \tag{17}$$

where

$$\vec{Q}_\perp = \hat{q} \times (\vec{s} \times \hat{q}), \tag{18}$$

and $r_e = \mu_0 e^2/4\pi m_e$ is the classical radius of the electron. $\hat{q}$ is a unit vector in the direction of $\vec{q}$.

It can be seen that $\vec{Q}_\perp$ is the vector projection of $\vec{s}$ on to the plane perpendicular to $\vec{q}$. Thus b_{mag} is zero when $\vec{q} \parallel \vec{s}$. This result may be seen on physical grounds. When $\vec{q} \parallel \vec{s}$, the phase difference between the waves scattered by two atoms in a plane perpendicular to $\vec{s}$ is zero (Fig. 1). The contribution to the integral

$\int \vec{\mu}_n \cdot \vec{B} \exp(i\vec{q}\cdot\vec{r}) d^3r$ over a plane perpendicular to $\vec{s}$ is therefore proportional to $\int \vec{B} \, dS$, where dS is an element of area in the plane, and the integral is taken over the whole plane. Now the contribution to the integral from the component of B perpendicular to the plane is zero, because div $\vec{B} = 0$, and the contribution from the component of $\vec{B}$ in the plane is likewise zero, from the symmetry of the field due to a dipole.

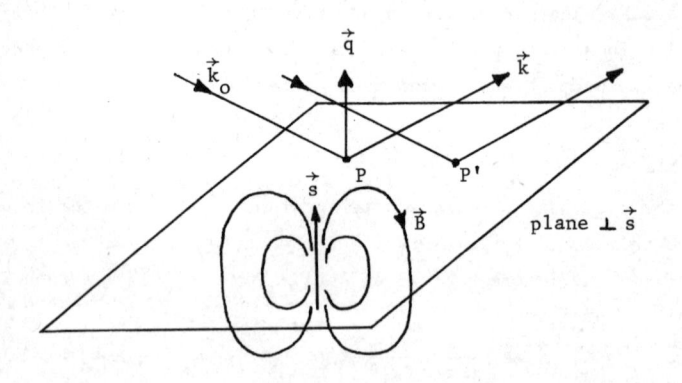

Fig. 1 Diagram for the case $\vec{q} \parallel \vec{s}$. The phase difference between the waves scattered at P and P' is zero.

It can be seen that the expression for b_{mag} is more complicated than its nuclear counterpart. The nuclear force is short-range and central. Thus the nuclear scattering length, which is proportional to the Fourier transform of the potential, is a constant, independent of $\vec{q}$. The magnetic force is long-range and non-central. Thus the magnetic scattering length depends on $\vec{q}$, and has the geometrical dependence given in (17) and (18).

For a single atom with L-S coupling the spins of the electrons combine to form a resultant spin angular momentum $\vec{S}$ for the atom, and b_{mag} becomes

$$b_{mag} = -\gamma r_e \vec{\sigma} \cdot \{\hat{q} \times (\vec{S} \times \hat{q})\} F(\vec{q}). \tag{19}$$

The magnetic form factor $F(\vec{q})$, which arises from the phase difference between the waves scattered by the different electrons in the atom, is the Fourier transform of $s(\vec{r})$, the normalised spin density of the unpaired electrons, i.e.

$$F(\vec{q}) = \int s(\vec{r}) \exp(i\vec{q}\cdot\vec{r}) \, d^3r. \qquad (20)$$

For a scattering system of many atoms we add the amplitudes of the scattered waves from each atom just as in the nuclear case. Coherent scattering occurs in systems with magnetic order, e.g. ferromagnets and antiferromagnets, and incoherent scattering occurs in paramagnets.

Elastic and inelastic scattering

For crystalline systems the scattering may be divided into elastic and inelastic processes. Coherent elastic scattering, both nuclear and magnetic, is the exact analogue of Bragg scattering of X-rays. For nuclear scattering and magnetic scattering from ferromagnets, the Bragg peaks occur at the same reciprocal lattice points as for X-rays. But for antiferromagnets and helimagnets the peaks are, in general, at other points in reciprocal space.

Inelastic scattering involves exchange of energy between the neutron and excitations such as phonons and magnons in the scattering system. All the scattering apart from coherent elastic scattering is sometimes referred to as thermal diffuse scattering, as it does not give peaks at points in reciprocal space. The term derives from X-ray scattering where it is used to denote the inelastic (phonon) scattering. This type of scattering produces a negligible fractional change in the energy of the X-ray quantum (in sharp contrast to the neutron case), but a significant change in the momentum, and thus gives rise to diffuse spots and streaks on X-ray photographs.

Applications

Coherent elastic scattering of thermal neutrons gives information on the location of atoms (neutrons are particularly useful for locating hydrogen) on the degree of structural order (e.g. in binary alloys), and on magnetic structures. Inelastic scattering gives information on phonon and magnon frequencies and lifetimes, time-dependent pair-correlation functions, and spin-correlation functions. Considerable work has been done in a number of fields, such as liquids, polymers, liquid crystals, molecular crystals, diffusion in solids, critical phenomena, and so on.

Additional information may be obtained by polarising the neutrons and analysing their spin states after scattering. Interference between nuclear and magnetic scattering provides a method for polarising neutrons and is also used to enhance the accuracy of some measurements, for example the determination of spin densities.

REFERENCES

Bacon, G. E. 1975. Neutron Diffraction, 3rd ed., Oxford: Clarendon Press. A comprehensive account of the experimental techniques and applications of thermal neutron scattering.

Koester, L. 1977. Neutron Scattering Lengths and Fundamental Neutron Interactions. Springer Tracts in Modern Physics, Vol. 80, Springer-Verlag. An account of the methods of measuring scattering lengths and tabulations of the results.

Lovesey, S. W. and Springer, T. 1977. Editors, Dynamics of Solids and Liquids by Neutron Scattering. Topics in Current Physics, Vol. 3, Springer-Verlag. An up-to-date account of a number of applications.

Squires, G. L. 1978. An Introduction to the Theory of Thermal Neutron Scattering, Cambridge University Press. Chapters 1 - 4 and 7 develop the theory for the results given in this paper.

STRUCTURE INFORMATION RETRIEVAL FROM SOLUTION
X-RAY AND NEUTRON SCATTERING EXPERIMENTS

Vittorio Luzzati

Centre de Génétique Moléculaire, Centre National de la
Recherche Scientifique, 91190 Gif-sur-Yvette, France.

The angular distribution of the intensity of X-ray and neutron beams scattered by macromolecules in solution usually displays two distinct regions : one, at "small" angles, is specifically sensitive to the long-range organization, the other, at "high" angles, to the short-range structural features. Most often this distinction, which mirrors the contrast between the sharp distribution of the shortest interatomic distances (in the 1-5 Å range) and the even distribution of the longer distances, is reasonably clear-out. Besides, if the macromolecules are large, the intensity decreases very rapidly with increasing scattering angle. We are mainly concerned here with the information relevant to the long-range structure, and thus with the "small" angle scattering ; we show below (see eq. 4) how to take into account, at least to a first approximation, the "small" angle effects of the short interatomic distances.

A typical solution scattering experiment consists of recording the intensity scattered by the solution, subtracting the intensity scattered by the solvent and by the instruments (sample cell, slits, etc.) and correcting for instrumental distortions, mainly collimation and polychroma-

---

Notation. Terminology and notation will be those commonly used in X-ray scattering studies[1] ; the results can be extended easily to neutrons. r (in Å) and s (in Å^{-1}) ($s = 2\sin\theta/\lambda$, 2θ being the scattering angle, λ the wavelength) specify positions in real and reciprocal space. $i(s)$ is the distribution of the scattered intensity ; $si(s) = \int_{-\infty}^{+\infty} rp(r)\sin 2\pi rs\, dr$, where $p(r)$ is the autocorrelation function, namely the spherical average of the convolution $\{\rho(\underline{r}) * \rho(-\underline{r})\}$, $\rho(\underline{r})$ being the electron density distribution.

tism. The result, usually expressed in a digit form, is a set of intensities associated to a number of channels (see fig. 1). It is clear that the intensities recorded at the different channels may well not be statistically independent, especially when the number of channels is large; it is worth while to note, in this respect, that some of the most common operations performed on the intensity curves - smoothing, interpolations, extrapolations - are based upon the very presence of such correlations. The nature of these correlations is closely linked to the information content of the data; information content and data analysis are the themes of this paper [2, 3].

It is hardly possible to tackle these problems without making some assumption on the structure of the sample. The following conditions define the framework of our treatment [1, 2, 3]:
- the sample is an ideal solution of discrete particles;
- the particles are all identical;
- the particles are globular, or more precisely none of their dimensions is large with respect to $(s_{min})^{-1}$, s_{min} being the lower limit of the interval of s explored experimentally.

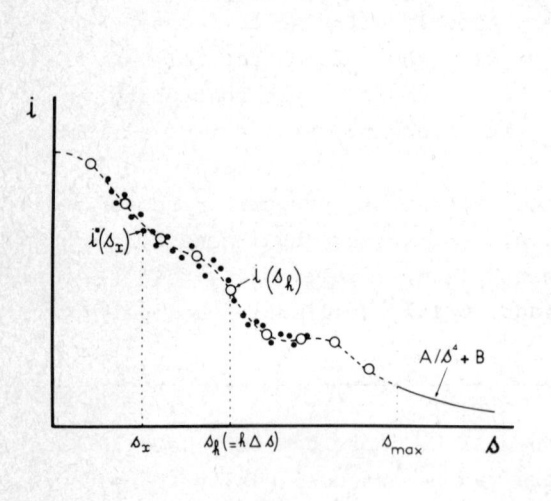

Fig. 1

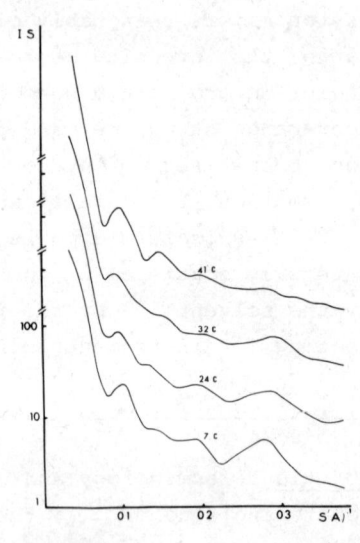

Fig. 2

Fig. 1 - Schematic representation of one intensity curve. The dotted line represents the correct curve, the small full dots the experimental points, the large open dots the intensities at the points of the lattice $s_h = h\Delta s$, the full line the asymptotic trend.

Fig. 2 - A few experimental curves, corrected for experimental distortions, obtained with simian low density serum lipoprotein, as a function of temperature [3].

Ideality is usually met by extrapolation to infinite dilution : monodispersity is a convenient simplification although the conclusions remain formally correct if the sample is heterogeneous. Globular shape is a strict requirement ; yet some of the results can be extended to rod-like and lamellar particles.

Within the framework of these hypotheses, the curvature at the origin of the intensity curve is porportional to one structural parameter, the radius of gyration R[4] :

$$i(s) = i(0)\{1 - (4/3)\pi^2 R^2 s^2 + ...\} \tag{1}$$

If, moreover, a few additional parameters are known - absolute scale, concentration, partial specific volume - the intensity at the origin yields the molecular weight M. Therefore, extrapolation to the origin - which involves some assumptions on the regularity of i(s) - can yield the values of R and M. Besides, it can often be assumed that the electron density inside the particle is fairly uniform ; in this case the asymptotic trend of i(s) takes the form [1, 5] :

$$\lim_{s \to \infty} s^4 i(s) \propto S \tag{2}$$

where S is the area of the outer surface of the particle. Equation 2 can be tested against the data : if it is fulfilled the value of S can be determined. In this case it is also possible to determine the volume V of the particle [4]:

$$V = i(0) \Big/ 2\pi \int_0^\infty s^2 i(s) ds \tag{3}$$

These four parameters - M, R, S, V - constitute the bulk of the information traditionnally retrieved from solution scattering experiments[4]; yet it is obvious, even upon visual inspection, that the experimental curves sometimes contain more information than that (see fig. 2).

It is particularly easy to discuss the problem of the information content when the particles are of globular shape. In this case the autocorrelation function vanishes beyond the maximal diameter of the particle, D_{max}. Since the function $si(s)$ is the Fourier transform of $rp(r)$ (see Notation), if $p(r) \equiv 0$ for $r > D_{max}$, then i(s) is completely defined by its values at the lattice points $h\Delta s$, with $\Delta s \leq (2D_{max})^{-1}$ (sampling theorem). Since the number of lattice points is infinite, a finite set of data is insufficient to define completely the function i(s). This problem can be circumvented if some assumption can be made about the mathematical form of i(s) at large s. For systems of biological interest (proteins, nucleic acids, lipids, etc.) the function

$$i(s) = A/s^4 + B \quad s > s_{max} \tag{4}$$

has been shown to provide a good empirical description of the asymptotic trend [1, 6]. Equation 4 can also be justified from a structural viewpoint ; the term A corresponds to the average long-range electron density distribution, the term B to the short-range fluctuations around the average. If eq. 4 is adopted, then the entire function i(s), from s = 0 to infinity, is defined by the intensity at the lattice points of the interval $0 < s < s_{max}$ plus the values of A and B. The number of these parameters is :

$$J = s_{max} \, 2D_{max} + 2 \tag{5}$$

Thus, we reach the conclusions :
- within the restrictions specified above - ideal solution of identical globular particles - and with the assumption that the high angle form of the intensity is known, the whole of the experimental information provided by one solution scattering experiment is ideally equivalent to a finite number J of independent measurements ;
- the maximal structural information which can be retrieved from that experiment is expressed by the same number J of independent parameters.

The explicit expression for the intensity at any point s, in terms of the intensity at the lattice points in the interval $0<s<s_{max}$ and of the parameters A and B, is :

$$si(s) = \sum_{h=1}^{h_{max}} h\Delta s \, i(h\Delta s)\Phi(s,h) + A \sum_{h_{max}+1}^{\infty} (h\Delta s)^{-3} \Phi(s,h) + B \sum_{h_{max}+1}^{\infty} (h\Delta s)\Phi(s,h) \tag{6}$$

where :

$$\Phi(s,h) = \frac{\sin\pi(s/\Delta s - h)}{\pi(s/\Delta s - h)} - \frac{\sin\pi(s/\Delta s + h)}{\pi(s/\Delta s + h)} \tag{7}$$

Each experimental value of the intensity ($i^*(s_x)$ in fig. 1) is equivalent to one equation (6) ; if the number of experimental points is equal to, or larger than, the number of unknowns J, then the system can be solved. The function i(s) is completely defined by the values of those J parameters.

We can make a few comments :

a) - Such analysis of the information content is to some extent idealized, and in any event incomplete, if the limitations in the accuracy of the data are ignored. The treatment above can be completed by a statistical analysis of the propagation of the experimental errors.

b) - The mathematical treatment is heavily dependent upon the properties of the functions $p(r)$ and $i(s)$ when r and s are large. The assumption that $p(r)$ vanishes beyond D_{max} is a logical consequence of the globular shape of the particles. The hypothesis that the asymptotic trend is a universal property of the intensity curves is more questionable, since, at all values of s, $i(s)$ depends upon the precise structure of the sample. It must be stressed that the postulated asymptotic form is meant to be an empirical approximation, acceptable within the limits of the experimental errors ; besides, the errors due to this approximation can be estimated.

c) - Equations 6 is not very sensitive to the precise value of Δs (and thus of D_{max}), provided the value chosen is not too large. One possible way of choosing D_{max} is to solve the system of equations 6 for different values of D_{max}, using for example a least squares algorithm. As D_{max} increases the residual can be expected to drop first, and then to level off : the breaking point defines the minimum value of D_{max}. Other experiments (for example electron microscopy) can provide independent information on this point.

d) - The solution of the system of equations 6 does not require data recorded at a regular interval of s, or even obtained in one single experiment. In fact the algorithm is well suited for matching together experiments performed under different conditions, for example different sample-detector distances, different wavelengths.

e) - The raw intensities, before corrections for polychromatism and collimation distortions, are related to the same J parameters by equations similar to eq. 6, in which $\Phi(s,h)$ (see eq. 7) is replaced by an operator which takes into account the experimental distortions. These equations provide a convenient algorithm for the correction of polychromatism and collimation distortions.

f) - The treatment we sketch here provides a rational approach to operations like smoothing, interpolations, extrapolations, and also to the determination of structural parameters (for example M, R, V, S, see above). Indeed, these operations can be expressed in terms of the J experimental parameters ; these expressions are more accurate than those based upon local properties of the data (for example Guinier's plots[4].

g) - It may be wondered why worry, here and now, about these formal problems. The reason must be sought in the recently revived interest in some theoretical aspects of the solution scattering phenomenon prompted by a variety of technical developments. With regard to X-rays, first the introduction of position sensitive detectors[7], later the use of synchrotron radiation[8] has had the effect, over the span of a few years, to shorten exposure time by a factor larger that 10^5. At almost the same time, the use of high flux reactors and of position sensitive detectors[9] have transformed neutron scattering into a powerful tool for structural studies of macromolecules in solution.

h) - It is common practice, in solution scattering studies, to perform experiments at variable solvent density; this is achieved by adding electron dense compounds (salts, sucrose, ..) in the case of X-rays and by varying the D_2O/H_2O ratio in the case of neutrons. These experiments are usually interpreted within the framework of the invariant volume hypothesis : a volume can be associated to each macromolecule in solution, inside of which the electron density distribution is independent of the density of the solvent. In this case,[1,3] the intensity scattered at any solvent density is a linear combination of three functions called the characteristic functions. In other words, the information which can be retrieved from any number of experiments performed under these conditions is contained in three independent intensity curves. Therefore the information content is equal to $3J$[3].

i) - We can illustrate the previous results with a few examples. The first is an ideal protein, of spherical shape, which we assume to be 30% hydrated, and whose partial specific volume is 0.74 $cm^3 g^{-1}$. If M is its molecular weight, its diameter is equal to 1.78 $M^{1/3}$ and $J = 3.56\ M^{1/3} s_{max} + 2$. Assuming that $s_{max} = (25\ \text{Å})^{-1}$ one can expect, at best, to retrieve the value of five structural parameters - for example M, R, V, S, B - when $M \simeq 10,000$ Daltons. The information content increases (rather slowly, in fact) with M, and is also greater for particles which are anisometric (D_{max} increases in this case). Another example is low density serum lipoproteins[3]. A few intensity curves are shown in fig. 2. For these particles D_{max} and s_{max} are close to 300 Å and to $(25\ \text{Å})^{-1}$ respectively. Therefore for one intensity curve the number of independent parameters is $J = 26$. Since this system has been studied systematically as a function of variable solvent density, and the invariant volume hypothesis has been shown to be fulfilled, the total information content is equivalent to $3J = 78$ parameters. Such a wealth of information is quite unexpected for solution scattering studies : indeed in this case the experiments have been analyzed in terms of an elaborate model[3].

References

(1) - Luzzati, V., Tardieu, A., Mateu, L. & Stuhrmann, H.B. (1976), J. Mol. Biol., 101, 115-127

(2) - Luzzati, V. (1979), Ann. Rev. Biophys. Bioengin., 9, in press

(3) - Luzzati, V., Tardieu, A. & Aggerbeck, L.P. (1979) J. Mol. Biol., in press

(4) - Guinier, A. & Fournet, G. (1955). Small-angle scattering of X-rays, Wiley, New York

(5) - Porod, G. (1951), Kolloidzschr. 124, 83-114.

(6) - Luzzati, V., Witz, J. & Nicolaïeff, A. (1961), J. Mol. Biol., 3, 367-378

(7) - Gabriel, A. & Dupont, Y. (1972), Rev. Scient. Instr. 43, 1600-1603

(8) - Stuhrmann, H.B. (1978), Quart. Rev. Biophys., 11, 71-98

(9) - Ibel, K. (1976), J. Appl. Cryst., 9, 296-309

GUIDED WAVES PROPAGATION AND INTEGRATED OPTICS

Y. LEVY

INSTITUT D'OPTIQUE

CENTRE UNIVERSITAIRE D'ORSAY

ORSAY

The current art of fiber optics and integrated optics has been reported in several papers and different symposiums. A brief review is presented here in order to describe some principal aspects of the energy light guiding realised by the optical fibers. The area of integrated optics is based on the phenomena involving light guided inside thin films. With the use of semiconductors, it is possible to realise integrated optical circuits. The advantages are the much larger bandwidth and negligible sensibility to interferences of electromagnetic fields of lower frequencies.

The rapid developement of optical communications systems started in 1970, when it was possible to make optical fibers having low attenuation of 20 dB/km. In last years, fibers with attenuation less than 1 dB/km have been realised. Associated with the new suitable sources and detectors, fiber optical transmission systems will find widespread commercial and military applications.

THE TRANSMISSION MEDIUM : PROPAGATION OF THE GUIDED LIGHT

The optical fiber consists of a core of a dialectrical material with refractive index n_1 and a cladding of another material which refractive index n_2 is less than n_1. The optical fiber is represented in figure 1.

The propagation of the guided light inside the fiber can be described by the modal theory. Generally, for the usual optical fibers, the refractive index n_1 of the core is slightly higher than the index of the cladding :

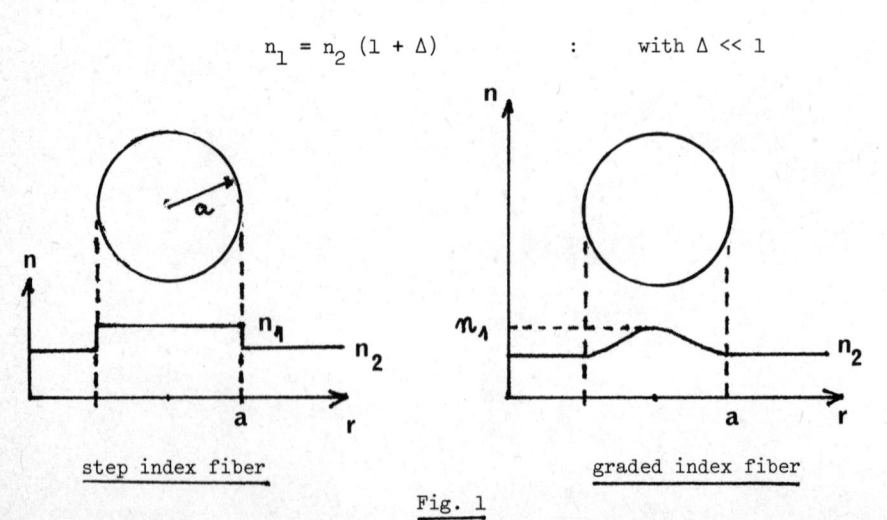

$$n_1 = n_2 (1 + \Delta) \qquad : \qquad \text{with } \Delta \ll 1$$

step index fiber   graded index fiber

Fig. 1

Another class of optical fibers is concerned by the graded index fibers for which the refraction index varies as a nearly parabolic function as one goes away from the center of the fiber (fig. 1).

$$n_1(r) = n_2(1 + \Delta(r/a)^\alpha)$$

where α has a value close to 2 to maximum fiber bandwidth and is the graded index power coefficient.

For the step index fiber, the propagation of the light in a meridional plane can be sketched in figure 2. The light propagates in the z direction and the physical picture of guided light propagation in then, that of light travelling in zig-zag fashion through the film. In the case of graded index fibers, the beam propagates on a curved way sketched in figure 3. Due to the variation of the refraction index as function of the radial coordinate r, the fiber acts as a continuous lensing medium.

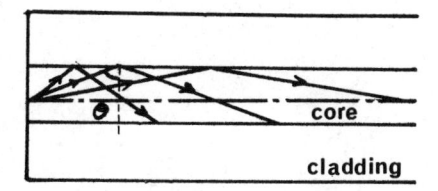

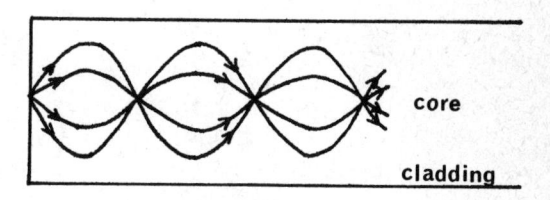

Ray picture in a meridional plane for a top index fiber.

Ray picture for a graded index fiber

Figure 2   Figure 3

Each zig-zag way can be defined with the propagation angle θ. However, all angles θ are not allowed. Only a discret set of angles corresponds to the propagation of the light. These angles are associated to the guided modes. The modal axpressions of the electromagnetic field are obtained from the Maxwell equations. Propagation equation in the cylindrical coordinates is given as:

$$\frac{d^2\psi}{dr^2} + \frac{1}{r}\frac{d\psi}{dr} + \frac{1}{r^2}\frac{d^2\psi}{d\phi^2} + (k^2 - \beta^2)\psi = 0 \qquad (1)$$

r is the radial coordinate, ϕ is the azimuthal coordinate, ψ is the wavefunction of the guided light, k is the wave vector, β is the wave vector component along the fiber axis.

Assuming a wavefunction of the form: $\psi = A F(r) e^{j\nu\phi}$ \qquad (2)

where A is a constant, ν is an integer, equation (1) can be written as :

$$\frac{d^2 F}{d r^2} + \frac{1}{r}\frac{dF}{dr} + \left(k^2 - \beta^2 - \nu^2/r^2\right) F(r) = 0 \qquad (3)$$

For step index fibers, solutions of equation (3) are given by the Bessel functions and the longitudinal component of the electric field has the form:

$$E_z = \begin{cases} A J_\nu(ur/a) \exp(j\nu\phi) \;;\; r < a \\ B H_\nu(wr/a) \exp(j\nu\phi) \;;\; r > a \end{cases}$$

where H is the Hankel function and a is the fiber radius. The boundary conditions require finite solution on axis (r=0) and that the fields disappear at infinity.

$$u^2 = (k_1^2 - \beta^2) a^2 \quad;\quad k_1 = 2\pi n_1/\lambda_0$$
$$w^2 = (\beta^2 - k_2^2) a^2 \quad;\quad k_2 = 2\pi n_2/\lambda_0$$

λ_0 is the wavelength in vacuum.

$$u^2 + w^2 = (2\pi a/\lambda_0)^2 (n_1^2 - n_2^2) = V^2$$

The parameter V is a characteristic of the fiber. The eigensolutions to u and w are determined from the boundary conditions. For ν = 0, the modes are T.E. and T.M. polarized and are radially symmetric. For $\nu \neq 0$, the modes are hybrid and the electromagnetic field possesses six components. The modes are denoted by HE_{mn} or EH_{mn}, depending on whether, the field has a more electric or more magnetic character. In the case of graded index fibers, the radial field equation can be solved approximately by the Hermite gaussian functions. While the mathematical expressions are more complex, the considerations on the guided modes are the same as those for the step index fibers. The modal description is described in figure 4, which represents the effective modal index β_{mn}/k_0 as function of the parameter V for the step index fiber. For V less than 2.405, the fiber has just a single mode denoted by HE_{11}. For high values of V, numerous guided modes can propagate inside the fiber. Single mode propagation is realised with fibers of a few wavelengths in cross sectional dimension and by having small refraction indices differences between the core and the cladding.

The ray theory is useful to define the numerical aperture of the fiber. The numerical aperture is related to the refraction indices of the core and the cladding by the following axpression:

$$\sin \theta = (n_1^2 - n_2^2)^{1/2}$$

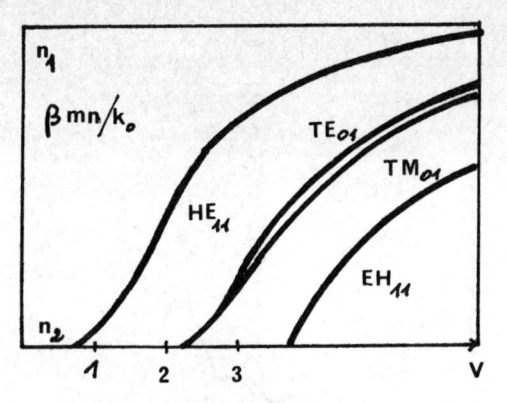

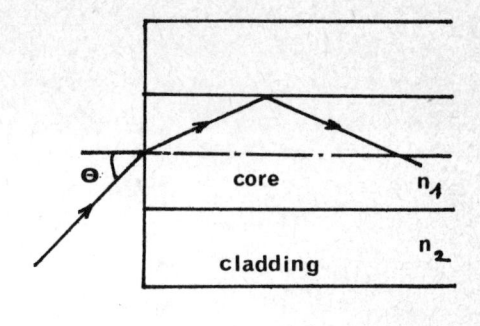

Effective modal index β_{mn}/k_o as function of the characteristic modal parameter V.

Figure 4   Figure 5

where θ is the incidence angle of the ray on the end of the fiber (figure 5). The numerical aperture, i.e., the maximum acceptance angle is used to calculate the source-fiber coupling efficiencies. When measurements of the numerical aperture are made, the values obtained depend on the length of the fiber. This effect is due to the excitation of the guided modes and leaky modes for short fibers. For long fibers, the power of the leaky modes is lost, after propagating on a long distance. As the higher order modes are lost, the effective numerical aperture decreases, causing the dependance of the numerical aperture on the length. For graded index fibers, the problem is more complex, because the numerical aperture is not a constant across the core as in the case of step index fibers. It is necessary to define a local numerical aperture NA(r) as function of the position r:

$$NA(r) = NA(0) \left(1 - (r/a)^\alpha\right)^{1/2}$$

PROPERTIES OF FIBERS

a) Attenuation

The principal fiber characteristic of interest is attenuated due to the intrinsic and extrinsic absorptions of the material which constitutes the fiber. Other contributions of attenuation are brought by inhomogeneities of the index and fiber shape. Impurity absorption arises from metal ions such as iron, copper and cobalt. Another important absorbing ion is OH^-. Figure 6 shows that there is a maximum absorption loss due to the OH^- of fused silica in the 0.9-1 μm region. The attenuation is also due to the scattering of the light. The propagating light can be coupled out of the fiber either by Rayleigh scattering or by guide inhomogeneities such as irregularities of the guide shape or curvature of the guide axis. Attenuation due to scattering is principally attributed to Rayleigh scattering ($\simeq \lambda^{-4}$). The magnitude of the Rayleigh scattering represents the lower limit. Attenuation is yet due to the microbending losses caused by fiber cabling. These losses result from the coupling between guided modes and radiation modes.

Figure 6

Figure 7

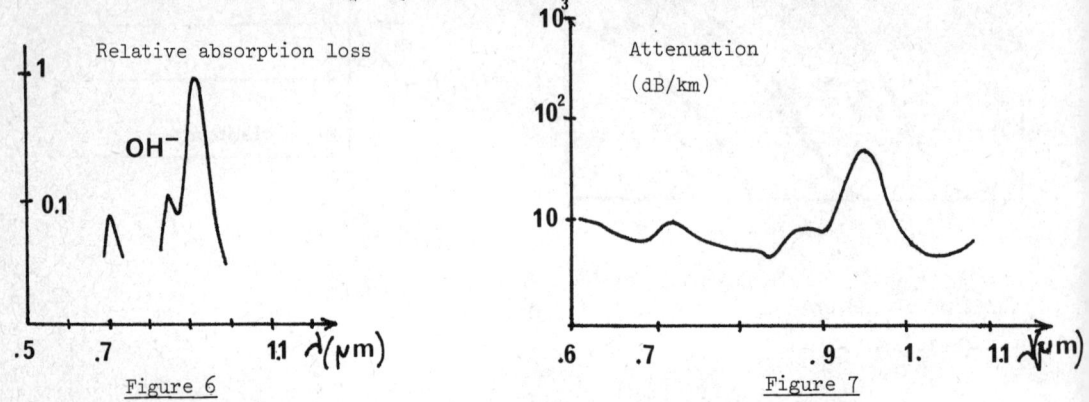

Figure 9 gives the loss of a fused silica multimode fiber versus wavelength. This fiber made by Corning, has a numerical aperture of 0.14 and consists of a doped fused silica core and fused silica cladding whose diameters are 91 and 125 μm respectively. The graph shows that the attenuation is the lowest at 0.85 μm and 1.05 μm, which are precisely the wavelengths chosen for the optical telecommunications.

b) Fiber information capacity

We will comment here the information capacity of the single mode and multimode fibers. The information capacity is considered from the standpoint of digital transmission. The important point is to analyse the different causes which produce the broadening of the pulses when they travel along the transmission fiber. Generally, the spreading is the result of the dispersion characteristic of the fiber.

For single mode fibers, it has been shown that the pulse spreading depends on the spectral width of the source. For multimode fibers, the spreading is due to group velocity differences between all the modes. Generally, there are three causes of dispersion in a fiber:

1)- Waveguide dispersion which produces a delay effect versus wavelength in each propagating mode.

2)- Material dispersion: for the most glasses, the refractive index depends on the wavelength.

3)- Multimode dispersion: at single freqency, the group velocities are different for the various modes.

For multimode fibers, all three effects must be considered.

With broad band sources, material dispersion and multimode dispersion must be considered. With narrow band laser and single mode fiber, the material dispersion is the most important cause. The evaluation of the dispersion characteristics of a fiber is obtained from the specific group delay (second per meter).
It has been shown by Glodge that the group delay has the following expression, with the assumption that the core-cladding dispersion characteristics are similar:

$$\tau = \frac{1}{c}\left\{ N_1 + (N_1 - N_2)\frac{m}{M}\right\}$$

where $N_1 = d(kn_1)/dk$; $N_2 = d(kn_2)/dk$; m is a mode group number and M is its maximum value. The parameters n_1, n_2 are the phase refractive indices of the mode and N_1, N_2 are the group refractive indices of the mode.
They are related to the phase velocity by the following expressions:

$$\frac{\omega}{k} = \frac{c}{n} \qquad \frac{d\omega}{dk} = \frac{c}{N} = \frac{c}{d(k_0 n)/dk_0}$$

From the above expression of τ, it is easy to calculate the multimode group delay distorsion. With $\Delta = 0.01$ and $n_1 = 1.46$, τ is about 50 ns/km.
Material dispersion is particularly significant for single mode fibers. The pulse spreading over a length L is given approximately by:

$$\tau = \frac{L}{c}\frac{\Delta\lambda}{\lambda}\lambda^2 \frac{d^2 n_1}{d\lambda^2}$$

where $\Delta\lambda/\lambda$ is the spectral width of the source, $\lambda^2 d^2 n_1/d\lambda^2$ is the material dispersion. This spreading appears because the group velocity of the mode is a function of the wavelength. It has been shown for a fused silica single mode that the material dispersion cancels exactly the wavelength dispersion when the wavelength λ equals 1.32 µm. At this wavelength, the bandwidth is very high (more than 100 GHz/km). The used source is an injection laser having a narrow linewidth of about 0.1 Å. The modal dispersion is the strongest cause of the delay spread. The maximum delay spread can be calculated easily from geometrical optics considerations leading to the expression:

$$\tau = \frac{(NA)^2}{2n_1 c} \quad ; \text{ for step index fibers.}$$

$$\tau = \frac{(NA(o))^4}{8n_1^3 c} \quad ; \text{ for graded index fibers.}$$

It appears from the above relations that the dispersion corresponding to the parabolic profile is $\Delta/2$ times smaller than the dispersion of the step index fibers. The width of a pulse propagating in a graded index fiber with the index profile $n(r) = n_1 (1 - (r/a)^\alpha)$, has been calculated as a function of the profile shape fector. For fused silica type, the optimum profile is nearly parabolic: $\alpha \simeq 2$.

The optimal value of α can be approximated by $\alpha \simeq 2(1 - 1.2\Delta)$.
Small deviations from the optimal value produce rapidly degradations on the capacity of the fiber. It is difficult to obtain practically the optimal profile.
We must notice the mode coupling effect on the capacity of multimode fibers. During the propagation, higher order power is lost by the coupling effect with the radiation modes. In this case, the numerical aperture is smaller and produces a reduction of the dispersion.
Mode mixing is also an effect which tends to equalize the modal velocities. Modal coupling reduces the spread of velocities and the coupled modes tend to possess a common velocity. Modal mixing is due to the fiber material inhomogeneities, diameter fluctuations and refractive index variations.

INTEGRATED OPTICS

We give now a brief tutorial introduction and a review of the growing research field of "Integrated Optics". We have given above the different properties of multimode fibers because fiber systems, under study today use multimode fibers in which a thousand modes can propagate. On the contrary, integrated optical circuits and devices are single mode structures. Single mode fibers are just needed for higher transmission speeds and longer transmission distances.
Integrated optics covers all the guided wave techniques used to realise new optical devices and waveguides. The waveguide allows to confine the light energy to a very small cross section, over long distances. The purpose of integrated optics is to construct optical guided wave devices with a very good thermal and mechanical stability. It is allowed to think that one will be able to realise on the same substrate, different integrated devices like integrated sources, integrated modulators in analogy with integrated circuits in electronics.
The planar waveguides used in integrated optics are dielectric waveguides which can be represented by a thin film of higher refractive index than the surrounding medium. They act like filters, wavelength multiplexes, directional couplers or detectors. The planar waveguide confines the light in one dimension.

- Dielectric waveguide.

The dielectric waveguide is shown in figure 8 and 9 where n_f, n_s, and the refractive indices of the film, substrate and cover materials. For the propagation of a guided wave, it is necessary to choose:

$$n_f > n_s, n_c$$

The light is guided inside the film if the energy is confined in the guide or near both interfaces. The guide supports radiation modes when the light is spread outside the film.

If the propagation angle θ is greater than the critical angle θ_ℓ

$$\sin \theta_\ell = n_s/n_f$$

the wave undergoes total reflection on the upper and lower interface. In this case, the modes are guided. Figures 8 and 9 represent the different types of modes.

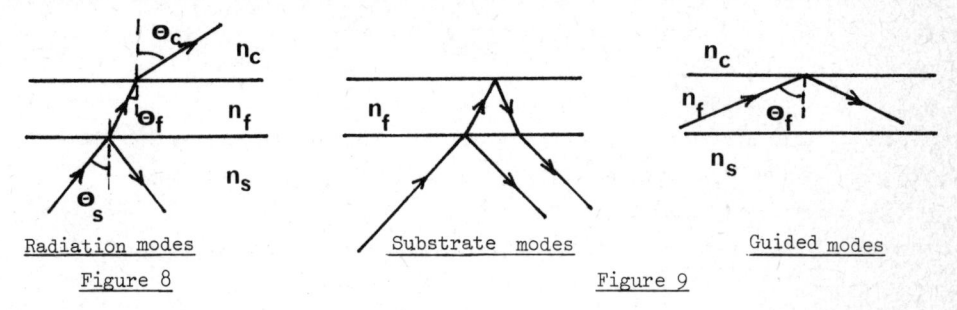

Radiation modes — Figure 8

Substrate modes — Figure 9

Guided modes

Strip guides confine the light in two dimensions. The refractive index n_f is always larger than the surrounding. Different strip guides are described in figure 10.

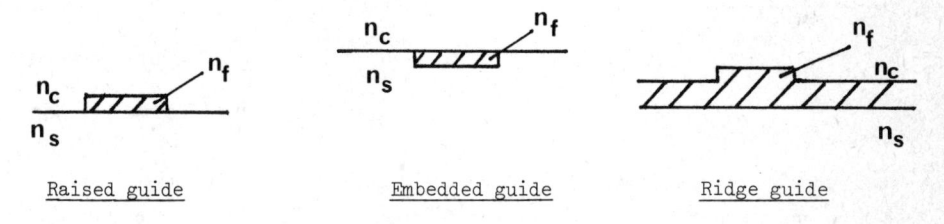

Raised guide — Embedded guide — Ridge guide

Figure 10

Strip guides are used to form various circuit patterns. If two identical strip waveguides are parallel and close to each other over a length L, they will couple energy due to the evanescent field between them. In this case, all the energy can be transferred from one waveguide to another if the coupling length satisfies the relation: $K L = (2m + 1)\ \pi/2$; $m = 0,1,2,\ldots$ This principle is the basis of the directional coupler described in figure 11.

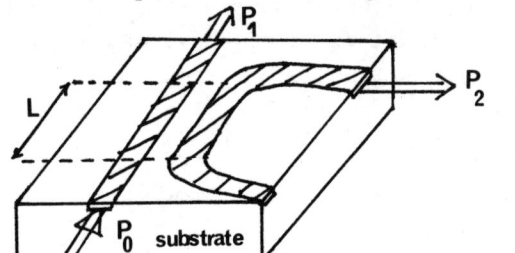

Figure 11
Directional coupler made of strip guides.
The incident energy P_o in guide 1 can theoretically be transferred to guide 2.

The thicknesses and the refractive indices of the strip guides depend on the method of fabrication. Typically, most guides have index differences with the substrate of the order of 10^{-3}. The thicknesses are of the order of a few microns.
The physical picture of the guided light propagation is that of light travelling in

zig-zag path. The electromagnetic field is the superposition of two plane waves which are totally reflected on the interfaces limiting the guide. For a guided wave, the propagation is allowed only for a discret set of angles θ_f. This propagation is described by the well known transverse resonance condition:

$$2k\, n_f\, d\, \cos\theta_f - \psi_s - \psi_c = 2m\pi$$

where $k = \omega/c$; ω is the angular frequency: c is the light velocity; θ_f is the propagation angle of the mode; ψ_s and ψ_c are the phase shifts imposed on the reflected waves; m is an integer (0,1,2, ...) which defines the mode number; d is the guide thickness (Figure 12).

The guided wave propagates in the z direction with the propagation constant

$$\beta = k\, n_f\, \sin(\theta_f)$$

It is often convenient to use the effective guide index defined by:

$$N_f = \beta/k = n_f \sin\theta_f$$

with $n_s < N < n_f$

Figure 12

The transverse resonance condition is also the dispersion relation giving the propagation constant β as function of the frequency ω and guide thickness d. From this condition, one can obtain the typical $\omega - \beta$ diagram of a dielectric waveguide (figure 13).

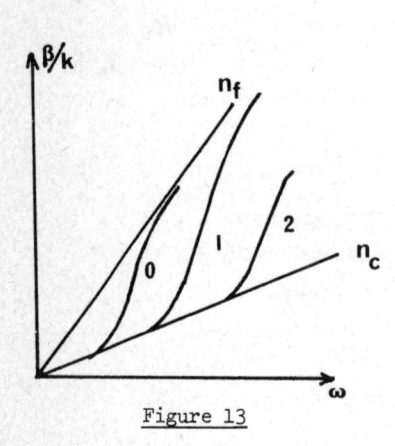

When ω or d increse, β/k tends to the upper bound b_f and more and more modes are guided. On the contrary, when ω (or d) decrease, β/k tends to the lower limit n_s and for small values of the frequency ω or the thickness d, only the first mode (order 0) propagates.

It is convenient to define a normalized frequency and a guide thickness V by:

$$V = k\, d(n_f^2 - n_s^2)^{1/2}$$

a normalized guide index

$$b = (N^2 - n_s^2)/(n_f^2 - n_s^2)$$

Figure 13

a parameter od asymetry of the waveguide

$$a = (n_s^2 - n_c^2)/(n_f^2 - n_s^2); \text{ for T.E. modes}$$

With the above defined parameters, the dispersion relation can be written in the form:

$$V(1-b)^{1/2} - \tan^{-1}\sqrt{b/(1-b)} + \tan^{-1}\sqrt{(b+a)/(1-b)} = m\pi.$$

This last relation allows to obtain the dependence of the guide index b as a function of the normalized thickness V for different values of the asymmetry parameter a. By setting b = 0, one determines the cutoff frequencies of the propagation modes.

- Switches and modulators

We have described above the characteristics of the light propagation in a passive dielectric waveguide. These waveguides serve the function of transporting light energy in the same way that conduits carry currents in integrated electronics circuits. In addition to the above passive dielectric waveguides, it is necessary to make active components like switches, modulators and integrated laser sources, such as transistors in integrated electronics. To obtain the light modulation, the electro-optic effect was applied to achieve modulation in a GaAs epitoxial film on a more heavily doped GaAs substrate. Considerable progress was made recently in the technology of guided wave modulators and switches.

The modulator or switch, based on the electro-optic effect, is shown schematically in figure 14. Two single mode strip guides are deposited on a common substrate. These guides are made with an electro-optic material. The two guides are parallel and separated by a gap g for a length L. Outside this length, the guides separate. An electric field is applied on the electrodes as shown in the figure. These electrodes are split in the middle to permit the application of voltages of reversed polarities. Without voltage, we have the configuration of the directional coupler. In this case, all the energy in one guide can be coupled to the other guide, if the length L satisfies the relation: $KL = (2m+1)\pi/2$, where K is the coupling constant between the two guides. With identical guides, the transfer is maximum if the propagation constants are equal: $\beta_1 = \beta_2$

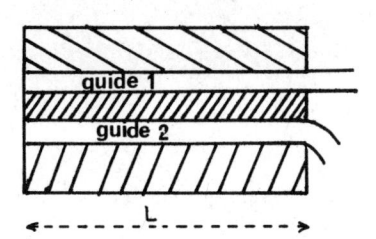

Figure 14

Electro-optic modulator

The light flowing in guide 1 will couple into guide 2 and vice versa. If the propagation constants are not matched: $\beta_1 \neq \beta_2$, all the light will not be coupled and the coupling length l will be shortened. The modulation of the light can be explained from these considerations. Without applied eletcric field, $\beta_1 = \beta_2$, and the light in guide 1 is totally coupled in guide 2. If a voltage is applied on the electrodes, the effective index of guide 1 will change in the opposite direction from that of guide 2. In this case, β_1 will be different from β_2 and the light will not be coupled in guide 2 and will propagate in both guides. The reader will find in numerous publications the detailed calculations

of the electro-optic modulator performances. We report here the results concerning
a modulator made with $LiNbO_3$. With a gap of 2 μm and a coupling length of 1.4 mm,
the energy transfer is maximum when V = 2 volts. An approximate calculation shows
that the specific energy is 18×10^{-6} W/MHz. As can be seen, high transfer is possible
with low voltage and power.

We may note at ounce that switches and optical modulators have to be important
components of any optical communication or data transmission system.

By connecting several such switches, one can build networks such as by applications
of proper voltages, the light entering in one of the input guides can be switched
in one of the output guides. Figure 15 shows an experimental 4x4 switching network.

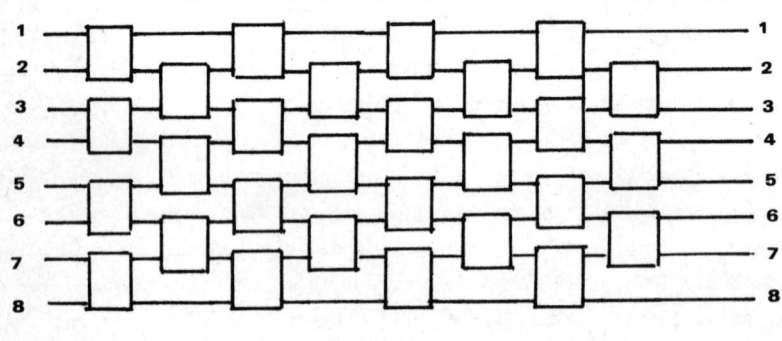

Figure 15

- Filters

The realisation of filters is used for application such as wavelength
multiplexing of different channels. Grating structures have been used as filters.
A corrugation is made into the surface of a film. The grating has a very short
period Λ. It has been shown that a filter has a bandwidth of approximately

$$\Delta\lambda/\lambda \simeq \Lambda/L$$

where L is the length of the periodic structure. The bandwidth is centered at the
wavelength λ_o with: $\lambda_o = 2N\Lambda$.
N is the effective index of the guided mode.

CONCLUSIONS

The ramarkable progress made up to date in developing components for use in optical fiber systems has led to a new guided light technology. Propagation loss in optical fibers has been reduced several orders of magnitude. Multimode graded fibers are now produced with low loss characteristics (few dB/km). Injection lasers and LED lifetimes have been improved by several orders of magnitude. Reliability of 10^5 hours is necessary for most practical applications and extrapolated lifetimes in this range have been demonstrated. The age of telecommunications appears to be a reality. The demand for communications has grown considerably over the past decade and continues to increase. One can conclude that optical fibers will have a significant impact on future data transfer applications.

REFERENCES

N.S. KAPANY, Fiber Optics, Academic Press, New York, (1967).
N.S. KAPANY and J.J. BURKE, Optical Waveguides, Academic Press, New York, (1972).
D. MARCUSE . Light Transmission Optics, Van Norstrand Remhold, Princeton, New Jersey,1972.
D. MARCUSE, Theory of dielectric waveguides, Academic Press, New York, (1974).
S.E. MILLER, E.A. MARCATILI and T. LI, Research toward optical fiber transmission systems, Proc. IEEE, 61, 1703, (1973).
R.D. MAURER, Introduction to optical fiber waveguides, in Integrated Optics, M. BARNOWSKI, Ed. New York, Plenum Press, (1974).
T.G. GIALLORENZI, Optical communications research and technology: Fiber optics, Proc. IEEE, 66, 744, (1978).
D. GLODGE, Multimode theory of graded core fibers, Bell Syst. Techn. J., 52, 1563, (1973)
R. OLSHANSKY, M.G. BLANKSHIP and D.B. KECK, Length-dependent attenuation measurements in graded index fibers, Proc. 2nd Eur. Conf., Paris, France, sept. 1976.
A.W. SNYDER, Power loss in optical fibers, Proc. IEEE, 60 757, (1972).
D.A. PINNOW, T.C. RICH, F. OSTERMAYER and M. DIDOMENICO, Fundamental optical attenuation limits in liquid and glassy state with application to optical fiber waveguides. Appl. Opt., 22, 527, (1973).
D.B. KECK, R.D. MAURER and P.C. SCHULTZ, On the ultimate lower limit of attenuation in glass optical waveguides, Appl. Phys. Letters, 22, 307, (1973).
D. MARCUSE, Power distribution and radiation losses in multimode dielectric waveguides Bell Syst. Techn. J., 51, 429, (1972).
D. GLODGE, Dispersion in weakly guiding films, Appl. Opt., 10, 2442, (1971).
D. MARCUSE, Impulse response of clad optical multimode fibers, Bell Syst. Techn. J., 52, 801, (1973).
D. MARCUSE, The impulse response of an optical fiber with parabolic index profile, Bell Syst. Techn. J., 52, (1973).
C.K. KAO and J.E. GOELL, Design process for fiber optical systems, Electronics,p.113, (1976).
R. OLSHANSKY and D.B. KECK, Pulse broadning in graded index optical fibers, Appl. Opt. 15, 483, (1976).
J.E. GOELL and R.D. STANDLEY, Integrated optical circuits, Proc. IEEE, 58, 1504, (1970).
S.E. MILLER, Integrated Optics: an introduction, Bell Syst. Techn. J., 48, 2059, (1969).
P.K. TIEN, Light waves in thin films and integrated optics, Appl. Opt. 10, 2395, (1971).
S.E. MILLER, A. Survey of integrated optics, IEEE J. Quantum Electronics, QE-8, 199, (1972).
M.K. BARNOSKI, Introduction to integrated optics, Plenum Press, New York, (1973).
P.K. TIEN, Integrated optics, Scientific American, 230, 28, (1974).
H.F. TAYLOR and A. YARIV, Guided waves optics, Proc. IEEE, 62, 1044, (1974).
W.S. CHANG, W.M. MULLER and F.J. ROSENBAUM, Integrated optics, laser applications, vol. 2, Academic Press, New York, (1974).
J.M. HAMMER and W. PHILLIPS, Low loss single-mode optical waveguides and efficient high speed modulators on $LiTaO_3$, Appl. Phys. Lett., 24, 545, (1974).
J.M. HAMMER, Modulation and Switching of light in dielectric waveguides; Integrated Optics, T. TAMIR, Ed., Springer, Heidelberg, (1975).
H. KOGELNIK, Theory of dielectric waveguides, T. TAMIR, Ed. Springer, Heidelberg, 1975.
T. TAMIR, Ed., Integrated Optics, Topics in applied physics, Springer, Heidelberg, 1975.
I.P. KAMINOV, Optical waveguide modulators, IEEE Transactions on microwave theory and technology, MTT 23, 57, (1975).
R.V. SCHMIDT and L.L. BUHL, Experimental 4x4 optical switching network, Elect. Lett. 12, 575, (1976).
R.V. SCHMIDT, D.C. FLANDERS, C.V. SHANK and R.D. STANDLEY, Narrow band grating filters for thin film optical waveguides, App. Phys. Lett., 25, 458, (1975).

PROSPECTS FOR LONG-WAVELENGTH X-RAY MICROSCOPY AND DIFFRACTION

D. Sayre
IBM Research Center
Yorktown Heights, New York 10598, U.S.A.

Our purpose in the first part of this talk is to call attention to certain advantages of the soft (λ = 10-100A) x-ray photon in the imaging of biological material, which are not shared by other particles. The second part will briefly survey the status of some of the problems which arise in the use of these photons for this purpose.

I. Properties of the Long-Wavelength X-Ray Photon as a Compositional Probe.

Examination of the reaction cross-sections[1] for photons in the soft x-ray region (Fig. 1) shows that these particles are well suited for the mapping of compositional features in intact, wet, unstained, and possibly living single biological cells or organelles. This arises as follows:

(1) Total reaction cross-sections correspond to mean free paths in biological materials of the order of 1 μm;

(2) Total cross-sections vary abruptly with specimen composition, due to existence of absorption edges;

(3) Photon absorption is the dominant reaction (absence of multiple scattering).

Sayre et al.[2] examined these effects in detail and concluded that an adequate signal/noise ratio for the imaging of 100A diameter features in 1 μm-thick wet unstained biological materials can be obtained at exposure levels of ~10^4 J/g. For electrons of the energies employed in electron microscopy, they concluded that exposure must rise to levels of ~10^7 J/g to achieve the same signal/noise ratio in the same materials. (Studies of radiation damage[3] indicate that exposures of 10^4 J/g, although destroying function in biological material, do not seriously affect structure visible at 100A resolution. Such structure is destroyed by exposures of 10^7 J/g.) Similar conclusions[4] apply to the mapping of the concentration of a particular atomic species Z in intact wet biological materials, except that for $Z \lesssim 11$ the method of electron energy-loss analysis can achieve exposures similar to those with soft x-rays.

High-resolution structural input to cell biology today comes mainly from studies on non-intact cellular material. Although this information is of extreme value, there seems to be agreement among cell biologists on the desirability of having techniques for the imaging of untreated cellular material.

II. Problems Attendant on the Use of These Photons.

A. Sources.

Until fairly recently, there existed no high-intensity laboratory sources of photons in the wavelength range under discussion. This problem has now been solved in most respects through the development of synchrotron radiation and plasma[5] sources. These allow ~10^4 J/g of fairly monochromatic radiation to be put on a sample, in seconds or minutes in the case of synchrotron radiation, and in 10^{-8} to 10^{-7} seconds with a

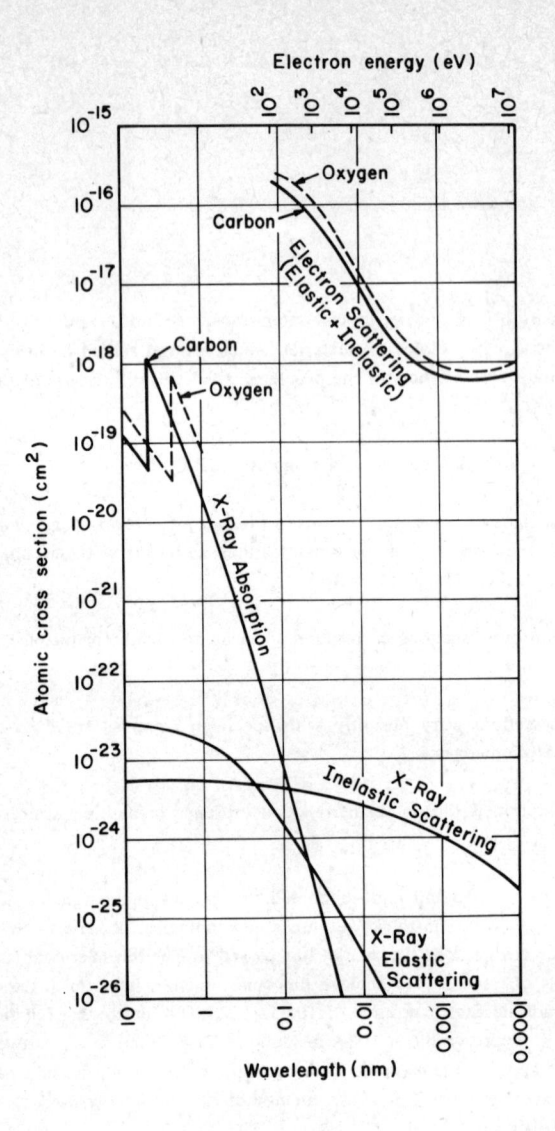

Fig. 1. Cross-sections for reactions of photons and electrons with carbon, as functions of particle wavelength λ. For comparison, portions of the corresponding curves for oxygen are also shown (dashed curves).

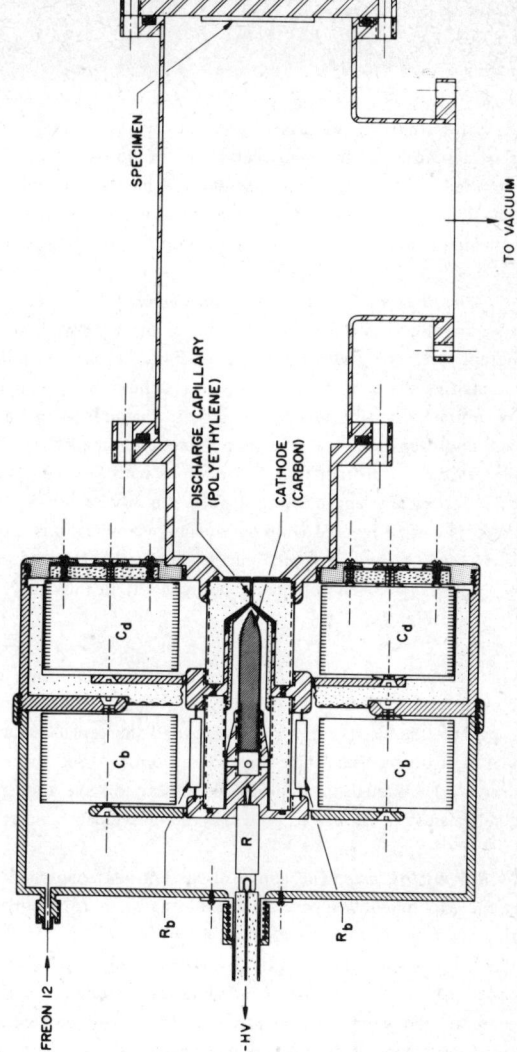

Fig. 2. The McCorkle-Vollmer pulsed plasma source. Upper left, schematic diagram of the source. Below, diagram showing some of the construction details. Upper right, photograph of the source. (Courtesy R.A. McCorkle.)

pulsed plasma source. The plasma source (Fig. 2) is also sufficiently compact and inexpensive to become widely available should the need develop. The synchrotron radiation source has the advantages of continuous tunability, narrow beam divergence, etc.

B. Imaging.

In the above study[2] it was assumed that every event of interest (absorption or non-absorption of the photon; elastic scattering, inelastic scattering, or non-scattering of the electron) could be detected and correctly assigned to the appropriate resolution element of the specimen; i.e., that the laws of optics and the quality of available components will allow an ideal image-forming system to be constructed. Somewhat similar idealizing assumptions were made in the study[4] on the mapping of atomic species.

Contact Microradiography. For thin specimens, or for resolutions which do not approach λ too closely, the above assumptions can be quite well met at present by the use of high-resolution detectors in contact microradiography[6]. In this method (Fig. 3) irradiation produces an image of the specimen in a "grainless" film of photon-sensitive material with which the specimen has been placed in contact. (To read out the full information contained in the image, the image is examined in an electron microscope.) To date the most widely used photosensitive material is polymethylmethacrylate (PMMA), which undergoes a change in solubility upon exposure to radiation. For photons with $\lambda = 40A$, the inherent resolution of PMMA is ~50A, with quantum efficiency approaching 1 (see the review by Spiller and Feder[7]). Accordingly, whenever diffraction effects can be ignored (thin specimens or applications in which resolutions do not approach λ), the conditions assumed above[2] are approximately met in this form of microscopy. Images of thin specimens with resolutions $\lesssim 100A$ have been obtained experimentally in this way[8]. Some biology is now being done with the technique[9,10] (see Fig. 4).

For high-resolution work with thick specimens the high-angle portion of the diffraction pattern must be processed more correctly than is done by the simple contact method. Proposals for doing this include deblurring by holographic or other processing and the replacement of the contact technique by scanning or conventional microscopy using wide-angle x-ray optics. (See the recent review of soft x-ray microscopy by Kirz and Sayre[11].) Although these proposals appear to have good potential, it may be some time before any have been sufficiently developed to produce an improvement over the simple PMMA-based contact method.

Soft X-Ray Diffraction. The above discussion has concerned microscopy, in which an image is formed directly by the experimental apparatus. An alternative is to adopt the practice of x-ray diffraction, in which the experiment is asked only to capture as much of the diffraction pattern as possible, and the image formation is carried out subsequently by computation. At $\lambda = 1.5A$, this technique, although slow, is extremely successful in the imaging of 3-dimensional structures at the full theoretical resolution of $\lambda/2$. There appears to be no fundamental reason why the same technique cannot be applied at $\lambda = 30A$. There would be a general scaling upward of distances in the image, i.e. the minimum resolvable distance would scale upward from ~0.75A to ~15A and the diameter of the structures imaged (imagible field of view) would similarly move upward from its present practical maximum of ~200A to perhaps 4000A. The method would thus cover a size-range of great interest in cell biology. One quantity which would scale downward is specimen size, because of the larger total reaction cross-sections at 30A; the appropriate diffracting structure would be the single biological cell or organelle. The diffraction would thus be micro-diffraction. It is not yet clear to the author what difficulties of instrumentation would result.

Physically the mechanism of diffraction changes from photon scattering at 1.5A to photon absorption at 30A. Fortunately the mathematics remains largely unchanged, with the diffraction pattern still being approximately a Fourier transform. (The function transformed is no longer $\rho(x)$ but $|m(x)-1|$, where m is the complex index of refraction, with $Im(m) = \lambda/4\pi s$; here s is the mean free path for absorption of the photons. For the approximation to be good, $|m-1|$ must be small (Rayleigh-Gans theory). For biological materials, m for $\lambda = 30A$ is of the order of 0.999-0.001i.) Thus the general scheme of structure analysis at

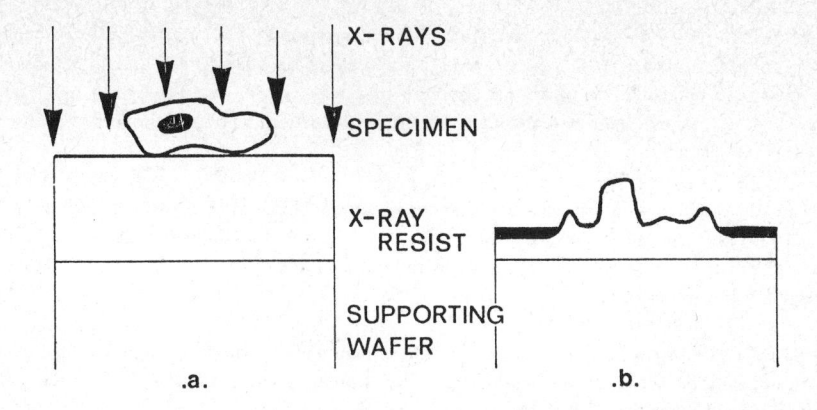

Fig. 3. Contact microradiography. (a) Irradiation of the x-ray resist through the specimen. (b) The resist after development. The projected image of the specimen is recorded in the resist profile and can be read out, after light metallization, in an SEM.

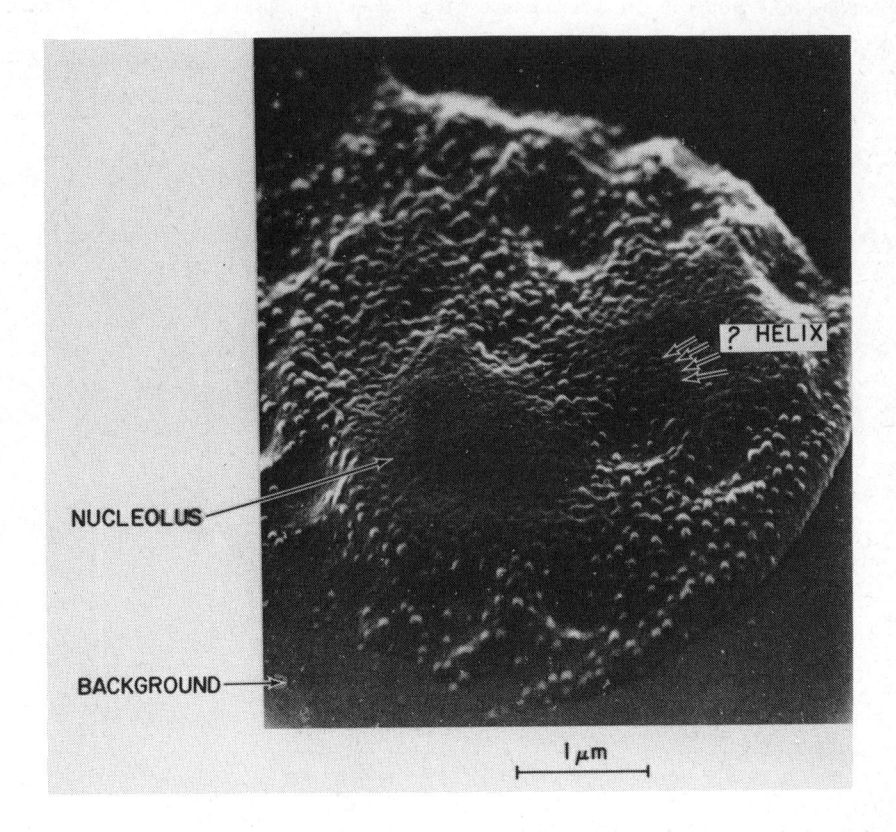

Fig. 4. Contact microradiograph of human interphase nucleus from glioblastoma tissue culture in PMMA, made with carbon Kα radiation. The x-ray source was operated at 5kV and 40mA, with a target-to-specimen distance of 18 cm and exposure time of 40 hours. (Courtesy L. Manuelidis (Yale University), J. Sedat (University of California at San Francisco), and R. Feder (IBM Research Center).)

1.5A carries over to the 30A case. In particular, the central problem of the analysis is that of supplying the phase of the diffraction pattern. It is suggested that at the scale of sizes involved, it should be possible to use artificially fabricated micro-objects of known structure to act as phasing references in a modified form of heavy-atom phasing. It should be noted also that the diffraction pattern will be continuous (not discrete), the specimen being non-crystalline.

Experimentally, diffraction patterns using synchrotron radiation at λ = 46.8A have recently been obtained[12] from 1 μm latex spheres. Exposures for collecting the small-angle diffraction pattern from assemblages of spheres were 2 minutes. The demonstration that the large-angle pattern can be collected from single objects remains to be done.

Imaging in 3 Dimensions. The thickness of the contemplated structures implies that 3-dimensional imaging will normally be necessary for comprehensibility. Contact microradiography lends itself naturally to the taking of stereo projections through tilting of the specimen-film pair (or more generally to the collection of n tomographic views), but is subject to the limitation on resolution noted above. Three-dimensional imaging without this limitation would also be provided by diffraction, as well as by microscopy with wide-angle optics (focussing on successive layers), or by holography.

In principle the exposure of $\sim 10^4$ J/g noted above is sufficient to establish the structure in 3 dimensions. However, the above techniques (holography excepted) require an increase in total exposure, and thereby in damage to the specimen. It is not clear as yet whether a 3-dimensional non-holographic method operating at 10^4 J/g can be realized physically or not.

C. Specimen Chambers.

To realize the full potential of soft x-ray imaging, the specimen should be maintainable in a normal environment at 1 atmosphere. Fortunately, mean free paths in air at STP are several mm. for $\lambda > 31$A (absorption edge of nitrogen), so that the design of suitable specimen chambers employing thin windows or differentially pumped small apertures should not be difficult.

REFERENCES

(1) B.L. Henke and E.S. Ebisu, in: Advances in X-Ray Analysis (C.L. Grant, C.S. Barrett, J.B. Newkirk, and C.O. Ruud, eds.), Vol. 19, pp. 150-213, Plenum Press, New York (1974). Also Wm. J. Viegele, Photon cross sections from 0.1 keV to 1 MeV for elements Z=1 to Z=94, Atomic Data Tables 5, 51-111 (1973).

(2) D. Sayre, J. Kirz, R. Feder, D.M. Kim, and E. Spiller, Transmission microscopy of unmodified biological materials: comparative radiation dosages with electrons and ultrasoft x-ray photons, Ultramicroscopy 2, 337-349 (1977). See also same authors, Potential operating region for ultrasoft x-ray microscopy of biological materials, Science 196, 1339-1340 (1977).

(3) M. Isaacson, in: Principles and Techniques of Electron Microscopy (M.A. Hayat, ed.), Vol. 7, pp. 1-78, Van Nostrand Reinhold, New York (1976).

(4) J. Kirz, D. Sayre, and J. Dilger, Comparative analysis of x-ray emission microscopies for biological specimens, Ann. N.Y. Acad. Sci. 306, 291-303 (1978). See also reference 11.

(5) R.A. McCorkle, Soft x-ray emission by an electron-beam sliding-spark device, J. Phys. B11, L407-408 (1978).

(6) R. Feder, D. Sayre, E. Spiller, J. Topalian, and J. Kirz, Specimen replication for electron microscopy using x-rays and x-ray resist, J. App. Phys. 47, 1192-1193 (1976).

(7) E. Spiller and R. Feder, in: Topics in Applied Physics (H.-J. Queisser, ed.), Vol. 22, pp. 35-92, Springer-Verlag, Berlin (1977).

(8) R. Feder, E. Spiller, J. Topalian, A.N. Broers, W. Gudat, B.J. Panessa, Z.A. Zadunaisky, and J. Sedat, High-resolution soft x-ray microscopy, Science 197, 259-260 (1977).

(9) J. Wm. McGowan, B. Borwein, J.A. Medeiros, T. Beveridge, J.D. Brown, E. Spiller, R. Feder, J. Topalian, and W. Gudat, High resolution microchemical analysis using soft x-ray lithographic techniques, J. Cell. Biol. (in press).

(10) L. Manuelidis, J. Sedat, R. Feder, and E. Spiller, Three-dimensional information on polytene and diploid interphase nuclei, to be presented at Conference on Ultrasoft X-Ray Microscopy, N.Y. Acad. Sci., June 1979.

(11) J. Kirz and D. Sayre, in: Synchrotron Radiation Research (S. Doniach and H. Winick, eds.), Plenum Press, New York (in press).

(12) K. Wakabayashi, A. Kazizaki, Y. Siota, K. Namba, K. Kurita, M. Yokata, H. Tagawa, Y. Inoko, T. Mitsui, E. Wada, T. Ueki, I. Nagakura, and T. Matsukawa, Soft x-ray small-angle scattering by polystyrene latexes using synchrotron radiation, J. Phys. Soc. Japan 44, 1314-1322 (1978).

KINEMATICAL AND DYNAMICAL DIFFRACTION THEORIES

A. AUTHIER

Laboratoire de Minéralogie-Cristallographie
Associé au C.N.R.S., Université P. et M. Curie
4, Place Jussieu, 75230 Paris Cedex 05

INTRODUCTION

Dynamical theory is the name given to the theory of diffraction of a wave by a perfect crystal. It differs from the kinematical, or geometric, theory used in structure determination problems in that it takes into account the influence of matter on the wave. Although it is as old as the discovery of X-ray diffraction, it has only been widely used for the past twenty years or so. This is due partly to the fact that before that time no perfect enough crystals were available to test it and partly to the fact that it became necessary in order to interpret the contrast of defect images in electron microscopy and X-ray or neutron topography. These techniques were, at least for the former two, developed more than twenty years ago, the latter beeing more recent. Very interesting optical problems related to the energy propagation arise at the vicinity of BRAGG's condition, both in perfect crystals for fundamental reasons and in crystals containing isolated defects since they enable these defects to be visualized. A concice bibliography on geometrical and dynamical theories is given in (1).

I.- PRINCIPLE OF THE KINEMATICAL, OR GEOMETRICAL THEORY

The principle of the kinematical or geometrical theory is to consider a spatial distribution of identical scattering centers and to assume that the amplitude of the incident wave is constant at each of these centers. The total amplitude diffracted in a given direction is therefore obtained by summing the amplitudes scattered by each center taking simply into account the phase relationships between them. This is only possible if the interaction between the incident wave and the scattering centers is so weak that partial reflections can be neglected. This approximation is no more valid for large perfect crystals and it is precisely the aim of the dynamical theory to take the above mentioned interaction into account.

Let O and P be two identical scattering centers and $\vec{k}_o = \vec{s}_o/\lambda$ the wave vector of an incident plane wave (fig. 1). Each center scatters this incident wave in every

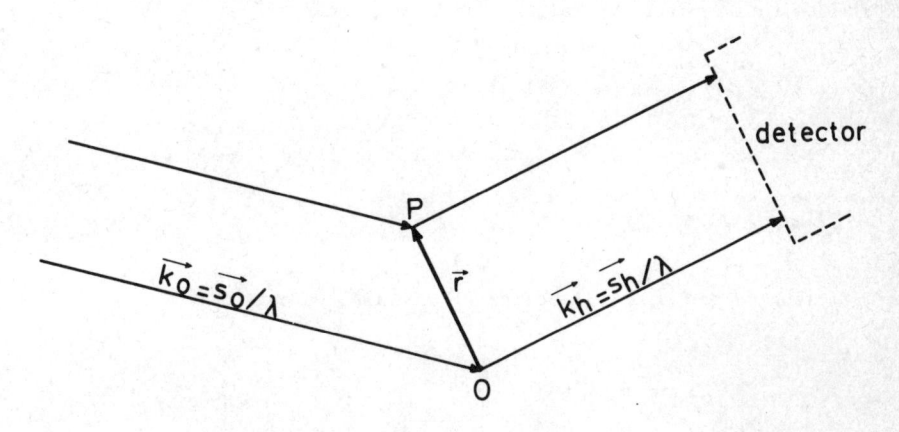

Fig. 1

direction. Let us consider one of them and place a receptor at a distance which is very large with respect to $\vec{OP} = \vec{r}$. We can assume that the receptor receives plane waves emitted at O and P and let $\vec{k}_h = \vec{s}_h/\lambda$ be their wave vector. Their phase difference is given by :

(1) $\quad \phi = \frac{2\pi}{\lambda} (\vec{s}_h - \vec{s}_o).\vec{r} = 2\pi (\vec{k}_h - \vec{k}_o).\vec{r} = 2\pi \vec{\Delta k}.\vec{r}$

If a is the amplitude scattered by each center in the direction of $\vec{k}_h$, the total amplitude diffracted by n centers is equal to :

$$A(\vec{\Delta k}) = a \sum_i^n e^{-2\pi i \vec{\Delta k}.\vec{r}_i}$$

where $\vec{r}_i$ is the position vector of the ith center.

If we now consider a continuous distribution of centers and if $\rho(\vec{r})\,d\tau$ is the number of centers contained in volume element $d\tau$, the total diffracted amplitude is equal to :

(2) $\quad A(\vec{\Delta k}) = a \int \rho(\vec{r})\, e^{-2\pi i\, \vec{\Delta k}\cdot\vec{r}}\, d\tau$

If the distribution of centers extends over an infinite volume, this amplitude is equal to :

(3) $\quad A_\infty(\vec{\Delta k}) = a\, \mathcal{F}[\rho(\vec{r})]$

where $\mathcal{F}[\rho(\vec{r})]$ is the Fourier transform of $\rho(\vec{r})$.

If, furthermore, the distribution $\rho(\vec{r})$ is triply periodic, that is if we are dealing with a perfect crystal, it can be written :

(4) $\quad \rho(\vec{r}) = \rho_0(\vec{r}) * \sum_{u_i,v_i,w_i} \delta(\vec{r}-\vec{r}_i)$

where $\rho_0(\vec{r})$ is the distribution within the unit cell and

$$\vec{r}_i = u_i\vec{a} + v_i\vec{b} + w_i\vec{c}$$

is the position vector of a cell. Using (3) and (4), we obtain :

(5) $\quad A_\infty(\vec{\Delta k}) = aF(\vec{\Delta k}) \sum_{h,k,\ell} \delta(\vec{\Delta k}-\vec{h})$

where $\vec{h} = h\vec{a}* + k\vec{b}* + \ell\vec{c}*$ is a reciprocal lattice vector and $F(\vec{\Delta k})$ is the Fourier transform of $\rho_0(\vec{r})$. It is called the structure factor. The amplitude distribution A_∞ is concentrated at the reciprocal lattice nodes. It is therefore only necessary to consider the values $F_{hk\ell}$ of $F(\vec{\Delta k})$ at these points. They are equal to

(6) $\quad F_{hk\ell} = \int_{\text{one cell}} \rho(\vec{r})\, e^{-2\pi i\, \vec{h}\cdot\vec{r}}\, d\tau$

and

(7) $\quad A_\infty(\vec{\Delta k}) = a \sum_{h,k,\ell} F_{hk\ell}\, \delta(\vec{\Delta k}-\vec{h})$

The fact that the diffracted amplitude is only different from zero if $\vec{\Delta k}$ is a reciprocal lattice vector is equivalent to BRAGG's law. (fig. 2)

$\overrightarrow{OH}$: reciprocal lattice vector

reciprocal space

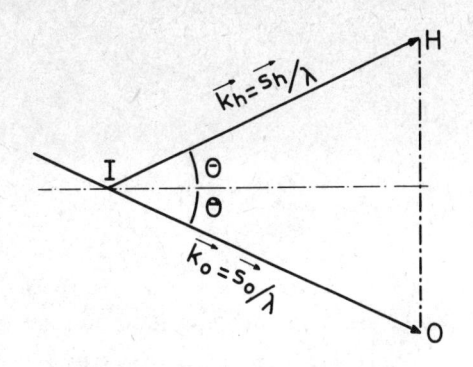

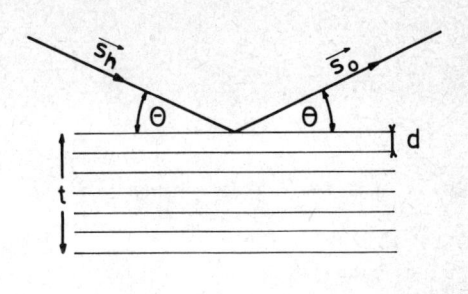

d : lattice spacing
t : crystal thickness
θ : Bragg angle

direct space

$$\frac{OH}{2} = \frac{1}{\lambda} \sin \theta \; ; \; OH = n/d \qquad 2d \sin \theta = n\lambda$$

<u>Fig. 2</u>

Actually, the crystal is finite. If $y(\vec{r})$ is a step function equal to 1 inside the crystal and to zero outside, the diffracted amplitude is equal to :

(8) $\quad A(\overrightarrow{\Delta k}) = a \; \mathcal{F} \; [\rho(\vec{r}) \; y(\vec{r})]$

If $Y(\vec{k})$ is the Fourier transform of $y(\vec{r})$, the amplitude distribution in reciprocal space is given by :

(9) $\quad A(\overrightarrow{\Delta k}) = A_\infty (\Delta k) * Y(\overrightarrow{\Delta k})$

Equations (7) and (9) show that the amplitude distribution is now a continuous function, triply periodic in reciprocal space. Around each reciprocal lattice point h,k,ℓ, it is equal to

(10) $\quad A(\overrightarrow{\Delta k}) = a \; F_{hk\ell} \; Y(\overrightarrow{\Delta k} - \vec{h})$

and the corresponding intensity to

(11) $\quad I = |a|^2 |F_{hk\ell}|^2 |Y|^2$

It can be shown, for instance, that in the case of an infinite crystal slice of thickness t (fig. 2), the interference function $|Y|^2$ is proportional to

$$\frac{\sin^2(\pi X t/d)}{\pi^2 X^2}$$

where $X = 2d\Delta\theta \cos\theta/\lambda$, d is the lattice spacing and $\Delta\theta$ the departure from BRAGG's law of the incident plane wave.

If the crystal is oscillated in the incident beam around the normal to the $\vec{k}_o$, $\vec{k}_h$ plane and the variations of the diffracted intensity are recorded in a detector, the shape of the rocking curve obtained (Fig 3) and in particular its half width only depends on the crystal thickness. The value of the maximum is proportional to the square of the modulus of the structure factor.

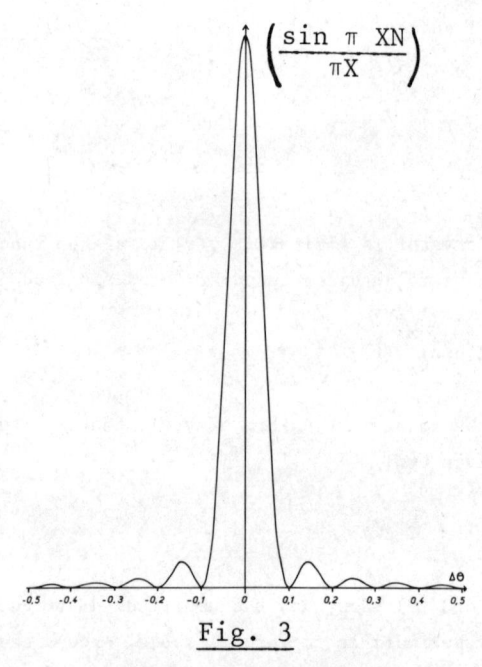

$$\left(\frac{\sin \pi X N}{\pi X}\right)$$

Fig. 3

Rocking curve for a thin crystal - Geometrical theory

The quantity which is actually used in practice is the total intensity received in the detector as the crystal is rocked : it is called <u>integrated intensity</u>. It is proportional to the area under the curve of figure 3. Its calculation shows that it is proportional to $|F_{hk\ell}|^2$ and to the crystal volume bathed in the incident beam. Its expression is only applicable to small enough crystals or to the so called "ideally imperfect" crystals which are made of small incoherent domains which can be considered to diffract the incident beam independantly. How "small" the crystal should be for the geometrical theory to be a good approximation is determined by the dynamical theory.

II.- PRINCIPLES OF THE DYNAMICAL THEORY

Dynamical theory is really a part of general optics and started out from optics. Its foundations were laid by P.P. EWALD even before the discovery of X-ray diffraction by M. von LAUE, W. FRIEDRICH and P. KNIPPING(2) in 1912 : the topic given him for his thesis by SOMMERFELD was the interpretation of double refraction in terms of the diffraction of an electromagnetic wave by a triply periodic anisotropic arrangement of isotropic dipoles. His work led to the formulation of the dynamical theory published in 1917 (3).

Independantly, another approach to the dynamical theory was developped by C.DARWIN in 1914 (4). It is based on the resolution of recurrence equations which state the balance between partially transmitted and partially reflected waves at each lattice plane. Its results are the same as those of the EWALD theory but it is not quite so convenient for the study of the energy propagation. It is very useful for the interpretation of optical phenomena related to dynamical diffraction as will be shown further on.

The formulation of the dynamical theory of X-ray diffraction which is widely used nowadays is that due to M. von LAUE and was derived in 1931 from EWALD's theory (5). It is based on the solution of Maxwell's equations in a medium with a triply periodic continuous electric susceptibility. Since the interaction with protons can be neglected, only electrons need be taken into account and a classical calculation shows that the electric susceptibility is equal to :

$$(12) \quad \chi(\vec{r}) = - \frac{R\lambda^2}{\pi} \rho(\vec{r})$$

where R is the classical radius of the electron and $\rho(\vec{r})$ the electronic density.

In a crystal, $\rho(\vec{r})$ and $\chi(\vec{r})$ are triply periodic and can be expanded in Fourier series :

$$(13) \quad \chi(\vec{r}) = \sum_{h,k,\ell} \chi_h \exp 2\pi i \, \vec{h}.\vec{r}$$

with $\chi_h = - \frac{R\lambda^2}{\pi V} F_{hk\ell}$

where V is the volume of the unit-cell and the structure factors $F_{hk\ell}$ are proportional to the coefficient of the Fourier expansion of the electronic density (eqs. (7) and (8)).

By eliminating $\vec{E}$, $\vec{B}$ and $\vec{H}$ in Maxwell's equations, the propagation is obtained :

(14) $\quad \Delta \vec{D} + \text{curl curl } \chi \vec{D} + 4\pi^2 k^2 \vec{D} = 0$

This equation is actually very similar to the wave equation used in the case of particle waves, since Schrödinger's equation can be written

(15) $\quad \Delta \psi + 4\pi^2 k^2 (1+\chi) \psi = 0$

where χ is proportional to the crystal electric potential, for instance.

Equations (14) and (15) are both second order linear partial derivative equations and their solutions have very similar properties. They differ in the particular properties of the wave : electrons, neutrons or X-rays. Equation (14) is more complicated since electromagnetic waves are vector waves and this leads to polarization effects which are discussed in P. SKALICKY's paper (5). They will not be considered here and the discussion will be limited to that of equation (15).

The theory of electrons in solids also considers solutions of equation (15), but with the boundary conditions that the electrons should be limited to the inside of the crystal, which is written through BORN's cyclic conditions. The solutions provide the various possible values of the energy of the electrons for given values of their wave-vectors. In the case of diffraction, the boundary condition is that the wave inside the crystal should be matched at the surfaces with the incident and reflected waves, and the solutions provide the various possible wave vectors for a constant energy. The latter is equal to that of the incident wave. The solutions in both cases have great similarities since they are subsets of the same general solution, but with different boundary conditions. In particular, to the energy gap in band theory corresponds a gap between the branches of the dispersion surface in diffraction theory. The dispersion surface is the locus of the extremities of the wave-vectors which can propagate inside the crystal.

The solution of equation (15) when the interaction term χ is triply periodic is a combination of Bloch waves :

(16) $\quad \psi = \exp - 2\pi i \, \vec{k}_o \cdot \vec{r} \sum_h \psi_h \exp 2\pi i \, \vec{h} \cdot \vec{r}$

It can also be written in the following way :

(16') $\quad \psi = \sum_h \chi_h \exp - 2\pi i \, \vec{k}_h \cdot \vec{r}$

(17) with $\vec{k}_h = \vec{k}_o - \vec{h}$

It shows that each particular solution of the wave equation is a wave field, that is a sum of plane waves of amplitude ψ_h whose wave vectors can be deduced from one another by reciprocal lattice translations (fig. 4).

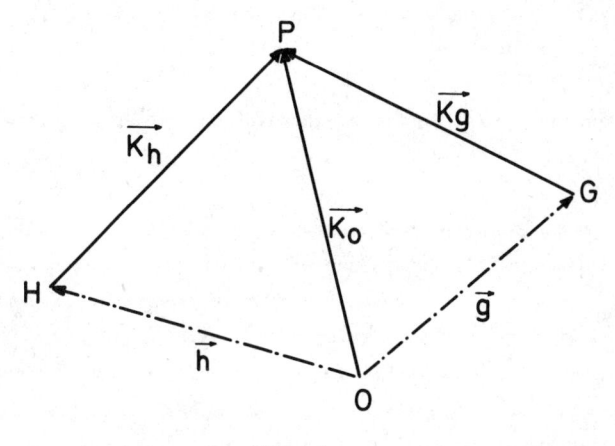

Fig. 4

The tie-point P characterizing a wave-field is the extremity of wave vectors $\vec{k}_o$, $\vec{k}_h$, $\vec{k}_g$ etc ... of the waves which constitute the wave field.

The propagation properties of X-rays in a perfect or nearly perfect crystal can be interpretated in terms of the optical paths of these wave fields inside the crystal. It is no more true, however, in the case of a highly distorted crystal, as is shown in the paper by F. BALIBAR (7). The wave field treatment is also only valid for an incident plane wave. It can be extended, with limitations, to incident wave packets or spherical waves by considering their Fourier expansion as has been done by N. KATO, but not to any kind of incident wave. This important point is also discussed by F. BALIBAR (7).

III.- FUNDAMENTAL EQUATIONS OF THE DYNAMICAL THEORY - PROPERTIES OF WAVE-FIELDS

If we put expansions (16') and (13) in equation (15), we obtain an equation with an infinite number of terms which is equivalent to an infinite set of linear equations :

$$(18) \quad \psi_h = \frac{k_h^2}{k_h^2 - k^2} \sum_{h'} \chi_{h-h'} \psi_{h'}$$

when $\chi_{h-h'}$ is the Fourier coefficient corresponding to the $\vec{h}-\vec{h}'$ reciprocal lattice point.

Although it cannot be proved rigorously it is clear that only a limited number of terms of expansion (16') takes non negligible values, those which correspond to wave numbers k_h very close to the wave number in vacuum, k, in other words, those which are associated with reciprocal lattice points lying simultaneously close to the Ewald sphere. This number is usually big in the case of electron diffraction since the wave length is to small and the radius of the Ewald sphere so large, but small in the case of X-rays or neutrons.

The system (18) is linear and homogenous in ψ_h. For it to have a non trivial solution, its determinant should be equal to zero. This provides a relation between the values of the wave vectors of the plane wave components of the wave field and shows that the extremity, called tie-point, of these wave vectors, drawn from the various reciprocal lattice points (eq. 17, fig. 4) should lie on a certain surface which is the dispersion surface.

We shall limit ourselves to the two beam cases and consider what happens in the plane parallel to the wave vectors, $\vec{k}_o$ and $\vec{k}_h$, of the two waves. The system (18) reduces in this case to two equations :

$$(19) \quad \begin{cases} 2X_o \psi_o - k \chi_{\bar{h}} \psi_h = 0 \\ -k\chi_h \psi_o + 2 X_h \psi_h = 0 \end{cases}$$

where X_o and X_h are respectively the distances of the tie-point P from the two spheres centered at the reciprocal lattice points O and H and with radius n/λ, $n = (1 + \chi_o/2)$ is the index of refraction (fig. 5)

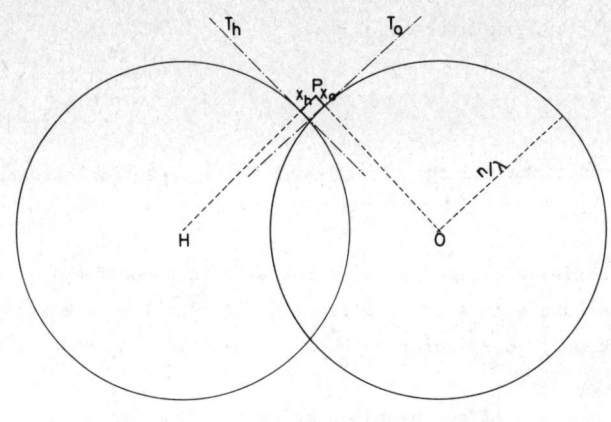

Fig. 5

If we equate the determinant of system (19) to zero, we obtain the following equation :

(20) $\quad X_o X_h = k^2 \chi_h \chi_{\bar{h}}/4$

Since the values of χ_h and $\chi_{\bar{h}}$ are very small, of the order of 10^{-5}, X_o and X_h are very small with respect to the radii of the spheres which can therefore be replaced by their tangential planes.

Equation (20) shows therefore that if we consider the $\vec{k}_o$, $\vec{k}_h$ plane, the tie-point lies on a hyperbola asymptotic to the tangents to the circles centered in O and H and with radii $k(1 + \chi_o/2)$ (fig. 6).

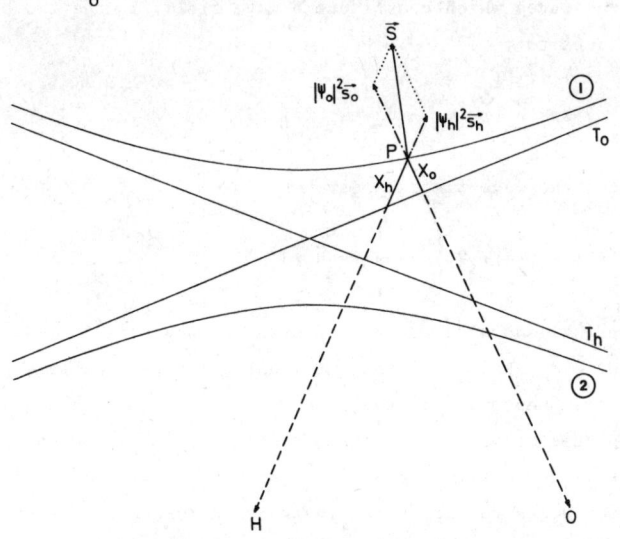

Fig. 6 : Dispersion surface

The diameter of this hyperbola is equal to :

(21) $\quad \Lambda_o^{-1} = k \sqrt{\chi_h \chi_{\bar{h}}} / \cos \theta = R \lambda F_{hk\ell} / V \cos \theta$

it is directly proportional to the strength of the interaction between the waves and matter.

The wave-fields defined by equation (16') are not simply a mathematical concept. They also have a physical reality and can actually be observed. Their physical reality is shown in particular by their propagation and absorption properties.

It can be shown either by calculating the Poynting vector or the group velocity of a wave packet that the common direction of energy propagation of all the waves which constitute a wave field is along the normal to the dispersion surface at the tie-point associated with the wave field. In the two beam case, the propagation direction is given by that of the vector (fig. 6)

(22) $\quad \vec{S} = |\psi_o|^2 \vec{s}_o + |\psi_h|^2 \vec{s}_h$

where ψ_o and ψ_h are respectively the amplitudes of the refracted and reflected wave components of the wave-field.

A very important property of the wave fields is that of anomalous transmission or BORRMANN effect (8). It can be readily understood by considering the interferences between the waves which constitute a wave field. In the two beam case, expansion (16) reduces to :

(23) $\quad \psi = \exp - 2\pi i \, \vec{k}_o \cdot \vec{r} \, [\psi_o + \psi_h \exp 2\pi i \, \vec{h} \cdot \vec{r}]$

and the intensity of the wave-field is given by

(24) $\quad I = |\psi_o|^2 [1 + |\frac{\psi_h}{\psi_o}|^2 + 2 \frac{\psi_h}{\psi_o} \cos 2\pi \, \vec{h} \cdot \vec{r}]$

Equation (24) shows that a set of standing waves is formed. The nodes lie on planes such that $\vec{h} \cdot \vec{r}$ should be constant, that is having the same spacing as that of the lattice planes. Depending on the sign of ψ_h/ψ_o either the nodes or the antinodes lie on the planes of maximum electronic density. It can be shown that it is the nodes for wave fields where tie-points lie on branch 1 of the dispersion surface (fig. 6) and the antinodes for branch 2. There is therefore a minimum of electric field on the atoms for branch 1 wave-fields and thus a minimum of photoelectric absorption and anomalous transmission. It is just the opposite for branch 2 wave fields. This effect is very sensitive to displacements of the atoms from their triply periodic equilibrium position. It is therefore reduced when temperature increases because of thermal vibrations, or when a high density of microdefects is present in the crystal.

IV.- BOUNDARY CONDITIONS FOR AN INCIDENT PLANE WAVE - RELATION BETWEEN GEOMETRICAL AND DYNAMICAL THEORIES

There are two boundary conditions at the surfaces of the crystal which relate the waves inside and outside the crystal :

1/ - continuity of the tangential components of the wave vectors
2/ - continuity of the tangential components of the electric field and of the normal components of the electric displacement.

The first condition provides a geometrical construction to determine the wave fields which are actually excited inside the crystal by the incident wave : their tie-points lie at the intersections of the dispersion surface and the normal to the crystal surface drawn from the extremity of the wave vector in vacuum.

In transmission geometry (LAUE case), these tie-points lie on the two different branches of the dispersion surface and two wave fields are excited which propagate through the crystal. These wave fields interfere giving rise to fringes which were called "Pendellösung" fringes by P.P. EWALD because they are associated with periodic exchanges of energy between the refracted and reflected components of the wave-fields. The period of these fringes in direct space is equal to the parameter Λ_o defined by equation (21) multiplied by a geometrical factor depending on the relative orientation of the normal to the crystal surface and the reflecting planes.

In reflection geometry (BRAGG case), the normal to the crystal surface intersects the dispersion surface either at two real points of the same branch or at two imaginary points. Two cases are to be considered : thick and thin crystals. In the thick crystal case, it can be shown that only one intersection point should be taken into account and there is total reflection of the incident wave. The width of the rocking curve is inversely proportional to $k \Lambda_o$ and, from equation (21), proportional to $\lambda^2 F_{hk\ell}$. In the thin crystal case, both intersection points must be taken into acount and interferences occur between the corresponding wave fields. The total reflection domain disappears and oscillations appear in the rocking curve. The shape and width of the rocking curve tend asymptotically towards those given by the geometrical theory as the ratio t/Λ_o becomes much smaller than one (t is the crystal thickness). Equation (21) shows that this ratio is proportional to $t \lambda F_{hk\ell}$. For a given crystal thickness and a given reflection, this ratio is very small for a very small wavelength, for instance for γ-rays, and geometrical theory is a good approximation, while, for longer wave lengths such as CuKα, the ratio may be large enough for dynamical theory to be necessary. In a similar way, for a given crystal thickness and a given wave length, dynamical theory may be necessary for the strongest, low order reflections, while geometrical theory is in general a good enough approximation for the weak high order reflections. Since the latter form the majority of reflections used for structure determinations, this explains why geometrical theory suffices for this purpose.

V.- DIFFRACTION OF A SPHERICAL WAVE

The above discussion applies to an incident plane wave. Actually, it is not possible to generate a true plane wave for X-rays. Pseudo plane waves can be obtained with particular settings, but normally, X-rays are produced as spherical waves.

The diffraction of a spherical wave may be treated by generalizing the solutions (16) of the wave equation. This was done by S.TAKAGI (19) who considered that the amplitudes ψ_h are slowly varying functions of position. The fundamental linear equations (18) are then replaced by partial derivative equations which admit analytical solutions for a spherical wave incident on a perfect crystal. Another method was used by N.KATO (10). He summed the diffracted waves associated to the plane wave components of the Fourier expansion of the spherical wave. This approach enables to keep the geometrical description provided by the dispersion surface. Each of its points is now simultaneously excited and wave fields are generated which propagate in all the directions which lie between the incident and reflected directions : they fill out a triangle, called the BORRMANN triangle (11) (fig. 7)

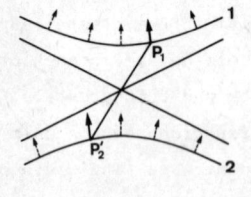

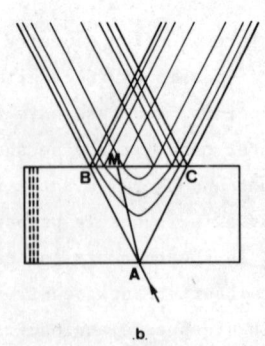

a.
Reciprocal space

b.
Direct space

Fig. 7 - Diffraction of a spherical wave in transmission geometry

Among the optical properties related to the diffraction of a spherical wave, two important ones may be selected out :

1/ - <u>Pendellösung fringes</u> : two wave fields propagate along any path AM within the BORRMANN triangle (fig. 7) These tie-points lie on different branches of the dispersion surface and are actually at the extremities of a diameter (fig. 7).These wave fields give rise to interference fringes first observed by N.KATO and A.R. LANG (12) and interpreted by N.KATO (10). These fringes are only visible for non too absorbing crystals since branch 2 wave-fields are anomalously highly absorbed. They are sensitive to the presence of defects and microdefects and their presence is a good test of crystalline perfection.

2/ - <u>Angular amplification</u> : a narrow wave packet of angular width $\delta\theta$ within the incident wave generates inside the crystal a packet of wave-fields. If the angular width of the paths of these wave fields is $\delta\alpha$, the ratio $\delta\alpha/\delta\theta$, or amplification ratio varies greatly within the BORRMANN triangle : it is equal to one on the sides and becomes very large, of the order of Λ_o/λ in the center. The consequence is that for all the optical problems related to the propagation of wave fields in the central part of the BORRMANN triangle it is Λ_o which should be considered as the wave length rather than λ.

1.- W. ZACHARIASEN, 1945, Theory of X-ray Diffraction in Crystals, J. Wiley (N.Y.)

M. v. LAUE, 1961, Röntgenstrahl-Interferenzen, IIIrd ed. Frankfurt-a-Main

R.W. JAMES, 1963, Solid State Physics, 15, 53

B.W. BATTERMAN and H. COLE, 1964, Rev. Mod. Phys., 36, 681

B.E. WARREN, 1969, X-ray Diffraction, Addison-Wesley

A. AUTHIER, 1970, Advances in Structure Research by Diffraction Methods, 3, 1, ed. Vieweg, Braunschweig

L.V. AZAROFF, R. KAPLOV, N. KATO, R.J. WEISS, A.J.C. WILSON, R.A.YOUNG, 1974, X-ray Diffraction - Mc Graw Hill (N.Y.)

O. BRÜMMER, H. STEPHANIK, 1976, Dynamische Interferenztheorie, Akademie Verlagsgesellschaft, Leipzig

2.- W. FRIEDRICH, P. KNIPPING und M. v. LAUE, 1912, Sitzungsberichte der Bayerische Akademie der Wissenschaften, 303

3.- P.P. EWALD, 1917, Analen der Physik, 54, 519

4.- C.G. DARWING, 1914, Phil. Mag., 27, 675

5.- M. von LAUE, 1931, Ergebnisse exakten Naturwissenschaften, 10, 133

6.- P. SKALICKY, 1979, Imaging Processes and Coherence in Physics, Springer.

7.- F. BALIBAR, 1979, Imaging Processes and Coherence in Physics, Springer.

8.- G. BORRMANN, 1941, Physik Z., 42, 157

G. BORRMANN, 1950, Z. Physik, 127, 297

9.- S. TAKAGI, 1962, Acta Cryst., 15, 1311

10.- N. KATO, 1961, Acta Cryst., 14, 527 and 627

11.- G. BORRMANN, 1959, Beiträge zur Physik und Chemie des 20 Jahrhunderts, Vieweg und Sohn, 262

DYNAMICAL THEORY OF X-RAY PROPAGATION IN DISTORTED CRYSTALS

F. BALIBAR
Laboratoire de Minéralogie-Cristallographie
Université P. et M. Curie, 4 place Jussieu,
75230 PARIS CEDEX 05

The Ewald-Laue theory is exact (it is a solution of Maxwell equations of propagation) ; but its use is limited because it deals with a very special type of experimental situation, seldom encountered in reality, namely :
- an incident plane wave, in vacuum,
- an ideally perfect crystal, extending to infinity in the x and z directions (see fig. 1).

I.- CHARACTERISTIC PARAMETERS

In order to evaluate the degree of ideality (or reality) of these assumptions, it is important to realise what are the orders of magnitude of the characteristic parameters involved in the problem. Bragg diffraction occurs because the crystalline field induces a split of degeneracy in the dispersion relation $E(\vec{K})$. If there were no crystalline field, the dispersion surface would consist of two spheres of radius $K = n/\lambda$, λ being the wave-length in vacuum of the radiation and n the index of refraction, centered on the 2 reciprocal lattice points O and H : the crystalline field introduces a gap in the region where these spheres intersect, of dimension $(K\chi_h)$ (χ_h = h-component of the crystal polarisability).

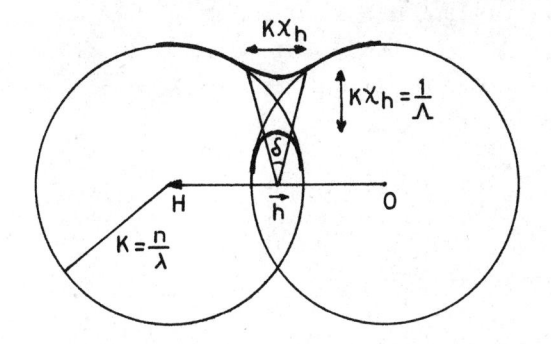

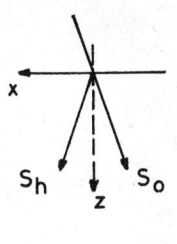

Fig. 1.

($K\chi_h$) is the characteristic dimension of the region of reciprocal space involved. A characteristic length Λ can be defined through the basic relation:

$$\Delta \vec{K} \cdot \Delta \vec{r} \sim 1 \qquad (1)$$

$$\Lambda = 1/(K\chi_h) \qquad (2)$$

Since $\chi_h \sim 10^{-5}$, $\Lambda = 10^{+5} \cdot \lambda \sim 10^5 \text{Å}$, for a typical value $\lambda = 1 \text{ Å}$. It is sometimes convenient to use the angular characteristic parameter:

$$\delta = (K\chi_h)/K \sim 10^{-5} \qquad (3)$$

which is the "width" of the Bragg reflection (or rocking curve).

The physical meaning of the characteristic length Λ is the following:
- along the direction of propagation, Λ is the distance over which appreciable changes in the amplitude occur; it is a "modulation" length;
- perpendicular to that direction (i.e. on a wave-front), Λ is the minimum size of aperture for the wave to propagate without alteration: it is clear from (1), that collimation by a slit smaller than Λ introduces some additional components in the wave-vector spectrum and therefore modifies the structure of the wave.

II.- LANG TECHNIQUE - PLANE WAVE OR SPHERICAL WAVE

Under these circumstances, a real wave can be considered as a plane wave if, and only if, its spread in wave-vector, specified by δK_x and δK_z, is much smaller than ($K\chi_h$):

$$\text{Plane wave} \Leftrightarrow \delta K_x \text{ and } \delta K_z \ll (K\chi_h) \qquad (4)$$

In the most popular imaging technique (the so-called "Lang topography"), the incident wave is produced by the point like focus of an X-Ray tube at a distance L from the crystal; the wave thus emitted is monochromatic and spherical. A plane wave can be extracted out of it by introducing a collimating slit (of size Δx) on the beam trajectory, along the entrance surface of the crystal to be investigated; the wave thus selected is characterized by:

$$\delta K_z \sim 0 \text{ (the incident wave is almost perfectly monochromatic)}$$

$$\delta K_x \sim K (\Delta x/L) \qquad (5)$$

From (4), it is clear that this wave cannot be considered as "plane" unless :

$$K \Delta x/L \ll K\chi_h \qquad (6)$$

a condition which imposes an upper limit to the width of the slit :

$$\Delta x \ll L \chi_h \qquad (7)$$

i.e., for a typical value $L = 10$ cm : $\Delta x \ll 10^4$ Å, a value which is less than the characteristic length Λ (10^5 Å). Such a collimated wave, though "plane" at the entrance surface, does not propagate as a "plane" wave in the crystal ; its wave-vector distribution suffers an inescapable spread in K_x, which can be easily evaluated, from (1) and (7) :

$$\Delta K_x > 1/(L\chi_h) \simeq 10 \; (K\chi_h) \qquad (8)$$

We therefore come to the conclusion that the ideal plane wave of the Ewald-Laue theory has no reality whatsoever in Lang topography. As a matter of fact it would be more relevant to speak of a "spherical" wave, since the spread in K_x extends on the whole range (and even more) of K_x available in the problem.

III.- RAY THEORY FOR A PERFECT CRYSTAL (wave optics based on ray trajectories)

It is remarkable that the amplitude of this "spherical" wave can be calculated in the conceptual frame of the Ewald-Laue theory... once it has propagated far enough in the crystal (still assumed to be perfect).

Let us expand the wave entering in the crystal as a superposition of plane waves, each being specified by its departure from the exact Bragg angle $\Delta\theta$. Each component excites 2 Bloch waves in the crystal (represented by 2 points P_1 and P_2 on each branch of the dispersion surface), according to the wave matching condition at the entrance surface (fig. 2).

At a given point Q in the crystal, the phase and amplitude of these Bloch waves are known from the Laue theory and the amplitude at Q due to the considered incident wave is just (!) the sum of all of them, with their proper phase relationships.

Such a calculation would be very complicated if, precisely, the spread ΔK_x were not so large ; under these circumstances, the range of phases involved in the summation,which is just the product of ΔK_x by the lateral width Λ of the beam, is larger than 1. Which means that the contribution of any component in the summation can be cancelled out by that of another one, so that the total amplitude finally adds to zero. This is true for all components, except those for which the phase is stationary . Imagine that some value of the deviation parameter $\Delta\theta$, call it $\overline{\Delta\theta}$, makes the phase of the Ewald-Laue theory stationary ; then those components which correspond to a value of $\Delta\theta$ very close to $\overline{\Delta\theta}$ are nearly in phase and add constructively and the only contribution to the amplitude at Q comes from those components which correspond to a value of $\Delta\theta$ close to $\overline{\Delta\theta}$. The calculation which looked, at first sight, very complicated, becomes much simpler.

It has been performed exactly by Kato in the case of a "perfect" spherical wave, i.e. assuming that the entrance slit is restricted to a single point $\Delta x=0$. The result (see fig. 2) is that only 2 wave-fields contribute to the amplitude at a given point Q, those which propagate along the OQ direction.

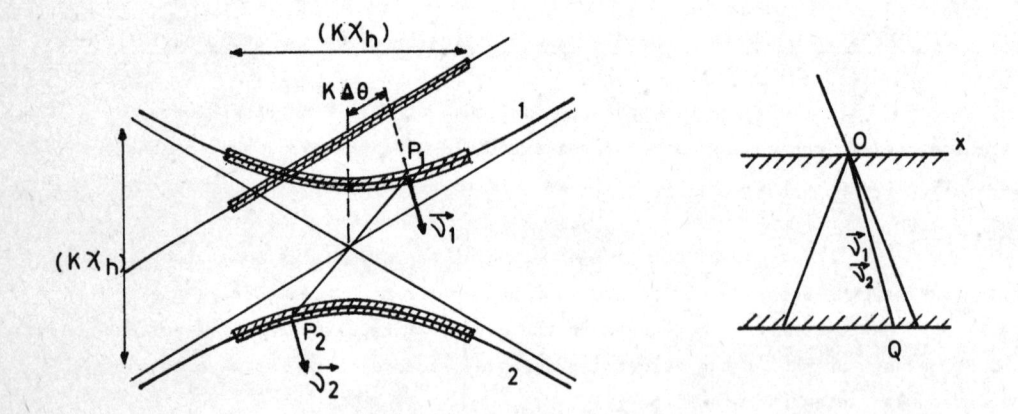

Fig. 2

Up to now, we have assumed a perfect crystal. As a matter of fact, Kato's "ray" theory can be easily extended in order to include the case of a crystal containing some kind of <u>planar defect</u>, a stacking fault, for instance (fig. 3).

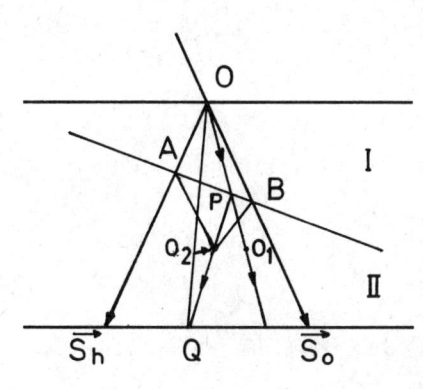

Let P be a point on the fault line AB ; the energy arriving at P travels along OP in part I of the crystal. Viewed from part II this same point P can be reached either along O_1P parallel to OP or O_2P, O_2 being such that $O_2B//OA$ and $O_2A//OB$. This means that at a given point Q on the exit surface, the amplitude is the sum of that corresponding to the direct trajectory OQ and that corresponding to the indirect trajectory OPQ. The exact calculation of the amplitude distribution on the exit surface of a crystal containing a stacking fault has been performed by several authors [2] and is in good agreement with the experimental observations.

Fig. 3

IV.- DISTORTED CRYSTALS - THE EIKONAL THEORY

The classical theory is not restricted to planar defects and can be extended in order to account for the contrast of other types of defects. Nevertheless, if we try to apply the technique of the previous section without any change (i.e. expand the incident wave in plane components and then make an argument about the real trajectory being that for which constructive interference between neighbouring paths occurs), then we come across a difficulty which is that the Ewald-Laue theory cannot be applied to the calculation of the phase along a given path, since this theory assumes a perfect and infinite crystal, which is no more the case. Generally speaking, it is well known that plane wave (Fourier) analysis is well adapted only to those systems which exhibit invariant translational properties (in space, or in time, for instance). Such an invariance exists for a perfect and infinite crystal and this is the reason why the (plane) Bloch-waves of the Ewald-Laue theory are the "normal" modes of the problem in that case ; but when the presence of a defect breaks this translational invariance, the plane wave analysis looses its relevance.

The situation here is very similar to that encountered in ordinary optics [3]. When the medium in which the light propagates is characterized by a varying index of refraction, the plane waves

$$A \exp 2\pi i \, \vec{k}.\vec{r} \quad (\vec{k} \text{ fixed and } A = Cste) \qquad (10)$$

are no more the "normal modes" of the problem and have to be replaced by "modified plane waves" of the type :

$$A(\vec{r}) \exp 2\pi i \, S(\vec{r}) \qquad (11)$$

where $A(\vec{r})$ is a slowly varying function of $\vec{r}$ and $S(\vec{r})$ is called the <u>Eikonal</u> function. A local wave-vector can then be defined through the relation

$$\vec{k}(\vec{r}) = \overrightarrow{\text{grad}} \, S(\vec{r}) \qquad (12)$$

Applying the same type of argument to the propagation in a non-perfect crystalline medium, we replace the Bloch-waves by "modified Bloch-waves" [4] :

$$A_o(\vec{r}) \exp 2\pi i \, S_o(\vec{r}) + A_h(\vec{r}) \exp 2\pi i \, S_h(\vec{r}) \qquad (13)$$

and define two local wave-vectors :

$$\vec{K}_o(\vec{r}) = \overrightarrow{\text{grad}} \, S_o(\vec{r}) \text{ and } \vec{K}_h(\vec{r}) = \overrightarrow{\text{grad}} \, S_h(\vec{r}) \qquad (14)$$

such that

$$\vec{K}_h(\vec{r}) = \vec{K}_o(\vec{r}) + \vec{h}(\vec{r}) \qquad (15)$$

$\vec{h}(\vec{r})$ is the so-called local reciprocal lattice vector ; it depends on the atomic displacement $\vec{u}(\vec{r})$ due to the defects and is related to the reciprocal lattice vector of the perfect crystal $\vec{h}$:

$$\vec{h}(\vec{r}) = \vec{h} - \vec{\nabla}(\vec{h}.\vec{u}(\vec{r})) \qquad (16)$$

With these definitions, it is then possible to calculate the phase along a given path and determine the real trajectory by a condition of stationarity along the same lines as those developped in the previous section. As a matter of fact, this procedure is just an extension of Fermat's principle to a crystalline medium. The main result is that the two wave-fields which would propagate along the same direction in the perfect crystal (see fig.2) now separate ; since their separation depends on $\vec{u}(\vec{r})$, and thus on the local defects, it is clear that the changes in amplitude distribution on the exit surface give some information on the defects themselves ; this is commented by C. Malgrange in this issue (5).

V.- LIMITS OF VALIDITY OF THE EIKONAL THEORY - GENERAL TREATMENT

V.1.- In order for the Eikonal approximation (13), and its consequence (15), to be valid, it is necessary that :

$$\left(\begin{array}{l}\text{the variation } \Delta_\Lambda \vec{K}_h \text{ of } \vec{K}_h \text{ over} \\ \text{a segment } \Lambda \text{ of ray trajectory}\end{array}\right) << \left(\begin{array}{l}\text{characteristic length} \\ \text{in } \vec{K} \text{ space: } (K\chi_h)\end{array}\right) \quad (17)$$

The left hand side of this inequality involves the derivative of $\vec{K}_h$; since $\vec{K}_h$ itself involves (see (16)) the first derivative of $\vec{h}.\vec{u}(\vec{r})$, it is clear that the left side of (17) contains the second derivative of $\vec{h}.\vec{u}(\vec{r})$. Exact calculation (6) leads to :

$$\frac{\partial^2[\vec{h}.\vec{u}(\vec{r})]}{\partial s_o \, \partial s_h} \cdot \Lambda << K\chi_h \quad (18)$$

($\vec{s}_o$, $\vec{s}_h$: unit vectors along the incident and reflected direction).

This is usually written in the form :

$$f = \frac{\partial^2[\vec{h}.\vec{u}(\vec{r})]}{\partial s_o \, \partial s_h} \cdot \Lambda^2 << 1 \quad (19)$$

Condition (17) can also be expressed in an alternative and equivalent manner as :

$$|\Delta_\Lambda \vec{K}_h / K| << \chi_h \quad (20)$$

Since $|\Delta_\Lambda \vec{K}_h / K|$ is just the disorientation of the reflecting planes over a distance Λ, the conclusion is that the Eikonal approximation holds only as long as this disorientation is less than δ, the width of the rocking curve.

V.2.- When condition (17) is not fullfilled, neither the crystal wave nor the normal modes have something to do with plane waves. Rather than find out, for each type of deformation, the new normal modes along which the amplitude $\psi(\vec{r})$

should be expanded, it is more rational to solve Maxwell equations of propagation directly in the non-perfect crystal. Such a theory has been developped by Takagi [7] and is, at the present time, the basic and general theory of X-Ray propagation in crystals.

In this treatment, nothing particular is assumed concerning the structure of the crystal wave, except that it has an "o" and an "h"-component. $\psi(\vec{r})$ is written as :

$$\psi(\vec{r}) = \sum_h \psi_h(\vec{r}) \exp - i\vec{k}_h \cdot \vec{r} ; (h = o, h) \qquad (21)$$

$\vec{k}_h \cdot \vec{r}$ represents a fast oscillation of constant periodicity λ , (wavelength in the perfect crystal) ; superimposed on that fast oscillation, $\psi(\vec{r})$ exhibits a modulation of its amplitude, extending over a range of order Λ and depending on the local state of deformation. Introducing (21) in Maxwell equations, Takagi obtains a set of partial differential equations of second order and hyperbolic type, which can be solved (in principle) by the Green function technique.

V.3.- The Green function (or propagator) $G(O,Q)$ represents the effect of a unique point source at a further point Q ; since the total amplitude at Q, is the superposition of all the wavelets emitted by the different field sources (Huyghens's principle), it can be calculated by performing the convolution product of the source distribution by the Green function.

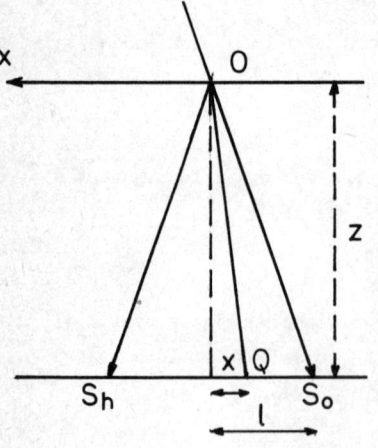

It is enlightening to see how the Ewald-Laue theory comes out of this general treatment as a special case (perfect crystal and plane incident wave). Let $\lambda_x = 2\pi/K_x$ be the periodicity of a plane incident wave along the entrance surface Ox ; in the present case, the source distribution consists of all the points of this entrance surface which are "lightened" by the incident wave (each of them acts like a point source). On the other hand, Takagi has shown that the Green function corresponding to the propagation in a perfect crystal is :

$$G(O,Q) = J_o\left[\frac{z}{\Lambda} \sqrt{1-(x^2/l^2)} \right] \quad \text{(see fig.4)}.$$

Fig. 4

At a given depth z in the crystal, this Green function exhibits oscillations of variable "wave-length" $\lambda'(x)$ (fig. 5). Convolution of this Green function with the periodic source distribution on the entrance surface, will amount to zero except at those points Q_1 and Q_2 where $\lambda_x = \lambda'(x)$. These points correspond to the two wave-fields of the Ewald-Laue theory.

Note that the Green function for a perfect crystal is just the amplitude distribution induced in the crystal by a "perfect" spherical wave, already calculated by Kato (see § III).

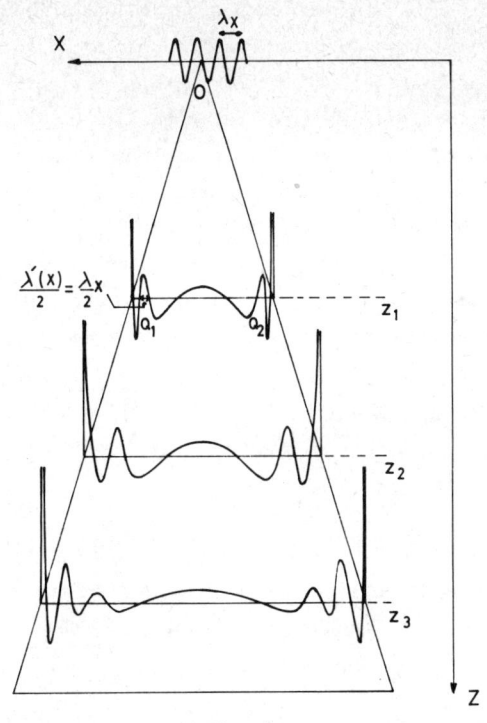

Fig. 5

V.4.- For a non perfect crystal, the determination of the Green function is not so easy as in the case of a perfect crystal ; as a matter of fact, it is a problem still to be solved. Nevertheless, it can be shown[6] (on the basis of the mathematical theory of distributions[8] that it is not necessary to calculate the proper Green function for each case of imperfection and that the Green function J_o of the perfect case is sufficient, at least for a first order approximation. It can be shown, that the crystal wave can be calculated by assuming a perfect crystal and replacing the deformations by an extra distribution of fictive sources which are to account for the local deformations[6].

As an example of this technique, let us examine the effect on

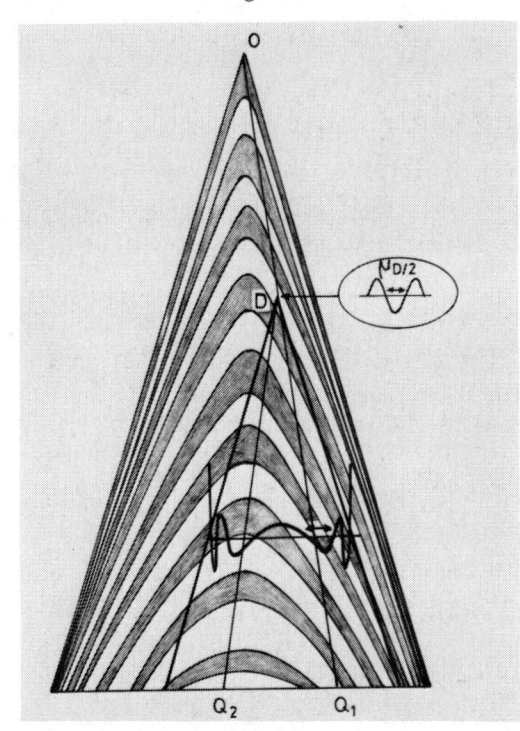

Fig. 6

a section topograph (i.e. assuming a "perfect" spherical wave at the entrance surface) of the region which lies in the immediate vicinity (by this we mean at a distance $<\Lambda$) of the core D of a dislocation line. It has been shown that the strength of the equivalent "fictive" sources depends both on the local deformation and on the amplitude which would exist at the considered point if the crystal were perfect. It is clear on fig. 6 that the effect of the dislocation will be more or less sensible according to whether D lies inside a dark fringe or right in the middle of a bright one. In the present case, it is even possible to predict along which direction the disturbance in amplitude will be the most sensible. The argument is of the same type as in section V.3 : the distribution of fictive sources around D can be characterized by a certain periodicity, call it μ_D. Convoluting this distribution by the Green function J_o will give a null result except along those lines (originating at D) where μ_D matches the periodicity of J_o, i.e. along two lines DQ_1 and DQ_2. In other words, we expect that two wave-fields will stem out of the highly distorted region around D.

Since it would seem that an "extra" wave-field has been created by the distorted region, this phenomenon is generally referred to as that of the "creation of new wave-fields" ; it is of great importance in the interpretation of the contrast of isolated defects such as dislocations.

REFERENCES

(1) N. KATO - Acta Cryst. 14, 526 (1961)
               14, 627 (1961)

(2) A. AUTHIER, D. MILNE and M. SAUVAGE - Phys. Stat. Solidi, 27, 77 (1968).

   N. KATO - X-Ray Diffraction edited by L.V. Azaroff, Mc Graw-Hill, New-York (1974), ch. 5.

(3) A. SOMMERFELD - Lectures on theoretical Physics, Vol. IV Optics, Academic Press, N.Y. (1964).

(4) N. KATO - J. Phys. Soc. Japan, 18, 1785 (1963)
   N. KATO - (ibid.) 19, 67 (1964)
   N. KATO - (ibid.) 19, 971 (1964)
   see also P. PENNING and D. POLDER, Philips Res. Rep., 16, 419 (1961).

(5) C. MALGRANGE in "Imaging Processes and Coherence in Physics", Springer (1979).

(6) A. AUTHIER, F. BALIBAR, Acta Cryst., A26, 647 (1970).

(7) S. TAKAGI, J. Phys. Soc. Japan, 27, 1239 (1969).

(8) L. SCHWARZ - Théorie des distributions, Herman Paris (1950).

POLARIZATION PHENOMENA IN X-RAY DIFFRACTION

P. Skalicky

Institute of Applied Physics
Technical University, Vienna

INTRODUCTION

The dielectric properties of most materials in the X-ray frequency region are such that the polarization phenomena which are observed for visible light (such as birefringence and optical activity) are not found. This is due to the fact that K or L electrons which are important in X-ray scattering are shielded from crystalline fields. The problem was already considered theoretically as early as 1939 by Moliere (1,2) who concluded that polarization phenomena are expected to be very small exept very close to an absorption edge. Hojo, Ohtsuki & Yanagawa (3) showed that what they call "polarization mixing" is expected to be theoretically exactly zero as long as hydrogen - like eigenfunctions are used to describe the scattering of X-rays.

The situation is different for a diffracting crystal. The Thomson electron scattering depends on the relationship between the polarization vector and the wave vectors. We can therefore expect that polarization phenomena would be marked when elastic (Bragg)-scattering is important.

THEORY

The influence of the dielectric properties of crystals on the polarization of the crystal waves was explored in detail in two papers by Moliere (1,2). The results shall be briefly discussed here. The fundamental equations of the dynamical theory which is discussed by Authier (4) have the following form

(1) $$\left(1 - \frac{k^2}{K_m^2}\right)\vec{D}_m = \sum_q \left[C_{mq}\vec{D}_q\right]_{\perp \vec{K}_m}$$

The C_{mq} have generally tensor character (in x,y,z, not in m,q!). They consist of the Fourier coefficients of the scalar dielectric susceptibility χ_{m-q} and a complex term A_{mq} which can be split into a scattering and an absorption term:

(2) $$C_{mq} = \chi_{m-q} + A_{mq}^{scat} + A_{mq}^{abs}$$

These A_{mq} terms become important only near an absorption edge. They have tensor character and thus determine the tensor properties of C_{mq}. For X-ray frequencies far from an absorption edge A_{mq} is neglegibly small and C_{mq} is equal to the Fourier coefficient of the dielectric susceptibility χ.

Following Moliere (1,2), we introduce N different coordinate systems (one for each beam) for a "mixed representation" of the tensor C_{mq}. The mth coordinate system has its z-axis parallel to the wave vector $\vec{K}_m$. The amplitudes $\vec{D}_m$ are perpendicular to the corresponding $\vec{K}_m$. Taking the index $\vec{K}_m$ on the right hand side of equation (1) into account, we obtain a two-dimensional reduced tensor C'_{mq} which contains no z-components. With the introduction of the excitation errors ε_m as usual (4):

(3) $$2\varepsilon_m = \left(1 - \frac{k^2}{K_m^2}\right)$$

a system of two dimensional matrix equations is obtained

(4) $$\sum_{q}^{N} \{C_{mq} - 2\varepsilon_m \delta_{mq}\} \vec{D}_q = 0$$

where δ_{mq} is Kronecker's delta. This can alternatively be written as a 2N dimensional matrix equation

(5) $$([C] - 2[\varepsilon])\vec{D} = 0$$

where $[\varepsilon]$ contains the excitation errors and the two-dimensional unitary matrices E_2:

(6) $$[\varepsilon] = \begin{bmatrix} \varepsilon_o E_2 & 0 & 0 & \cdot \\ 0 & \varepsilon_m E_2 & 0 & \cdot \\ 0 & 0 & \varepsilon_q E_2 & \cdot \\ \cdot & \cdot & \cdot & \end{bmatrix}$$

Equation (5) can be regarded as an eigenvalue problem. The 2N-dimensional eigenvectors $\vec{D}$ contain the components of the electric vectors of the crystal waves. Their polarization depends on the properties of the matrix $[C]$. In the most general case, this is a non-hermitean matrix and the system of crystal waves is elliptically polarized with no definite relationship between the axes and the eccentricities

of the ellipses. When the crystal has a center of symmetry and no
absorption the matrix [C] is real and symmetric and the system of
crystal waves is linearly polarized with mutually perpendicular planes of polarization. This is the case which is generally assumed in
the dynamical theory of diffraction. The following Table 1 summarizes
the polarization of the crystal waves as a function of the properties
of the matrix [C].

Table 1

center of symmetry	no center
$A_{oo}^{abs} = 0$	
[C] real, symmetric $\vec{D}_{o1} = \begin{pmatrix} a \\ 0 \end{pmatrix}$ $\vec{D}_{o2} = \begin{pmatrix} 0 \\ a \end{pmatrix}$ linearly polarized	[C] complex, hermitean $\vec{D}_{o1} = \begin{pmatrix} a \\ ib \end{pmatrix}$ $\vec{D}_{o2} = \begin{pmatrix} ib \\ a \end{pmatrix}$ elliptically polarized
$A_{oo}^{abs} \neq 0$	
[C] A_{oo}^{scat} + A_{oo}^{abs} ↓ ↓ real imaginary symm. symm. $\vec{D}_{o1} = \begin{pmatrix} a \\ ib \end{pmatrix}$ $\vec{D}_{o2} = \begin{pmatrix} -ib \\ a \end{pmatrix}$	[C] complex, non-hermitean $\vec{D}$ elliptically polarized no relation between a,b

For simple transmission (no Bragg reflection excited) the optical
properties of the crystal are determined by the quantity $\chi_o + A_{oo}$.

Following Moliere (1,2) we can better understand the nature
of the terms making up the [C] matrix when we consider the electric
polarization $\vec{P}(\vec{r},t)$ in the crystal which is induced by the X-ray
field $\vec{E}(\vec{r},t)$. It has the following form:

(7) $$\vec{P}(\vec{r},t) = \int_{crystal} \left[\chi(\vec{r}') \cdot \delta(\vec{r}-\vec{r}') + \underline{A}(\vec{r},\vec{r}') \right] \vec{E}(\vec{r}',t) \, dV$$

The quantity $\underline{A}(\vec{r},\vec{r}')$ has tensor character and its Fourier coefficients are the A_{mq} in equation (2). From this equation we see that the
polarization $\vec{P}(\vec{r},t)$ depends not only on the electric field at the

same position but depends on the electric field throughout the crystal. These non-local terms cause non-diagonal elements in the matrix [C] and consequently complex eigenvectors which describe polarization phenomena. The quantity $\underline{A}(\vec{r},\vec{r}')$ has the following form:

$$(8) \quad \underline{A}(\vec{r},\vec{r}') = \frac{1}{\varepsilon_o} \frac{e^2}{hmc^2} \frac{\lambda^2}{4\pi^2} \sum_n \left[\frac{\{\vec{j}_{on}(\vec{r}) \cdot \vec{j}_{on}^+(\vec{r}')\}}{\gamma - \gamma_{no}} + \frac{\{\vec{j}_{on}^+(\vec{r}) \cdot \vec{j}_{on}(\vec{r}')\}}{\gamma + \gamma_{no}} \right]$$

where the $\vec{j}_{on}$ are the current matrix elements between the unperturbed $|0\rangle$ state and the various other states $|n\rangle$ of the many electron solid. The $\{\}$ brackets denote tensor products. The γ_{no} are the resonance frequencies. There are some important points to the behaviour of $\underline{A}(\vec{r},\vec{r}')$. With increasing frequency $\underline{A}$ falls off as γ^{-3} whereas χ falls off as γ^{-2}. From a certain γ_{no} we can therefore neglect A_{mq} in equn.(2). Also, $\underline{A}$ falls off very rapidly with increasing difference between $\vec{r}$ and $\vec{r}'$, having nearly the character of a delta function such as the first term in equn.(7). This would cause $\underline{A}$ to loose its tensor character and to become essentially a scalar correction term for χ_{m-q}. The reason for this behaviour is that the current (indicated by $\vec{j}_{on}$) which flows in atom Nr. i when an electron in its K-shell is excited flows to a good approximation only in that one atom, causing the non-local character of the second term in eqn.(7) to be small. The argument is similar to the one that justifies the use of the random phase approximation in the treatment of many electron systems which neglects, in the electron field and electron electron interactions, the coupling between interactions of different momentum transfers (5,6). Since K or L electrons (which are important at X-ray frequencies) are rather well shielded from crystalline fields we can expect this to be the case in X-ray scattering.

It is therefore not surprising that up to now no polarization effects have been found for X-rays in simple transmission. Careful measurements by Hart (7) performed for $\lambda = 1,54 \text{ Å}$ were able to detect neither significant optical rotatory power, nor circular dichroism in quartz, nor linear dichroism in sheet polaroid.

BRAGG REFLECTING CRYSTALS

Even if the matrix [C] is real and symmetric (Table 1) a wide variety of polarization phenomena exists in the case where a strong Bragg reflection is excited. We restrict the discussion to the well-known two-beam case (4). Figure 1 shows the four-branch dispersion surface for this case which lies in real k-space when absorption is neglected. The excitation points (tie-points) for waves with their electric vectors perpendicular and parallel to the plane of incidence (the plane of the drawing) lie on the σ and π -branches, respectively. The important points are that the wave field propagation (wave fields (1) and (2)) is perpendicular to the dispersion surface at the tie-point of the wave vectors and that the absorption is different for both wave fields (1) and (2) and for both states of polarization. A diffracting crystal will therefore exhibit linear dichroisms (8,9). It is also birefringent as was shown in ray tracing experiments. At the exit surface of the crystal we expect a fringe system (the "Pendellösung-fringes") for each state of polarization due to the wave vector difference $\Delta \vec{k}_{1,2}$. When natural light (i.e. light where all states of polarization are equally present) is diffracted this will result in the superposition of two fringe systems with different spacings Λ_σ and Λ_π for the two polarization states, respectively. This causes a "beat-pattern" or a periodic fading of the fringes which was first explained by Hart and Lang (11) and by Hattori et.al.(12) for spherical waves. The amplitude of the fringes is modulated by the beat of the two patterns with the factor

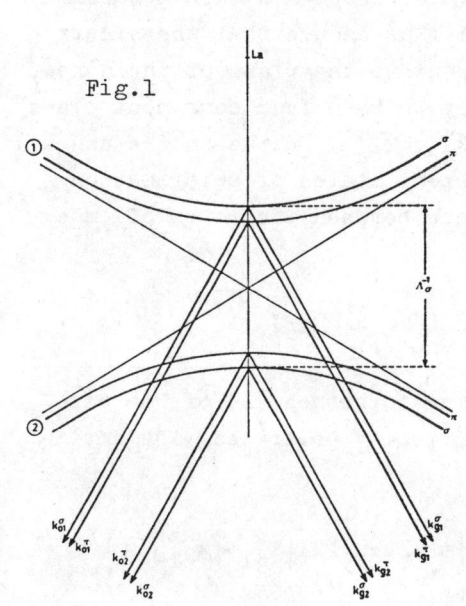

(9) $$\cos \left\{ \pi \left(\frac{1}{\Lambda_\sigma} - \frac{1}{\Lambda_\pi} \right) \cdot t \right\}$$

where t is the crystal thickness. For the exact Bragg condition and the symmetrical Laue-case, the wave vector difference is given by

Fig.1

$$(\vec{k}_{g1} - \vec{k}_{g2})_{\sigma,\pi} = \frac{k}{\cos\theta} C |\chi_g| = \frac{1}{\Lambda_{\sigma,\pi}} \tag{10}$$

with the polarization factor C which is 1 for σ-polarization and $C = \cos 2\theta$ for the π-polarization. The quantity χ_g is again the Fourier coefficient of the susceptibility χ. Both the transmitted and the diffracted waves at the exit surface remain unpolarized.

When the incident wave is already linearly polarized we must consider a possible interference between all crystal waves which are now coherent. As was shown by Skalicky and Malgrange (9) the crystal will produce elliptically polarized crystal waves and turn the plane of polarization of the incident wave. Let us assume that the electrical vector is inclined at 45° with respect to the plane of incidence. If we consider the diffracted waves only we have four component plane waves with the wave vectors $\vec{k}_{g1}^{\sigma}$, $\vec{k}_{g1}^{\pi}$, $\vec{k}_{g2}^{\sigma}$, $\vec{k}_{g2}^{\pi}$. Unlike in the case of natural light these waves with different states of polarization are not independent. The phase difference between the waves of index (1) is given by

$$2\pi(\vec{k}_{g1}^{\sigma} - \vec{k}_{g1}^{\pi})\cdot\vec{r} = \pi(\frac{1}{\Lambda_{\sigma}} - \frac{1}{\Lambda_{\pi}})\cdot t \tag{11}$$

This means that the wave fields belonging to branch (1) of the dispersion surface will generally be elliptically polarized. The amplitude ratio is

$$\frac{\vec{D}_{g1}^{\sigma}}{\vec{D}_{g1}^{\pi}} = \frac{a_1^{\sigma}}{a_1^{\pi}} \cdot \exp(-\frac{1}{2}[\mu_1^{\sigma} - \mu_1^{\pi}]t)\cdot \exp(2\pi i[k_{g1}^{\sigma} - k_{g1}^{\pi}]\cdot t) \tag{12}$$

where μ_1^{σ} and μ_1^{π} are the absorption coefficients for the two states of polarization. The ratio of the amplitudes of the plane wave components is $a_1^{\sigma}/a_1^{\pi} = a_2^{\sigma}/a_2^{\pi} = \cos 2\theta$. The wave fields belonging to branch (1) of the dispersion surface are of right hand polarization while the branch (2) waves are of opposite hand. The state of polarization of the crystal waves and the result of their interference are schematically summarized in Fig.2. We can introduce a polarization periodicity Ω. It is given by:

$$\frac{1}{\Omega} = \frac{1}{2}(1 - |\cos 2\theta|)\frac{1}{\Lambda_{\sigma}} \tag{13}$$

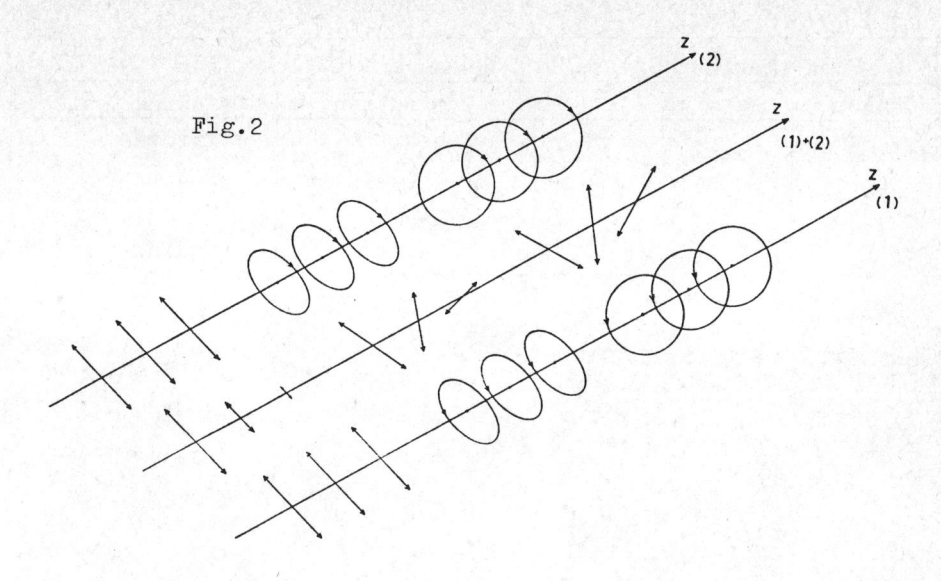

Fig.2

The phase difference between the σ and the π components can thus be expressed as $\exp(2\pi i t/\Omega)$. This means that the same state of polarization is repeated with a depth periodicity of Ω. Circular polarization occurs for a thickness of $\Omega/4 + n\Omega/2$. A crystal with such a thickness is therefore a quarter wave plate for both wave fields (1) and (2). The thickness of a silicon quarter wave plate for copper K_α radiation using the 220 Bragg reflection is 24,2 µm.

If the two wave fields (1) and (2) overlap, they will interfere and give fringes at the exit surface. The interference between two elliptically polarized waves with the same amplitude and eccentricity but opposite sense of turning gives a linearly polarized wave. The amplitude and orientation of this linearly polarized resultant wave depends on the phase difference between the two elliptically polarized waves. The contrast of the fringes depends therefore on the state of polarization. It is strongest where both wave fields have linear polarization and fading is expected where both wave fields have circular polarization. The results are summarized in Table 2 and Fig.2.

An elegant way of doing experiments with polarized X-rays is the use of synchrotron radiation which has been used to verify the above results (13). The advantage is a high intensity for any given orientation of the plane of polarization with respect to the plane of incidence which is always a problem in many crystal settings (7,9). The two photographs in Fig.3 were both recorded with the same exposure time of 2 min at the D.C.I. storage ring at L.U.R.E. (Orsay, France).

Table 2 Crystal thickness and polarization			
thickness	$\delta = |k^\pi - k^\sigma|$	polarization	fading
0	0	linear	fringes
$\Omega/4$	$\pi/2$	circular	fading
$\Omega/2$	π	linear	fringes
$3\Omega/4$	$3\pi/2$	circular	fading

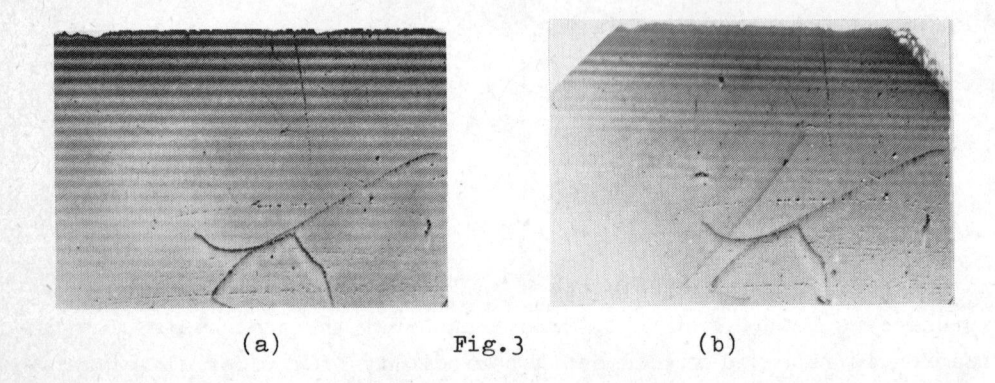

(a) Fig.3 (b)

A silicon wedge shaped crystal was oriented for a 220 reflection of $\lambda = 0,9 Å$ with the plane of polarization normal to the plane of incidence (Fig.3a) and inclined at 45° with respect to the plane of incidence at nearly the same wavelength $\lambda = 1,1 Å$ (Fig.3b). According to the theory a periodic fading of the fringes separated by 5,2 fringes is observed thus indicating the excitation of component waves on all four branches of the dispersion surface.

When a beam of circularly polarized X-rays is desired, the diffracting crystal must either be very thick and set away from the exact Bragg condition so that the wave fields (1) and (2) will propagate in different directions which are given by the normals to the dispersion surface or the Borrmann effect may be employed to damp wave field (2) which has the higher absorption, out. Such a setting was used by Hart (7) with a three quarter wave plate which is superior to the quarter wave plate because it is thicker.

In principle thus all polarization experiments which are possible with visible light can be performed with X-rays with the help of diffracting crystals.

REFERENCES

1. G. Molière, Ann.Phys. 35, 272, (1939)
2. G. Molière, Ann.Phys. 35, 297, (1939)
3. A. Hojo, Y.H. Ohtsuki & S.Yanagawa, J.Phys.Soc.Japan, 21, 2082 (1966)
4. A. Authier, Imaging Processes and Coherence in Physics, Springer: (1979)
5. M.von Laue, Röntgenstrahlinterferenzen, Frankfurt: Akademische Verlagsgesellschaft, (1960)
6. H.Y. Fan, Photon Electron interaction, Handbuch der Physik XXV/2a, Springer (1967)
7. M.Hart, Phil.Mag. 38, 41, (1978)
8. H. Cole, F.W. Chambers & C.G. Wood, J.Appl.Physics 32, 1942, (1961)
9. P. Skalicky and C. Malgrange, Acta crystallogr. A28, 501, (1972)
10. A. Authier, Bull.Soc.fr.Miner.Cristallogr. 84, 51, (1961)
11. M. Hart and A.R. Lang, Acta crystallogr. 19, 73, (1965)
12. H. Hattori, H. Kuriyama & N.Kato, J.Phys.Soc.Japan, 20, 1047, (1965)
13. M. Sauvage, J.F. Petroff & P. Skalicky, phys.stat.sol. (a) 43, 473, (1977)

PERFECT CRYSTAL NEUTRON OPTICS

A. Zeilinger
Atominstitut der Österreichischen Universitäten
A-1020 Wien, Austria
and
Massachusetts Institute of Technology Cambridge, MA 02139, USA

1. Introduction and Fundamentals

In recent years neutron optics started to gain increasing attraction despite the handicap this field experiences because the index-of-refraction (compare /1-3/) of all materials and of magnetic fields for thermal neutrons

$$n = 1 - \frac{1}{2E}(V_N(0) \pm \mu B) \tag{1}$$

is very close to 1. This is because the kinetic energy E of thermal neutrons ($\sim 10^{-2}$ eV) is very much larger than both the mean nuclear interaction potential $V_N(0)$ ($\sim 10^{-8}$ to 10^{-7} eV) and the interaction potential of the neutron magnetic moment μ with a magnetic field B ($6.03 \cdot 10^{-8}$ eV for B = 1 T).

This limitation was surpassed in two different ways. On the one hand the use of very cold and ultra cold neutrons (see /1/) makes it possible to work in a neutron energy range where the index-of-refraction is high.
On the other hand the introduction of perfect crystals allows the construction of optical components which provide large deflection angles even for thermal neutrons. This feature has finally lead to the successful development of neutron interferometers /4/.

The results of the dynamical theory of neutron diffraction /5/ as obtained from the Schroedinger equation turns out to be closely analogous to the X-ray case, even the size of some effects is the same. Thus we can refer to the corresponding articles in this volume /6,7/.

The reason for this feature is, that the time-independent Schroedinger equation is formally quite similar to the Helmholtz-equation one obtains from Maxwell's equations. The analog to the spatially dependent polarizability in the X-ray case is the spatially dependent potential in the neutron case. For more detailed comparisons we refer to review papers on neutron dynamical diffraction /8,9,10/.

The phenomenological peculiarities of the neutron case are
1. With few exceptions, the absorption for neutrons is very low, it is particularly negligible for Si-crystals of usual laboratory thickness.
2. The neutron velocity is low ($\sim$ 1000 m/s for thermal neutrons) as

compared to X-rays.
3. The neutron is a spin-$\frac{1}{2}$ particle with magnetic moment and experiences considerable magnetic interaction.

This latter property leads to nontrivial extensions of the X-ray dynamical theory results /11/, if perfect magnetic crystals are employed. Another extension concerns the incorporation of inelastic neutron scattering into dynamical diffraction theory /12/. The theoretical work mentioned above has in general been restricted to the 2-beam case, but there has already been some work done on the neutron 3-beam case /13/. The present paper will be restricted mainly to those aspects which have found experimental verification.

Two large and important fields of applications, i.e. neutron topography and neutron interferometry are not covered here since they are the topics of two other papers of this volume /14,15/.

2. Laue-Case Reflectivity Experiments

Knowles /16/ was the first one to demonstrate neutron dynamical diffraction effects by showing that the intensity of neutron-capture γ-radiation varies in a characteristic way with a variation close to the exact Bragg-angle of the direction of incidence of the incoming neutrons for $CdSO_4$-crystals. This effect is related to the anomalous transmission phenomenon which is generally encountered in X-ray dynamical diffraction. More experimentation on neutron anomalous transmission was done by Sippel et al. /17/ using InSb-crystals and by Shilstein et al. /18/ using CdS. A rather interesting phenomenon which is characteristic for neutrons and has no X-ray counterpart is anomalous spin-incoherent scattering. This effect was demonstrated by Sippel et al. /19/ for a KH_2PO_4-crystal. In that crystal the planes containing the H-atoms are arranged between the planes containing the K and P atoms. Thus, by varying again the angle of incidence in an angular region close to the exact Bragg-angle the amplitudes of the wave-fields at the positions of the H-atoms can be varied. This leads to a strong variation of the transmitted intensity because H shows a very strong spin-incoherent scattering which reduces the coherent scattering.

The validity of the integrated reflectivity predictions of dynamical diffraction theory for nonabsorbing crystals has also been shown first by Sippel et al. /20/.

3. The Bragg-Case Double-Crystal Diffractometer

For the absorption-free case the reflectivity of a perfect crystal in symmetric Bragg-geometry is obtained to be unity over an angular range

$$\Delta\theta = \frac{2|F_N(\vec{G})|\lambda^2}{\pi v_c \sin(2\theta_B)} \qquad (2)$$

where $F_N(\vec{G})$ is the appropriate structure factor, λ is the neutron wave length, v_c is the volume of the crystal unit cell and θ_B is the Bragg-angle. The reflection width is typically a few seconds of arc, the reflection curve was measured in detail by Kikuta et al. /21/.
One employs this high angle resolution feature for small angle scattering experiments in the double-crystal spectrometer where two perfect crystals in parallel arrangement are used with the sample between them, the angular changes of the beam due to the sample being measured. This arrangement has been used, among others, in experiments on a limit of the neutron charge /22/, on single-slit diffraction of neutrons /23/ and refraction of neutrons at magnetic boundaries /24,25/. In the backscattering spectrometer /26/ the double-crystal arrangement provides an energy resolution of about 10^{-7} eV.

4. Ray-Optical Aspects

In the following we restrict our considerations to the absorption-free symmetric Laue-case in the 2-beam approximation. Then the solutions to the Schroedinger equation are the two wave fields (see e.g. /9/)

$$\psi_{1,2} = u_{1,2}(0) \exp(i\vec{K}_{1,2}\cdot\vec{r}) + u_{1,2}(\vec{G}) \exp[i(\vec{K}_{1,2}+\vec{G})\cdot\vec{r}] \qquad (3)$$

with

$$\vec{K}_{1,2} = \vec{k} + \frac{\pi}{\Delta}(-y \pm \sqrt{y^2+1})\hat{n} - \frac{kV(0)}{2E\cos\theta_B}\hat{n} \qquad (4)$$

$$y =: (\theta_B - \theta)\frac{E}{|V(\vec{G})|}\sin(2\theta_B), \quad \Delta = \lambda\frac{E}{|V(\vec{G})|}\cos\theta_B \qquad (5)$$

$V(\vec{G})$ is the Fourier-component of the neutron-crystal interaction potential and Δ is the pendellösung length. The amplitudes u are also functions of y and given elsewhere /9/. The angle Ω of the mean neutron current with the lattice planes is given through

$$\Gamma \equiv \frac{\tan \Omega_{1,2}}{\tan \theta_B} = \frac{\mp y}{\sqrt{1+y^2}} \qquad (6)$$

This equation defines the Borrmann-fan, which is excited if a fine entrance slit is used (Fig.1). The intensity distribution within the Borrmann-fan,

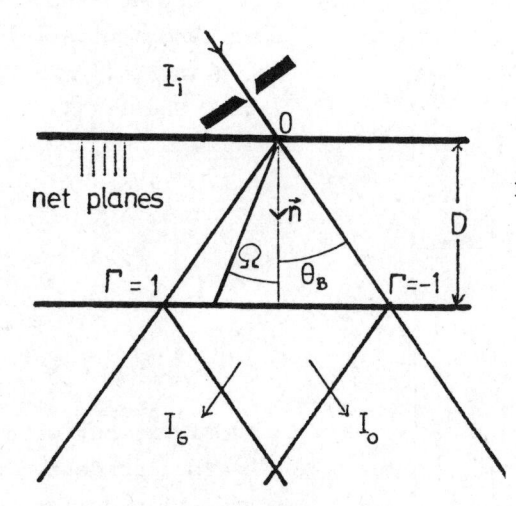

Fig.1: Borrmann-fan neutron propagation in the symmetric Laue-case

the so-called "Kato-profile" was measured in detail by Shull and co-workers /27/. With no coherence between the +y and -y components we obtain

$$I_G(\Gamma) = I_i/2 \sqrt{1-\Gamma^2} , \qquad (7)$$

which, if smeared out with the experimental resolution gives the result of Fig.2, which was obtained with white radiation and poor collimation. If the collimation is improved, the wave length becomes well defined and thus interferences between +y and -y components are observable which are the characteristic spherical wave interference oscillations as shown in Fig.3. This feature was used by Shull et al. for a precision determination of the scattering lengths of Si and Ge.

An interesting question concerns the optical length of the path given by Equ.6, which is different from its geometrical length because the momentum eigenstates $\hbar \vec{K}_{1,2}$ and $\hbar(\vec{K}_{1,2}+\vec{G})$ are not parallel to the path. Thus one defines an infinitesimal zig-zag path (Kato /28/) with components along the momentum eigendirections, whose length is

$$D_{opt} = D/\cos \theta_B \qquad (8)$$

independent of y in the symmetrical Laue-case. This path length was measured in a neutron time-of-flight experiment /29/.

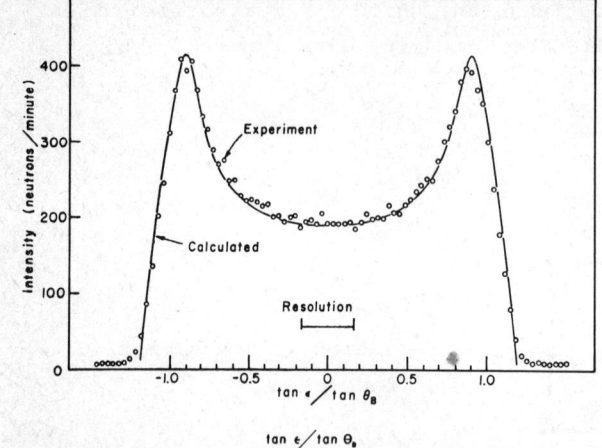

Fig.2: Distribution of the intensity I_G released from the backface of an arrangement as shown in Fig.1 (after Shull /11/).

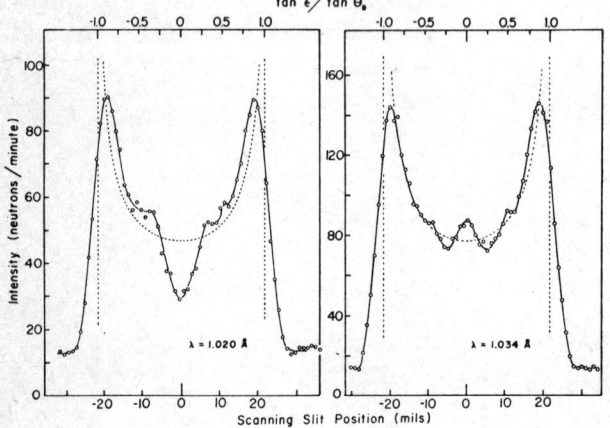

Fig.3: As Fig.2, but with better incident radiation collimation, such that spherical wave interferences occur (after Shull /11/).

5. Applications of Ray Optics

5.1 Diffraction Focussing of Neutrons

There are different types of focussing effects to be expected in dynamical diffraction /30/. One of them occurs if two crystals of equal thickness are arranged parallel in symmetric Laue-geometry, which is most easily seen from symmetry considerations (Fig.4). This double Laue-crystal arrangement is very sensitive to the relative angular position of the crystals /31,32/, because the double crystal rocking curve exhibits a narrow peak of width

$$\delta\theta = \lambda/W \qquad (9)$$

on top of the broader peak whose width is approximately that of Equ.2. As W is the width of the beam between the crystals $\delta\theta$ is typically in the range of a few 10^{-3} arc sec (Fig.5).

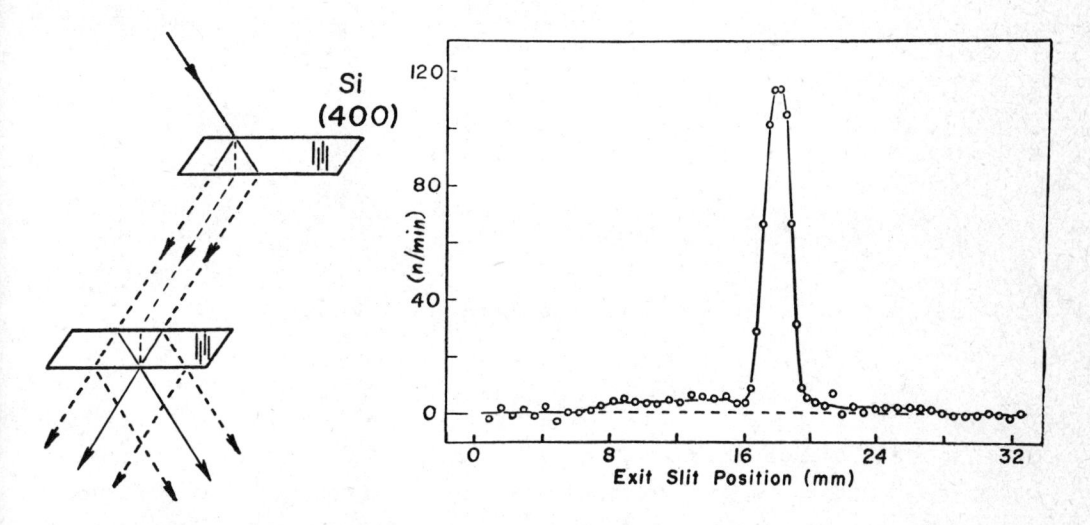

Fig.4: Diffraction Focussing of Neutrons:
Ray paths in a double Laue-case arrangement showing focussing action (left) and intensity released from the backface of the second crystal into the double diffracted direction as scanned with an exit slit (right) /31/.

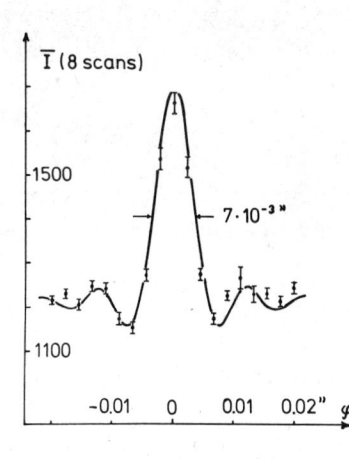

Fig.5: The very narrow central peak occuring in the rocking curve of two Laue-crystals with equal thickness /32/.

The physical reason for this effect is the property that such a double crystal arrangement shows interference action in the central focus /31/ if the crystals are exactly parallel. This constructive interference is deteriorated by optical path length changes due to angular changes, which leads directly to Equ.9. An alternative interpretation can be given in terms of the oscillatory type Laue-case reflection curves /32/.

5.2 The Angle Amplification Feature

For small y we obtain from Equs.5 and 6

$$\Omega \approx (\theta - \theta_B) \frac{E}{|V(\vec{G})|} 2 \sin^2 \theta_B \qquad (10)$$

which for e.g. Si (400)-reflection and $\lambda = 1.865$ Å gives

$$\Omega \approx 4.4 \cdot 10^5 (\theta - \theta_B). \qquad (11)$$

Thus, a small change of the incoming radiation direction results in a large change of in-crystal neutron propagation direction. This feature was first used by Authier /33/ for X-rays and later exploited by Kikuta et al. /21/ to measure the small directional changes of a neutron beam due to prism refraction (Fig.6). In that experiment neutron absorbing slits in front and on the backface of the first crystal plate act as a "crystal-collimator" selecting only those neutrons which are travelling through the crystal along paths close to the lattice planes. These neutrons travel in the second crystal plate again in the same directions. If now the beam is deflected on its way between the two crystals, its path in the second crystal plate splits according to the equations above (Fig.6).

The same phenomenon occurs if not the direction but the wave length of the neutrons is changed on their way between the crystal plates because a change of wave length results in a change of Bragg-angle according to

$$\delta \theta_B = \frac{\delta \lambda}{\lambda} \tan \theta_B. \qquad (12)$$

This feature was utilized /34/ in a direct measurement of the wave length change of a neutron in a magnetic field (Fig.7)

$$\frac{\delta \lambda}{\lambda} = \mp \frac{\mu B}{2E}. \qquad (13)$$

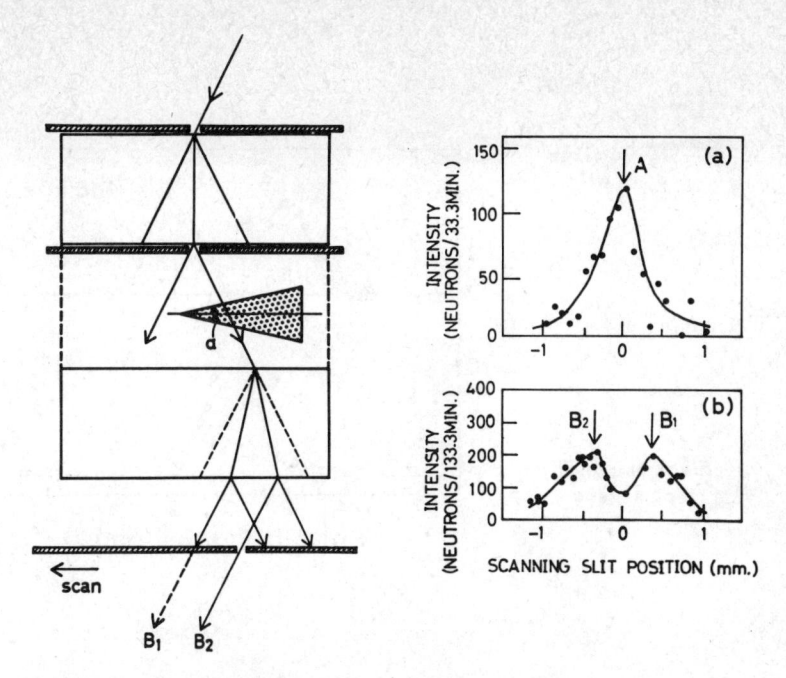

Fig.6: Measurement of small directional changes utilizing the angle amplification feature. Principle of the setup (left) and backface intensity profiles (right) without (a) and with (b) a 0.032 arc sec refracting prism in beam.

The results demonstrate a sensitivity of about 10^{-8} eV. This experiment can also be regarded as the first verification of dynamical diffraction theory predictions because the experimental situation corresponds to the case of a purely nuclear reflection in a ferromagnet. Additionally, by simultaneous magnetic field action and prism refraction a separation of the spin states could be achieved within the crystal /34/.

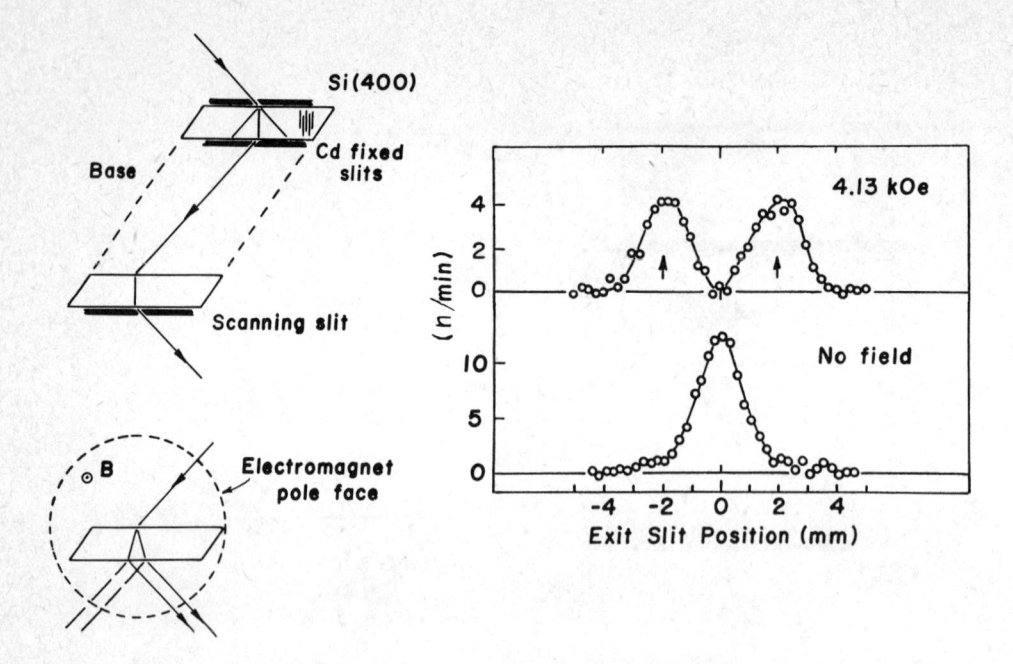

Fig.7: Measurement of the wave length change of a neutron in a magnetic field /34/. Principle of the arrangement (top left) and ray path in the second crystal plate with magnetic field applied (bottom left). The experimental results (right) demonstrate the expected separation effect.

6. Concluding Comments

The neutron's magnetic dipole interaction introduced to dynamical diffraction opens an exciting field of new experimental possibilities. The extension of the experiments mentioned in the last paragraph will be to use inhomogeneous rather than homogeneous magnetic fields applied to, say, a nonmagnetic crystal.
For such an experiment and for a similar one where the potential gradient is due to the gravitational interaction of the neutron /35/ an extension of dynamical diffraction theory to the case of a force acting on the particle while in the crystal is needed, a situation not encountered in the X-ray case. It turns out that approaches very similar to diffraction by a crystal with a gradient in lattice spacing /7/ can be used. For example one obtains in the Laue-case curved in-crystal neutron paths /36/ with curvatures immensly larger than in a free-space Stern-Gerlach experiment. Creation of new wavefields can be

expected if the gradient of the mean potential V(0) does not fulfil the inequality

$$\Delta_o \vec{G} \cdot \text{grad}(V(0)) \ll k(V(\vec{G})).$$

anymore. It seems feasible that such field gradients can be achieved experimentally. This offers the possibility of directly testing the different approaches used in topography /7,37/.

The field of new effects expected is even more extended if the polarisation of the neutron is taken into account explicitely. So, e.g. there exist spin rotation phenomena if the two polarisation states travel as different wave fields /38/, which effect is related to the coherence phenomena in neutron polarisation /39/. Finally, the application of magnetic fields to perfect nonmagnetic crystals /40/ or the use of perfect ferromagnets /41/ in off-set Bragg-geometry permits the construction of new types of highly efficient neutron polarizers. Here, with inhomogeneous magnetic fields an increase in reflectivity is to be expected.

I wish to thank Profs. H.Rauch and C.G.Shull for invaluable support and encouragement and Prof.M.A.Horne, Dr.G.L.Squires and Prof.S.A.Werner for useful discussions. The following institutions are acknowledged for granting permission to reproduce material in the present paper: Int.Union of Crystallography (Figs.2 and 3), Oxford University Press (Figs.4 and 6), North-Holland Publ.Co. (Fig.5) and Phys.Rev. (Fig.7).

References

/1/ A.Steyerl: Propagation of Neutron Beams (this volume)
/2/ J.Sivardiere: Formalism of Wave-Matter Interaction (this volume)
/3/ G.L.Squires: Interactions of Neutrons with Matter (this volume)
/4/ H.Rauch, W.Treimer, U.Bonse: Phys.Lett. $\underline{47A}$ (1974) 271
/5/ M.L.Goldberger, F.Seitz: Phys.Rev. $\underline{71}$ (1947) 294
  Yu.Kagan, M.Afanasev: Soviet Phys. JETP $\underline{22}$ (1966) 1032; Z.Naturforsch. $\underline{28a}$ (1973) 1351
  P.Hiismäki: Acta Cryst. $\underline{A25}$ (1969) 377
  R.Lenk, H.Solbrig: phys.stat.sol. $\underline{B46}$ (1971) 273
  H.Solbrig: phys.stat.sol. $\underline{B47}$ (1971) 143 and $\underline{B51}$ (1972) 555
/6/ A.Authier: Kinematical and Dynamical Theories of X-ray Diffraction (this volume)

/7/ F.Balibar: Geometric and Diffraction Approaches in X-Ray Diffraction (this volume)
/8/ F.Eichhorn: Dynamische Neutroneninterferenzen und ihre Anwendungen, in: Dynamische Interferenztheorie (O.Brümmer, H.Stepanik, Eds.), Akademische Verlagsgesellschaft, Leipzig, 1976
/9/ H.Rauch and D.Petrascheck: Dynamical Neutron Diffraction and Its Application, in: Neutron Diffraction (H.Dachs, Ed.) Vol.6 of Topics in Current Physics, Springer Verlag, Berlin, 1978
/10/ V.F.Sears: Canad.J.Phys. 56 (1978) 1261
/11/ C.G.Shull: J.Appl.Cryst. 6 (1973) 257
 C.Stassis, J.A.Oberteuffer: Phys.Rev. B10 (1974) 5192
 J.Sivardiere: Acta Cryst. A31 (1975) 340
 H.H.Schmidt, P.Deimel: J.Phys.C: Solid State Phys. 8 (1975) 1991
 S.K.Mendiratta, M.Blume: Phys.Rev. B14 (1976) 144
 V.A.Belyakov, R.Ch.Bokun: Soviet Phys. Solid State 18 (1976) 1399
 O.Schärpf: J.Appl.Cryst. 11 (1978) 626 and 631
/12/ S.K.Mendiratta: Phys.Rev. B14 (1976) 155
/13/ W.Treimer: Phys.Lett. 68A (1978) 162 and Z.Naturforsch. 33a (1978) 1432
/14/ W.Graeff: X-ray and Neutron Interferometry (this volume)
/15/ M.Schlenker: Neutron Topography (this volume)
/16/ J.W.Knowles: Acta Cryst. 9 (1961) 61
/17/ D.Sippel, K.Kleinstück, G.E.R.Schulze: phys.stat.sol. 2 (1962) K104
/18/ S.Sh.Shilshtein, V.A.Somenkov, V.P.Dokashenko: JETP Lett. 13 (1971) 214
/19/ D.Sippel, F.Eichhorn: Acta Cryst. A24 (1968) 237
/20/ D.Sippel, K.Kleinstück, G.E.R.Schulze: Phys.Lett. 14 (1965) 174
/21/ S.Kikuta, I.Ishikawa, K.Kohra, S.Hoshino: J.Phys.Soc.Japan 39 (1975) 471
/22/ C.G.Shull, K.W.Billman, F.A.Wedgewood: Phys.Rev. 153 (1967) 1415
/23/ C.G.Shull: Phys.Rev. 179 (1969) 752
/24/ S.Sh.Shilshtein, V.A.Somenkov, M.Kalanov: Sov.Phys.JETP 36 (1973) 1170
 N.O.Elyutin, A.O.Bubleinik, V.A.Somenkov, S.Sh.Shilshtein: JETP Lett. 18 (1973) 186
/25/ O.Schärpf, H.Strothmann: Vol.II, p.713 in Proc.Conf.Neutron Scattering, Gatlinburg 1976, publ. by U.S.Dept. of Commerce, CONF-760601-P2
/26/ M.Birr, A.Heidemann, B.Alefeld: Nucl.Instr.Meth. 95 (1971) 435
/27/ C.G.Shull: Phys.Rev.Lett. 21 (1968) 1585
 C.G.Shull, J.Oberteuffer: Phys.Rev.Lett. 29 (1972) 871

C.G.Shull, W.M.Shaw: Z.Naturforsch. $\underline{28a}$ (1973) 657
/28/ N.Kato: J.Appl.Phys. $\underline{39}$ (1968) 2225
/29/ C.G.Shull, A.Zeilinger, G.L.Squires, M.A.Horne: to be published,
preliminary result presented at "Int.Workshop on Neutron Interferometry", Grenoble 1978
/30/ V.L.Indenbom, I.Slobodetskii, K.G.Truni: Sov.Phys.JETP $\underline{39}$ (1974) 542
E.V.Suvorov, V.I.Polovinkina: JETP Lett. $\underline{20}$ (1974) 145
A.M.Afanasev, V.K.Kon: Sov.Phys.Solid State $\underline{19}$ (1977) 1035
V.V.Aristov, V.I.Polovinkina, I.M.Shmytko, E.V.Shulakov: JETP Lett. $\underline{28}$ (1978) 4
/31/ A.Zeilinger, C.G.Shull, M.A.Horne, G.L.Squires: Proc.Int.Workshop on Neutron Interferometry, Grenoble 1978, U.Bonse and H.Rauch (Eds.), Oxford University Press, 1979
/32/ U.Bonse, W.Graeff, H.Rauch: Phys.Lett. $\underline{69A}$ (1979) 420
/33/ A.Authier: Compt.rend.Acad.Sci.Paris $\underline{251}$ (1960) 2502
/34/ A.Zeilinger, C.G.Shull: Phys.Rev. B (in press)
/35/ S.A.Werner: private communication
/36/ A.Zeilinger: paper to be presented at Int.Conf. Polarized Neutrons in Condensed Matter Res., Zaborow 1979
/37/ C.Malgrange: X-ray Topography: Principles (this volume)
F.Balibar, Y.Epelboin, C.Malgrange: Acta Cryst. $\underline{A31}$ (1975) 836
/38/ A.Zeilinger: Proc.Int.Workshop on Neutron Interferometry, Grenoble 1978, U.Bonse and H.Rauch (Eds.), Oxford University Press 1979
/39/ F.Mezei: Coherent Approach to Polarized Neutrons (this volume)
/40/ S.Funahashi: Nucl.Instr.Meth. $\underline{137}$ (1976) 99
/41/ M.Hart: X-ray Optics (this volume)

COHERENT APPROACH TO NEUTRON BEAM POLARIZATION

F. MEZEI
Institut Laue Langevin, 156 X
38 Grenoble Cedex, France

and

Central Research Institut for Physics
Budapest, 114 POB 49, Hungary

I. INTRODUCTION

The classical approach to polarized neutron beams consists of considering the probabilities of finding a neutron in the "up" or "down" spin state with respect to the magnetic field direction, often referred to as quantization direction. For a spin wave function

$$|\chi\rangle = \alpha |\uparrow\rangle + \beta |\downarrow\rangle, \quad |\alpha^2| + |\beta^2| = 1 \quad (1)$$

(which is often written in a matrix form $|\chi\rangle = \binom{\alpha}{\beta}$) these probabilities are:

$$P_\uparrow = |\alpha|^2 \quad\quad P_\downarrow = |\beta|^2$$

and the polarization is defined as

$$P = \overline{P_\uparrow} - \overline{P_\downarrow} = \overline{|\alpha|^2} - \overline{|\beta|^2}$$

where the bar means the ensemble average for the neutron beam. To make clear why this approach can be called "incoherent" let us recall the general notions of "interference" and "coherence".

We can consider the solution $|\psi\rangle$ of the equation of motion for an experimental situation as a superposition of a set of (quasi stationary) solutions

$$|\psi\rangle = \alpha_1 |\psi_1\rangle + \alpha_2 |\psi_2\rangle + \ldots$$

The expectation value of an arbitrary physical quantity $\hat{A}$ is then given as (the hat ^ stands for operators)

$$\langle\psi|\hat{A}|\psi\rangle = |\alpha_1|^2 \langle\psi_1|\hat{A}|\psi_1\rangle + |\alpha_2|^2 \langle\psi_2|\hat{A}|\psi_2\rangle + \ldots + \\ + \alpha_1^* \alpha_2 \langle\psi_1|\hat{A}|\psi_2\rangle + \alpha_2^* \alpha_1 \langle\psi_2|\hat{A}|\psi_1\rangle \ldots$$

The measured value of A in a repeated experiment, i.e. over a thermodynamical ensemble of particle states determined by the experimental conditions (e.g. a beam of particles coming from a given source) reads:

$$\overline{A} = \overline{|\alpha_1|^2} \langle\psi_1|\hat{A}|\psi_1\rangle + \overline{|\alpha_2|^2} \langle\psi_2|\hat{A}|\psi_2\rangle + \ldots + \\ + \overline{\alpha_1^* \alpha_2} \langle\psi_1|\hat{A}|\psi_2\rangle + \overline{\alpha_2^* \alpha_1} \langle\psi_2|\hat{A}|\psi_1\rangle + \ldots$$

We will, rather subjectively, talk of interference if

a) we can think of the solutions $\psi_1, \psi_2 \ldots$ as known, easy to materialise and to visualise entities (e.g. by closing one or the other slit in the case of Young interference);

b) we measure a quantity $\hat{A}$ for which some of the $<\psi_1|\hat{A}|\psi_2>$ type matrix elements are not zero.

In this case the expectation value of $\hat{A}$ contains not only the weighted sum of the familiar values for ψ_1, ψ_2..... (terms with $|\alpha_1|^2$, $|\alpha_2|^2$,...) but also the cross-terms, which did not appear for ψ_1, ψ_2...... alone (terms with $\alpha_1^*\alpha_2$, etc.). These cross-terms are called interference.

The interference effects can only be observed in a real experiment i.e. for the pertinent ensemble average, if in addition the "coherence" condition is met: i.e. some of the $\overline{\alpha_i^*\alpha_j}$ ensemble averages are not zero. The degree of partial coherence can be obviously given by the expression

$$P = \frac{|\overline{\alpha_i^* \alpha_j}|}{|\alpha_i||\alpha_j|} .$$

In view of these the above definition of beam polarization is obviously incoherent: it tacitly assumes that $\overline{\alpha^*\beta} = 0$. The coherent approach is concerned with the possible effects of the interference terms between "up" and "down" spin states. Let us remember that the three spin components for spin $\frac{1}{2}$ particles are related to the Pauli matrices

$$\hat{\sigma}_x = \begin{pmatrix} 0 & 1 \\ 1 & 0 \end{pmatrix} \qquad \hat{\sigma}_y = \begin{pmatrix} 0 & -i \\ i & 0 \end{pmatrix} \qquad \hat{\sigma}_z = \begin{pmatrix} 1 & 0 \\ 0 & -1 \end{pmatrix}$$

and thus the expectation values of the spin components are

$$<X|\hat{S}_x|X> = \tfrac{1}{2}\hbar(\alpha^*\beta + \beta^*\alpha)$$
$$<X|\hat{S}_y|X> = \tfrac{i}{2}\hbar(\beta^*\alpha - \alpha^*\beta) \qquad (2)$$
$$<X|\hat{S}_z|X> = \tfrac{1}{2}\hbar(|\alpha|^2 - |\beta|^2)$$

Notice furthermore that introducing the angles $0<\vartheta<\pi$ and φ by the definitions

$$\cos\tfrac{\vartheta}{2} = |\alpha| \qquad e^{i\varphi} = \frac{\alpha}{|\alpha|} \Big/ \frac{\beta}{|\beta|} \qquad (3)$$

eqs. (2) become

$$<\hat{S}_x> = \tfrac{1}{2}\hbar\left(\sin\tfrac{\vartheta}{2}\cos\tfrac{\vartheta}{2}e^{i\varphi} + \sin\tfrac{\vartheta}{2}\cos\tfrac{\vartheta}{2}e^{-i\varphi}\right) = \tfrac{1}{2}\hbar\sin\vartheta\cos\varphi$$
$$<\hat{S}_y> = \tfrac{1}{2}\hbar\sin\vartheta\sin\varphi \qquad (4)$$
$$<\hat{S}_z> = \tfrac{1}{2}\hbar\left(\cos^2\tfrac{\vartheta}{2} - \sin^2\tfrac{\vartheta}{2}\right) = \tfrac{1}{2}\hbar\cos\vartheta$$

Thus there is a unique correspondence between a spin $\frac{1}{2}$ wave function $|\chi\rangle = |\chi(\vartheta,\varphi)\rangle$ and the classical notion of a spin direction unit vector $\vec{S}$, defined by the polar coordinates ϑ and φ.

Eqs. (2) show that the interference term $\alpha^*\beta$ appears in the expectation value of the x and y spin components. Thus the coherent approach to neutron spin polarization simply means the study of the spin polarization in three dimensions (in particular the observation of Larmor precessions) as opposed to the usual one dimensional "up" and "down" approach.

This new concept of polarised neutron work was initiated around 1970 in three laboratories[1][2][3] on very different grounds, and using different techniques. Of course, it did not grow out of anything like a methodical search for something more general. Rather it was suggested, as it is often the case with "novelties", by simple minded semiclassical analogies taken from other fields of science, which in this case was mostly the nuclear magnetic resonance[4].

In this lecture, however, I will follow the methodical way advocated by Levy-Lebond in his talk. Simplified analogies, classical or not, can be helpful both in producing new ideas and in popularization of science. But only the rigorous quantum mechanical treatment can show their validity and limitations.

In the next section a coherent, fully quantum mechanical theory of the spin $\frac{1}{2}$ particle beam polarization is presented, to my knowledge, for the first time. Section III. is devoted to a short description of the basic experimental concepts and to a summary of the applications of the new technique of three dimensional "coherent" neutron spin polarization analysis.

II. QUANTUM THEORY OF POLARIZATION FOR SPIN $\frac{1}{2}$ PARTICLE BEAMS

1.) Polarization of a single particle

The notion of spin polarization is obvious for (the non-existant abstraction of) a point-like classical particle with an arrow-like spin. Its tempting simplicity is in strong contrast to the quantum mechanical reality, especially in two respects.

Firstly, the quantum mechanical particle wave function has to describe both the spatial and the spin behaviour; in other words a particle is not point-like. The general form of wave function is:

$$|\psi\rangle = \varphi_+(\vec{r}, t)|\uparrow\rangle + \varphi_-(\vec{r}, t)|\downarrow\rangle \qquad (5)$$

with the normalization

$$\int (|\varphi_+(\vec{r}, t)|^2 + |\varphi_-(\vec{r}, t)|^2) d^3\vec{r} = 1$$

Secondly, when we talk of spin polarization as an observable quantity, we have to define exactly with what kind of a measurement we are concerned with, i.e. define the corresponding operator which properly takes into account the spin and space variables. In other words the spin is not an arrow with an obvious direction.

The simplest situation is that of the nuclear magnetic resonance: the study of the spin states as a function of time, independently of the spatial variables. The corresponding spin component operators are thus given as

$$\hat{S}_i = \tfrac{1}{2} \hbar \int d^3\vec{r}\, \hat{\sigma}_i \qquad (6)$$

To proceed, we will expand the wave function (5) in terms of the free particle eigenfunctions, where, for simplicity, we assume for the moment that the magnetic field B is homogeneous, and parallel to the Z ("up" - "down") axis. (This is convenient, though not necessary):

$$\varphi_\pm(\vec{r}, t) = \frac{1}{(\sqrt{2\pi})^3} \int a_\pm(\vec{k}) e^{i(\vec{k}\vec{r} - \omega_\pm(\vec{k})t)} d^3\vec{k} \qquad (7)$$

where

$$\hbar \omega_\pm(\vec{k}) = \frac{\hbar^2 k^2}{2m} \mp \mu B$$

(m is the mass, μ is the magnetic moment of the neutron). The expectation values of the polarization operators (6) are given as e.g.

$$\langle \hat{S}_x(t)\rangle = \frac{\hbar}{2(2\pi)^3} \int \{a_+^*(\vec{k}) a_-(\vec{k}') e^{i[(\vec{k}-\vec{k}')\vec{r} - (\omega_+(\vec{k}) - \omega_-(\vec{k}'))t]} + c.c.\} d\vec{k}\, d\vec{k}'\, d\vec{r}$$

where c.c. stands for the complex conjugate. Integrating over $d\vec{r}$ (using δ function normalization $^{(5)}$)

$$\langle \hat{S}_x(t)\rangle = \frac{\hbar}{2} \int a_+^*(\vec{k}) a_-(\vec{k}') e^{i(\omega_+(\vec{k}) - \omega_-(\vec{k}'))t} \delta(\vec{k} - \vec{k}') d\vec{k}\, d\vec{k}' + c.c.$$

$$= \frac{\hbar}{2} e^{-i \frac{2\mu B}{\hbar} t} \int a_+^*(\vec{k}) a_-(\vec{k}) d\vec{k} + c.c.$$

Similarly

$$\langle \hat{S}_z(t)\rangle = \frac{\hbar}{2} \int (|a_+(\vec{k})|^2 - |a_-(\vec{k})|^2) d\vec{k}$$

These expressions can be parametrized similarly to eqs. (4):

$$\langle \hat{S}_x(t) \rangle = p \sin\vartheta \cos\varphi(t)$$
$$\langle \hat{S}_y(t) \rangle = p \sin\vartheta \sin\varphi(t) \qquad (9)$$
$$\langle \hat{S}_z(t) \rangle = \cos\vartheta$$

where

$$p = \frac{\left| \int a_+^*(\vec{k}) a_-(\vec{k}) d\vec{k} \right|}{\sqrt{\int |a_+(\vec{k})|^2 d\vec{k} \int |a_-(\vec{k})|^2 d\vec{k}}}$$

and

$$\varphi = \varphi_0 - \frac{1}{\hbar} 2\mu B t, \qquad \vartheta, \varphi_0 = \text{const.}$$

We observe that the $\langle S_x \rangle$, $\langle S_y \rangle$ interference terms are reduced by the factor p which is essentially the (momentum) overlap of the ↑ and ↓ spatial wave functions. No interference is observable if these wave functions are orthogonal (p=0), quite the same way as no spatial interference is possible between opposite spin states. Furthermore the time dependence of φ corresponds to the well known classical Larmor precession, whose frequency for neutrons is given by the constant

$$\gamma_L / 2\pi = -2\mu/h = 2916.4 \text{ Hz/Oe}$$

(μ is negative).

In practice $\langle \hat{S}_i(t) \rangle$ can be conveniently measured for the nuclear spin ensemble in solids or liquids (NMR). On the other hand, for propagating particles, and particularly neutron beams, the space variables always play an essential role, which seriously limits the applicability of this "time only" approach. Nevertheless in special cases such a nuclear magnetic resonance type technique has been successfully used in neutron beam experiments [6][7].

For particle beams what is really measured is the particle flux impinging on the detector. That is why the polarization measured on a beam is in reality the spin flux defined by the operator

$$\hat{S}_i(\vec{r}) = \frac{1}{2} \hbar \hat{\sigma}_i \hat{J}(\vec{r}) \qquad (8)$$

where $\hat{J}$ is the familiar current operator [8] ($\nabla_i = \partial/\partial x_i$)

$$\hat{J}(\vec{r}) = \frac{\hbar}{2mi} \left(\vec{\nabla} \delta(\vec{r}) + \delta(\vec{r}) \vec{\nabla} \right)$$

In order to evaluate the polarization described by eq. (8) for a wave packet (in one dimension, for the sake of simplicity) we have to calculate the time integral over the particle passage:

$$\langle \hat{S}_i(r) \rangle = \tfrac{1}{2}\hbar \int \langle \psi(t) | \hat{\sigma}_i \, \hat{f}(\vec{r}) | \psi(t) \rangle \, dt$$

which gives the following results by a straightforward, but lengthy algebra:

$$\langle \hat{S}_x(r) \rangle = \tfrac{1}{2}\hbar \int [a_+^*(k+\delta k) a_-(k) e^{-i\delta k \cdot r} + c.c.] dk$$

$$\langle \hat{S}_z(r) \rangle = \tfrac{1}{2}\hbar \int [|a_+(k)|^2 - |a_-(k)|^2] dk$$

Here δk is the wave number shift related to the Zeeman energy $2\mu B$:

$$\delta k = \sqrt{k^2 + \frac{4\mu B m}{\hbar^2}} - k \sim \begin{cases} 2\mu B m/\hbar^2 k, & \text{if } |k|>|\delta k| \\ \sqrt{4\mu B m/\hbar^2}, & \text{if } |k|<|\delta k| \end{cases}$$

For, say, $B = 1\,kOe$, $2\mu B \sim 10^{-8}$ eV, and for ordinary thermal neutrons $\hbar^2 k^2/2m \sim 10^{-2}$ eV ($|k| \sim 2$ Å^{-1}) so δk can vary between $\sim 10^{-6}$ and 10^{-3} Å^{-1}.

The basic point of our considerations is now the following: if, and only if, $|a_+(k)|$ and $|a_-(k)|$ are "smooth" functions, i.e. slowly varying on the scale of δk:

$$|a_\pm(k+\delta k)| \simeq |a_\pm(k)| \qquad (11)$$

then eqs. (10) can be simply interpreted in terms of classical notions. (Note that this relation has to hold in the three dimensional space for any direction of $\vec{k}$. The best known example to the contrary is the classical Stern-Gerlach experiment, where, for the very finely collimated beam, (11) does not hold for k perpendicular to the beam propagation).

Indeed, using assumption (11), we can rewrite eqs. (10) in the following parametrised form (for the case of $|\delta k| \ll k$)

$$\langle \hat{S}_x(r) \rangle = \tfrac{1}{2}\hbar \int f(v) \sin\vartheta(v) \cos\varphi(v)\, dv$$

$$\langle \hat{S}_y(r) \rangle = \tfrac{1}{2}\hbar \int f(v) \sin\vartheta(v) \sin\varphi(v)\, dv \qquad (12)$$

$$\langle \hat{S}_z(r) \rangle = \tfrac{1}{2}\hbar \int f(v) \cos\vartheta(v)\, dv$$

where $v = \frac{\hbar k}{m}$, and $f(v), \vartheta(v), \varphi(v)$ are defined by the equations

$$f(v) = \frac{\hbar}{m}\left(|a_+(k)|^2 + |a_-(k)|^2\right)$$

$$\cos\frac{\vartheta(v)}{2} = \frac{|a_+(k)|}{\sqrt{|a_+(k)|^2 + |a_-(k)|^2}}, \qquad 0 \leq \vartheta(v) \leq \pi$$

$$\varphi(v) = \frac{\gamma_L B r}{v} + \varphi_0(v), \qquad e^{i\varphi_0(v)} = \frac{a_+^*(k) a_-(k)}{|a_+(k)||a_-(k)|}$$

Notice that f(v) can be considered as the classical velocity distribution function corresponding to the wave packet as it would be measured e.g. in a time-of-flight experiment.

Consequently eqs. (12) have the form of averages over a velocity distribution of classical pointlike particles displaying classical Larmor precessions in the field B.

The generalization of these results for time dependent magnetic fields is obtained by differentiation, which gives the classical result

$$\frac{d}{dt}\vec{S}(v) = \gamma_L [\vec{S}(v) \times \vec{B}(t)]$$

where $\vec{S}(v) = \vec{S}(\vartheta(v), \varphi(v))$ is the classical spin direction unit vector. For spatially inhomogeneous fields the plane waves are no longer stationary solutions of the Hamiltonian. However, the variations of the fields we are interested in occur on a macroscopic scale, i.e. they correspond to wave numbers $< 10^{-6}$ Å^{-1}, as compared to which our wave packets are assumed to be "smooth" and "broad" in the momentum space. Therefore we can consider the solutions over small, but still macroscopic, homogeneous field regions, and join them together to approximate the global result.

Thus in conclusion we find that the time-space evolution of the polarization of a "well behaved" spin ½ quantum particle in a magnetic field $\vec{B}(\vec{r},t)$ can be formally obtained as the average over a corresponding ensemble of pointlike classical particles, for which the spin precession is described by the classical equation

$$\frac{d}{dt}\vec{S} = \gamma_L [\vec{S}(t) \times \vec{B}(\vec{u}(t), t)] \tag{13}$$

where $\vec{u}(t)$ is the classical particle trajectory. The term "well behaved" means that the three dimensional wave packet describing the quantum particle is "smooth and broad" in momentum space on the scale defined by the Zeeman energy $\hbar\gamma_L|B|$, in the sense given in detail above.

It seems that in practice this "well behavedness" can always be assumed, if there is no specific reason to the contrary (e.g. deliberate beam collimation). Especially the observation of Larmor precessions of classical character for neutron beams in fields of order of 100 Oe imply that the incoming "untreated" neutron wave functions can only correspond to wave packets not narrower than about 10^{-5} Å^{-1}: (for average wave vectors of 1 Å^{-1}). This is by no means a trivial statement, since a priori we know little about the initial wave functions of particles coming from a complex source like a reactor. On the other hand, the best known case of not "well behaved" particles, the Stern-Gerlach experiment, not surprisingly, displays "real quantum effects" as opposed to the usual "classical" Larmor spin precessions. It is interesting to note that for very low energy (ultra cold) neutrons[9] and for very monochromatic beams involved in single crystal neutron interferometry[10] it would be particularly easy to produce "non-well-behaved" particle states and, consequently, "non-classical" spin polarization effects.

2.) Polarization of an ensemble of particles

As we have seen above, the complete quantum mechanical analysis of spin $\frac{1}{2}$ polarization for cases of interest in neutron beam scattering results, paradoxically enough, in a classical type picture, as opposed to the superficially quantum mechanical, commonly used "up" - "down" picture. As a consequence, this "classical" description is obviously applicable for an ensemble of particles too, with a more general distribution function $f_{\vec{r},t}(\vec{v},\vartheta,\varphi)$ which is the probability of finding a pointlike classical particle of velocity $\vec{v}$ and spin vector $\vec{S} = \vec{S}(\vartheta,\varphi)$ at position $\vec{r}$ and time t. The evolution of ϑ and φ i.e. $f_{\vec{r},t}$ in space and time is governed by eq. (13) and the vector polarization at a given point $\vec{r}$ and instant t is obviously given as (dΩ stands for integration over the polar angles)

$$P_x(\vec{r},t) = \int f_{\vec{r},t}(\vec{v},\vartheta,\varphi) \cos\varphi \sin\vartheta \, d\vec{v} \, d\Omega$$

$$P_y(\vec{r},t) = \int f_{\vec{r},t}(\vec{v},\vartheta,\varphi) \sin\varphi \sin\vartheta \, d\vec{v} \, d\Omega$$

$$P_z(\vec{r},t) = \int f_{\vec{r},t}(\vec{v},\vartheta,\varphi) \cos\vartheta \, d\vec{v} \, d\Omega$$

This picture has, however, a fundamental quantum mechanical limitation: while a velocity distribution can be directly measured, a distribution of spin directions vectors $\vec{S}(\vartheta,\varphi)$ can not. To show this let us consider the result of the measurement corresponding to an operator $\hat{O}$ over the ensemble of quantum particles described by the distribution $f(\vartheta,\varphi)$ of the spin states $|\chi(\vartheta,\varphi)\rangle$ (cf. eqs. (1)(2) and (3)).

$$\langle \hat{O} \rangle = \int f(\vartheta,\varphi) \langle \chi(\vartheta,\varphi)| \hat{O} |\chi(\vartheta,\varphi)\rangle \, d\Omega$$

It is readily seen that this becomes:

$$\langle \hat{O} \rangle = \langle \uparrow | \hat{O} | \uparrow \rangle \int f(\vartheta, \varphi) \cos^2 \frac{\vartheta}{2} d\Omega +$$
$$+ \langle \downarrow | \hat{O} | \downarrow \rangle \int f(\vartheta, \varphi) \sin^2 \frac{\vartheta}{2} d\Omega +$$
$$+ \left(\langle \downarrow | \hat{O} | \uparrow \rangle \int f(\vartheta, \varphi) \cos \frac{\vartheta}{2} \sin \frac{\vartheta}{2} e^{-i\varphi} d\Omega + c.c. \right)$$

We see that, whatever $\hat{O}$ may be, the result of the measurement only depends on the four integrals on the right hand side (which we denote in order $\varsigma_{\uparrow\uparrow}$, $\varsigma_{\downarrow\downarrow}$, $\varsigma_{\uparrow\downarrow}$ and $\varsigma_{\downarrow\uparrow}$ and it is independent of any further details of $f(\vartheta,\varphi)$. Thus from a measurement we can not have more direct information about $f(\vartheta,\varphi)$ than these four integrals; still we can often consider $f(\vartheta,\varphi)$ as known in more detail by calculating its evolution starting from an a priori (more or less) known situation. (E.g. for a beam which has not been in contact with anything magnetized, we can assume $f(\vartheta,\varphi) = 1/4\pi$)

Note for completeness that obviously

$$\varsigma_{\uparrow\uparrow} = \tfrac{1}{2} + \tfrac{1}{2} P_z \qquad \varsigma_{\downarrow\downarrow} = \tfrac{1}{2} - \tfrac{1}{2} P_z$$
$$\varsigma_{\uparrow\downarrow} = \tfrac{1}{2}(P_x + i P_y) \qquad \varsigma_{\downarrow\uparrow} = \varsigma_{\uparrow\downarrow}^*$$

The matrix $\hat{\varsigma}$ formed by these four quantities is called the spin $\tfrac{1}{2}$ density matrix and formally one finds that

$$\hat{\varsigma} = \tfrac{1}{2} + \tfrac{1}{2} \sum_i P_i \hat{\sigma}_i \qquad \text{and} \qquad \langle \hat{O} \rangle = \text{Tr}(\hat{O}\hat{\varsigma})$$

III. BASIC EXPERIMENTAL FACTS AND APPLICATIONS

A polarised beam can be produced e.g. by reflection of neutrons from an appropriate, ferromagnetic crystal or ferromagnetic thin film structure which is magnetically polarised to saturation by an applied field. Within the strict framework of the coherent description of polarization, the polarizer should be described by a quantum mechanical operator (so called $\hat{S}$ matrix) (cf. ref.(11)). For most practical purposes however, we can limit ourselves to the traditional approach, within which the polarizer is characterized by the two reflectivities, $R_\uparrow$ and $R_\downarrow$, for neutrons with the spin respectively parallel and antiparallel to the magnetizing field. This implies that for an unpolarized incoming beam the reflected beam polarization is:

$$P_x = P_y = 0 \qquad P_z = \frac{R_\uparrow - R_\downarrow}{R_\uparrow + R_\downarrow} = P$$

where P is called the polarizer's efficiency. On the other hand, the reflectivity for a polarized incoming beam is obviously:

$$R(P_z) = \frac{1}{2}(R_\uparrow + R_\downarrow) + \frac{1}{2} P_z (R_\uparrow - R_\downarrow) \tag{14}$$

Thus the polarization dependence of the reflectivity permits the measurement of P_z (spin analysis) too.

To produce a beam with a vector polarization P_x, $P_y \neq 0$, one has to turn the polarization (initially P_z) with respect to the field. For this purpose different techniques have been developed using the evolution of the neutron spin direction $\vec{S}$ in inhomogeneous, time independent magnetic fields, as given by eq. (13), which has two simple limiting cases:

a) Adiabatic case: if the direction of $\vec{B}$ changes slowly, as seen by the neutron, as compared to the Larmor frequency i.e.

$$\gamma_L |\vec{B}(\vec{u}(t))| \gg \frac{1}{|\vec{B}(\vec{u}(t))|} \frac{d}{dt} \vec{B}(\vec{u}(t))$$

the angle ϑ between $\vec{B}$ and $\vec{S}$ stays constant i.e. the precession cone of $\vec{S}$ around $\vec{B}$ rigidly follows $\vec{B}$.

b) Majorana field flip: if $\vec{B}$ changes direction very rapidly as compared to $\gamma_L |B|$, $\vec{S}$ has no time to "follow" and the angle between $\vec{S}$ and $\vec{B}$ changes.

One scheme[2] of turning $\vec{S}$ from the field direction Z to e.g. the perpendicular direction X is illustrated in the figure.

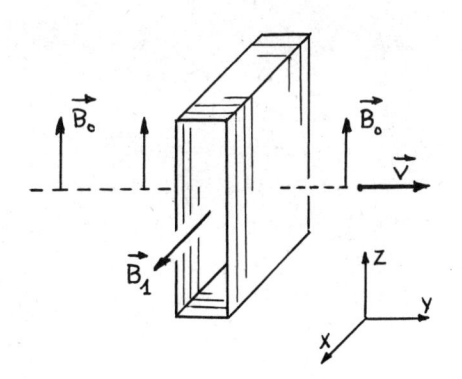

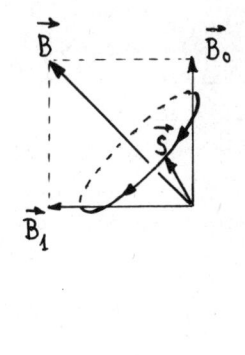

The neutron with velocity $\vec{v}$ goes through a rectangular coil, whose field $\vec{B}_1 || X$ confined to its interior adds to the external homogeneous field $\vec{B}_o || Z$. When the neutron enters the coil $\vec{B}(\vec{u}(t))$ jumps from $\vec{B}_o$ to $\vec{B}_o + \vec{B}_1$. Assume that the neutron spin was initially parallel to $\vec{B}_o$, and $|B_o| = |B_1|$. Then inside the coil $\vec{S}$ will start to precess around $\vec{B}$ on a cone of 45°. If the thickness of the coil is such, that the neutron leaves the coil after one half precession, this will happen at the point where $\vec{S}$ is parallel to X. Outside the coil again, $\vec{S}$ will now of course precess around $\vec{B}_o$. More generally, the Larmor rotation inside the coil (or in a field limited the same way) can be described by a rotation matrix transformation[12][13] which e.g. in the present case gives

$$P_x \to P_z \quad P_y \to -P_y \quad P_z \to P_x$$

i.e. $\vec{P}(P_x, P_y, P_z) \to \vec{P}'(P_z, -P_y, P_x)$

This coil device can be used to initiate Larmor precessions by turning the polarization perpendicular to the field, but also to analyse them by turning P_x into P_z which then can be measured directly (cf. eq. (14)).

A detailed review of the applications of the novel technique of three dimensional polarization analysis can not be given here. These new possibilities will only be pointed out.

The change of neutron polarization in a magnetic field provides a quite sensitive microprobe for the study of the field itself. E.g. thermal neutrons (velocity ≃ 2000 m/sec) in 20 kOe field (≈ saturation moment of iron) make a full, 360° Larmor precession within about 30 μ distance. Thus the change of the neutron beam vector polarization in the course of transmission through a ferromagnet carries a great deal of information about the magnetic domain structure of the sample on the scale of a few μ. The method has been developped and successfully used by Rekvelt[1][12] and the Leningrad group[3].

Another field which can benefit of the use of vector polarization analysis is the study of complicated magnetic crystal structures. It has been shown by Blume[14] that there is in general a tensorial relation between the polarization vector of the incoming and the Bragg reflected neutron beams, which becomes particularly complex e.g. for non-centrosymmetrical or non-collinear structures. The existence of such tensorial effects has been demonstrated by Alperin[15] and a vector polarization analysis instrument for such crystallographic use is being developped at the ILL[16].

Finally we will consider a special use of the Larmor precessions in inelastic neutron scattering[17]: the neutron spin echo, which was introduced in 1972[2]. The general principle of the method[18] will be discussed in a form somewhat simplified by the omission of the momentum dependence of the scattering effects.

In inelastic neutron scattering we are concerned with the change E of the neutron energy in the course of the scattering process, which is related to the atomic dynamics of the scatterer[17]:

$$E = \frac{1}{2} m v_1^2 - \frac{1}{2} m v_0^2$$

where v_0 and v_1 are the incoming and outgoing neutron velocities, respectively. The usual inelastic scattering methods are based on the determination of the average of v_0 and v_1 for the whole beams, and the measured value of E becomes

$$\bar{E} = \frac{1}{2} m (\bar{v}_1)^2 - \frac{1}{2} m (\bar{v}_0)^2$$

This is a difference of two quantities measured separately and the resolution in $\bar{E}$ is limited by both of these measurements, especially by the monochromatization of the incoming beam.

By the use of the Larmor precession technique it became possible to observe the beam average of the energy change of each individual neutron instead, i.e.

$$\overline{E} = \overline{\frac{1}{2} m v_1^2 - \frac{1}{2} m v_0^2}$$

The main interest of such a direct measurement of the difference as compared to the above classical approach is that the resolution is no longer limited by the monochromatization of the (incoming) beam. This permits better resolutions with still acceptable beam intensities.

As we have seen above (cf. eqs. (12)) the angle of the Larmor precession in a constant, homogeneous magnetic field B over a distance ℓ can be considered as a measure of the neutron velocity v:

$$\varphi = \gamma_L \frac{B \ell}{v}$$

In a neutron spin echo scattering experiment we make both the incoming neutrons precess before the scattering (field B_0 of length ℓ_0), and the scattered neutrons precess in the opposite sense after the scattering (field B_1 of length ℓ_1). Thus the total Larmor precession angle is:

$$\varphi = \gamma_L \left(\frac{B_0 \ell_0}{v_0} - \frac{B_1 \ell_1}{v_1} \right)$$

Let us take $v_0 = \bar{v}_0 + \delta v_0$ and $v_1 = \bar{v}_1 + \delta v_1$ where $\bar{v}_0$ and $\bar{v}_1$ are the respective average values, and use the approximations

$$\varphi = -\gamma_L \left(\frac{B_0 \ell_0}{\bar{v}_0^2} \delta v_0 - \frac{B_1 \ell_1}{\bar{v}_1^2} \delta v_1 \right) + \bar{\varphi}$$

and

$$E = m (\bar{v}_1 \delta v_1 - \bar{v}_0 \delta v_0) + \bar{E}$$

where $\bar{\varphi} = \gamma_L(B_0 \ell_0/\bar{v}_0 - B_1 \ell_1/\bar{v}_1)$ and $\bar{E} = \frac{1}{2}m(\bar{v}_1^2 - \bar{v}_0^2)$
are constants. φ will become a practical measure of E if for each combination of the variables δv_0 and δv_1

$$\varphi - \bar{\varphi} = t(E - \bar{E})$$

with a constant t. This is fulfilled if and only if:

$$\frac{\gamma_L}{m} \frac{B_0 \ell_0}{\bar{v}_0^3} = \frac{\gamma_L}{m} \frac{B_1 \ell_1}{\bar{v}_1^3} = t$$

This is the spin echo condition for this case. Now, if the scattering probability to be determined is S(E), the measured P_x polarization will be given as ($\int S(E)dE = 1$)

$$P_x = \int S(E) \cos(\varphi(E)) dE = \int S(E) \cos(Et) dE$$

i.e. $P_x(t)$ is the Fourier transform of S(E). The Fourier parameter t is proportional to the magnetic field which can be varied in an experiment.

The approximations we have used above (and those used in the more general case) are more or less precise according to the range of δv_0 and δv_1 used (monochromatization). In practice, depending on the experimental details, energy resolutions of $10^{-3} - 10^{-5}$ can be achieved with modest monochromatizations of $10^{-1} - 10^{-2}$. This is precisely the most interesting feature of the neutron spin echo from the point of view of applications. Until now it has been successfully used to study both quasielastic scattering effects, like the diffusion in polymer solutions[19] or the magnetic dynamics of spin glass systems[20], and scattering on elementary excitations, like phonons and rotons in superfluid He liquid[21] with a highly improved energy resolution (up to 50 times).

IV. CONCLUSION

We have seen that the interference effects between the familiar "up" and "down" spin states can be observed in a particle beam by observing all three vector components of the spin polarization. As usual, interference can only be observed if the coherence can be maintained over the whole particle ensemble. In this particular case the decay mechanism of the coherence is particularly conspicuous. In equations (12) we find in the argument of the sine and cosine function the Larmor precession phase angle $\varphi = \frac{\gamma_L B r}{v}$, which can achieve very high values: e.g. for B = 100 Oe field, r = 1 m distance and v = 2000 m/sec (thermal neutron velocity) $\varphi \simeq$ 900 radians. Thus both inhomogeneities of the field, and a distribution of the velocity result a scatter of φ, i.e. a dephasing of the precessions. If this becomes of the order of 2π, the coherence is completely lost and $P_x = P_y = 0$. This explains why it is more difficult to work with a three dimensional polarization than to restrict attention to the phase insensitive P_z, as it was the rule earlier. As we have seen above, in the neutron spin echo the neutron velocity differences are explicitly taken care of, thus the only remaining dephasing factor is the inhomogeneity of the magnetic fields, which is an important feature for further applications[22]. This is why the neutron spin echo represents to the highest degree the coherent approach to neutron beam polarization.

References

(1) M. Th. Rekvelt, J. de Physique 32 (1971) C579
(2) F. Mezei, Z. Physik 255 (1972) 146
(3) A.I. Okorokov, V.V. Runov, V.I. Volkov and A.G. Gukasov, Zh. Eks. Teor. Fiz. 69 (1975) 590
(4) F. Bloch, Phys. Rev. 70 (1946) 460
(5) cf. L.I. Schiff, Quantum Mechanics (McGraw Hill, New York 1968) p. 59
(6) N.F. Ramsey, Phys. Rev. 76 (1949) 996
(7) A. Abragam, in Trends in Physics (European Physical Society, Geneva, 1973) p.177
(8) cf. L.I. Schiff, ibid. p. 27
(9) cf. A. Steyerl, in this volume
(10) cf. A. Zeilinger, in this volume
(11) F. Mezei, Comm. on Physics 1 (1976) 81
(12) M. Th. Rekveldt, Z. Physik 259 (1973) 391, J. of Magnetism and Magn. Mat.
(13) J.B. Hayter, Z. Physik B, 31 (1978) 117
(14) M. Blume, Phys. Rev. 130 (1963) 1670
(15) H. Alperin, Proc. ICM-73, 111 (Moscow, 1973) 128
(16) F. Tasset, P. Gardner, K. Ben Saidane, private communication
(17) cf. G.L. Squires, in this volume
(18) F. Mezei, in Neutron Inelastic Scattering 1977 (IAEA, Vienna 1978) p. 125
(19) D. Richter, J.B. Hayter, F. Mezei and B. Ewen Phys. Rev. Lett. 41 (1978) 1484
(20) F. Mezei and A.P. Murani, to be published
(21) F. Mezei, to be published
(22) O. Schärpf and H. Warneck, J. of Magnetism and Magn. Mat. 2 (1976) 113

X-RAY AND NEUTRON INTERFEROMETRY

W. Graeff

Deutsches Elektronen-Synchrotron DESY

2000 Hamburg 52, Notkestrasse 85

Fed. Rep. of Germany

The extension of light optical interferometry to the Angström range required the development of new techniques for coherent beam splitting and handling since the refractive index for X-rays or matter waves like neutrons is too close to unity to allow the efficacious application of usual optical components.

Interferometry with such a small wavelength started with electrons. Marton et al.[1] tested in 1953 an interferometer composed of three separate crystal lamellae for beam splitting, deflection and superposition. Alignment and stability problems resulting in a rather poor interference contrast did not encourage to follow up this way. One year later Möllenstedt and Düker[2] therefore used the principle of Fresnel's biprism interferometer to realize the first electron interferometer with electrostatic lenses. Again Fresnel's biprism arrangement was chosen for the first attempt to neutron interferometry made by Maier-Leibnitz and Springer[3] in 1962. The very small beam separation of about 60 µm made the insertion of a sample rather difficult.

In 1965 Bonse and Hart[4] developed the first powerful interferometer in the Angström range with X-rays. They solved the problem of the Marton interferometer by making the three crystal lamellae out of the same perfect crystal leaving a rigid connection between the crystal slabs. The coherent beams were widely separated and easily accessible to macroscopic samples. A similar perfect crystal interferometer marked in 1974 the break through of neutron interferometry (Rauch, Treimer, Bonse).[5]

During this lecture we want to restrict ourselves to the discussion of this interferometer type. A description of other interferometer designs may be found in one of the recent review articles and workshop proceedings[6,7,8]. Furthermore, due to space limitations it is not possible to give a detailed reference list which may be found in [7] and [8].

The perfect crystal interferometer

Working principle

The geometry of components and beam paths is sketched in Fig. 1. The incoming wave is split, deflected and superposed through three consecutive Bragg reflections within the perfect crystal lamellae. This arrangement is known as triple Laue case (LLL) interferometer.

Fig. 1:
Triple Laue case interferometer made out of a perfect single crystal.

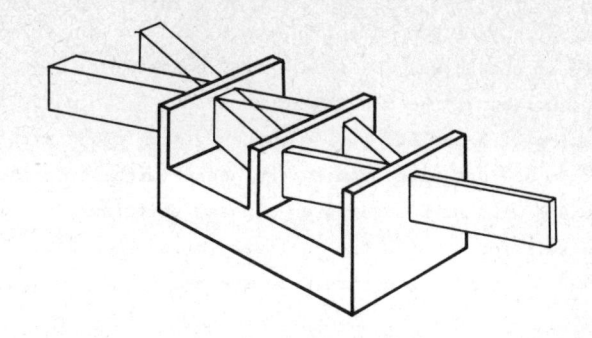

The superposition of the two coherent waves in front of the last crystal lamella, the so-called analyzer, forms a standing wave pattern with the same spacing as the reflecting net planes. Its relative position determines the intensities in the outgoing beams.

The theory of the LLL interferometer dealing both with incident plane and spherical waves can be found in the literature[7,9]. With perfect crystals dynamical theory has to be applied. As this theory is subject of a different lecture we only mention that for every incident plane wave two internal propagation modes, the so-called wavefields, are excited. Their slightly different wave vectors cause a beating of the intensity in the outgoing beams known as Pendellösung effect[10]. For a single reflection and without absorption, as it is the case with thermal neutrons in silicon, the intensity T of the transmitted and R of the reflected beam are complementary and have a constant sum, the incident intensity (here set to 1).

The values of T and R strongly depend on angle of incidence and crystal thickness. A quick calculation shows that the intensities of the interfering beams are in forward direction (0-beam) $T_S R_M R_A$ and $R_S R_M T_A$ but for the diffracted direction (h-beam) $T_S R_M T_A$ and $R_S R_M R_A$. For maximum interference contrast we find the conditions of ideal geometry:

(a) parallel setting of all crystal components (in most cases guaranteed by the common base)
(b) focussing of the coherent beams on to the entrance surface of the third crystal
(c) equal thickness of first and third crystal lamella.

Interference contrast

If these conditions are met the 0-beam theoretically has always 100 % interference contrast whereas the h-beam has an averaged contrast of 43 % with no absorption and all three lamellae equally thick. With increasing absorption the Pendellösung effect vanishes and with strong absorption ($\mu t \gg 1$) the Borrmann effect (anomalous transmission) dominates. Here the R- and T-curves nearly coincide so that condition (c) can be omitted and the contrast in the h-beam can reach 100 %, as well.

Simply speaking with low absorption a phase shift in one partial beam switches the intensity from one outgoing beam to the other (with an incoherent background in the h-beam) whereas with high absorption a phase shift switches between anomalously high and low transmission through the analyzer for both outgoing beams. In the intermediate range ($\mu t = 0.3 - 0.4$) the contrast in the h-beam can even drop to 10 %.

Geometrical tolerances

Deviations from the ideal geometry may occur with different crystal thicknesses of splitter and analyzer or different gap widths between the crystals. The latter results in a defocussing Δz of the partial beams. With low absorption a tolerable value of Δz is $\Delta_e/6$ where Δ_e is the extinction length. With high absorption the incident spherical wave gets more and more a plane wave and $\Delta z \approx \Delta_e$ may be tolerated. But this does not help so much because with increasing absorption the extinction length at least with X-rays decreases.

Absolute values of the defocus Δz which causes 50 % loss of contrast range between 10 and 16 μm[7].

Applications

The interferometer as a sample

As already mentioned a misfit in either spacing or position of the standing wave pattern in front of the analyzer and the netplanes inside alters the intensity of the outgoing beams analogue to the moiré technique used in optics for comparison of two similar gratings. Hence the interferometer is a very sensitive tool to detect minute strains and misorientations in one of the crystal components.

This technique was used to study inhomogeneous distributions of point defects, strain fields of single dislocations and dislocation aggregates, radiation damage due to ion implantation and stresses exerted by a thin metal film evaporated on the analyzer.

Absolute lattice parameter determination

By shifting the analyzer crystal with respect to the rest of the interferometer sideways every passage of one net plane corresponds to one oscillation period of the outgoing intensity. Thus simply by counting fringes and measuring the total shift with a light optical interferometer directly operated with the length standard lamp it was possible to establish a secondary standard in the Angström range with an accuracy of better than 10^{-6} (Deslattes et al., 1974)[11].

With such a standard at hand the redefinition of Avogadro's constant, several ratios of fundamental constants involving Planck's constant and the recalibration of γ-ray reference lines are possible.

Determination of refractive index

Measuring the refractive index of a sample is the obvious task in interferometry. For X-rays the refractive index is directly related to the forward scattering factor f(o) and for neutrons to the coherent scattering length b_c.[12]
The interesting part of the scattering for X-rays is the anomalous dispersion correction f' in the neighbourhood of absorption edges. The exact knowledge of f' with varying wavelength is important for phase determination in crystallography. Furthermore the short wavelength side of the absorption edge displays a fine structure which is closely related to the structure of absorption spectra, known as EXAFS. Such measurements have both been made with synchrotron radiation[13] (see Fig. 2) and Bremsstrahlung radiation.

Fig. 2

Anomalous dispersion correction f' for Ni near the K-edge (courtesy G. Materlik)

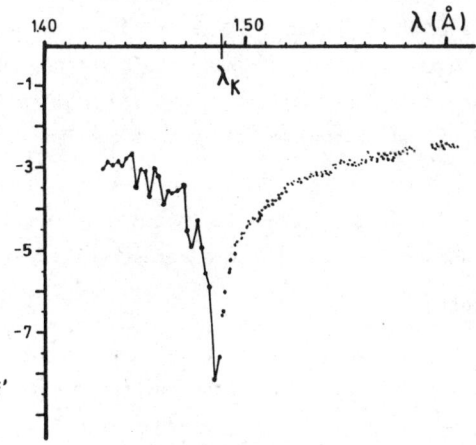

For neutrons three kinds of interactions are traceable with interferometric methods: the nuclear scattering, the magnetic interaction (if present) with the electron shell and to some extent the Foldy interaction between the atomic Coulomb field and the charge distribution of the neutron. Besides the measurements of the forward directed scattering for a variety of solid, liquid and gaseous samples experiments have been performed with crystalline samples near Bragg reflection to reveal the change in refractive index described by the dispersion surface in dynamical theory[14] (see Fig. 3).

The quantities f and b_c are determined via the measurement of the λ-thickness t_λ of a sample.(A plate of thickness t_λ shifts the phase of a transmitted wave about 2π). In addition the wavelength, the density and the sample thickness must be known with sufficient precision. The precision of t_λ-measurements is of the order of 10^{-3}. Phase shifts are induced either by rotation of a plane parallel sample[15] or by moving the sample in and out and detecting the phase shift by means of the scanning interferometer[16] where the analyzer is shifted sideways. By working simultaneously at two wavelengths the sample geometry can even be eliminated.

Fig. 3:
Interference pattern caused by rotating a perfect Si sample inside the interferometer through a Bragg reflection.

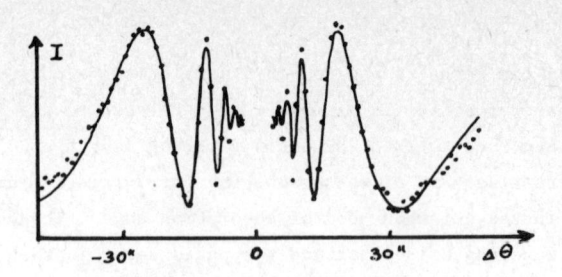

Phase contrast topography

By inserting a sample in the interferometer and detecting the outgoing beams on high resolution photographic film, in case of neutrons an additional converter like Gd is necessary, a direct phase contrast image is obtained, provided the absorption length $t_a = \mu^{-1}$ is much larger than the λ-thickness t_λ of the material under consideration. This is true for X-rays and light elements and might be useful for imaging of biological samples as it was demonstrated by Ando and Hosoya[17].

In case of thermal neutrons the condition $t_\lambda \ll t_a$ is quite often encountered. In addition phase contrast may be obtained from magnetically ordered regions in the sample. Experiments on neutron phase contrast imagery are under way[18].

Special experiments with neutrons

Two experiments should be mentioned which aimed at the particle properties of the neutron, namely its spin and mass. From quantum mechanics it is known that a spinor wave function of a spin 1/2 particle produces a phase factor -1 under 2π rotation, hence has a 4π-periodicity. This could clearly be demonstrated in several interferometric experiments[19].

Another experiment performed by Colella et al.[20] impressively showed the sensitivity of interferometric methods. They measured the phase difference of two coherent neutron beams travelling in earth's gravitational potential a few centimeters apart. Suggestions have been made to use the scanning interferometer as a high frequency neutron chopper[21] for neutron spectroscopy. Chopping rates of 10 to 100 MHz seem feasible.

Summary

Since the invention of the perfect crystal interferometer X-ray and neutron interferometry have found a widely spread field of applications. Scattering factors for X-rays and coherent scattering lengths for neutrons are measured with high precision. Experiments answering questions in fundamental physics, metrology, crystallography and material science, as well, have been performed.

Theories have been developed to predict intensity profiles of the interfering beams, contrast and image formation for both absorbing and non-absorbing interferometers.

However, the influence of crystal quality on phase blurring and interference contrast needs some clarification.

References

1) L.A. Marton, J.A. Simpson, J.A. Suddeth: Phys. Rev. 90, 410 (1953)
2) G. Möllenstedt, H. Düker: Naturwiss. 42, 41 (1954)
3) H. Maier-Leibnitz, T. Springer: Z. Physik 167, 386 (1962)
4) U. Bonse, M. Hart: Appl. Phys. Lett. 6, 155 (1965)
5) H. Rauch, W. Treimer, U. Bonse: Phys. Lett. 47A, 369 (1974)
6) M. Hart: Proc. Roy. Soc. A346, 1 (1975)
7) U. Bonse, W. Graeff: X-ray and Neutron Interferometry, in "X-Ray Optics", ed. H.J. Queisser (Springer Verlag Berlin Heidelberg New York 1977) pp. 93-143.
8) Proceedings of the Internat. Workshop on Neutron Interferometry, ed. by U. Bonse, H. Rauch (Oxford University Press, in press).
9) D. Petrascheck, R. Folk: phys. stat. sol. (a) 36, 147 (1976)
10) see also A. Authier in this volume
11) R.D. Deslattes, A. Henins, H.A. Bowman, R.M. Schoonover, C.L. Carrol, I.L. Barnes, L.A. Machlan, L.J. Moore, W.R. Shields: Phys. Rev. Lett. 33, 463 (1974)
12) See also L. Gerward, G. Squires, A. Steyerl in this volume
13) U. Bonse, G. Materlik: Z. Physik B24, 189 (1976)
14) W. Graeff, W. Bauspieß, U. Bonse, H. Rauch: Acta cryst. A34, S238 (1978)
15) W. Bauspieß, U. Bonse, H. Rauch: Nucl. Instrum. Meth. 157, 495 (1978)
16) C. Cusatis, M. Hart: Proc. Roy. Soc. A354, 291 (1977)
17) M. Ando, S. Hosoya: An Attempt at X-Ray Phase Contrast Microscopy, in Proc. of the 6th Int. Conf. on X-Ray Optics and Microanal., ed. by G. Shinoda, K. Kohra, T. Ichinokawa (University of Tokyo Press 1972) p. 63.
18) W. Graeff, W. Bauspieß, U. Bonse, M. Schlenker, H. Rauch: Acta Cryst. A34, S239 (1978).
19) H. Rauch, A. Wilfing, W. Bauspieß, U. Bonse: Z. Physik B29, 281 (1978)
    S.A. Werner, R. Colella, A.W. Overhauser, C.F. Eagen: Phys. Rev. Lett. 35, 1053 (1975)
20) R. Colella, A.W. Overhauser, S.A. Werner: Phys. Rev. Lett. 34, 1472 (1975).
21) M. Hart: Multiple Wavelength Interferometry with X-Rays and Neutrons, in (8).

X-RAY TOPOGRAPHY : PRINCIPLES

C. MALGRANGE

Laboratoire de Minéralogie-Cristallographie,
Université P. et M. Curie, 4 place Jussieu
75230 PARIS CEDEX 05 - FRANCE

1.- INTRODUCTION

Quite a lot of very good crystals are now available and their use is widening rapidly in modern technology.

Most physical properties depend on the defects of the crystal and interact with them. X-Ray topography methods (which are non destructive) give a two dimension projection image of the bulk defects of the crystal. The contrast on the topographs arises from the difference between the intensity diffracted by perfect and imperfect areas (section 3) and the propagation properties of X-Rays in slightly distorted crystals (section 4). Consequently, these methods give an image of the strain field in the crystal. The position of the defects in the crystal and their strain tensor can thus be determined in principle.

There are now quite a lot of different topographic methods with different and complementary fields of applications (for review see 1 to 6). A crude characterization of the defects (position for example) very helpful for a lot of applications (A. Mathiot conference) is quite easy but a careful interpretation leading to the value of the strain tensor around the defects is more difficult and requires a good knowledge of dynamical theory of diffraction in perfect and imperfect crystals.

Topographic methods can be divided in two main groups :
- reflexion methods : incident and diffracted X-Rays enter and leave the crystal surface at the same surface (fig. 1)
- transmission methods : X-Rays enter at a crystal surface and diffracted rays leave at the opposite surface.

As a first approximation one can say that crystal thicknesses which can be studied by X-Ray topographic methods are such that the product of the linear absorption coefficient by the length of X-Ray path is of the order of 1. Bragg

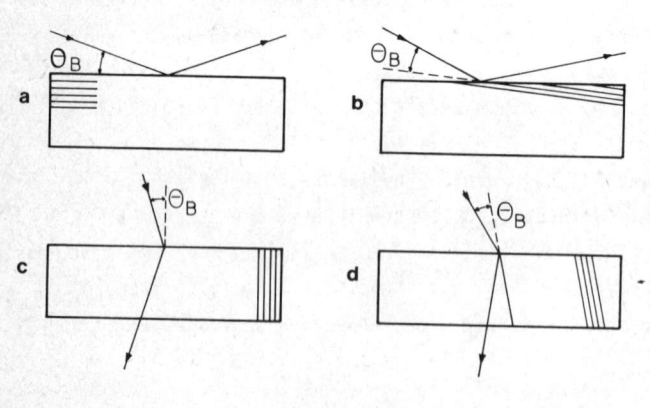

Fig. 1
Reflexion case : a) symmetric case, b) non symmetric case. Transmission case c) symmetric case, d) non symmetric case.

angles being of the order of 10° this implies that crystal thicknesses can be greater in the case of transmission.

Typically, crystal thicknesses which can be studied by transmission vary from about 100 μm for high absorbing crystals to several millimeters for low absorbing crystals (containing light atoms). In the case of reflexion methods, the thicknesses which contribute to diffraction and thus which give images vary from about 1 to 20 μm.

One transmission method will be studied here in more detail : Lang method (7) , in order to show the basic principles of X-Ray topography. Then, synchrotron white radiation topography and double crystal topography will be described briefly.

2.- LANG METHOD

The incident beam coming from a point focus (∼ 100 μm x 100 μm) is collimated by a thin vertical slit (∼ 10 μm) S perpendicular to the plane of the figure 2. The crystal is adjusted so that the mean incident ray is at exact Bragg incidence for a characteristic radiation of the X-Ray tube and a given family of parallel (h,k,l) planes which we assume, for simplicity, to be perpendicular to the crystal surface. If the crystal is perfect, the divergence δ of the incident beam which contributes to the reflected beam is of the order of a few seconds. The corresponding X-Ray wave-fields propagate inside a triangle, the so-called Borrmann fan giving rise to interference phenomena. Afterthis, the wave-fields split into two beams, one in the incident direction which we call the transmitted or refracted beam, the other in the reflected direction, the reflected beam. Due to interference phenomena the beams present fringes (Pendellösung or Katò fringes (8)) which are vertical lines if the crystal is parallel sided and hyperbola if it is a wedge shaped crystal.

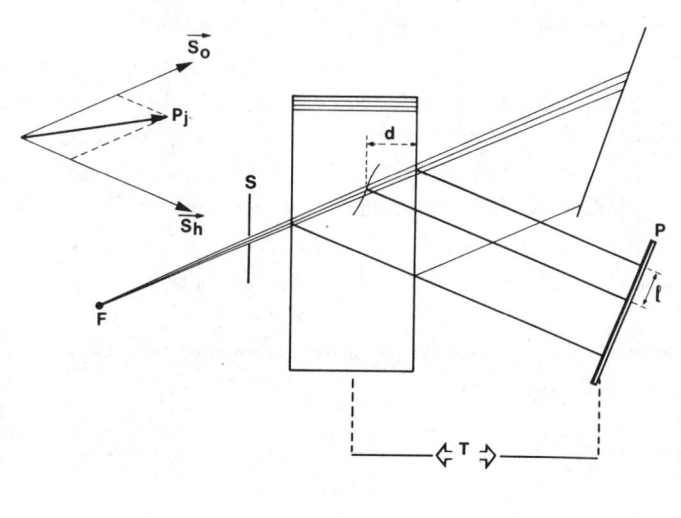

Fig. 2
Principle of Lang method

Let us notice that the locus of the rays which are at exact Bragg incidence is a cone the apex of which is the focus, the axis being the normal to the reflected planes. The focus-crystal distance being of the order of 40 cm and the height of the crystal of the order of 1 cm, the rays at exact Bragg incidence form a sheet nearly perpendicular to the plane of the figure.

The divergence of the incident beam is of the order of 1 minute so that, for a perfect crystal, most of the incident beam is not reflected and propagates straightforward through the crystal, interacting with the crystal through photoelectric absorption only. If this part of the beam, called the direct beam, falls on a defect, a dislocation line for example, it can be diffracted by the distorted area surrounding the line and gives a black image on a photographic plate or film P perpendicular to the reflected beam. The position of the image in the beam is directly related to the depth of the defect which can then be measured (1 = 2 d tgθ in fig. 2). One thus obtains an image of the defects contained in a section of the crystal by the vertical sheet of incident X-Rays. This is the reason why Lang called this a "section topograph". If the crystal and the photographic plate are simultaneously translated along direction T one obtains an image which is a projection of the defects of the whole crystal. This image is called a traverse topograph. Of course the quantitative information on the depth of the defects in the crystal is lost with the translation and qualitative information only can be obtained by looking simultaneously at two "stereo images" (h k l and $\bar{h}\,\bar{k}\,\bar{l}$ reflections).

The resolution of the method depends on two factors :
a) the broadening of the image due to experimental factors : in particular the focus size and the natural wavelength spread of the characteristic radiation which induce a divergence α of the beam diffracted by a point inside the crystal. The size s of the image is then equal to the product of this divergence α and the distance h crystal-film. With good experimental conditions s is of the order of 1 μm (details in ref. 4).
b) the intrinsic size of the image related to the strength of the strainfield around the defect as will be discussed just below.

3.- INTRINSIC WIDTH OF THE DIRECT IMAGE

The distorted area which contributes to the direct image corresponds to a local misorientation of the reflecting planes larger than something of the order of the angular width δ of the beam which can be reflected by a perfect crystal (2, 9). This region behaves as a small ideally imperfect crystal embedded in the perfect crystal and reflects X-Rays according to kinematical theory. The reflected intensity is thus proportionnal to the volume of the distorted area (Authier conference). The size of this area depends on :
1) the value of δ which is

$$\delta \sim \frac{2|x_h|}{\sin 2\theta_B} \propto \frac{\lambda^2 F_h}{\sin 2\theta_B}$$

x_h is the h Fourier coefficient of the electrical susceptibility.
δ depends on the wavelength and on the structure factor F_h of the considered reflection. Its order of magnitude is a second of arc.

Fig. 4
a) Section topograph of a KDP crystal, (020), MoKα. A,B : images of stress relaxation distortions at the intersections of two different dislocations ; one cuts the entrance surface (A), the other the exit surface (B).
b) the corresponding traverse topograph.

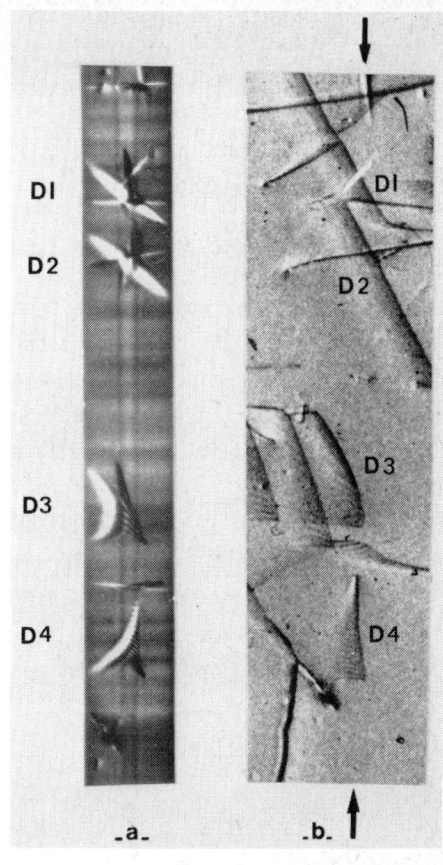

Fig. 5
a) Section topograph of a silicon crystal, (220), MoKα.
b) the corresponding traverse topograph. Note the blurring of the image as compared to the section.

Several cases may be considered and can occur simultaneously :
a) the two regions are translated with respect to one another through a translation which is not a translation of the lattice (stacking fault (21,22), antiphase boundary (23))...
b) the reciprocal lattice vector is different in the two regions (I and II are rotated with respect to one another (21) or/and have different lattice parameters. In each case characteristic fringes appear which have been experimentally observed and theoretically explained either using stationnary phase method (24) or by computer simulations (25, 26).

5.- SYNCHROTRON WHITE RADIATION TOPOGRAPHY (review articles 27, 28)

The source has very different characteristic features as compared to X-Ray tubes (Coïsson conference) :
- high flux of photons ;
- white radiation ;
- small emission divergence (order of 20 sec. for $\lambda \sim 1\text{Å}$ DCI ring in Orsay France) ;
- larger focus (order of 1,5mmx6mm) but located far (order of 20 m) from the sample.

Consequently, the divergence of the beam irradiating a point in the sample is of the same order as for classical X-Ray tube arrangements.
- due to the curvature of the electron orbit in the ring, the lateral extension of the beam is wide enough to cover the whole sample, every point of it being at Bragg incidence for slightly different wavelengths if the crystal is distorted. Consequently :

a) one obtains, without any translation, an image of the whole crystal even if it is made of several grains and several reflections alltogether.
b) the analysis of the contrast is different from classical Lang method (29, 30).
c) due to the high flux of photons, time exposures are reduced by a factor of the order of 50 allowing dynamic experiment (exposure time $\sim$ 20 sec.)(31, 32).

6.- DOUBLE CRYSTAL TOPOGRAPHY

The principles of such arrangements are given in (33, 34, 35).

A first monochromator crystal is designed to obtain a monochromatic beam with an angular divergence $\Delta\theta$ smaller than the intrinsic width of the sample reflexion profile δ. Consequently there are no more "direct images" but a high sensitivity to small local distortions and the possibility to determine the sign of the distortion (34).

If the crystal is adjusted far from the maximum of the rocking curve, only those parts of the crystal which are very distorted (dislocation core for example) will contribute to the image leading to very thin images and thus a high resolution (37).

REFERENCES

[1] U. BONSE, M. HART, J.B. NEWKIRK : Advan. X-Ray Anal. 10, 9 (1967).
[2] A. AUTHIER : Advan. X-Ray Anal. 10, 9 (1967).
[3] A.R. LANG : Advan. X-Ray Anal. 10, 91 (1967).
[4] A.R. LANG in Modern Diffraction and Imaging Techniques in Material Science edited by S. Amelincks and al. North Holland p. 407 (1970).
[5] A. AUTHIER in Modern Diffraction... p. 481.
[6] A. AUTHIER in X-Ray Optics edited by H.J. Queisser, Springer-Verlag, p. 145 (1977).

[7] A.R. LANG : J. Appl. Phys. 29, 597, (1958).
[8] N. KATO : Acta Cryst. 14, 526 and 627 (1961)
[9] J. MILTAT, D.K. BOWEN : J. Appl. Cryst. 8, 657 (1975).
[10] P. PENNING, D. POLDER : Philips Res. Rept. 16, 419 (1961)
[11] P. PENNING : Thesis Delft (1966)
[12] N. KATO : J. Phys. Soc. Jap. 18, 1785 (1963).
[13] N. KATO : J. Phys. Soc. Jap. 19, 67, 971 (1964).
[14] C. MALGRANGE : International Summer School on X-Ray dynamical theory and topography. Limoges, France (1975).
[15] E. DUNIA, C. MALGRANGE, J.F. PETROFF : Phil. Mag. (1979), to be published.
[16] J.F. PETROFF, A. AUTHIER : Phys. Stat. Sol. 13, 373 (1966).
[17] N. KATO, Y. ANDO : J. Phys. Soc. Jap. 21, 964 (1966).
[18] Y.M. FISHMAN and V.G. LUTSAU : Phys. Stat. Sol. (a), 18, 443 (1973).
[19] N. KATO, J.R. PATEL : J. Appl. Phys. 44, 965, 971 (1973).
[20] Y. ANDO, J.R. PATEL, N. KATO : J. Appl. Phys. 44, 4405 (1973).
[21] A. AUTHIER, M. SAUVAGE : J. Physique 27, C3, 137 (1966).
[22] A. AUTHIER, J.R. PATEL : Phys. Stat. Sol. 27a, 213 (1975).
[23] B. CAPELLE, C. MALGRANGE : Phys. Stat. Sol. 20a, K5 (1973).
[24] A. AUTHIER : Phys. Stat. Sol. 27, 77 (1968).
[25] A. AUTHIER, Y. EPELBOIN : Phys. Stat. Sol. 41a, K9 (1977).
[26] T. KATAGAWA, H. ISHIKAWA , N. KATO : Acta Cryst. A31 , 246 (1975).
[27] B.K. TANNER : Progress Cryst. Growth 1, 23 (1977).
[28] M. SAUVAGE, J.F. PETROFF in "Synchrotron Radiation Research" edited by S. Doniach and H. Winick, Plenum Press, New-York (1979).
[29] B.K. TANNER, D. MIDGLEY, M. SAFA : J. Appl. Cryst. 10, 281 (1977).
[30] M. HART : J. Appl. Cryst. 8, 436 (1975).
[31] J. GASTALDI, C. JOURDAN : Phys. Stat. Sol. 49a, 529 (1978).
[32] J. MILTAT , D.K. BOWEN : J. Physique, 40, 389, (1979).
[33] U. BONSE, E. KLAPPER : Z. Naturforsch 13a, 348 (1958)
[34] U. BONSE : Z. Physik, 153, 278 (1958).
[35] H. HASHIZUME, K. KOHRA, T. YAMAGUCHI : Appl. Phys. Letters, 18, 213 (1971).
[36] U. BONSE : Z. Physik, 184, 71 (1965)
[37] J.F. PETROFF, M. SAUVAGE, P. RIGLET : Phil . Mag. To be submitted.

EXAMPLES OF X-RAY TOPOGRAPHIC RESULTS

A. MATHIOT
Departement de Metallurgie
Centre d'Etudes Nucleaires
85X - 38041 GRENOBLE CEDEX - FRANCE

INTRODUCTION

X ray diffraction by crystals has been extensively described former in this book and the aim of this contribution is, once the principles have been recalled, to look at the possible uses for the X ray topographic tool. We shall first outline the most important characteristics of the topographic image and then try to show some typical applications of these techniques. This review cannot be exhaustive, because the fields of application as well as the users are becoming increasingly numerous, since its invention by Lang in 1958(1). In particular the advent of synchrotron radiation facilities opened recently a new field for this tool, owing to the extremely intense white spectrum available. This paper will be confined to classical transmission topography, and the reflection techniques as well as the double crystal assessments will not be reviewed here. However the interested reader can have a look at the comprehensive reviews on that subject by Lang(2), Authier(3) or Tanner(4). Some specific applications will be described in order to show that X ray topography can be used either as a characterization technique for the study of defects in nearly perfect crystals or as a tool for a deeper investigation of elastic fields inside the crystal by means of models and simulation.

CHARACTERISTICS OF THE X RAY TOPOGRAPHIC IMAGE

The origin of the image lies in a perturbation of the ray trajectories inside the crystal due to the presence of elastic displacements. Those displacements arise from singularities in the crystal lattice configuration. So the contrast of the image is mainly due to defects and the size of the image is strongly dependent on the amplitude and derivatives of the elastic displacement.
As has been shown by Malgrange(5) the kinematical image is created from the direct beam in strongly distorted regions so that its size is usually quite small (a few microns). The dynamical image, due to perturbations inside the Borrmann fan of the wave field propagation in less distorted regions, is generally much larger than the kinematical one and can reach 100μm or more.
From the size of the images two important limitations arise for the study of defects: the resolution limit cannot be better than 1μm and practically will be of the order of 5 to 10μm. This is why it will not be possible by this technique to look conveniently at crystals containing more than a few defects (maximum dislocation density $10^4 cm^{-2}$). This means that applicability of X ray topography is restricted to quite perfect crystals.
To balance these disadvantages the interests of the tool are that it is non-destructive and that it can manage with quite macroscopic samples (a few cm^2 in surface and 50μm to few mm in thickness).

The exposure time is strongly dependent on the source used as well as on the detector (6). As an example the exposure time required for a Lang traverse topograph with a high brilliancy rotating anode X ray generator on a moderately absorbing crystal can be of the order of 10h/cm, if recorded on a nuclear plate. The same image can be obtained in 1mn with the white synchrotron radiation at LURE-DCI (1.8GeV, 100mA) and can also be recorded directly with poor resolution on a TV detector (exposure time 0.04sec).

X RAY TOPOGRAPHY AS A CHARACTERIZATION TOOL

The few following examples will try to show the variety of fields open to the topographic techniques in materials science.

1. CRYSTAL GROWTH

Looking at the local strains inside the crystal allows the experimentalist to follow the growth habit from the nucleation, to understand the growth mechanisms and to rebuild the growth history of the sample.

A-Naturally occurring crystals(7,8,9)

The non-destructive study of diamond by Lang(10) is one of the most beautiful applications in that field. X ray topography is able to show the strain field inside the crystal without slicing it off. Fig. 1 shows a section topograph through a slightly rounded natural diamond of about 5mm edge length, with axis [001] vertical. The natural growth faces in diamond correspond usually to octahedral growth sectors with {111} facets and cuboid rounded surfaces with mean orientation {100}. The sample presented here was enveloped in a layer of coat which prevented optical observation. The section passes 0.75mm distant from the center of the crystal. The areas appearing very dark on the print represent zones of non-faceted growth which are densely populated by minute bodies of unidentified composition. Due to local strain these bodies produce intense diffraction contrast from the matrix surrounding them.

Fig.1: Section topograph through a natural diamond; reflection $\overline{4}04$, MoKα

B-Synthetic crystals grown from aqueous solution(11,12)

X ray topography is widely used in order to determine the optimal growth parameters in crystals. Fig. 2 to 4 show the results obtained by Gits-Leon & al.(13) on the effect of stirring on the quality of solution grown potash alum. The crystals are obtained by the temperature decrease of a saturated mother solution of potash alum (from 40 to 25°C). The seeds are (001) platelets cut from a previous crystal,

which are kept moving in the bath to obtain a constant flux velocity. Fig. 2 shows the main characteristics of the growth and the crystallographic orientation of the samples:
-The {111} and {001} facets appear simultaneously.
-The growth rate of the {001} facets is governed by the existence of screw dislocations. These dislocations are nucleated on local stress concentrations due to impurity segregation on the growth front (on the seed or on inclusions as at I).

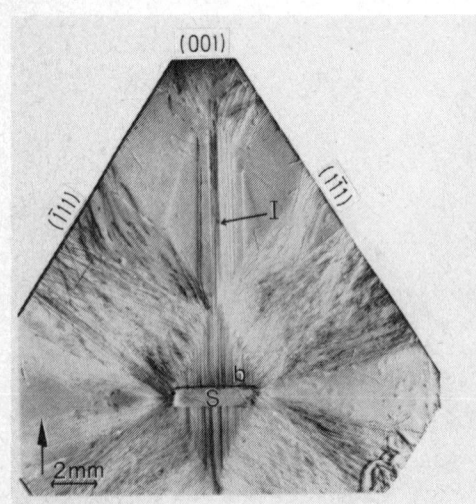

Fig.2: Potash alum
220 topograph, MoKα

Fig.3: Potash alum; 2̄2̄0 topograph MoKα showing the influence of flux direction

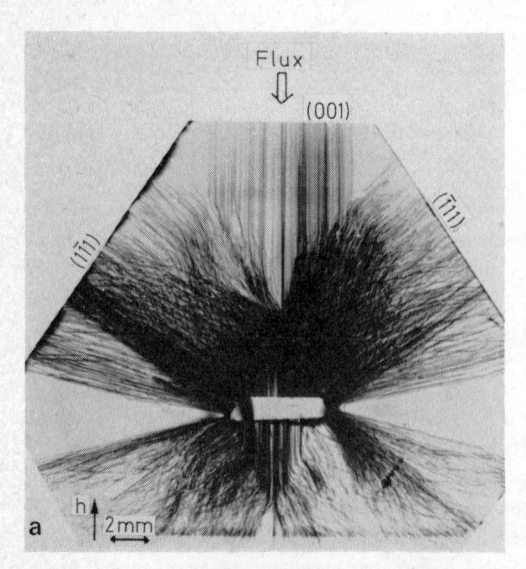

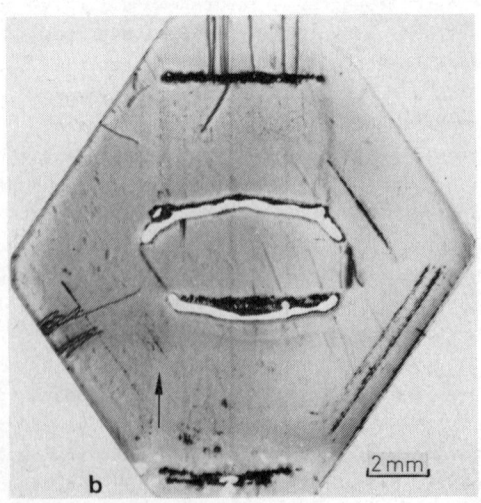

Fig.4: Influence of flux velocity on perfection. 2̄2̄0 topographs, MoKα. a) V=14 cm/sec   b) V=0

—When these dislocations meet the limit between two growth sectors they are "refracted" in order to follow a normal to the growth front.

The variations in growth rate explain the fact that the {001} facets alternatively appear and disappear on the growth habit of the crystal. Fig. 3 shows the influence of the flux direction on the growth habit. The dislocations are mostly visible in front of the flux. In the back sectors, due to turbulence inside the liquid, the supersaturation may vary so that liquid inclusions can be embedded in the solid. They stop the dislocation propagation, and the crystal perfection looks better with the {001} facets growing slower and wider. Fig. 4 shows the influence of flux velocity on the quality of the crystals. On fig. 4a the relative velocity is 14cm/sec, when on fig. 4b it is zero. For high speeds the dislocation density is higher, many among them being of edge character do not contribute to the growth. For low speeds fewer dislocations and more inclusions are visible, due to a reduction of the solution exchange near the growing crystal, resulting in a better crystalline quality.

C-Dynamic study of recrystallization(14).

Fig. 5 shows an example of in-situ experiment by mean of synchrotron radiation. A slice of high purity aluminium, after a 3% critical strain, is heated directly in front of the beam, between 300 and 400°C. The rearrangement of dislocations gives rise to small nuclei of perfect crystal. The recrystallization proceeds by the growth of those seeds at the expense of the deformed matrix. These grains are of random orientations and the fact that the incident beam is polychromatic allows to get Laue spots for some of them, corresponding to appropriate Bragg reflections. So that it is possible to follow the growth by recording these spots which are in fact topographic images on a film behind the furnace. Topographs of fig. 5 have been obtained by exposing X ray films 10sec every 3mn. They show the expansion of the grain by low indices facets and the interaction of these facets with neighbouring recrystallized grains, which results in a deformation of the growth habit (fig. 5c & d).

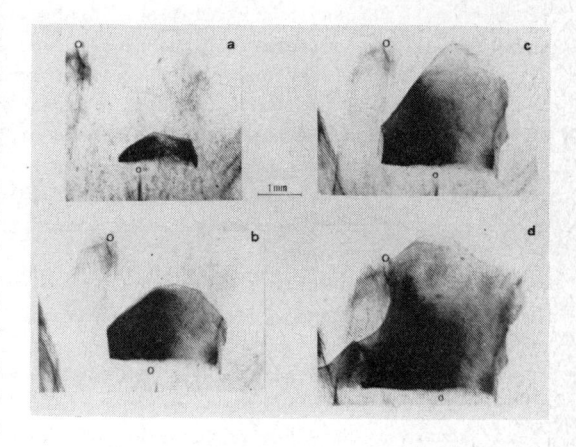

Fig.5: Synchrotron radiation topographs of Al recrystallization showing the first stages of a crystal growth

2. SYSTEMATIC STUDY OF DEFECTS IN CRYSTALS
A-Grain boundary dislocations(15).

Germanium bicrystals are a good model for a systematic study of grain boundaries. X ray topography is used here as a complement of electron microscopy, then it allows the examination of macroscopic samples on a larger scale. Apart from the so-called primary grain boundary dislocations of the classical Frank model for small angle grain boundaries, two types of dislocations are visible on fig. 6:

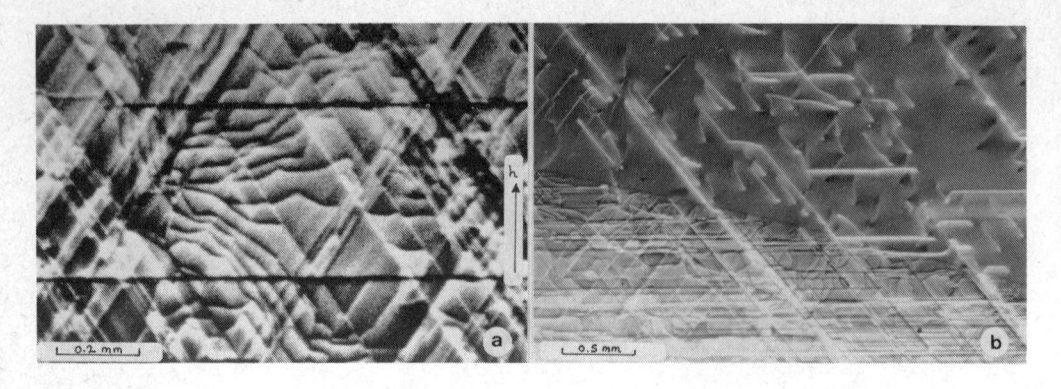

Fig.6: Grain boundary dislocations in Germanium. 400 reflection, MoKα
a) Secondary dislocations   b) Extrinsic and matrix dislocations

-For special disorientations between the two crystals the total elastic energy is minimum; these particular respective orientations correspond to high densities of sites in coincidence position for both lattices. The crystals shown here can be superposed by a rotation of angle $38.94°$ around [110] which correspond to a twin index $\Sigma=9$. The so-called secondary dislocations account for a small deviation from the exact coincidence orientation in the same way as for primary dislocations in a small angle boundary. Their spacing D is related to their Burgers vectors b (which are quite smaller than those of primary dislocations) and to the deviation angle $\Delta\theta$ by the well known Frank relation: $D \times \Delta\theta = b$. When $\Delta\theta$ is very low as in fig. 6a ($\Delta\theta \leq 0.005°$) these dislocations can be resolved on X ray topographs.

-The extrinsic dislocations are matrix dislocations which end up on the grain boundary and then follow a straight direction which is the intersection of the boundary plane and the glide plane of the dislocation. This trapping by the boundary is clearly visible on fig. 6b. The identification of these dislocations is in progress(16) but some difficulties arise from the numerous reactions which take place on the boundary plane between secondary and extrinsic dislocations.

B-Magnetic domain walls.

In two adjacent domains of a single crystal two different magnetization directions produce two different magnetostrictive deformations. The domain wall corresponds to a discontinuity in the magnetoelastic field(17). When the angle between the two magnetization directions is not 180° this elastic singularity results in a contrast on X ray topographs(18), which allows a study of the domain configuration in magnetic samples.

Fig. 7 shows the evolution of domain structure in a (110) slice of ferrimagnetic Terbium-Iron garnet ($Tb_3Fe_5O_{12}$, TbIG) at low temperatures(19). In TbIG magnetization, magnetic anisotropy and magnetostriction increase notably when the temperature is lowered. This comes from a progressive reorientation of the magnetic moments in the ferrimagnetic lattice(20). As the domain number is roughly proportional to the square of the magnetization, it can be seen on fig. 7 that the wall images (vertical straight lines) become increasingly numerous for low temperatures. Moreover the

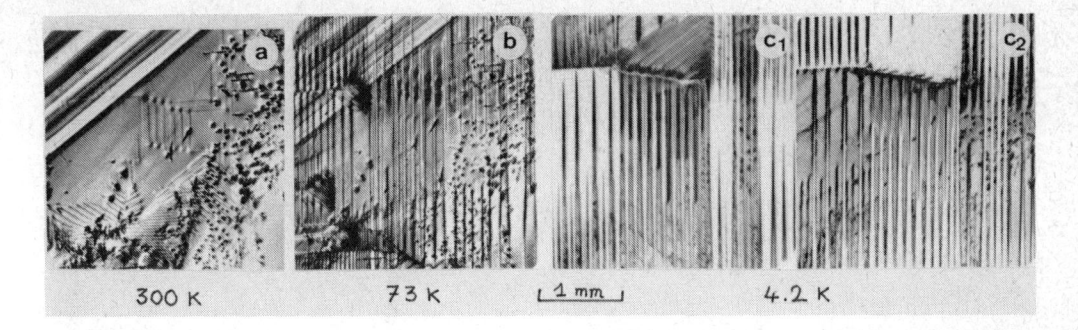

Fig.7: Evolution of domain structure in TbIG with temperature $\overline{444}$ reflection, MoKα

magnetostrictive deformation increases so that the departure from Bragg angle(5) may become larger than the width of the rocking curve. For such cases (fig. 7c) the Bragg peak splits into two parts, so that one can obtain two distinct complementary topographic images of the crystal, each of them corresponding to one set of domains with parallel magnetization. The measurement of this splitting allowed a direct determination of the magnetostriction coefficient for this garnet(21).

Another striking example of domain wall study is shown on fig. 8. This is a dynamic experiment performed at LURE-DCI(22). By means of an alternative magnetic field(80Hz) the domain walls in this Fe-3%Si crystal are vibrating and a stroboscopic topographic image is obtained by integration on a fixed phase. The wall images appear thin enough in the whole sample, which seems to prove the reversibility of the wall motion. An instability can be noticed at point P, where a precipitate is embedded in the crystal. This indicates an elastic interaction between the precipitate and the oscillating wall. The systematic study of such interactions between walls and defects is in progress.

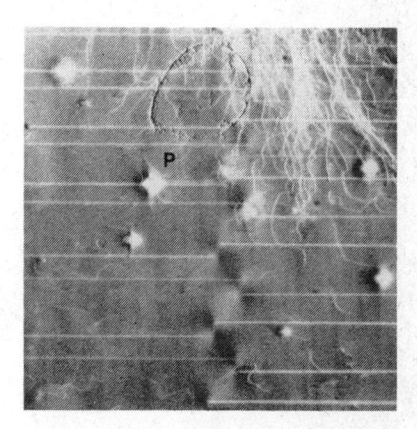

Fig.8: Synchrotron radiation stroboscopic image of oscillating walls in Fe-Si crystal Interaction with precipitate at P.

3. APPLICATION TO ELECTRONICS

As a non-destructive method X ray topography is widely used during the processing of electronic devices and correlated with electrical fault analysis.

Topograph of fig. 9 represents a (111) silicon wafer during the fabrication of integrated circuits(23). The surface has been oxidized and the oxide layer has been removed by etching on selected square areas. Then a phosphorus diffusion has been performed at high temperature(1050°C). Three characteristic sets of defects are visible and can be correlated with the electrical properties of the device:
-Emitter edge dislocations due to stress relaxation at the edge of the oxide layer, which run parallel to the surface from one edge to another one.

-Straight dislocations (lines along <110> and <112> directions) which limit stacking faults in the {111} planes, running likewise parallel to the surface.
-Extended defects inside the oxide windows which have been shown to correspond to local phosphorus overconcentrations.

Fig. 10 corresponds to another application in the field of optoelectronics(24). Deuterated potassium dihydrogenphosphate (D-KDP) is used as electro-optical modulator in optical relays which allow the transfer of an electronic, X ray or visible image by means of local variations of the refraction index due to an electric charge pattern. This device is mentioned by Clair in this book. The topographic image of the D-KDP crystal (fig. 10a) shows some dislocations and four growth sectors corresponding to growth along [001] and [010] directions (vertical and horizontal) through {110} facets. Those defects are electrically active and produce a damage in the stored image shown on fig. 10b after a few tens of seconds.

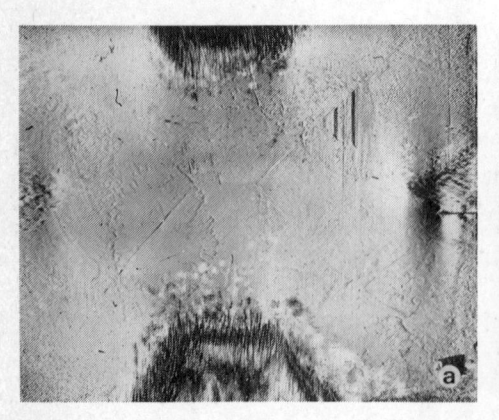

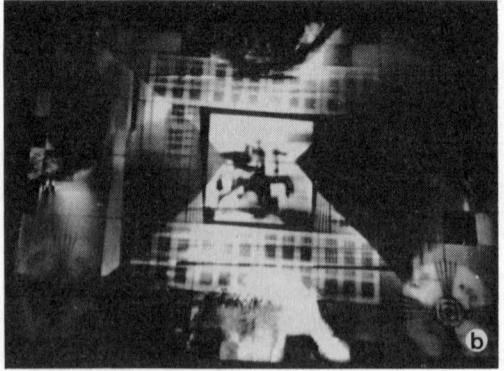

Fig.9: Silicon with oxide windows and P diffusion, 440 reflection, MoKα

Fig.10: a) X ray topography of D-KDP single crystal with growth defects
b) Damage on the TV image electrically stored in the crystal due to the defects

X RAY TOPOGRAPHY AS A TOOL FOR INVESTIGATION

Three examples are given to show that this technique can help for the quantitative investigation of elastic fields in the crystal when it is possible to give a model for the distortion inside the crystal.

1. SILICON WITH SUPERPOSED OXIDE FILM (25)

As it has been mentioned above the oxide layer on silicon single crystals produces local stress concentrations which can be relaxed during heat treatment by emission of dislocations. These stresses are due to a lattice difference between layer and substrate. A model of this deformation has been built and section topographs show a perturbation of the Pendellösung fringes with appearance of new elliptical fringes. This perturbation arises from the small strain gradient introduced in the substrate by the oxide film. These dynamical images can be analyzed in terms of the Eikonal theory, by computing the wave-field trajectories inside the Borrmann fan. Starting from the elastic model of the deformation this computation gives the exact phase integral for each wave-field and results in the fringes positions at the exit surface of the crystal. A comparison with the experimental positions, as measured on the section topographs for different oxide film thicknesses allowed a determination of the force per film unit length, and an evaluation of the stress inside the film.

2. CONTRAST OF A Y SHAPED MAGNETIC DOMAIN WALL JUNCTION (26)

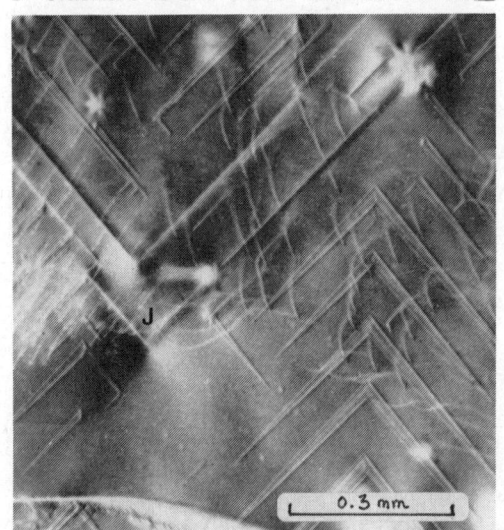

Fig.11: Domain walls in Fe-Si single crystal. Y shaped junction at J. 0$\bar{2}$0 reflection, MoKα.

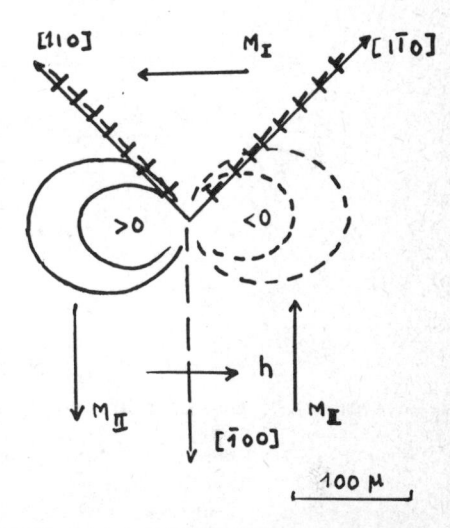

Fig.12: Representation of the junction, same orientation as fig.11; M_I & M_{II}: magnetization; solid and dashed curves are equi-G; h= diffraction vector

The magnetic domain wall configuration in Fe-3%Si at equilibrium exhibit a lot of junctions between two or three walls(fig. 11). The Y shaped junction consists in two 90° {110} walls and one 180° {100} wall (point J on fig. 11, fig. 12). The dynamical contrast of such junctions can be described with the help of continuous theory of dislocations(27) as the intersection of two planar quasi-dislocations densities (with infinitesimal Burgers vector) along the 90° walls. These walls produce an incompatible state of deformation (due to the differences in magnetostrictive deformation) which is equivalent to the presence at the junction of a wedge disclination of angle $3\lambda_{100}$, where λ_{100} is the magnetostriction coefficient. Starting from this model it is possible to compute the total displacement field and the strain gradient G which governs the dynamical contrast for such small deformations. Taking into account the surface relaxation the equi-G curves have been plotted (fig. 12) and correspond fairly well to the image's shape. Thus the disclination model of the junction is in good agreement with the experimental results.

3. CROSS SLIP OF ISOLATED DISLOCATIONS ON THE SURFACE OF SILICON CRYSTALS (28)

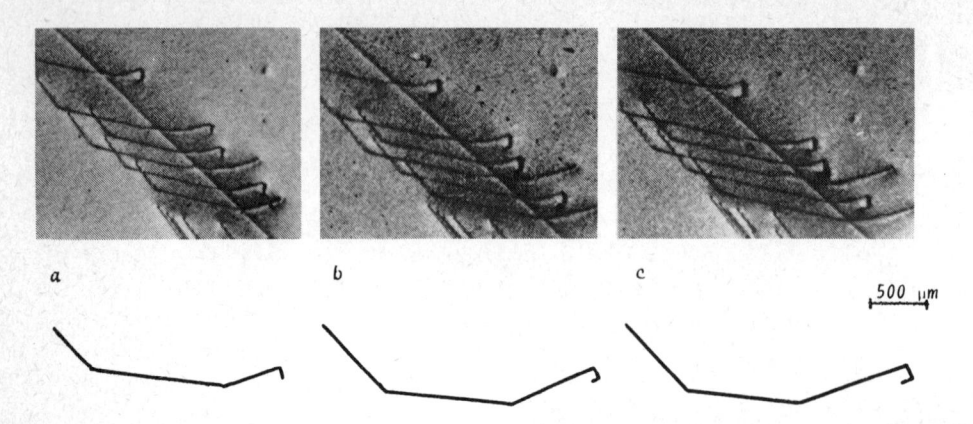

Fig.13: Cross slip of isolated dislocations on silicon surface. $2\bar{2}0$ reflection, MoKα.

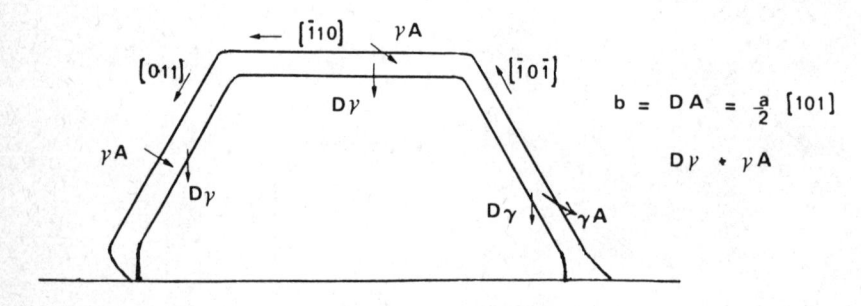

Fig.14: Geometry of the dislocation loop when dissociated

The study of plastic deformation and dislocation movement is one of the very interesting applications of X ray topography. On the surface of a dislocation free silicon crystal a scratch can act as a source of hexagonal half loops of dislocations during a high temperature tensile experiment. The movement of such dislocations can be followed by X ray topography. Moreover in-situ experiments are now in progress with the synchrotron radiation at LURE-DCI. Under specific conditions one can observe, as shown on fig. 13, that the half loops exhibit cross-slip near the surface. This phenomenon can only be interpreted in terms of dissociated dislocations. Fig. 14 shows the geometry of the loop in its $(11\bar{1})$ glide plane. The Burgers vector is 1/2[101], one segment is a pure screw, the two others are 60° dislocations. If the loop is dissociated in two Shockley partials with Burgers vectors 1/6[112] and 1/6[2$\bar{1}$1], surrounding an intrinsic stacking fault ribbon, the geometry is that of fig. 14. Electron microscopic observations have shown on Cu-Al that, near the surface, the partials tend to a screw orientation(29), which results in a narrowing or widening of the fault ribbon as shown on fig. 14. This configuration has been shown to be energetically favourable. This explains that the segment where both the dissociation width is narrower and the two partials have a dominant screw component can easily cross-slip. This last example shows that a lot of microscopic informations can be obtained on the basis of the macroscopic results from X ray topographs.

CONCLUSION

This brief review of X ray topography applications to materials science was not complete at all. But the aim was, above all, starting from various examples, to show the most important fields where it can help. As it is mostly sensitive to elastic fields, this technique has first been used as a characterization tool for the defects' study. But the current developement of simulation procedures allows deeper studies whenever elastic models can be built. Furthermore the recent appearance of intense, tunable polarized sources allows to foresee a lot of new applications both for the investigation of diffraction processes in crystals and for the study of dynamic phenomena.

REFERENCES
(1) LANG A.R.(1958) J. Appl. Phys. 29, 597
(2) LANG A.R. Diffraction and Imaging Techniques in Material Science. Amelincks, Gevers, van Landuyt eds. North-Holland(1978)
(3) AUTHIER A. X Ray Optics. Queisser ed. Topics in Applied Physics Vol.22, Springer(1977)
(4) TANNER B.K. X Ray Diffraction Topography, Int. Ser. Sc. Sol. St. Vol.10, Pergamon(1976)
(5) MALGRANGE C. This book
(6) COISSON, MILTAT : this book
(7) LANG A.R.(1974) J. Cryst. Growth 24/25, 108
(8) SAUVAGE M. and AUTHIER A.(1965) Bull. Soc. Fr. Min. Crist. 88, 379
(9) ZARKA A.(1974) J. Appl. Cryst. 7, 453
(10) SUZUKI S. and LANG A.R.(1976) Diamond Research 1976 p.39
(11) AUTHIER A.(1972) J. Cryst. Growth 13/14, 34
(12) IZRAEL A., PETROFF J.F., AUTHIER A. and MALEK Z.(1972) J. Cryst. Growth 16, 131
(13) GITS-LEON S., LEFAUCHEUX F. and ROBERT M.C.(1978) J. Cryst. Growth 44, 345
(14) GASTALDI J. and JOURDAN C.(1978) Phys. Stat. Sol.(a) 49, 529
(15) BACMANN J.J., MATHIOT A., SILVESTRE G. and DE TOURNEMINE R.(1977) IVth Eur. Cryst. Meet., Oxford p.646
(16) MATHIOT A. and GAY M.O. (to be published)
(17) CHIKAZUMI S. Physics of Magnetism, Wiley(1964)
(18) POLCAROVA M. and LANG A.R.(1962) Appl. Phys. Lett. 1. 13
(19) MATHIOT A. and PETROFF J.F.(1976) J. Appl. Phys. 47,1639
(20) NEEL L., PAUTHENET R. and DREYFUS B.(1964) Prog. Low Temp. Phys. 4, 344
(21) PETROFF J.F. and MATHIOT A.(1974) Mat. Res. Bull. 9, 319
(22) MILTAT J. private communication
(23) ROLLAND G. These 3eme cycle, Grenoble(1973)
(24) BELOUET C., DUNIA E. and PETROFF J.F.(1974) J. Cryst. Growth 23, 243
(25) KATO N. and PATEL J.R.(1973) J. Appl. Phys. 44, 965
     PATEL J.R. and KATO N. ibid p.971
(26) MILTAT J.E. and KLEMAN M.(1973) Phil. Mag. 28, 1015
(27) KRONER E.(1958) Kontinuum Theorie der Versetzungen und Eigenspannungen. Springer
(28) GEORGE A. These d'Etat Nancy(1977)
(29) HAZZELDINE P.M. KARNTHALER H.P. and WINTNER E.(1975) Phil. Mag. 32,81

NEUTRON TOPOGRAPHY

M. Schlenker[1,2] and J. Baruchel[1]

(1) Laboratoire Louis Néel du CNRS, associé à
l'Université Scientifique et Médicale de
Grenoble
B.P. 166,
38042 Grenoble - France

(2) Institut Laue Langevin
B.P. 156,
38042 Grenoble - France

Introduction: The preceeding papers in this volume should make it very easy to understand neutron topography: just like its older sister, X-ray topography[1,2], it consists in imaging singularities in single crystals via local variations of reflectivity in Bragg reflections. Because the wavelengths and scattering amplitudes involved are alike[3], the diffraction processes both in perfect and in imperfect enough crystals are very similar for neutrons and for X-rays, and indeed some of the mechanisms responsible for the contrast of defects are the same in either case[1].

There are, however, three major differences[4], and they determine the limitations as well as the value of neutron topography. We will examine them in turn:

- neutrons are few;
- in most materials, their absorption is weak;
- the neutron has a magnetic moment.

1. Neutrons are few

There are rather few neutron beams in the world: this certainly is the reason for the late start[5] of neutron topography. Furthermore the intensities available in a small wavelength range even at a high-flux reactor[6] are small compared to what a conventional X-ray tube offers in a characteristic emission line; consequently, the resolution of neutron topography is poor - no better than $\sim 4 \cdot 10^{-2}$ mm typically - basically because exposure times cannot be increase above an order of magnitude of about a day, and therefore larger beam divergences and wavelength ranges have to be tolerated, leading to a geometrical loss of resolution.

On the other hand, this allows neutron topography to be instrumentally simple: a beam wide enough to illuminate the part of the crystal to be imaged is monochromatized, Bragg diffracted by the specimen and recorded on a neutron-sensitive photographic detector - typically X-ray dental film with a Gd converter foil. Using a thin beam and traversing the specimen, as in Lang's method of traverse X-ray topography [1], is a luxury that would increase the exposure time too much. The geometric resolution, determined by the specimen-detector distance and by the divergence of the Bragg-diffracted beam, can be tailored to a large extent by the choice of the relative setting of the monochromator and specimen as well as by collimation of the primary white beam. One possibility is to work with a low-divergence primary beam: the very clean beam emerging from the end of a curved neutron guide-tube is particularly suitable and, since the massive shielding surrounding the monochromator in "normal" neutron diffractometry becomes unnecessary, the instrument can be compact, flexible and cheap. The instrumentation can even be reduced to its very simplest form, nothing, by using the collimated white beam directly: each Laue spot is then a topograph, just like in synchrotron radiation X-ray topography[1]. Alternatively, neutron topography can be performed on an ordinary thermal neutron beam[7], or a set-up can even be installed behind an existing experiment by pushing back the beam-stop, and using some of the neutrons that would otherwise be wasted.

2. <u>Low absorption of neutrons</u>

The difficulty of low counting rates is of course also encountered in "normal" neutron scattering work, e.g. in structure determinations, and there it is often obviated simply by the use of large specimens, made possible because the absorption of neutrons by most materials is very small indeed[4]. Although it would be futile to hope to improve resolution and/or exposure times in neutron topography by using large crystals, this property of low absorption does make neutron topography very valuable in those situations where X-ray topography is impractical because the X-ray absorption factor, $\mu_x t$ where t is the specimen thickness, is large. This can obviously occur if μ_x is large and/or if t is large, and in both cases boils down to heavy crystals.

The situation is straightforward for thin crystals containing heavy elements (high μ_x): the growth history of natural crystals of lead carbonate $PbCO_3$, cerusite, could thus be, to some extent, reconstructed by observing the growth defects[8] - an investigation that just cannot be done in transmission with X-rays.

For thick crystals, there is an intrinsic difficulty in unravelling the complicated

3-dimensional arrangement of defects from a two-dimensional image. One way out is of course to restrict the study to just a slice of the specimen; neutrons offer the pleasant possibility of looking at a slice without cutting it, just by using the principle of section topography[1], originally developed by Lang for the X-ray case, and which acquires a new dimension with neutrons because really large (cm-thick) crystals can be investigated[9].

Fig. 1 shows a normal (projection) topograph and a section topograph of an as-grown crystal of terbium iron garnet, kindly given by J.P. Remeika (Bell Telephone Laboratories): considerably more detail can be seen on the section topograph, where the investigated region is limited to a slice 1 mm thick. In conjunction with a cruder but much faster Polaroid detector, this technique is now routinely used to select appropriate regions before cutting out monochromators from large crystals.

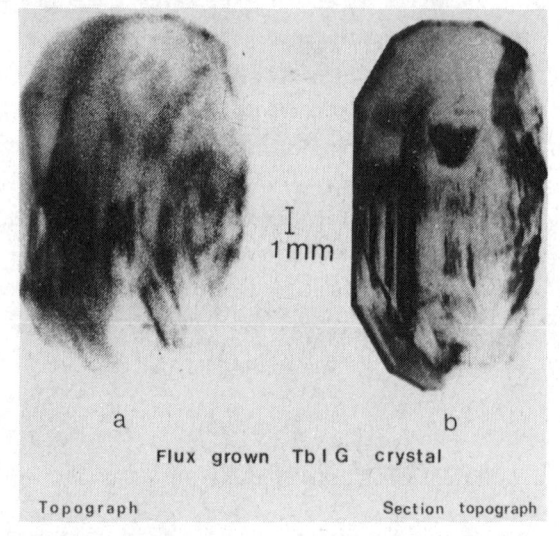

Fig. 1 a) Projection, b) Section topographs of flux-grown TbIG single crystal; λ=4.4 Å. 400 reflection; graphite monochromator using 0002 reflection, mosaic spread 80'. Exposure times a) 1 hour, b) 14.5 hours.

3. The neutron has a magnetic moment

Because neutrons can sense the distribution of magnetic moments they have become the major tool in magnetic structure determination work[4]. In the topographic approach, they can, in principle, provide <u>direct</u> imaging of magnetic domains of all kinds, ferro, ferri, or antiferromagnetic, because different arrangements and/or directions of magnetic moments will lead to different reflectivities in some Bragg reflections[10].

The pioneering work in this direction was a very daring and elegant investigation of spin-density-wave domains in antiferromagnetic chromium by Ando and Hosoya[11], who found a disagreement of several orders of magnitude between the domain sizes observed and those calculated from the then accepted theoretical model.

Later work focused on more classical systems, ferromagnetic and antiferromagnetic. In the case of ferromagnetic domains, the interest was mainly in the optics of neutron propagation in almost perfect magnetic crystals, the theory of which was recently outlined by several authors [references in (12)]; as an example, fig. 2 shows a polarized neutron topograph of ferromagnetic domains in Fe-3% Si.

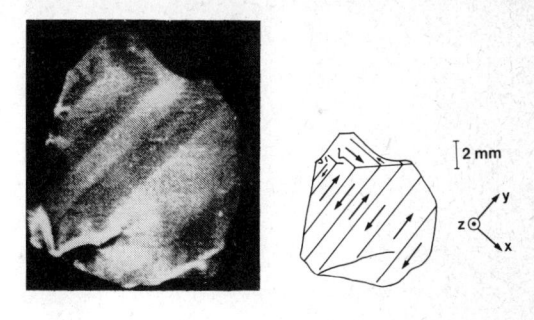

Fig. 2: a) Polarized neutron topograph of Fe-3% Si specimen, 0.1 mm thick. λ = 1.35 Å, 110 reflection; Cu$_2$MnAl monochromator-polarizer, using 111 reflection. b) schematic diagram of domain structure

In the case of antiferromagnetic domains, a study of the best-known material, NiO, provided direct confirmation[13] of the accepted model of the magnetic moment arrangement; fig. 3 shows the T-domains simultaneously imaged and characterized as to their propagation vector by use of magnetic (superstructure) reflections.

Recently it was possible using this technique, to directly eliminate one of the suggested possibilities for the magnetic moment direction[14].

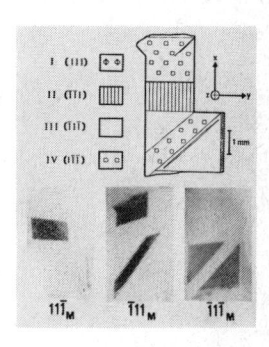

Fig. 3: a) 11$\bar{1}$, b) $\bar{1}$11, c) $\bar{1}$1$\bar{1}$ magnetic (superstructure) reflections from specimen of NiO; d) Diagram of T-domain structure. Note that the 111 reflection could not be found - in agreement with the absence of domains with propagation vector 111.

The investigation of MnF$_2$, a case where 180° antiferromagnetic domains appear to be physically distinguishable, is under way; these domains, shown in fig. 4 as seen

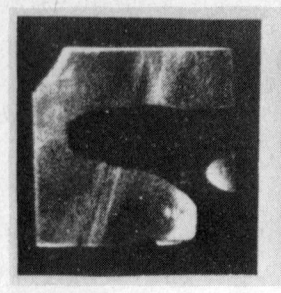

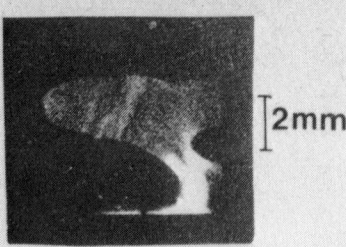

Fig. 4: Polarized neutron topographs of MnF_2 specimen at 20 K. $\lambda=1.35$ Å. 210 reflection, with incident neutron polarization reversed. Only one type of domains is visible on each.

by polarized neutron topography, are, to the best of our knowledge, not visible by any other technique; their physics is a completely new field, only poorly understood at the moment.

Conclusion

Neutron topography definitely has poor resolution - it is in this respect closer to the magnifying-glass than to microscopy. But it has the capability of investigating very large single-crystal specimens of most materials - centimeters thick - and to reveal their crystal defects. And it is an invaluable technique in magnetism because it can directly reveal the shape, size and arrangement of magnetic domains of all kinds and at the same time characterize the arrangement of magnetic moments in the domains. New physics is beginning to emerge from such investigations.

References

(1) C. Malgrange, X-ray topography: principles (this volume)
(2) A. Mathiot, Applications of X-ray topography (this volume)
(3) G.L. Squires, Interaction of thermal neutrons with matter (this volume)
(4) G.E. Bacon, Neutron diffraction, 3rd edition, Clarendon Press, Oxford, 1975
(5) K. Doi, N. Minakawa, H. Motohashi, N. Masaki, J. Appl. Cryst., 4, 528 (1971)
(6) J.B Forsyth, Intense sources of thermal neutrons for research (this volume) R. Coisson, X-ray sources (this volume)
(7) C. Malgrange, J.F. Petroff, M. Sauvage, A. Zarka, M. Englander, Phil. Mag. 33, 743 (1976)
(8) J. Baruchel, M. Schlenker, A. Zarka, J.F. Petroff, J. Cryst. Growth 44, 356 (1978)
(9) H. Tomimitsu, K. Doi, J. Appl. Cryst., 7, 59 (1974); J.B. Davidson, J. Appl. Cryst.,7, 356 (1974); M. Schlenker, J. Baruchel, R. Perrier de la Bathie, S.A. Wilson, J. Appl. Phys., 46, 2845 (1975)
(10) M. Schlenker, J. Baruchel, J. Appl. Phys., 49, 1996 (1978)
(11) M. Ando, S. Hosoya, Phys. Rev. Lett., 29, 281 (1972), J. Appl. Phys. 49, 6045 (1978)
(12) A. Zeilinger, Perfect crystal neutron optics (this volume)
(13) J. Baruchel, M. Schlenker, W.L. Roth, J. Appl. Phys., 48, 5 (1977)
(14) J. Baruchel, M. Schlenker, K. Kurosawa, S. Saito (to be published)

CRYSTAL DIFFRACTION OPTICS FOR X-RAYS AND NEUTRONS

Michael Hart

Wheatstone Laboratory, King's College, London

INTRODUCTION

In the wavelength range near $\lambda = 1$ Å Bragg reflecting crystals are the only satisfactory dispersion elements for spectroscopy. Modern commercially available single crystals of silicon and germanium are inexpensive, almost uninfluenced by radiation damage and, from the diffraction viewpoint, ideally perfect.

Although perfect crystals Bragg reflect x-rays only over a very small range of incident angles, they do so with very high efficiency so that multiple Bragg reflections can be used without a serious loss of intensity. Multiple Bragg reflections allow great freedom in optical design and we will explore here the control and measurement of spectral profile, phase and polarization.

DYNAMICAL DIFFRACTION THEORY

The theory of diffraction in perfect crystals is fully developed in a number of standard texts[1-9]. Two important cases are distinguished; in the Bragg-case (reflection) the diffracted wave $\underline{K}_h$ emerges from the same (entrance) surface as the incident wave $\underline{K}_o$ while in the Laue-case (transmission) the diffracted waves emerge from the lower surface. Only in the symmetric Laue-case when the Bragg planes are normal to the crystal surface does the Bragg law give the exact Bragg angle θ_L, so that

$$2 d \sin \theta_L = n \lambda$$

In all other cases the Bragg angle is shifted slightly by refraction by an amount which depends on the optical parameters and on the relative orientations of the crystal surface and the wavevectors. The principal features of multiple reflection systems can be understood in terms of the examples shown in Fig 1 which are drawn for the symmetric Bragg-case.

Fig 1(a) Graph of the reflectivity $R(\theta)$ for the fundamental (n=1) 220 Bragg reflection from silicon at $\lambda = 1.54\text{Å}$ and for the harmonics n=2 (440 at $\lambda = 0.77\text{Å}$) and n=3 (660 at $\lambda = 0.385\text{Å}$). The σ-state of polarization is assumed.

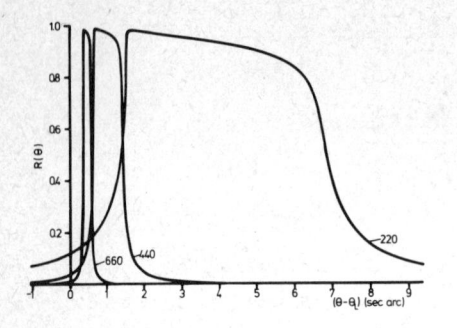

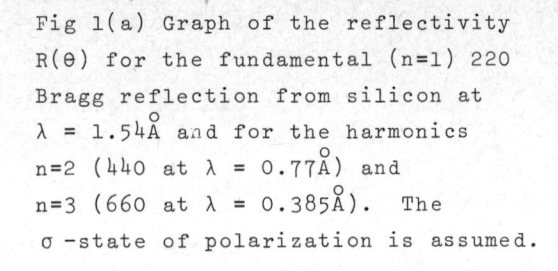

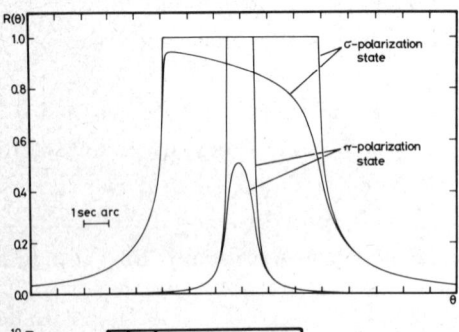

Fig 1(b) $R(\theta)$ for the two polarization states in the 440 Bragg reflection from germanium at $\lambda = 1.54\text{Å}$. The flat-topped curves show the result when absorption is ignored.

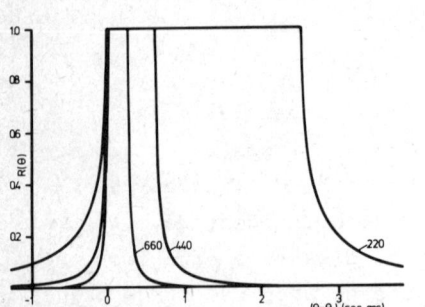

Fig 1(c) $R(\theta)$ for the fundamental (n=1) 220 Bragg reflection of neutrons from silicon at $\lambda = 3\text{Å}$ and for the harmonics n=2 (440 at $\lambda = 1.5\text{Å}$) and n=3 (660 at $\lambda = 1\text{Å}$).

The principal effect of refraction is to shift the Bragg peak and its harmonics away from θ_L. The shift of the centre of the Bragg peak is given by

$$\theta_o - \theta_L = \frac{|\chi_{ro}|}{2\sin 2\theta}\left(1 + \frac{\sin(\theta-\phi)}{\sin(\theta+\phi)}\right)$$

where ϕ is the angle between the crystal surface and the Bragg planes and $\chi_{ro} = -8r_e \lambda^2 (Z+f')/\pi a^3$ in diamond structure materials. For the harmonics the shift is proportional to λ^2 or n^{-2}. The peak widths $\Delta\theta_o$ are given by

$$\Delta\theta_o = \frac{2 |C| |\chi_{rh}|}{\sin 2\theta} \sqrt{\frac{\sin(\theta-\phi)}{\sin(\theta+\phi)}}$$

where C is 1 for the σ-polarized state and $\cos 2\theta$ for the π-polarized width $\Delta\theta_o$. Thus large sources can be used to gain intensity. In practice $\Delta\theta_o$ lies within the range $10^{-8} < \Delta\theta_o < 10^{-4}$ and the corresponding energy or momentum resolution is determined by the width of the convolution of the two crystal profiles. It is approximately

$$\Delta E/E \simeq \cot\theta \cdot \Delta\theta_o$$

MULTIPLE BRAGG REFLECTIONS

Since the peak reflectivity is almost 1 (Fig 1) multiple Bragg reflections can be made with little intensity loss. This design freedom allows one to control the shape of the Bragg reflection profile, to split and recombine beams and to control their paths through a system. For example, multiple reflections lead to the elimination of the tails of the Bragg reflection curve[12] and make possible interferometry with x-rays and neutrons[13-18]. In practice most of the multiple reflection diffraction optical systems are monolithic perfect crystals, but a new design freedom is added by the realisation that elastic adjustments are practicable if the crystal is suitably designed. In this way it has been possible to make ultra-stable scanning interferometers and fast interferometric choppers[19,20], harmonic-free monochromators[21] and tuneable polarizers [22].

HARMONIC FREE SYSTEMS FOR SPECTROSCOPY

There is a fundamental requirement in spectroscopy for "monochromatic" beams with a controlled spectral passband. In practice, with single crystal monochromators, the excellent selectivity of the Bragg reflection (Fig 1) is spoiled by the presence of harmonics and by the background intensity in the wings or tails of the Bragg reflection curve. The former problem has been tackled by a variety of means, shown in Fig 3, while both problems can be solved with suitable multiple Bragg reflection optical systems.

Although specular reflection at a mirror (usually metal-coated quartz, bent to focus the beam) has been widely used to reduce crystal monochromator harmonics at synchrotron radiation sources, there appears to be little information on their actual performance. It is straightforward (e.g. from James (1948)[1]) to show that the harmonic reflectivity at twice the critical angle is 0.4%. In practice, if we include the influence of absorption, the harmonic content will be higher.

state. $\chi_{rh} = r_e \lambda^2 F_h/\pi a^3$ where F_h is the structure amplitude for the active Bragg reflection. The peak widths are therefore proportional to $F_h n^{-2}$ for the harmonics. In the approximation of point stationary atoms – close to the neutron case at room temperature – Fig 1(c) represents a universal result in which changes of atom, wavelength or obliquity simply alter the scale of the abscissa. Comparisons between Fig 1(a) and Fig 1(c) show the influence of finite atoms on the peak widths and the (small) influence of thermal motion ensures that the left hand edges of the peaks in Fig 1(c) do not in fact quite coincide at $\theta = \theta_L$. In practice, x-ray absorption limits the peak reflectivity somewhat as Fig 1(a) and Fig 1(b) show and the diffuse nature of the electron density around atoms moves the near edge of the Bragg peak away from θ_L.

SPECTROSCOPIC DESIGN

There are only two generic designs for spectroscopic instruments[10,11].

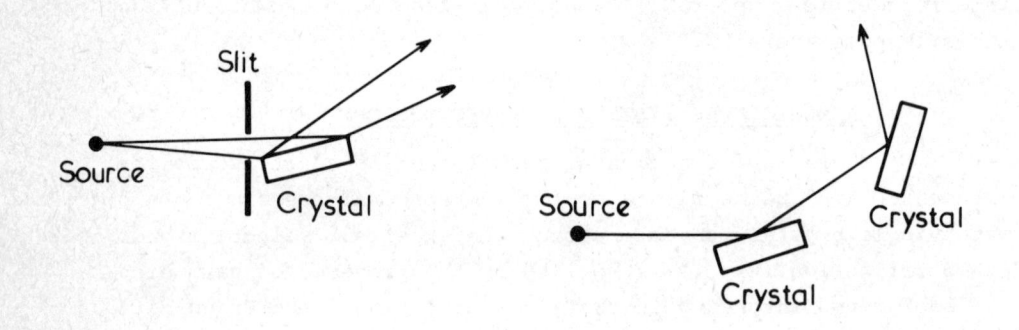

Fig 2(a) Bragg spectrometer (b) Compton or double crystal spectrometer

In the Bragg spectrometer the energy resolution is determined by the divergence of the incident beam $\Delta\theta_i$ which is fixed in practice by the geometry of the source and collimator slit. $\Delta\theta_i > 10^{-4}$ can be achieved in the laboratory with reasonable intensity and at synchrotron radiation sources it might be possible to work with $\Delta\theta_i < 10^{-5}$. The corresponding energy resolution follows from Bragg's law

$$\Delta E/E = \cot\theta \cdot \Delta\theta_i$$

Higher resolution can be achieved in the double crystal arrangement because the angular pass band is determined by the Bragg

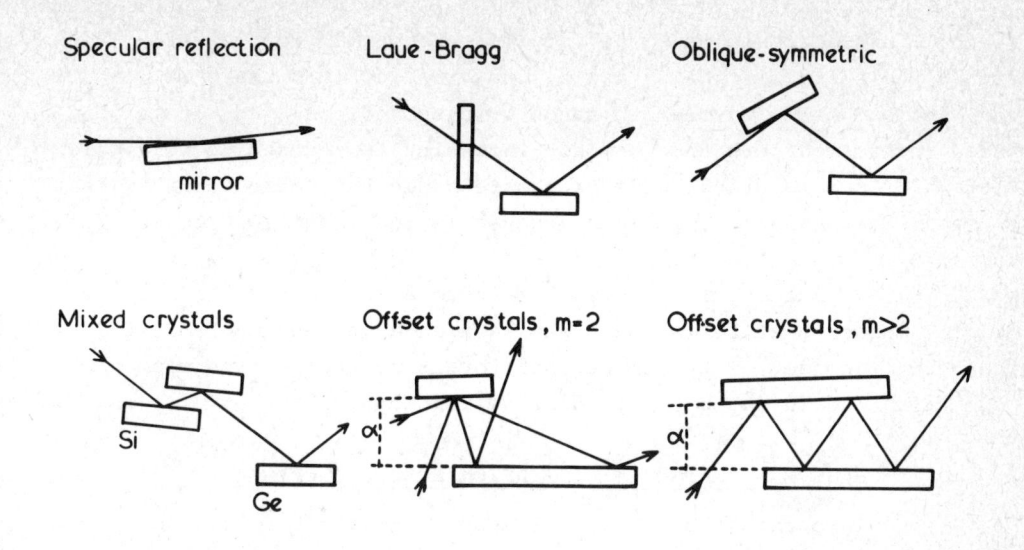

Fig.3 Systems used to control harmonic contamination in spectrometers.

Three of the harmonic reducing optical systems rely on shifting the relative positions of fundamental and harmonics by refraction (Fig 1). Bonse, Materlik and Schröder[23] found that the Laue-Bragg combination did not reduce the harmonic content below a few percent of the fundamental intensity. However, when the Laue-case device is an interferometer and the Bragg-case device is a double-reflection grooved crystal as it was in those experiments, harmonics can at least be conveniently reduced by this method. The refractive index shift can also be exploited by mixing oblique and symmetric Bragg reflections[24]. Unfortunately, the system cannot be scanned over a wide wavelength range unless several oblique crystal cuts are available. By mixing crystals with different refractive indices, for example, a double 220 reflection from silicon and a single 220 reflection from germanium, harmonic suppression to 0.3% can be achieved[23]. Such a system could with difficulty be scanned over wavelength but its passband is fixed.

The off-set grooved crystal systems suffer from none of these disadvantages. Their operation depends not on the refraction effect but on the fact that the Bragg reflection widths are different for the fundamental and the harmonics. The off-set angle α is larger than the width of the harmonic peak and much less than the width of the fundamental (Fig 1). With m = 2 Bragg reflections there is no restriction on the wavelength range or passband compared with a single Bragg reflection, but the harmonic intensity is reduced to well below 0.5% [21]. As more Bragg reflections are added the harmonic ratio is reduced substantially and can be chosen in the range 10^{-2} to (say) 10^{-10} by appropriate design.

An incidental advantage of these elastic monolithic off-set crystal monochromators is that they have an excellent mechanical frequency reponse. They can be used for x-ray and neutron beam modulation and in stroboscopic diffraction topography[33].

PHASE MEASUREMENT - DISPERSION SPECTROSCOPY

Since this topic is covered by Zeilinger[17] in the neutron case and by Graeff[18] for the x-ray case we will be concerned here only with crystal optical principles and with the results so far obtained.

Two different approaches have been adopted. Where film detection was used to determine phase shifts in nickel[25], harmonic control is necessary and was achieved as outlined in the previous section. On the DESY synchrotron source the complete nickel spectrum shown in Fig 4 was obtained in only three hours. By using energy selective detectors and a scanning interferometer[20,26,27] one may work simultaneously at two separate harmonic wavelengths which provide internal calibration of the sample thickness. With 0.8 kVA laboratory Bremsstrahlung sources each spectrum represents 600 hours of counting time but the resulting dispersion curves for zirconium, niobium and molybdenum are very encouraging as Fig 4 shows. With a 100 kW rotating anode x-ray generator these spectra could be obtained in only 5 hours.

In the corresponding neutron case the scattering is entirely characterised by a single parameter; the scattering length, which may be measured at any convenient wavelength. As complete spectra are not necessary, results have already been obtained[17,28] for many different elements.

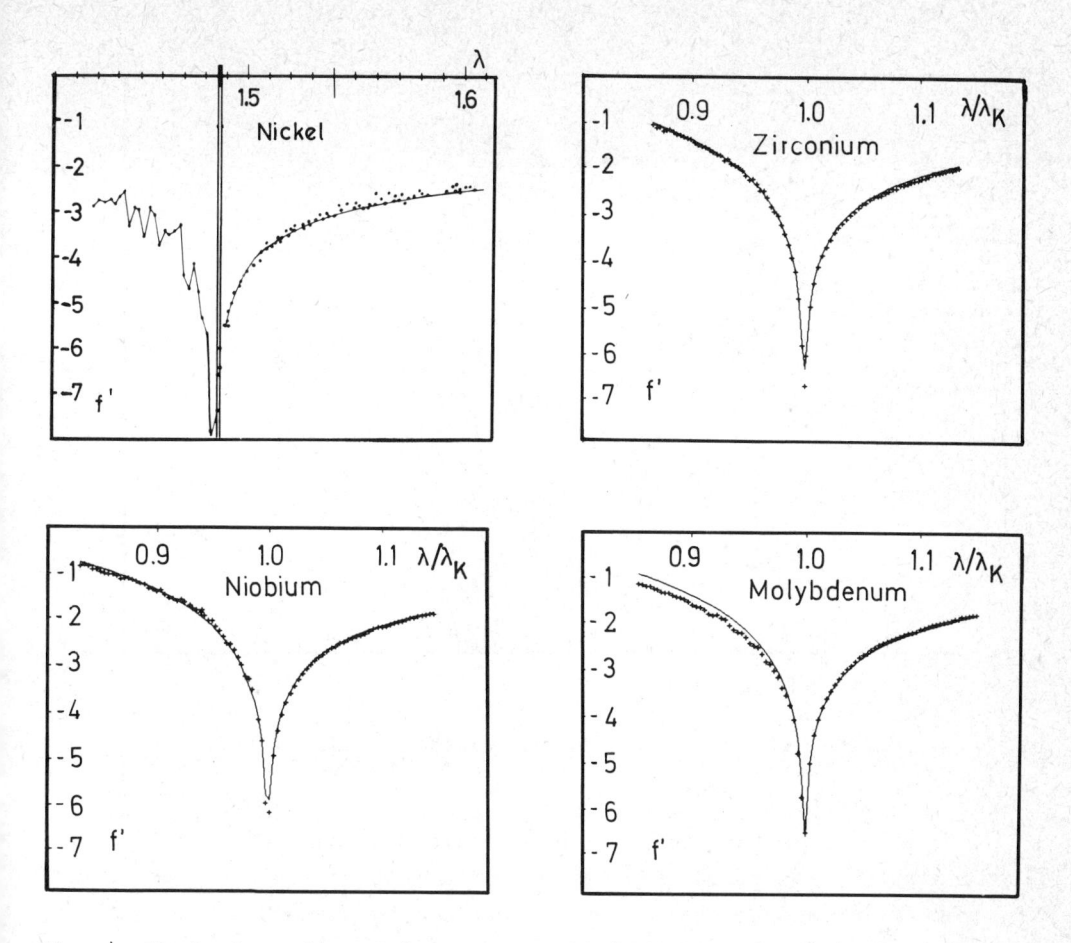

Fig. 4 K-electron dispersion curves obtained by x-ray interferometry.

POLARIZING OPTICAL SYSTEMS

Prior to the 1960's very little work had been done with polarized x-rays[29]. Simple Rayleigh scattering and coherent Bragg scattering at 90° were the only available methods for producing polarized beams. More recently partially polarized sources of synchrotron radiation have become available. Polarization phenomena are reviewed by Skalicky[30].

Recently a tuneable x-ray polarizer has been developed[22]. As Figure 1 shows, the peak reflectivity of a crystal is lower for the π-polarization state than for the σ-state. By making multiple Bragg reflections in a groove cut into a perfect crystal we can produce polarized beams. Figure 5 shows calculations of the polarizing ratio I_π/I_σ for $1 < m < 10$ in the case of the 440 Bragg reflection from germanium for the wavelength range near the polarizing Bragg angle.

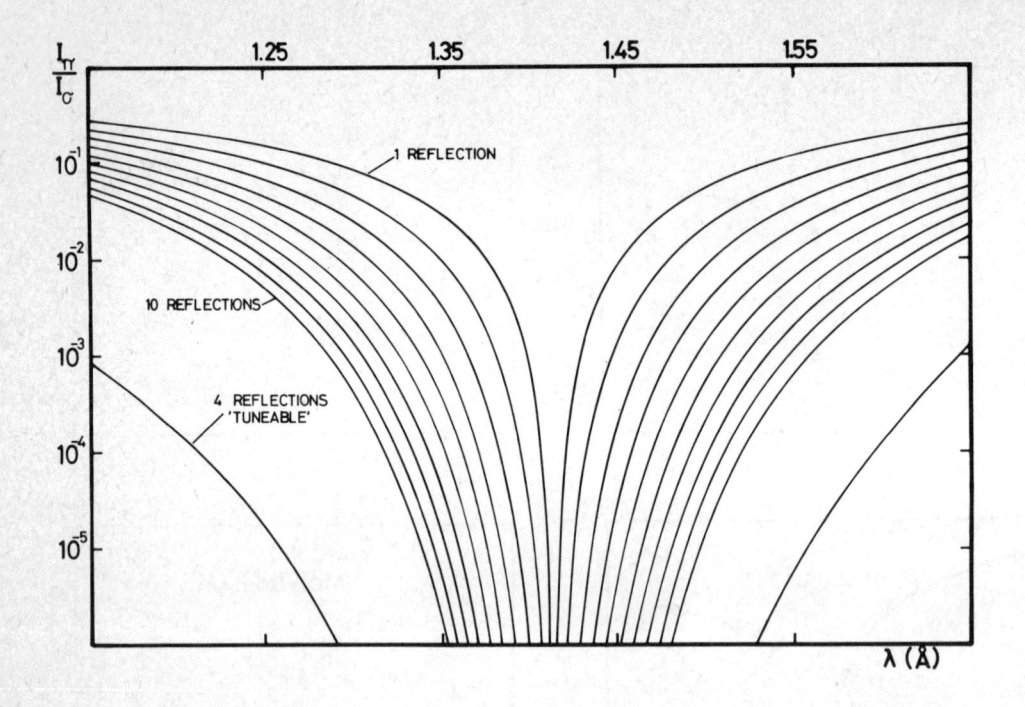

Fig. 5 Calculated polarization ratios for the 440 Bragg reflection from germanium with m-fold Bragg reflections.

As the π-state reflection curve is much narrower than the σ-state reflection curve (Fig 1), the polarization ratio can be further improved if one side of the groove is off-set by a small angle α with respect to the other as in the arrangement for harmonic elimination. The result, for a four-fold Bragg reflection system, is shown as the outermost curve in Fig 5. Detailed calculations in the Table below (for the silicon 422 Bragg reflection with m = 4 Bragg reflections at λ = 1.38Å) show how the same off-set grooved crystal acts both as a harmonic free monochromator and as a polarizer. As mentioned before, this general purpose device can also be used as a wavelength modulator and stroboscopic monochromator.

TABLE

α/sec.arc.	$I_\sigma(\alpha)/I_\sigma(o)$	$I_\pi(\alpha)/I_\sigma(\alpha)$	I_{844}/I_{422}
0.0	1.00	5.8×10^{-2}	9.6×10^{-2}
0.1	0.95	3.6×10^{-2}	1.0×10^{-3}
0.2	0.85	6.3×10^{-3}	5.0×10^{-6}
0.3	0.73	5.4×10^{-4}	1.5×10^{-7}
0.4	0.60	7.9×10^{-5}	2.5×10^{-8}
0.5	0.47	2.0×10^{-5}	7.5×10^{-9}

This same principle can also be used to polarize neutrons[22]. At $\lambda = 2\overset{o}{A}$ using the 110 Bragg reflection from saturated iron we find that the Bragg widths are approximately $\Delta\theta(+) \sim 7.4$ sec.arc and $\Delta\theta(-) \sim 2.1$ sec. arc. At an off-set $\alpha \sim 3$ sec.arc the polarization ratio is calculated to be $I_-(\alpha)/I_+(\alpha) = 5 \times 10^{-4}$.

BEAM STEERING AND CALIBRATION

Multiple Bragg reflections can also be used to simplify the experimental geometry. For example, the precision and range of goniometric control required in a classical double crystal spectrometer (Fig 2) is so daunting that few have been built. The arrangement shown in Fig 6a performs the same task but is far simpler to implement[31]. The double Bragg reflectors might each be off-set grooved crystals.

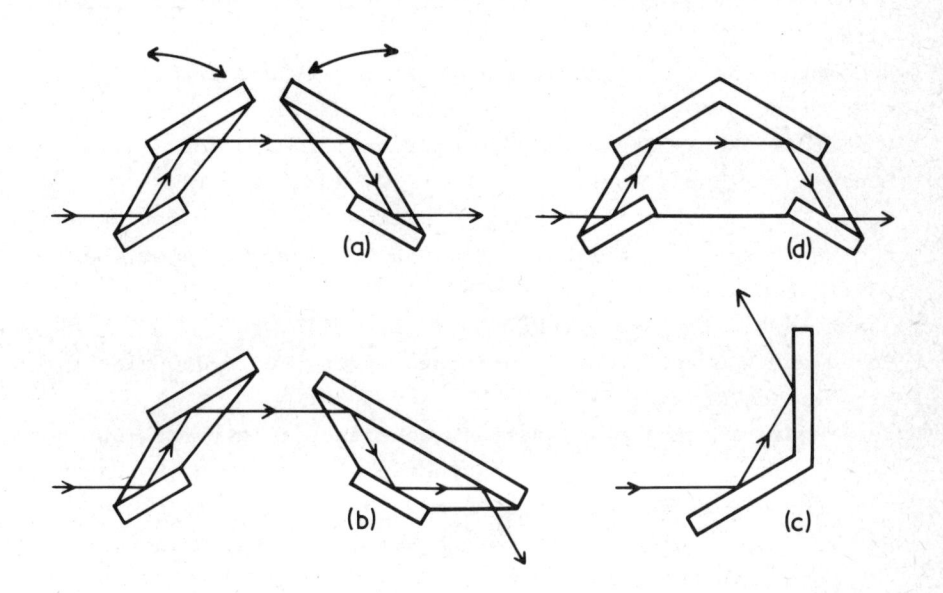

Fig.6 Beam steering and calibration by multiple Bragg reflections

If the final beam is required in the deviated direction then, with virtually no intensity penalty, another Bragg reflection can be added as in Fig 6b. In some situations only a fixed wave-length is required. Than can be prov ided by the monolithic double reflection monochromator shown in Fig 6c. A zero-deviation version, shown in Fig 6d has recently been constructed to calibrate synchrotron radiation spectrometers[32].

REFERENCES

1. R W James, The optical principles of the diffraction of x-rays, Bell, London, (1948)
2. W H Zachariasen, Theory of x-ray diffraction in crystals, Wiley: New York, (1945)
3. J C Slater, Rev. Mod. Phys. 30, 197, (1958)
4. G Borrmann, Röntgenwellenfelder in Beiträge zur Physik und Chemie des 20 Jahrhunderts, Vieweg und Sohn, Brunswick, (1959)
5. Laue M von, Röntgenstrahl-Interferenzen, Akademische, Verlag, Frankfurt, (1960)
6. R W James, Solid State Phys. 15, 55, (1963)
7. B W Batterman and H Cole, Rev. Mod. Phys. 36, 681, (1964)
8. L V Azaroff, K Kaplow, N Kato, R J Weiss, A J C Wilson and R A Young, X-ray Diffraction, McGraw-Hill, New York (1974)
9. Z G Pinsker, Dynamical scattering of x-rays in crystals. Springer, Berlin, (1978)
10. A H Compton and S K Allison, X-rays in theory and experiment, Van Nostrand, New York, (1935)
11. L V Azaroff, X-ray Spectroscopy, McGraw-Hill, New York (1974)
12. U Bonse and M Hart, Appl. Phys. Lett. 7, 238, (1965)
13. U Bonse, Proc. 5th International Conference on X-ray Optics and Microanalysis; Ed: G Möllenstadt & G H Gaukler, Springer, Berlin, (1969)
14. M. Hart, Rep. Prog. Phys. 34, 435, (1971)
15. M. Hart, Proc. Roy. Soc. Lond. A346, 1, (1975)
16. U Bonse and W Graeff, Ch. 4 in X-ray Optics, Ed: H-J Queisser, Springer: New York, (1977)
17. A Zeilinger, Imaging processes and coherence in physics, Springer: (1979)
18. W Graeff, Imaging processes and coherence in physics, Springer (1979)
19. M Hart and D P Siddons, Nature, Lond. 275, 45, (1978)
20. M Hart and D P Siddons, Workshop on x-ray and neutron interferometry, Ed: U Bonse and H Rauch, (1978)
21. M Hart and A R D Rodrigues, J. Appl. Cryst. 11, 248, (1978)
22. M Hart and A R D Rodrigues, Phil. Mag. (1979)
23. U Bonse, G Materlik and W Schröder, J. Appl. Cryst. 9, 223, (1976)
24. S Kikuta and K Kohra, J. Phys. Soc., Japan, 29, 1322, (1970)
25. U Bonse and G Materlik, Z. Phys. B24, 189, (1976)
26. C Cusatis and M Hart, Anomalous Scattering, Munksgaard, Copenhagen (1974)
27. C Cusatis and M Hart, Proc. Roy. Soc. London A354, 291 (1977)
28. W Bauspiess, U Bonse and H Rauch, Nucl. Instrum. & Meth. 157, 495, (1978)

29. M Hart, Phil. Mag. 38, 41 (1978)
30. P Skalicky, Imaging Processes and coherence in physics, Springer: (1979)
31. J H Beaumont and M Hart, J. Phys. E. 7, 823, (1974)
32. M Hart, Submitted to J. Phys. E. (1979)
33. J Miltat, Imaging processes and coherence in physics, Springer: (1979)

ELECTRON IMAGING TECHNIQUES

R.H. WADE

Section de Physique du Solide
Département de Recherche Fondamentale
Centre d'Etudes Nucléaires de Grenoble
85 X, 38041 Grenoble Cédex, France

1. INTRODUCTION

Our aim is to give an introductory description of the electron microscope suitable for those having no acquaintance with this field. We will describe how the instrument is used to obtain the maximum possible amount of information about different types of specimen with particular attention to biological specimens.

There are so many types of electron beam instruments that it will not be possible to describe them all. Particular attention will be paid to the high resolution instruments which are still being actively developped because of their importance for ultrastructural research.

It is well known that any optical instrument has an ultimate resolution limit (r) of the form

$$r = \lambda/\alpha$$

depending both on the angular aperture of the system α and on the radiation wavelength λ. The de Broglie wavelength associated with an electron of mass m moving with velocity v is

$$\lambda = h/mv$$

where h is Planck's constant. By equating the kinetic and potential energy we obtain the electron wavelength in terms of the accelerating potential,

$$\lambda_e = (150/V)^{1/2} \text{ Å, where } 1 \text{ Å} = 10^{-8} \text{ cm, is in volts.}$$

The wavelength of an electron accelerated through 100 keV is $\lambda_e \simeq 4 \times 10^{-10}$ cm which is well below atomic dimensions so a dramatic improvement in resolution is possible compared with photons of wavelength $\lambda \simeq 5 \times 10^{-5}$ cm. The best electron lenses so far devised have aberrations which limit of the resolution to $\sim 100\ \lambda_e$. These lenses consist of many turns of copper wire round a soft iron core in the gap of which specially shaped soft-iron pole pieces constitute the lens itself.

2. MAGNETIC LENSES AND THEIR ABERRATIONS

The quality of the lens depends on the pole piece geometry. The most important parameters being the spacing between the upper and lower pole pieces and the diameters of the bore of the two parts. Electrons travelling at a small angle to the optical axis are focussed by the magnetic field between the pole pieces. In the thin lens approximation the focal length (f) of the lens is given by :

$$\frac{1}{f} = \frac{V.75}{V} \int H_z^2 \, dz \; ; \; f = \text{const } V/I^2$$

where V is the electron energy in volts and H_z is the axial component of the field in amp/m , f is in cm, I is the excitation current of the lens [1].

Of the third order aberrations for a magnetic lens the spherical aberration does not vanish on the optical axis [2,3]. Since the focal length depends on the square of the field all electron lenses are positive so there is no way of correcting this defect in a rotationally symmetric, charge free system [4].

The rays incident on the lens at an angle θ are deflected through an angle $C_s \theta^3$; these rays are focussed nearer to the lens than the axial rays which form the Gaussian image (C_s the spherical aberration constant is in the range 0.07 - 0.15 cm for the best magnetic lenses). In fact, a caustic surface is associated with this aberration. It can be shown that the associated disc of least confusion has a diameter of $C_s \theta_m^3$ and is situated at $3 C_s \theta_m^3/4$ in front of the Gaussian image plane [5]. On this arguement the resolution limit is reached when the lens aperture is such that the Airy disc associated with the diffraction by this aperture is equal to the disc of least confusion due to spherical aberration. To within multiplicative constants this gives $C_s \theta^3 = \lambda/\theta$ which yields a resolution $r = (C_s \lambda^3)^{1/4}$.

The wave aberration associated with the lens is $W(\theta) = 2\pi/\lambda \, (-z\theta^2/2 + C_s\theta^4/4)$ where z is the defocussing distance [6-9]. This phase term is stationary for θ = 0 or $\sqrt{C_s/z}$. It can be shown that for maximum phase contrast $W(\theta) = \pi/2$ whence we find that for a defocus $z = (C_s \lambda)^{1/2}$ (Scherzer defocus) the resolution is given by $r = (C_s \lambda^3)^{1/4}$ in agreement with the semi-geometrical arguement above. In fact the resolution is more correctly discussed in terms of contrast transfer theory [6-9] which shows that by varying the defocus z we can "tune-in" to different spatial frequency bands.

The electrical supplies of the microscope are never perfectly stable. If δV is the fluctuation of the high tension supply V and δI that of the lens current we find a varying defocus $\delta f = C_c \{\frac{\delta V}{V} - 2\frac{\delta I}{I}\}$ where the constant C_c is called the chromatic aberration constant of the lens, and is in the range 0.1 to 0.2 cm. The beam itself has an energy spread equivalent to the voltage fluctuation ΔV part of which is thermal and part due to space charge - the Boersch effect [10 - 12]. Since all these fluctuations are independant it seems reasonable to take the RMS value as the defocus fluctuation :

$$\delta z = C_c \{ (\frac{\delta V}{V})^2 + 2 (\frac{\delta I}{I})^2 + (\frac{\Delta V}{V})^2 \}^{1/2}$$

The effect of this fluctuating defocus is to limit the resolution. It can be shown quite generally that for parallel illumination the image of an object of period d is identical in planes separated by defocus increments $\delta z = 2 d^2/\lambda$ and that the contrast reverses at half this distance. From this equality we find that in the presence of a defocus fluctuation δz the resolvable periodic components (r) in the object must satisfy the inequality $r \geq (\lambda \delta z)^{1/2}$.

In the contrast transfer theory the effect of this chromatic fluctuation is described by an envelope function which attenuates the higher spatial frequencies in the object giving a cut-off frequency in agreement with the above inequality to within a constant factor.

Astigmatism results from departure from exact cylindrical symmetry in the lens. Its effect is that different directions in the object do not fall in focus in the same plane. It is as if the lens has different focal lengths in two perpendicular planes. This defect is easily corrected by using compensating coils built into the lens regions.

3. TRANSMISSION ELECTRON MICROSCOPE : CTEM

The strong absorbtion of electrons by matter requires that the electron beam runs within a vacuum enclosure. The electrons emitted from a heated tungsten filament are accelerated to the operating voltage by the potential difference applied between the filament and an anode which is at the same potential as the microscope column. For obvious reasons, the filament is at a negative potential and the microscope column at earth ! An automatically biased triode system is usual. The bias voltage determines the total emission current. The brightness of such a source is $10^4 - 10^5$ A/cm^2/sterad. The optics of the triode produces a cross-over region between the filament and the anode. This minimum cross-section acts as the effective source and has a diameter of about 10^{-2} cm, [1].

A double condenser lens system enables the beam to be focussed at the specimen plane with varying cross-over sizes and beam currents. This enables the illuminated area to be limited to the field of view. The first condenser lens is strong and the second lens projects the highly demagnified image onto the specimen. The minimum spot diameter at the specimen is 2 μm. The use of the second lens allows an adequate working space between the condenser system and the object. Since the brightness is constant within an optical system the current density in the spot illuminating the object is $j_o = j_s (\emptyset_o/\emptyset_s)^2$ and depends on the solid angles $\emptyset_o$ and $\emptyset_s$ subtended by the condenser aperture at the object and the source respectively. Typically j_o is of the order 1 amp/cm^2 or less,[1, 2, 13].

The specimen is situated just above the objective lens of the microscope. In practice it may be within the magnetic field of this lens. The pre-field can be considered as part of the condenser system. The electron beam is scattered by the specimen with an angular distribution which can be visualised in the back focal plane of the objective lens, this is the Fraunhofer diffraction image of the object. Apertures placed in this plane can delimit which part of the scattering distribution contributes to the image (spatial frequency filtering). The focal length of the objective lens is in the range 0.1 - 0.2 cm so that a scattering angle of 2×10^{-2} rad will correspond to a diffraction spot about 30 μm off the axial unscattered beam. Objective aperture sizes used to exclude these scattered beams will be in the range 10 - 100 μm diameter.

The objective lens produces an image magnified in the range 10 - 50 x. This image is further magnified by a projector lens system of two or three lenses to produce a final image whose magnification can be varied over a wide range from below 5,000 up to 500,000 or more. The focal length of the lens below the objective lens can be varied so that the final image plane is conjugate either to the object or to the back focal plane.

The final image is viewed on a fine grain fluorescent screen which has an efficiency of about 25 % for the conversion of electrons to photons. This efficiency increases with accelerating voltage up to 100 keV and then falls off. Only about 2×10^{-4} of all the emitted light is collected by the eye. The resolving power of the screen is of the order 100 μm. The screen is viewed through a lead-glass port offering X-ray protection. Binocular viewing permits fine focussing. Images are recorded on photographic plates or film stored in a holder below the viewing screen.

4. SCANNING TRANSMISSION ELECTRON MICROSCOPE : STEM [14]

The electron source is a field emission electron gun consisting of a pointed tungsten filament and an extraction electrode. The filament tip is roughly hemispherical with a radius $R \sim 500$ Å. When the extraction electrode is held at a few kilovolts the field generated in the region of the tip is of the order of 1 volt/Å. This causes electrons to be drawn from the tip by the tunnel effect. These electrons are accelerated up to the working voltage 20 - 100 keV in one or several stages. Since the electrons are emitted radially from the tip the apparent source size r is much smaller than the tip, $r = R(V_T/V)^{1/2}$, where V_T is the energy spread of the emitted electrons ($V_T < 0.5$ eV) and V_o the accelerating potential ($V \sim 1$ keV). This gives an apparent source radius of 10 Å.

The gun operates as a stable source in the absence of adsorped layers and therefore requires a very high vacuum in the region of the tip. For this reason the gun chamber is usually differentially pumped to better than 10^{-10} torr. This type of gun has a brightness of the order 10^8 amp/cm^2/sterad.

The image of the source produced by the gun optics is about 100 Å in diameter. The simplest scanning microscopes consist of the gun and an "objective" lens ; an additional condenser lens is sometimes added. In either case the spot size δ at the specimen is about 5 Å and is essentially spherical aberration limited, $\delta \approx 0.7(C_s \lambda^3)^{1/4}$.

Since the energy spread associated with field emission is small (< .25 volts) the chromatic aberration of the objective lens does not limit the spot size. Recent results indicate that at high beam currents the energy spread may become much larger [32]. A beam current of about 10^{-8} amps can be obtained in a spot size of 10 Å in a commercially available instrument.

The specimen is situated after the objective lens. The illumination spot is scanned across the specimen in a raster generated by sets of coils placed between the source and the objective lens. Detectors placed after the specimen pick-up the scattered electrons and a cathode ray tube scanned synchronously is modulated according to the electron current. The magnification depends only on the size of the scanning raster at the specimen. There are no lenses after the specimen. This makes it easy to fit an energy analyser. The function of the detectors is described in the section on contrast mechanisms.

5. SPECIMENS

The specimens examined in metallurgy and solid state physics are prepared in a variety of ways. They must be thin which in practice can mean in the thickness range from a few tens to several thousand Å units. The final thinning operation may consist of electrolytic or chemical polishing, ion bombardement, cleavage, grinding and vacuum deposition. The aim is to obtain the largest possible electron transparent area. The specimens may be self supporting or mounted on copper grids, [13].

Biological materials are dehydrated by the vacuum within the microscope and are destroyed by electron irradiation. In addition most biological specimens have low contrast in bright field electron microscopy. Some materials are embedded in resins and sliced-up on an ultramicrotome whilst others can be placed on a thin carbon support. In both cases negative stains of heavy metal salts are used to preserve and enhance the contrast from structural irregularities. The stain engulfs the material and forms a glass-like mould. The best resolution of material so treated is found to be around 20 Å and is related to the grain of the staining material.

6. CONTRAST MECHANISMS

6.1. Scattering

6.1 (a) - **Elastic scattering.** The atomic scattering factor has the form :

$$f(\theta) = \frac{m_o e^2}{2h^2} \left(\frac{\lambda}{\sin \theta}\right)^2 (Z - f_x)$$

where $m_o e^2/2h^2 = 2.38 \times 10^6$ cm^{-1} [15]. The first term in the bracket is due to Rutherford scattering from the nucleus and the second term, the atomic scattering factor for X-rays, is due to scattering from the electron cloud. The differential scattering cross section is directly related to $f(\theta)$, $d\sigma_e(\theta)/d\Omega = f^2(\theta)$. The variation of $d\sigma_e/d\Omega$ as a function of θ is schematised in figure 1 for 100 keV electrons. The figure is equivalent to a radial section across the diffraction image obtained in the back focal plane of the objective lens. The total number of electrons scattered through an angle θ is obtained by integration over an annulus of radius proportional to θ. Since the cone of the incident radiation has typically an angle of 10^{-3} rad, it is seen that most electrons are scattered outside the cone of the transmitted beam. The total scattering cross section σ_e found by integrating $f^2(\theta)$ over all scattering angles is $\sigma_e \propto Z^{4/3}$ using an exponentially screened Coulomb atomic potential distribution. Experimentally σ_e would be obtained by collecting all the elastically scattered electrons.

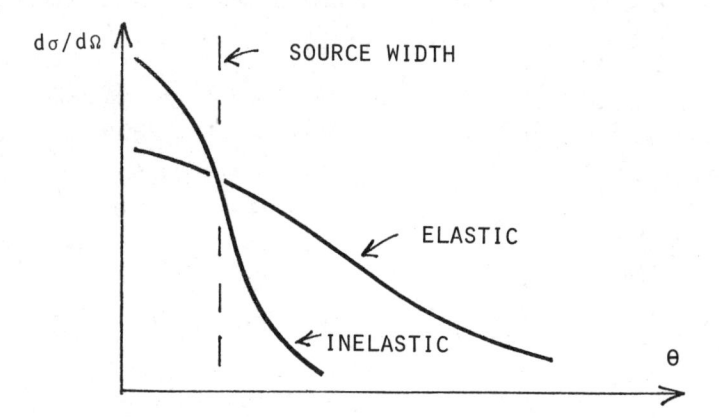

Figure 1 - The angular dependence of the elastic and inelastic differential scattering cross sections compared to the source width.

6.1 (b) - _Inelastic scattering_. There are many processes by which energy is transferred from the electron beam to the irradiated material [16]. For the light atoms present in biological material the probability of an inelastic scattering process is about twice that of an elastic one. The differential scattering cross section $d\sigma_i/d\Omega$ is given by

$$\frac{d\sigma_i}{d\Omega} = \text{const} \frac{1}{\theta^2 + \theta_E^2}$$

θ is the scattering angle and the screening angle $\theta_E = \Delta E/2 E_o$ where ΔE is the energy loss, typically in the range 1 - 40 eV, and E_o is the incident electron energy [16]. This distribution is sharply peaked in the forward direction for angles less than θ_E, ie $\theta_{max} < 10^{-3}$ rad. The total inelastic cross section varies according to the atomic model. In the Lenz theory [17] using a Thomas-Fermi model and for single electron excitation $\sigma_i \propto Z^{1/3}$.

The essentiel point is that the beam transmitted through a thin specimen consists of three parts. The unscattered beam N_o ; the elastically scattered component N_e ; the inelastically scattered component N_i.

6.2. Imaging

6.2 (a) - _CTEM_ : A perfect lens collecting all the scattered electrons would form an image directly representing the square modulus of the electron wave function at the exit face of the specimen. In the absence of electron absorbtion the image so formed would show no contrast. Image contrast can be generated either by introducing apertures into the back focal plane of the objective lens in order to select a particular angular component of the overall scattering distribution (Fraunhofer diffraction amplitude) or by using the objective lens defocus as a means of introducing phase shifts between the scattered and unscattered components.

1) Diffraction contrast : a small aperture selects the electrons corresponding to essentially one scattering angle. In the bright field method the aperture selects the unscattered beam. The image will show modulations of intensity depending on the local scattering at the specimen. The resolution, depending on the aperture size, is relatively low. The method is commonly used in work on single crystals. If the aperture is placed over a scattered beam a dark field image is produced. A great deal of information concerning the nature of crystal defects can be obtained by this method [13].

2) Bright field interference contrast : in this case the scattered and unscattered beams are allowed to pass into the image where they interfere. When $N_o \gg N_e$ the image contrast is proportional to the object exit surface wave function. For lattice

plane images the incident beam can be tilted so the unscattered and scattered waves make equal angles with the optical axis in which case the image is aberration free. Images combining many diffracted beams have been successfully exploited in the study of complex oxide crystals. In this case the lens aberrations and finite source size must be taken into account. This is usually done by a comparison of experimental and digitally generated images. The agreement can be quite impressive, a simple account is given in [18].

The method can also be applied to crystals of biological interest exposed under low dose conditions in order to minimise the radiation damage. The images consist of a periodic signal with a strong noise superposed. Digital treatment of the images enables the periodic component to be separated from the noise and since both the amplitudes and phases of the Fourier components are obtained a Fourier synthesis of the crystal structure can be made [19-21]. Image processing can also be carried out optically [22, 23].

3) Dark field : the image is formed with the scattered electrons only. The idea being that since it is only these electrons which carry information about the object they alone should be retained in preference to the unscattered electrons. In one method a central stop is used. Although difficult to realise experimentally, this gives a good collection efficiency for the elastically scattered electrons whilst a good deal of the inelastic scattering can be cut out by the stop along with the unscattered beam. Another method uses a circular aperture whilst the incident beam is tilted. This method gives less efficient electron collection and it is rather more difficult to take account of the effect of lens aberrations. Images of low molecular weight proteins may yield valuable high resolution (< 10 Å) information after some image averaging [24].

6.2 (b) <u>STEM</u> : There are no lenses after the object so the detection takes place essentially in the far-field or Fraunhofer diffraction region. Use is made of the different angular distributions of the elastic and inelastic electrons to separate all three beam components. An annular aperture collects the elastically scattered electrons, the unscattered and inelastic electrons pass through the central hole and are separated by an energy filter. The three signals can be detected simultaneously.

1. Dark field : The signal N_e (or N_e/N_o) corresponds to the dark field image with an extremely good efficiency since the annular aperture corresponds to a very large collecting angle for the elastic electrons. The signal is approximately proportional to σ_e

2. Z contrast : The ratio $N_e/N_i \simeq \sigma_e/\sigma_i$ gives a signal which is proportional to Z or to some power of Z depending on the form taken for the scattering potential. This method has no equivalent in the CTEM and has been greatly exploited by Crewe and his collaborators for the visualisation of heavy atoms [14].

3. Bright field : The detector angle must be very small and the illumination angle large. It is then possible to observe phase contrast and lattice images as in bright field CTEM [33]. The method compares unfavourably to the CTEM as far as the object irradiation is concerned.

7. RECIPROCITY

It is found experimentally that STEM images can be similar to CTEM images. Examples are Fresnel fringes, phase contrast, lattice images. From the description of the two instruments it is not at all obvious that the two instruments give similar results since the CTEM involves detection in a plane conjugate to the object whilst the STEM detection plane appears roughly conjugate to the illumination aperture.

However if we consider the two systems as they are represented in figure 2 we see that, in STEM, electrons from a point A of the source are focussed on the specimen and the intensity detected at B far below the specimen. In CTEM electrons from a distant point B' of the source are scattered by the specimen and focussed by the lens to the image point A'.

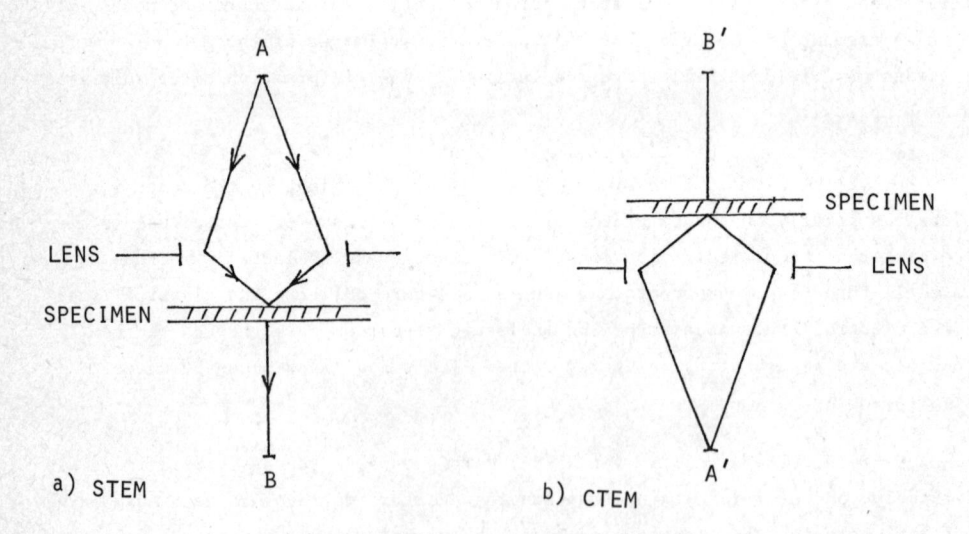

Figure 2 - The imaging process in STEM and CTEM are schematised in (a) and (b) respectively. Inversion of the ray directions in one diagram gives the other so the two methods are related by reciprocity.

The Helholtz reciprocity theorem would suggest that in fig. 2(a) the point source A will produce the same effect at B as a point source of equal intensity at B would produce at A. Comparison of fig. 2(a) and (b) shows that (b) is just an inversion of (a), provided the specimen is inverted. The two microscopes are therefore equivalent ; in STEM the detector size plays the same resolution limiting role for phase contrast images as does the source in CTEM [25, 26]. However the two microscopes are not equivalent as far as radiation dose is concerned.

8. RESOLUTION LIMITS

That part of the image intensity distribution directly related to the scattering due to the object represents the useful signal. The rest is noise and includes :

1) inelastic electrons which contribute essentially a low contrast low resolution background.

2) the elastic image of the specimen substrate, for example the phase contrast image of the carbon support used for biological materials.

3) electrons scattered through large angles whose contribution to the image is blurred out by the enveloppe functions associated with a finite source size and chromatic defocus fluctuations.

4) shot noise which depends on the number of electrons detected within a "resolution cell" and so on the electron dose at the specimen.

A useful relationship which indicates the number of quanta (n/unit area) necessary to image an object of contrast c at a resolution d is the Rose expression

$$n \geq (t/cd)^2,$$

where the threshold signal to noise ratio t will typically have a value of around 5 [27].

A thermal source imaged onto the specimen plane by the condenser lenses can easily give a radiation dose of 60 el/$\mathring{A}^2$/sec. For a total magnification of 4×10^4 an intensity at the fluorescent screen equivalent to about 2 el/$\mathring{A}^2$/sec at the object is required to allow visual observation and focussing. At the same magnification a photograph of unit average density requires a dose of ~ 1 el/$\mathring{A}^2$ at the object.

Since biological material is destroyed by doses in the range .05- several el/$\mathring{A}^2$ [28, 29] meaningful observations on biological material require very low total electron doses. The Rose expression then shows that the attainable resolution is severely limited. The resolution can be improved by averaging over a number of identical objects. This can be achieved by either using two dimensional crystal arrays [20], or adding images of isolated molecules [30]. Other possibilities are :

1) to increase the contrast using the dark field method [24]. The doses used destroy the structure but the remaining "skeleton" seems to be closely related to the original structure in the case of low molecular weight proteins ($\sim$ 1000 daltons). Image averaging is also required. Equivalent results should be attainable by subtracting the background intensity from bright field images.

2) Recent results suggest that the radiation resistance may be dramatically improved if the specimens are cooled to liquid helium temperatures [31]. If these results are confirmed a dramatic breakthrough in biological structure investigation by electron microscopy will be possible.

References

[1] M.E. HAINE, The Electron Microscope, Spon Ltd, (1961).
[2] C.E. HALL, Introduction to Electron Microscopy, Mc Graw Hill, (1953).
[3] M. BORN and E. WOLF, Principles of Optics, Permagon Press (1964).
[4] O. SCHERZER, Z. Physik 101, 593 (1936).
[5] D. GABOR, Proc. Roy. Soc. A 197, 454 (1949).
[6] O. SCHERZER, J. Appl. Physics, 20, 20, (1949).
[7] K.J. HANSSEN, in Advances in Electron and Optical Microscopy, ed. by R. Barer and V.E. Cosslett, vol. 4 pp. 1-84, Academic Press (1971).
[8] P.W. HAWKES, this book.
[9] F.A. LENZ, in Electron Microscopy in Materials Science, ed. by U. Valdrè, pp. 541-569, Academic Press (1971).
[10] G. FONTAINE, this book.
[11] H. BOERSCH, Z. Physik 139, 115 (1954).
[12] R.W. DITCHFIELD, M.J. WHELAN, Optik, 48, 163 (1977).
[13] P.B. HIRSCH, A. HOWIE, R.B. NICHOLSON, D.W. PASHLEY, M.J. WHELAN, Electron Microscopy of Thin Crystals, Butterworths (1965).
[14] A.V. CREWE, in Electron Microscopy in Materials Science, ed. by U. Valdrè, pp. 162-207, Academic Press (1971).
[15] J. SIVARDIERE, this book.
[16] Ch. COLLIEX, this book.
[17] F.A. LENZ, Z. Naturf. 9a, 185 (1954).
[18] J.M. COWLEY, S. IIJMA, Physics Today, 30, n° 3, 32 (1977).
[19] I.A.M. KUO, R.M. GLAESER, Ultramicroscopy 1, 53 (1975).
[20] P.N.T. UNWIN, R. HENDERSON, J. Mol. Biol. 94, 425 (1975).
[21] D. McLACHLAN, Proc. N.A.S. 44, 948 (1958).
[22] A. KLUG, J.E. BERGER, J. Mol. Biol. 10, 565 (1964).
[23] U. AEBI, P.R. SMITH, J. DUBOCHET, C. HENRY, E.KELLENBERGER, J. Supramol. Structure 1, 498 (1973).
[24] F.P. OTTENSMEYER, D.P. BAZETT-JONES, J. HEWITT, G.B. PRICE, Ultramicroscopy, 3, 303 (1978).
[25] J.M. COWLEY, Appl. Phys. Letters, 15, 58 (1969).
[26] E. ZEITLER, M.G.R. THOMPSON, Optik, 31, 258 (1970).
[27] A. ROSE, Adv. in Electronics, 1, 131 (1948).
[28] K. STENN, G.F. BAHR, J. Ultrastructure Res. 31, 526 (1970).
[29] R.M. GLAESER, J. Ultrastructure Res. 36, 466, (1971).
[30] J. FRANK, W. GOLDFARB, D. EISENBERG, T.S. BAKER, Ultramicroscopy, 3, 283 (1978).
[31] I. DEITRICH, H. FORMANEK, F. FOX, E. KNAPEK, R. WEYL, Nature 277, 380 (1979).
[32] M. TROYON, Optik 52, 401 (1979).
[33] A.J. CRAVEN, C. COLLIEX, J. Microscopie et Spectroscopie Electron. 2, 511 (1977).

SPECKLE AND INTENSITY INTERFEROMETRY. APPLICATIONS
TO ASTRONOMY

F. RODDIER - Université de Nice

I/ - STATISTICAL PROPERTIES OF THERMAL RADIATIONS

We shall mainly deal with the wave aspect of thermal radiations. A thermal radiation is classicaly considered as a random electromagnetic field. When dealing with steady sources, the field is described as a *stationary random process*. Thermal emission is incoherent, i.e. radiations emitted by individual atoms are statistically independant. The resulting field, being the sum of a large number of statistically independant processes, is therefore a *Gaussian process*, as a consequence of the central-limit theorem. A Gaussian process is a process which has Gaussian probability density functions (p.d.f.), i.e. the joint probability density function of the field at any number of points at any time is given by a multidimensional Gaussian law.

We shall only deal with quasimonochromatic fields for which the field bandwidth $\Delta\omega$ is much smaller that the average frequency ω. Using scalar theory, such a field can be expressed, in complex notation, as

$$Z(\vec{r},t) = \psi(\vec{r},t) \exp - i\omega t \qquad (1)$$

where $\vec{r}$ and t are space and time coordinates. ψ is a generalised complex amplitude with negligible variations during the period $2\pi/\omega$. ψ is often refered to as the analytic signal. It is a circular complex Gaussian process, i.e. the variance of the real part is equal to the variance of the imaginary part.

A Gaussian process is entirely defined by its first and second order moment. The first order moment $<\psi(\vec{r},t)>$ is zero for a circular process. Since our process is stationary, the second order moment $<\psi(\vec{r}_1,t_1)\psi^*(\vec{r}_2,t_2)>$ is a function $C(\Delta\vec{r},\Delta t)$ of the spacing $\Delta\vec{r}$ and of the time delay $\Delta\vec{t}$. It goes to zero when the time delay Δt is much larger than the correlation time τ, which is of the order of $1/\Delta\omega$.

Since there is a finite correlation time τ, our process is also ergodic. This means that a time average taken over a time interval, T, much larger than τ, can be considered as an ensemble average.

In the following, we shall be interested only in the spatial properties of the process $\psi(\vec{r},t)$, that is we shall only consider time delays Δt small compared with the correlation time τ, and we shall examine the properties of the spatial covariance $<\psi_1\psi_2^*>$ where ψ_1 and ψ_2 denote $\psi(\vec{r}_1)$ and $\psi(\vec{r}_2)$.

II/ - SPATIAL COHERENCE AND ZERNIKE'S THEOREM

Let us consider the fundamental Young's holes experiment (figure below). In this experiment a thermal light source illuminates a mask with two pinholes. Let ψ_1 and ψ_2 be the complex amplitudes of the field at each hole. Behind the mask, the two holes behave like point sources, and at some distance from the mask, near the axis of symmetry, the complex amplitude is proportional to

$$\psi = \psi_1 + \psi_2 \exp i\phi \qquad (2)$$

where ϕ is the phase difference due to the inequality in the optical path length between the two illuminating beams.

The observed illumination, given by the squared modulus of the field amplitude, is therefore proportional to

$$I = <|\psi|^2> = <|\psi_1+\psi_2 \exp i\phi|^2> \qquad (3)$$

where the time integration of the eye, made over a time interval much larger than the correlation time τ, has been replaced by an ensemble average

Expanding equation (3) leads to

$$I = <|\psi_1|^2> + <|\psi_2|^2> + 2\text{Re} <\psi_1\psi_2^*> \exp -i\phi \qquad (4)$$

in which the complex covariance, $<\psi_1\psi_2^*>$, appears. Let θ be the phase of this complex quantity, so that

$$<\psi_1\psi_2^*> = |<\psi_1\psi_2^*>| \exp i\theta \qquad (5)$$

and let us suppose that

$$\langle|\psi_1|^2\rangle = \langle|\psi_2|^2\rangle = \sqrt{\langle\psi_1^2\rangle\langle\psi_2^2\rangle} \qquad (6)$$

i.e. the irradiance output from each hole is equal. Equation (4) becomes

$$I = 2\left[\sqrt{\langle\psi_1^2\rangle\langle\psi_2^2\rangle} + |\langle\psi_1\psi_2^\star\rangle|\cos(\theta-\phi)\right] \qquad (7)$$

The cosine term of equation (7) describes Young fringes. Let $I\max$ and $I\min$ be the associated maxima and minima of I. The fringe visibility is given by the symmetrized expression

$$V = \frac{I\max - I\min}{I\max + I\min} = \frac{|\langle\psi_1\psi_2^\star\rangle|}{\sqrt{\langle|\psi_1|^2\rangle\langle|\psi_2|^2\rangle}} = |\Gamma| \qquad (8)$$

where Γ is the normalized complex covariance. $|\Gamma|$ is called the degree of coherence between ψ_1 and ψ_2.

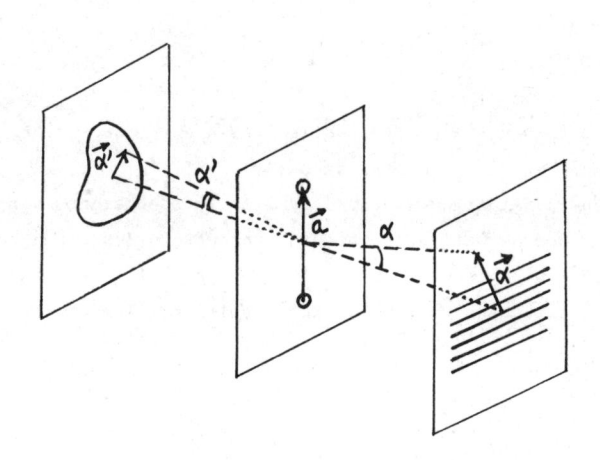

Young's hole experiment

Let us consider a point source illuminating the mask from a direction $\vec{\alpha}'$ (see figure) and let $J(\vec{\alpha}')$ be the irradiance output of each hole, produced by the spherical wave issuing from that point source. At the hole entrance the complex amplitudes are identical, but one is delayed with respect to the other, so that

$$<|\psi_1|^2> = <|\psi_2|^2> = J(\vec{\alpha}') \tag{9}$$

and $\quad <\psi_1\psi_2^*> = J(\vec{\alpha}') \exp - i\phi'$

The illumination in the observed fringe pattern in a direction $\vec{\alpha}$ (see figure) is given by equation (7) which leads to

$$I_o(\vec{\alpha},\vec{\alpha}') = 2J(\vec{\alpha}')\left[1 + \cos -(\phi+\phi')\right] \tag{10}$$

In the paraxial approximation,

$$\phi = \frac{2\pi}{\lambda} \vec{a}.\vec{\alpha} \quad \text{and} \quad \phi' = \frac{2\pi}{\lambda} \vec{a}.\vec{\alpha}' \tag{11}$$

where $\vec{a}$ is the position vector of hole 1 with respect to hole 2 (see figure), and equation (10) becomes

$$I_o(\vec{\alpha},\vec{\alpha}') = 2J(\vec{\alpha}')\left[1 + \cos - \frac{2\pi}{\lambda} \vec{a}(\vec{\alpha} + \vec{\alpha}')\right] \tag{12}$$

In this case, the fringe visibility equals 1, since the degree of coherence between ψ_1 and ψ_2 is also unity.

Let us now consider an extended, spatially incoherent source. Each point of the source will produce a fringe pattern described by equation (12) and — since the source is incoherent — all the fringe patterns will add in intensity so that the resulting illumination will be proportional to

$$I(\vec{\alpha}) = \int I_o(\vec{\alpha},\vec{\alpha}')d\vec{\alpha}'$$

$$= 2\int J(\vec{\alpha}')d\vec{\alpha}' + 2\int J(\vec{\alpha}') \cos - \frac{2\pi}{\lambda} \vec{a}(\vec{\alpha} + \vec{\alpha}').d\vec{\alpha}'$$

$$= 2\int J(\vec{\alpha}')d\vec{\alpha}' + 2\mathrm{Re}\, \exp-\frac{2i\pi}{\lambda}\vec{a}\vec{\alpha}\int J(\vec{\alpha}')\exp-\frac{2i\pi}{\lambda}\vec{a}\vec{\alpha}'.d\vec{\alpha}'$$

$$= 2\hat{J}(0) + 2\mathrm{Re}\ \exp -\frac{2i\pi}{\lambda}\vec{a}\vec{\alpha}\ \hat{J}(\frac{\vec{a}}{\lambda})$$

$$= 2\left[\hat{J}(0) + \left|\hat{J}(\frac{\vec{a}}{\lambda})\right|\cos(\theta - \frac{2\pi\vec{\alpha}\vec{a}}{\lambda})\right] \quad (13)$$

where $\hat{J}$ is the Fourier transform of J and θ is the phase of this Fourier transform.

A comparison between (13) and the general expression (7) shows that

$$\Gamma = \frac{\langle\psi_1\psi_2^\star\rangle}{\sqrt{\langle|\psi_1|^2\rangle\langle|\psi_2|^2\rangle}} = \frac{\hat{J}(\vec{a}/\lambda)}{\hat{J}(0)} \quad (14)$$

The normalized covariance Γ between ψ_1 and ψ_2, also called complex degree of coherence, is given by the Fourier transform of the intensity distribution in the extended incoherent source, normalized to unity at the origin. (Zernike's theorem).

III/- FOURTH ORDER STATISTICS AND INTENSITY INTERFEROMETRY

Let us now consider the statistical properties of the instantaneous intensity $|\psi(\vec{r},t)|^2$. They can be entirely deduced from the statistics of $\psi(r,t)$. Since ψ is a complex circular Gaussian process, the one-dimensional probability density function of $|\psi|$ is a Rayleigh density and the one-dimensional p.d.f. of $|\psi|^2$ is an exponential density (such statistics occur in any two-dimensional random walk process). This means that the instantaneous intensity $|\psi|^2$ undergoes large fluctuations with a high probability of complete extinction (the exponential distribution is infinite at the origin).

It can be shown that the correlation time of the intensity fluctuations is of the order of the correlation time τ of the fluctuations of ψ that is of the order of the inverse of the bandwidth $\Delta\omega$.

We shall rather focus our attention on the spatial correlation scale of the intensity fluctuations and show that it is also of the order of the correlation scale of ψ.

Let

$$\Delta I = |\psi|^2 - \langle|\psi|^2\rangle \quad (15)$$

be the intensity fluctuation and let us calculate the spatial covariance

$$<\Delta I_1 \cdot \Delta I_2> = <|\psi_1|^2|\psi_2|^2> - <|\psi_1|^2><|\psi_2|^2> \quad (16)$$

between the intensity fluctuations ΔI_1 and ΔI_2 at two points $\vec{r}_1$ and $\vec{r}_2$

Since ψ is a Gaussian process any fourth order moment can be expressed as a function of the second order moment. In the complex case the relation is

$$<\psi_1\psi_2^*\psi_3\psi_4^*> = <\psi_1\psi_2^*><\psi_3\psi_4^*> + <\psi_1\psi_4^*><\Psi_2^*\Psi_3> \quad (17)$$

so that

$$<|\psi_1|^2|\psi_2|^2> = <\psi_1\psi_1^*\psi_2\psi_2^*>$$
$$= <\psi_1\psi_1^*><\psi_2\psi_2^*> + <\psi_1\psi_2^*><\psi_1^*\psi_2>$$
$$= <|\psi_1|^2> \cdot <|\psi_2|^2> + |<\psi_1\psi_2^*>|^2 \quad (18)$$

Putting (18) into (16) leads to

$$<\Delta I_1 \Delta I_2> = |<\psi_1\psi_2^*>|^2 \quad (19)$$

showing that the covariance of the intensity fluctuations is the squared modulus of the covariance of the complex amplitude. The correlation scale of the intensity fluctuations is therefore of the order of the correlation scale of ψ.

Applying equation (19) to the same point, $\vec{r}_1 = \vec{r}_2 = \vec{r}$ leads to

$$<(\Delta I)^2> = |<|\psi|^2>|^2 = <I>^2 \quad (20)$$

showing that the standard deviation $\sqrt{<(\Delta I)^2>}$ of the intensity fluctuations is equal to the average intensity $<I>$.

Normalizing equation (19) leads to

$$\frac{<\Delta I_1 \Delta I_2>}{\sqrt{<(\Delta I_1)^2><(\Delta I_2)^2>}} = \frac{<\Delta I_1 \Delta I_2>}{<I_1><I_2>} = \frac{|<\psi_1\psi_2^*>|^2}{<|\psi_1|^2><|\psi_2|^2>} = |\Gamma|^2 \quad (21)$$

The normalized covariance of intensity fluctuations is equal to the squared modulus of the degree of coherence.

Therefore Zernike's theorem also applies to intensity fluctuations under the following form

$$\frac{\langle \Delta I_1 \Delta I_2 \rangle}{\sqrt{\langle (\Delta I_1)^2 \rangle \langle (\Delta I_2)^2 \rangle}} = \left| \frac{\hat{J}(\vec{a}/\lambda)}{\hat{J}(0)} \right|^2 \qquad (22)$$

The normalized spatial covariance of intensity fluctuations is given by the squared modulus of the normalized Fourier transform of the intensity distribution in the incoherent source. The measurement of this covariance is called 'intensity interferometry" by analogy with usual or amplitude interferometry.

The intensity fluctuations we have described are often referred to as thermal noise. In this classical description we have not taken into account quantum effects and the related quantum or photon noise. We shall say a word on it in part VI.

Since they are extremely rapid, thermal fluctuations are difficult to visualize. Photon noise is usually dominant in the visible range and at shorter wavelengths, whilst in the infrared or at longer wavelengths thermal noise dominates.

However the spatial properties of thermal intensity fluctuations can be easily simulated by putting a diffuser across a coherent beam of light, such as a laser beam. At some distance of the diffuser, the field is produced by superposition of a large number of statistically independant scattered beams, so that its complex amplitude obeys Gaussian statistics. The related intensity fluctuations are easily visible and called "speckle". Our mathematical description of thermal noise as well as Zernike's theorem applies equally to laser speckles.

A speckle pattern is an interference pattern. It can be considered as a superposition of interference fringes produced by any pair of points on the diffuser. Each Fourier component of the speckle noise therefore corresponds to pairs of points on the diffuser with a given spacing. The whole Fourier spectrum is given by the autocorrelation of the intensity distribution J on the diffuser, as shown by taking the Fourier transform of equation (22). The spatial correlation scale, or "speckle size", is proportional to the inverse of the width of the illuminated part of the diffuser.

Thermal noise can also be considered as an interference pattern, but in this case the phase relationships are random so that the life time of thermal speckles is very short, being equal to the life time of the random phase relations, i.e. the coherence time τ of the light beam.

IV/ - APPLICATIONS TO ASTRONOMY

In astronomical observations, the ultimate angular resolution is limited by the diameter d of the telescope. The smaller detectable angular separation of a double star is

$$\alpha \simeq \lambda/d \qquad (23)$$

By observing interference fringes between two telescopes at a distance a apart, information is obtained on much smaller details. It is equivalent to the Young fringes experiment described above. The fringe visibility is given by the normalized Fourier transform of the intensity distribution in the object (equation 14). For a double star the visibility is a cosine function of a. The first minimum occur at a distance a, when the angular spacing of the double star is

$$\alpha = \lambda/a \qquad (24)$$

an expression similar to equation (23), but in which a can be much larger than d.

The method is very powerful for measuring the angular size α of small objects. Again the fringe visibility goes down to zero at a distance a approximately related to α by equation (24).

Michelson and his coworkers were the first to successfully apply this technique, in 1920, to the measurement of stellar angular diameters. The major problem encountered is mechanical stability (the two optical path must stay equal to within a few microns). For this reason their baseline remained limited to 12 metres. Only a few close giant stars have been measured and their technique was subsequently discarded.

A similar technique was then extensively applied by radio-astronomers. Because of the much longer wavelength there was a need for very long baseline, in order to get the same resolution as in the visible. The mechanical stability was no longer crucial and large arrays of antennas have been built in order to map the Fourier transform $\hat{J}(\vec{a}/\lambda)$ of the intensity distribution in the object. Synthetic images have been obtained by computing the inverse Fourier transform (the technique is now called aperture synthesis).

In order to transmit signals more easily over kilometers, several techniques have been developed to reduce the signal frequency : Instead of correlating the signals from the antennas, radioastronomers correlate beats between each signal and a reference wave. This is heterodyning. It is now done over baselines as large as the earth's diameter. Another technique consists in correlating thermal noise as described above (part III). This is intensity interferometry. Thermal noise can be considered as natural beats between the different components of the signal itself.

In 1956 a radioastronomer, Handbury Brown and an opticist Twiss, suggested that intensity interferometry might work at optical wavelengths. That would solve the mechanical stability problem by correlating intensity fluctuations in the gigahertz range instead of optical frequencies. Tolerances become of the order of a few centimetres. However the normalized covariance is reduced by the ratio of the electronic to optical bandwidth.

They built large multimirrors light collectors (diameter : 6 m) and succeeded in measuring diameters of about 30 very bright stars with baselines up to 200 metres. The main drawback of this method is its very low sensitivity, which is due to photon noise. We shall come back later on this problem. Photon coincidence techniques were used to measure the correlation. The magnitude of the faintest star measured was 2.5 and the measurement required 100 hours of integration. A major advantage of this technique is its insensitivity to atmospheric turbulence which affects Michelson type interferometry. These effects will be discussed in part V.

Heterodyning is now also used in infrared interferometry by two groups : one directed by Townes in the U.S. and the other by Gay in France. However this technique also suffers from a lack of sensitivity because the bandwidth is again limited by electronics. Moreover it is sensitive to atmospheric turbulence.

Due to the work of a French opticist A. Labeyrie, there is now a considerable renewal of interest in Michelson type interferometry.

The mechanical stability problem has been solved by using concrete telescope mounting, and a small prototype instrument is already operating at C.E.R.G.A., near Grasse (France), with a baseline of the order of 20 metres. It will be increased to 40 metres in a near future. A larger instrument with two 1.5 metre telescopes is under construction. An original solution to the atmospheric turbulence problem has been developed by A. Labeyrie himself. The technique, called speckle interferometry, will be now examined.

V/ - ATMOSPHERIC TURBULENCE and SPECKLE INTERFEROMETRY

Atmospheric turbulence generates small scale temperature inhomogeneities. Due to the related fluctuations of the index of refraction, the earth atmosphere behaves like a diffuser. The angle of diffusion is of the order of a few arc-seconds.

In the visible range, it compares with the angular resolution of a lens of few centimeters in diameter. This means that the image of any telescope is blurred by such an amount. Large telescopes have been built because of their light-collecting capabilities. Through the atmosphere, their angular resolution is that of a small lens !

Due to this atmospheric problem, the size of the apertures of Michelson's stellar interferometer were limited to a few centimetres, so that the two superimposed stellar images were not blurred. However the optical path difference between the two beams was randomly varying causing a restless jigger of the observed fringes. The amplitude of this random fringe motion greatly exceeds the fringe spacing, so that fringes disapear when their motion is too fast for the eye. Happily, when the atmosphere is quiet enough, they remain visible to the observer.

Let us now examine what happens when using a single large aperture. An unresolved star is a spatially coherent light source. Observed through a diffuser it gives rise to a speckle pattern very similar to a laser speckle pattern. This speckle pattern can again be considered as a superposition of interference fringes produced by any pair of points on the telescope entrance pupil.

As in the Michelson experiment, these fringes hold high angular resolution information on the observed object but they move quite rapidly, so that the speckle pattern is always changing. On a few seconds photographic exposure time, speckles are entirely blurred out by their motion and all high angular resolution information is lost.

A. Labeyrie was the first to point out the information content of atmospheric speckles and their similarity with laser speckles. Indeed, the theory developed in part III, also applies to atmospheric speckles. As we have seen, the speckle size is proportional to the inverse of the width of the illuminated aperture. The covariance of the speckle pattern is given by equation (22) where $\hat{J}$ is now the average intensity distribution on the telescope entrance pupil, so that $\hat{J}(\vec{a}/\lambda)^2$ is the intensity distribution on the Airy disk, that one would observe if turbulence was removed. The speckle size is therefore the size of the Airy disk.

Let us now examine what happens when observing an extended astronomical object $O(\vec{\alpha})$. As far as the optical paths in the atmosphere are similar, each point $\vec{\alpha}$ in the object gives rise to the same speckle pattern $S(\vec{\alpha}' - \vec{\alpha})$ shifted by an amount $\vec{\alpha}$ so that the resulting instantaneous illumination in the image plane is

$$I(\vec{\alpha}') = \int O(\vec{\alpha}) S(\vec{\alpha}' - \vec{\alpha}) d\vec{\alpha} \tag{25}$$

This convolution relation, usually written,

$$I(\vec{\alpha}) = O(\vec{\alpha}) \star S(\vec{\alpha}) \tag{26}$$

shows that the random process $I(\vec{\alpha})$ is obtained by a linear filtering of the random process $S(\vec{\alpha})$. The covariance of $I(\vec{\alpha})$ is therefore given by [†]

$$<\Delta I_1 \cdot \Delta I_2> = AC[O(\vec{\alpha})] \star <\Delta S_1 \cdot \Delta S_2> \tag{27}$$

Since $<\Delta S_1 \cdot \Delta S_2>$ is an Airy function, the object autocorrelation $AC[O(\vec{\alpha})]$ is obtained with the ultimate resolution of the telescope by statistically measuring the covariance of the irradiance in its focal plane. This is the principle of stellar speckle interferometry.

The same technique can be applied to a Michelson type stellar interferometer with large apertures. Each aperture produces a speckle pattern. When the two patterns are superimposed, fringes appear with

---

[†] We assume here, as a first approximation, that the process $S(\vec{\alpha})$ is homogeneous (spatially stationary) so that its covariance is a function of a single variable.

random shifts on speckle overlap areas. The fringe contrast is deduced from the covariance of this random fringe pattern.

VI/ - QUANTUM NOISE AND SIGNAL-TO-NOISE RATIO LIMITATIONS

Let us now consider the quantum aspect of radiation. At very low light level, individual photons are detected. Photon events occur randomly in time and space. The probability of detecting a photon on a small area around a point $\vec{r}$ during a small time interval around time t is proportional to the irradiance $I(\vec{r},t)$. When $I(\vec{r},t)$ is a constant, such a random process is called a *Poisson process*. Here, $I(\vec{r},t)$ is itself a random function. Such a process, with a random parameter $I(\vec{r},t)$, is called a *compound Poisson process*. In a compound process there are two types of fluctuations. Averages must be taken over fluctuations of both types.

In an intensity correlation experiment, the amount of noise is given by the variance of the statistical estimation of the correlation. Assuming a compound Poisson process, detailed calculations show that, in the very low light level limit, the signal-to-noise ratio is of the order of the number of photons per coherence area and coherence time (or number N_s of photons per speckle) multiplied by the square root of the number M of independant measurements

$$\text{Signal/Noise} = N_s \sqrt{M} \qquad (28)$$

On such a basis, let us compare speckle interferometry with intensity interferometry. Assuming the same optical bandwidth (of the order of 200 Å) for both experiments, let n and n' be the related number of photons received per cm² and per second inside that bandwidth. For speckle interferometry

$$N_s = n\tau\sigma \qquad (29)$$

where τ is the life time and σ the area of atmospheric speckles. And

$$M = T/\tau \qquad (30)$$

where T is the duration of the experiment.

For intensity interferometry :

$$N'_s = n'\tau'\sigma' \qquad (31)$$

where τ' is the life time of thermal speckles. In the Handbury Brown and Twiss experiment the area of the light collector is smaller that the area of thermal speckles, so that σ' must be taken as the area of the collector. Assuming the same duration T of the experiment, the number of independant measurements is

$$M' = T/\tau'' \qquad (32)$$

where τ'' is the correlation time, much longer that the natural coherence time of thermal speckles due to the electronic low pass filtering.

The two signal-to-noise ratio are identical if :

$$n\tau\sigma\sqrt{\frac{T}{\tau}} = n'\tau'\sigma'\sqrt{\frac{T}{\tau''}} \qquad (33)$$

or

$$\frac{n'}{n} = \frac{\tau\sigma}{\tau'\sigma'}\sqrt{\frac{\tau''}{\tau}} = \frac{\sigma}{\sigma'}\frac{\sqrt{\tau\tau''}}{\tau'} \qquad (34)$$

With $\sigma = 10^{-3}$ m^2, $\sigma' = 30$ m^2, $\tau = 2 \times 10^{-2}$ s, $\tau' = 10^{-14}$ s and $\tau'' = 10^{-9}$ s, one gets :

$$\frac{n'}{n} \sim 2 \times 10^4 \qquad (35)$$

showing that intensity interferometry needs 2×10^4 times more photons than speckle interferometry in order to get the same signal-to-noise ratio. In other terms, the limiting magnitude difference between the two techniques is of the order of 11. Indeed, the limiting magnitude in the Handbury Brown experiment is 2.5 while it is of the order of 13.5 in a stellar speckle interferometry experiment.

Because of its low sensitivity, intensity interferometry is no longer used in optical astronomy. In a near future, speckle interferometry techniques will be applied to long base line Michelson-type interferometry allowing a similar angular resolution with a much higher sensitivity.

HOLOGRAPHY AND λ-CODING

J.P. Goedgebuer and J.Ch. Viénot

Laboratoire d'Optique, Associé au CNRS n° 214, "Holographie et Traitement Optique des Signaux", Université de Franche Comté, 25030 Besançon cedex France

INTRODUCTION :

In the 1950's, D. Gabor solves the basic problem of recording and retrieving the phase, as well the amplitude of a wave. His purpose is to record electron waves and to reconstruct them in the visible domain, the advantage being to remove the difficulties linked to the spherical aberrations of electron lenses. An improvement of the resolution is expected in the images obtained in electron microscopy. His method [1] is based on the "in-line" superposition of a uniform background generated by a reference wave to the electron wave-front carrying the information. In fact the method pioneered by D. Gabor leads to twin images of poor quality. These major drawbacks are overcome 10 years later by E.N.Leith and J. Upatnieks who introduce the "off-axis" method of holography [2]. The reference and object waves impinge the holographic plate at an angle, to form very fine interference fringes. The resulting hologram acts as a grating and generates non-overlapping orders of diffraction. Each 1st order term of diffraction produces an image, and its quality is largely improved by working with diffused light. These ideas form the basis of conventional holography. A tremendous number of applications have been developed, combining the holographic process and the coherence properties of the laser sources. Before going further with conventional holography (section I), it is worth noting that a new trend has appeared for some years, by introducing white light in holography. In this case, one generally deals with wavelength-coding (section II).

I.- Conventional holography

1.- *Recording and retrieving of a 3-D wavefront* :

Fig. 1a represents the off-axis method of recording a hologram. Simple mathematics illustrate the reasonning. Let A_o and A_r be the complex amplitudes of the

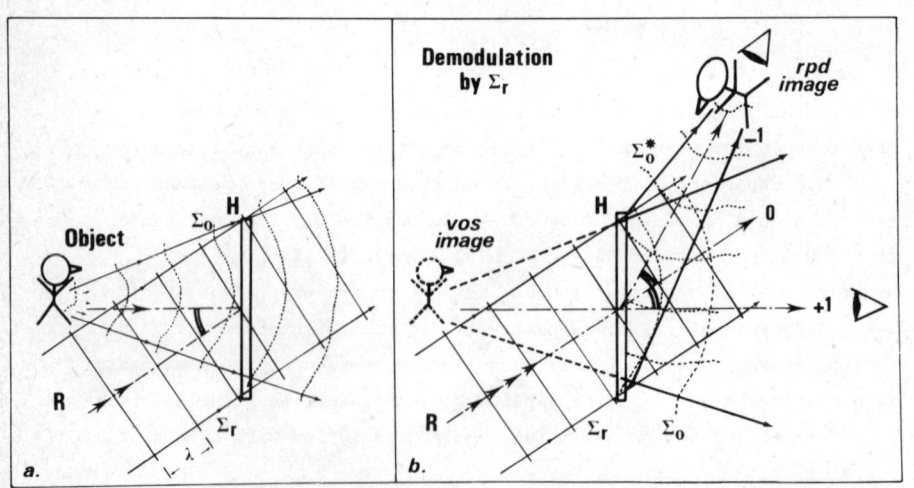

Fig. 1 : a - Recording set-up : R reference wave, O object, H : holographic plate (the dots indicate the emulsion), Σ_o object wave, Σ_r reference wave ;
b - Demodulation of the hologram illuminated by the reference wave Σ_R.

object wave and reference wave respectively. The energy E recorded on the plate is :

$$E = |A_o + A_r|^2 = |A_o|^2 + |A_r|^2 + A_r A_o^* + A_o A_r^* \qquad (1)$$

where * stands for a conjugate. This hologram is recorded under the conditions of linearity between the incident energy E and the transparency in amplitude H of the plate.

Fig. 1b illustrates the demodulation of the hologram illuminated with the reference wave. The amplitude of the light transmitted through the hologram is nothing but :

$$A_r \cdot H = A_r(|A_o|^2 + |A_r|^2) + |A_r|^2 \cdot A_o + A_r^2 \cdot A_o^* \qquad (2)$$

Two images corresponding to the last terms of eq. (2) are reconstructed. One of them is viewed in the direction of A_o : it is a *virtual, orthoscopic and stigmatic* image (vos) identical in position and in size to the object. The other one is generated by the second term of diffraction in the direction symmetrical of A_o with respect to A_r. This image is *pseudoscopic* and *distorted* (p d) and thus is usually of less interest.

From a general point of fiew, the conditions of geometry or/and wavelength of retrieving may be quite different from those of recording. Object, image, size and positions are no more similar and magnification effects may occur [3,4]. As an illustration, let us consider a conjugate beam Σ_r^* illuminating the hologram discussed so far. The conjugate beam Σ_R^* may be obtained by inverting the direction of propagation of Σ_r (Fig. 2). Eq (2) becomes :

$$A_r^* H = A_r^*(|A_o|^2 + |A_r|^2) + A_o A_r^{*2} + |A_r|^2 A_o^* \qquad (3)$$

The last term describes a conjugate object wave A_o^*. It generates a *real, pseudoscopic* and *stigmatic* image (r, p, s). Its position and size are identical to those of the object, but its relief is inverted (as in the previous arrangement, a distorted image is also obtained).

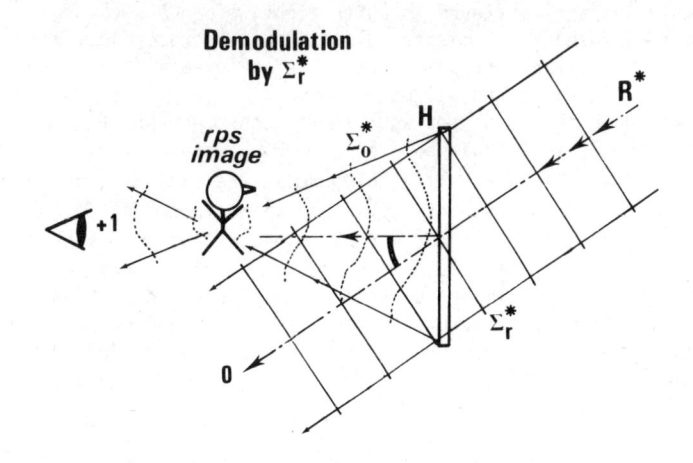

Fig. 2 : Demodulation of the hologram by a conjugate beam. A pseudoscopic image is viewed in front of the plate.

2.- *Coherence requirements* :

A hologram resulting from an interference phenomenon, the object and reference waves have to be spatially and temporally coherent. Practically two conditions are required when recording the hologram : (i) the two waves are emitted by a unique point source, (ii) the rays reflected from all parts of the object must interfere with the reference beam - that implies the coherence length (*) to be about twice the object depth. In general only the laser can meet this requirement, its coherence length ranging from some meters to several hundreds of meters.

One could expect that the same conditions apply to the demodulation of holograms. In fact, the demodulation requires less temporal coherence since the process does not involve any reflections, i.e. no large optical delays intervene. In these conditions, the image reconstruction can be carried out quite well with the 5461 Å line of a Hg source for instance.

II.- A cornerstone of modern holography : the introduction of white light

It would be an improvement to perform holography in white light. First, it would be advantageous in industrial and commercial applications since it would allow the hologram to be viewed anytime and any place without the use of a laser. Second, speckle effects would be smoothed, yielding a better resolution in the image.

1.- *Volume Holograms* [5]

The technique takes advantage of the thickness of the emulsion. The hologram is recorded in laser light. The object and reference beams are introduced from opposite sides of the plate, yielding fringes parallel to the emulsion and with a $\lambda/2$ spacing between them. When processing the plate, these fringes induce multiple reflective Ag layers ; their number may be thousands for very thick (a few mm) recording materials. The resulting hologram can be roughly described as an interference filter. Its wavelength selectivity allows the demodulation to be performed directly in white light.

2.- *Rainbow holography*

The recording is a two-step process based on conventional holography in laser light. The holographic images are retrieved in white light, along a λ-coding.

Fig. 3 indicates the procedure. A master hologram must be recorded first with the conventional technique (Fig. 1a). The demodulation is achieved with a conjugate beam, yielding a real, pseudoscopic and stigmatic image (r, p, s). The latter is projected through a slit and used as the object wave in the construction of a second hologram (Fig. 1b). The result is a "rainbow" hologram [6]. It can be viewed with a white light point source that illuminates the hologram at the conjugate angle of the reference beam (Fig. 1c). Each radiation of wavelength λ generates an image of the object. Each image is viewed through the corresponding reconstructed image of the slit acting as a gate. As a result, the technique provides a discrimination between the various images by displaying them individually in different colors (as a rainbow !). Unlike the volume holograms, the reconstructed image is brighter.

Last developments of this technique deal with holographic images of coloured [7] and animated [8] objects.

(*) coherence length = $1/\Delta\sigma$, where $\Delta\sigma$ is the spectral width of the source (in wavenumbers).

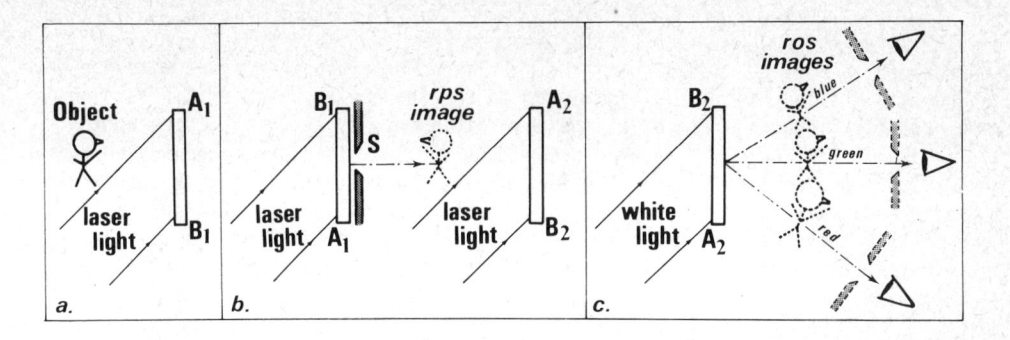

Fig. 3 : Principle of rainbow holography[6]. a - recording of a master hologram ; b - recording of a second hologram. The master hologram is illuminated by the conjugate reference beam (obtained by turning the hologram upside down) ; c - the reconstruction in white light yields polychromatic r, o, s images that are viewed through different "windows" - that prevent any color blurring.

3.- Chromatic or temporal holography

The fundamentals of holography (modulation of a spatial carrier) can be applied to the chromatic, or temporal, domain itself. It permits the recording of a so-called temporal hologram [9,10] in white light. The demodulation process is achieved in laser light. A rough idea of temporal holography can be given through the phenomenon of channelled spectrum [11]. A basic example is shown in Fig. 4. A Michelson interferometer is illuminated by a white light point source and a spectroscope analyzes the light emerging from the interferometer. The power-spectrum $B(\sigma)$ displayed at the output of the spectroscope is [11] :

$$B(\sigma) \propto 1 + \cos 2\pi\sigma\Delta \qquad (4)$$

where Δ is the path-difference between two white light waves incident onto the spectroscope (σ : wave-number). It is a channelled spectrum. The latter can be

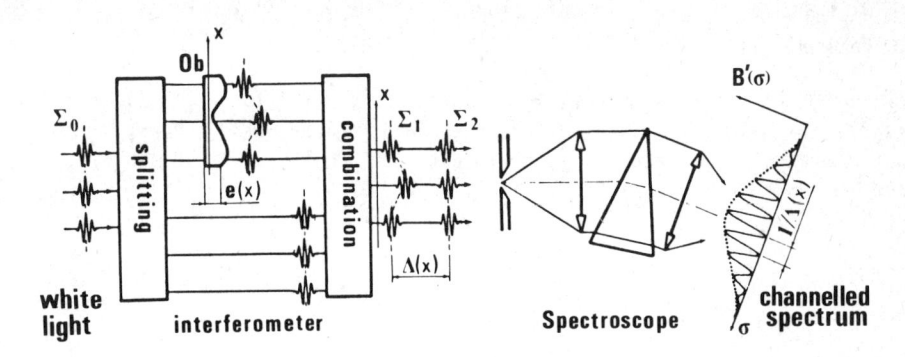

Fig. 4 : Recording of a channelled spectrum ; S white light source, Sp spectroscope. The channelled spectrum is a chromatic hologram.

considered as the superposition of the spectra of two white light waves Σ_1, Σ_2 separated by Δ, one of them playing the role of a reference with respect to the other. It comes to the description of a hologram in the temporal domain. By diffraction in monochromatic light, an "image" of the path-difference is actually reconstructed.

Unfortunately, the resolution in this type of hologram depends on that of the spectroscope and is generally low compared to that obtained with the previous methods. As a result, this method does not meet the requirements of visualization. However this technique has provided useful scientific applications, especially in metrology when measuring absolute path-differences [12]. It is why this technique is worth noting.

CONCLUSION :

A tremendous number of applications have been developed in various fields : pattern recognition, interferometry, contouring, information processing, optical memories (a volume hologram may store up to 10^9 bits/mm3), spectroscopy, optical reading, microscopy, detection of fatigue failure in materials or atmospheric pollutants, etc... Moreover the basic concepts of holography in the optical wavelengths can be generalized, and extended to other radiations (X-rays, electronwaves, acoustic waves, etc...) with applications in medicine, biology, astronomy... Some aspects of these applications are given in ref. 13.

REFERENCES :

1 - D. Gabor, Nature 161, 771 (1948)
2 - E. Leith, J. Upatnieks, J. Opt. Soc. Am. 52, 1123 (1962) ; 53, 1377 (1963) ; 54, 1295 (1965)
3 - e.g. R.J. Collier, C.B. Burckhardt, L.H. Lin, "Optical Holography" Academic Press ed., 1971
4 - e.g. J.Ch. Viénot, P. Smigielski, H. Royer, "Holographie optique ; développements; applications", Dunod ed., 1971
5 - Y.N. Denisyuck, Sov. Phys. Dokl., 7, 543 (1962) ; Opt. Spectrosc., 15, 279 (1963)
6 - S.A. Benton, J. Opt. Soc. Am. 59, 1545 A (1969)
7 - P. Hariharan, Opt. Eng., 16, 520 (1977)
8 - N. Aebischer, B. Carquille, 17, 23 (1978), 3698-3700
9 - C. Froehly, A. Lacourt, J.Ch. Viénot, Nouv. Rev. Optique, 4, 4 (1973)
10- J.Ch. Viénot, J.P. Goedgebuer, A. Lacourt, Appl. Opt., 16, 2 (1977)
11- e.g. H. Bouasse, Z. Carrière, "Interferences", Delagrave ed., Paris, 1923
12- J.P. Goedgebuer, A. Lacourt, J.Ch. Viénot, Opt. Comm. 16, 1, 99 (1976)
13- see for example "Le Courrier du CNRS", n° 28, April 1978, pp 11-19

ELECTRON HOLOGRAPHY

Juliet ROGERS

Section de Physique du Solide
Département de Recherche Fondamentale
Centre d'Etudes Nucléaires de Grenoble
85 X, 38041 Grenoble Cédex, France

Electron holography was introduced in electron microscopy thirty years ago by Dennis Gabor (1948). Although the technique is only just beginning to prove itself useful it has been widely applied in other fields (B.J. Thompson, 1978). The attraction of holography is the possibility of reconstructing the complex amplitude of a wavefront from an intensity recording with diffraction limited resolution. Holograms also have the advantage of being taken first and imaged later, which minimises object exposure. Comprehensive discussions of electron holography are given by K.J. Hanssen (1973) and R.H. Wade (1979).

The method relies upon the interference of two coherent wavefronts, one of which is known. This reference wavefront is then used to illuminate the hologram which reconstructs by diffraction the unknown wavefront. A thorough treatment of the subject is given by R.J. Collier, C.B. Burckhardt and L.H. Liu (1971). With appropriate scaling the reconstruction can be made at any wavelength or by digital processing. A conjugate wavefront is also generated because the interference pattern records only the modulus of the phase difference. Let $U_R(x) = A_R(x) \, e^{i\phi_R(x)}$ be the complex amplitude of the reference wave, with a similar expression for the unknown image wave U_I, then the intensity at the hologram plane is given by

$$I_H(x) = |U_R(x) + U_I(x)|^2$$
$$= A_R^2(x) + A_I^2(x) + 2A_R(x) \, A_I(x) \, \cos[\phi_R(x) - \phi_I(x)] \quad (1)$$

Interference fringes are made evident by the cosine term. In practice, there will be wavefronts within an angular spread and with a range of wavelengths. One may consider that wavefronts at each angle and each wavelength will form their own set of interference fringes, all overlapping (M.E. Haine and T. Mulvey, 1952, E.N. Leith and J. Upatnieks, 1967). The resolution of the hologram will be determined by the finest set of fringes that remain clearly defined when the different sets are added. This is quite comparable to the conclusions of the envelope approximation in contrast transfer theory (K.J. Hanssen and L. Trepte, 1971, J. Frank, 1973) where it is found, for small deviations from perfect coherence, that the effect is simply to multiply the transfer function by an envelope that attenuates higher spatial frequencies, thus eliminating finer fringes. The highest resolution is only attainable about a certain small defocus (R.H. Wade, 1978, J. Frank, 1978) in conventional transmission electron microscopy.

Before reconstruction the hologram is processed so that its transmittance, $t_H(x)$, is proportional to the intensity to which it was exposed, i.e. $t_H(x) = g I_H(x)$. Thus the reconstructed wavefront is given by $U_c(X) = U_R(x) t_H(x)$, or

$$U_c(x) = g|U_R(x)|^2 U_R(x) + g|U_I(x)|^2 U_R(x) + g|U_R(x)|^2 U_I(x) + g U_R^2(x) U_I^*(x) \qquad (2)$$

The last two terms are generated by the interference between the reference and image waves. The third term is directly proportional to the original image wave and the fourth to its conjugate. However, $t_H(x)$ may have both an amplitude and phase component and the type of hologram used determines the quantity that is linearly reconstructed (K.J. Hanssen 1970, R.H. Wade 1974). If g is purely real or purely imaginary the complex amplitude reconstructed is proportional to that of the image wave, whose components may be separately visualised by optical methods such as interferometry or phase contrast (G.W. Ellis 1966, K.J. Hanssen 1973). For the special case of a weak phase object the intensity reconstructed from a weak phase hologram will be proportional to the original phase distribution, and similarly for a weak amplitude object.

The quantity $U_R(x)$ is determined by the layout. For a plane, tilted reference wave, $U_R(x) = e^{i\beta x/\lambda}$, where λ is the wavelength, and equation 2 becomes :

$$U_c(x) = g (1 + A_I^2(x)) e^{i\beta x/\lambda} + g A_I(x) e^{i\phi_I(x)} + g A_I(x) e^{i(2\beta x/\lambda - \phi_I(x))} \qquad (3)$$

The inclination of the reference with respect to the image wave results in the spatial separation of the reconstruction and its conjugate from the illuminating beam.

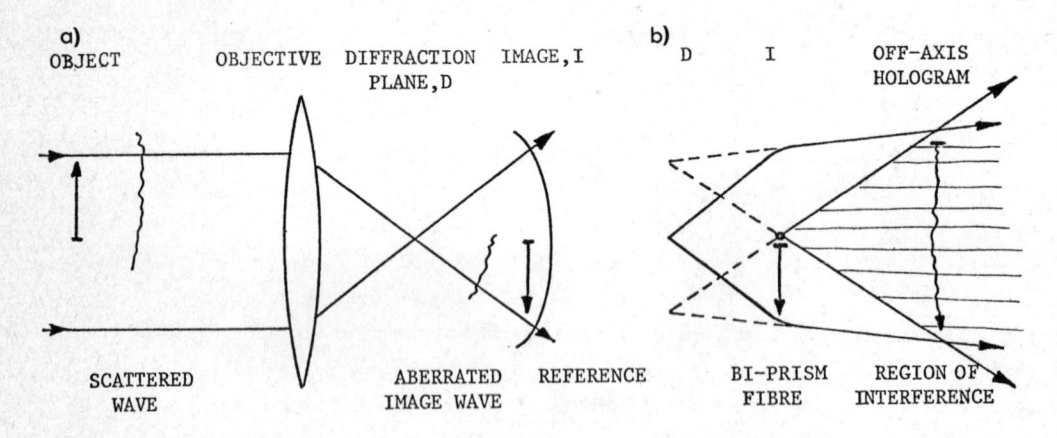

Figure 1 : Holographic layouts in electron microscopy. a) In-line. The unscattered illumination acts as a reference wave. b) Off-axis. The bi-prism produces interference between the image and an adjacent clear area.

This can be seen clearly from the spectrum of the hologram,

$$\tilde{I}_H(u) = \delta(u) + \tilde{U}_I(u) \otimes \tilde{U}_I(u) + \tilde{U}_I^*(u-\beta/\lambda) + \tilde{U}_I(u+\beta/\lambda) \qquad (4)$$

thus the reference beam inclination angle β should be large enough to separate the off-axis terms from the central auto-correlation. In general, this implies $\beta > 3\, u_o \lambda$, where u_o is the highest frequency in the image, but the special case of a weak object requires only that $\beta > 2\, u_o \lambda$ since the auto-correlation is limited to the bias.

If in recording the hologram the reference and image wave are on the same axis, i.e. $\beta = 0$, then the reconstructions will be in-line and all the terms of equation 4 will be centered on the origin. In-line holography is only appropriate for objects whose auto-correlation is negligeable and which transmit enough of the beam unscattered to act as a reference wave. Gabor (1949) originally proposed a projection method, using the change in wavelength of 10^5 between the electron microscope and the light reconstruction to provide magnification. The arrangement was too difficult to use at the time, although it can now be compared to S.T.E.M. imaging (L.H. Veneklasen, 1975) and was later shown by R.W. Meier (1965) to introduce additional aberrations. A typical brightfield micrograph however may also be interpreted as an in-line hologram, as shown in figure 1. To compensate for the original aberration and defocus the reconstruction is imaged backwards through a light optical model of the electron microscope (K.J. Hanssen, 1968, J. Rogers, 1978). At the original object plane is a perfect reconstruction and a conjugate having twice the image defocus and aberration, whose combined intensity may be described by an all positive transfer function which in general still contains zeros.

The first electron holograms were taken using the in-line method by M. Haine and T. Mulvey (1952). The specimens used were zinc oxide crystals and carbon black. Resolution at the time was about 10 Å and limited by mechanical and electrical instability. The resolution of the reconstructions did not match that of the holograms. Similar experiments were performed by T. Hibi (1956). The first optical reconstruction of an electron hologram that achieved the resolution of the original, 3 Å, was made by K.J. Hanssen and G. Ade (1976) using a carbon film as object with full com-

Figure 2 : Reconstructed image of gold particles on a carbon film from an in-line hologram recorded at a defocus of 2.1 μm and 10^5 magnification. The diffractogram extends to 7 Å. (from N. BONNET et al, 1978).

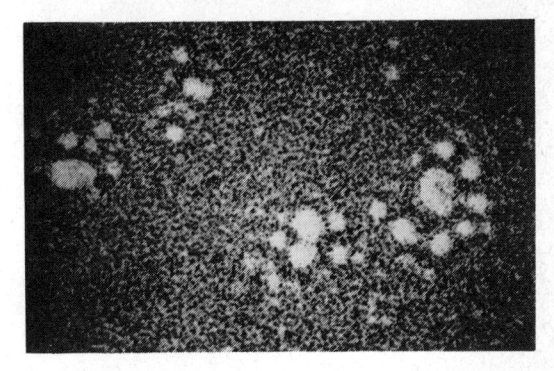

200 Å

pensation of spherical aberration and defocus. With contemporary microscopes the effects of spherical aberration are only noticeable at resolutions below about 5 Å.

When an in-line hologram is taken at a very large defocus, in the far-field of the whole image, the two reconstructions will be so far separated that one will be undisturbed by the other. The local resolution will be continuous, i.e. the transfer function will be constant and without zeros, although if the whole reconstruction is considered the effects of the superimposed conjugate image will still be apparent. Fraunhofer holography, as it is called, has been the subject of many experiments in electron microscopy e.g. A Tonomura et al. (1968), J. Munch (1975), M. Troyon (1977), N. Bonnet et al. (1978). Because of the need for high spatial coherence to record any detail at defocus values of the order of microns a field emission gun is essential. Even so the resolution will be limited by the source and only defocus need be compensated. The best reconstruction to date is that shown in figure 2 whose diffractogram extends to 7 Å.

Single-sideband holography uses a half-plane aperture in the diffraction plane of the microscope and has been thoroughly investigated despite great experimental difficulty, by K.H. Downing (1974). In general two complementary holograms are required, but the conjugate reconstruction is suppressed and in a fully compensated optical system a resolution of 4 Å was obtained. True off-axis holography is required to reconstruct a complex object from one hologram. The difficulty lies in introducing a coherent, tilted reference wave. The most practical method is to use the equivalent of the Fresnel bi-prism (T. Hibi and K. Yada, 1976) at the image plane, as shown in figure 1b. The first demonstrations of off-axis electron holography were by G. Mollenstedt and H. Wahl (1968) and by H. Tomita et al (1970) who

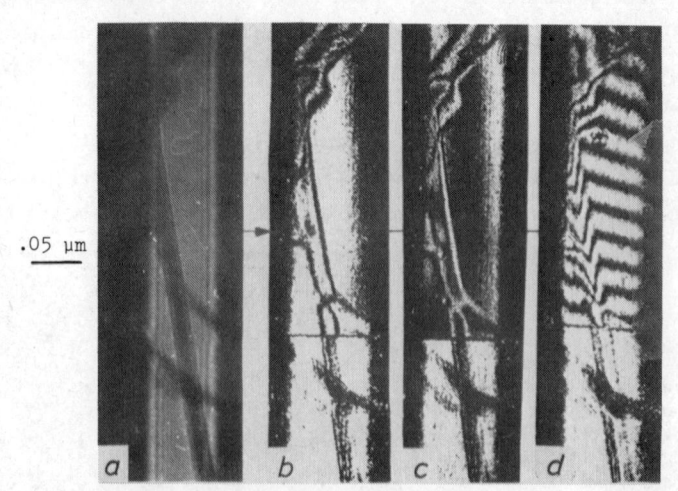

Figure 3 : Off-axis results from H. WAHL (1974). a) Hologram of ZnO crystals. b) -d) Reconstructions. The upper part is doubly illuminated for interference by a plane wave whose phase is b) 0 , c) π and d) slanted.

worked at a resolution of 500-1000 Å. An investigation of the method was made by G. Saxon (1972) with consideration of aberration correction. A more flexible microscope and bleaching the holograms before reconstruction enabled H. Tomita et al. (1972) to obtain an apparent resolution of 20 Å. Considerable development applied to improving the brightness and monochromaticity of the source has allowed A. Tonomura et al. (1979a) to obtain off-axis holograms in a microscope that has demonstrated .62 Å lattice fringes. A bi-prism working at up to 200 V produced a large field of 2 Å interference fringes at a defocus of 6000 Å. Since the objects were pentagonal gold crystals about 100 Å across the hologram is already in the Fraunhofer field. A through focus series of the optical reconstruction, which is not compensated for spherical aberration, clearly shows 2.4 Å lattice fringes and their half-spacing.

The three dimensional nature of the reconstruction was shown by J. Munch (1975). H. Wahl (1974, 1975) demonstrated with electrons interferometric techniques developed in light optics to produce the difference between two image waves or interference between reconstructions from two holograms, as shown in figure 3. The resolution was limited to about 50 Å and the specimens, as in most previous work were crystals of zinc or magnesium oxide. The technique has recently been applied to the study of magnetic domain walls (B. Lau and G. Pozzi, 1978). A Mach-Zehnder reconstruction of a hologram obtained by A. Tonomura et al. (1979b) clearly demonstrates the practicality of such methods for phase visualisation.

The continuous resolution obtained from most off-axis holograms has been much less than from in-line techniques. This is because a low magnification is used to balance the conflicting demands of a small source size and sufficient intensity for a short exposure. But such technical difficulties can be overcome, as shown by the work of A. Tonomura et al. Aberration compensation must be used in the reconstruction to obtain the best possible results. With an exposure time of a few seconds off-axis holography may now be applied to more delicate specimens. However there are not many microscopes that could meet this standard at present. Hence for most high resolution work in-line techniques offer the simplest approach to image interpretation.

Acknowledgement :

The author holds a Royal Society Fellowship at the Centre d'Etudes Nucléaires de Grenoble. The copyright of figure 2 resides with the Microscopical Society of Canada and of figure 3 with the Wissenschaftliche Verlagsgesellschaft mbH, Stuttgart.

References :

N. BONNET et al. 1978, 9th E.M. Toronto, 1, 222.
R.J. COLLIER, C.B. BURCKHARDT and L.H. LIU, 1971, "Optical Holography" Academic Press.

K.H. DOWNING, 1974 "Image Enhancement in Electron Microscopy by Single Sideband Methods", Ph. d. Thesis, Cornell Univ. New York.
G.W. ELLIS, 1966, Science 154, 1195.
J. FRANK, 1973, Optik 38, 819.
J. FRANK et al. 1978, Optik, 52, 49.
D. GABOR 1948, Nature 161, 778.
D. GABOR 1949, Proc. Roy. Soc. A197, 454.
M.E. HAINE and T. MULVEY 1952, J.O.S.A., 42, 763.
K.J. HANSSEN, 1968, Eur. Reg. Conf. on E.M., Rome 1, 153.
K.J. HANSSEN, 1970, Optik 32, 74.
K.J. HANSSEN, 1973 in "Image Processing and Computer Aided Design" ed. P. Hawkes, Academic Press.
K.J. HANSSEN and L. TREPTE, 1971, Optik 32, 519, 33, 182, 166.
K.J. HANSSEN and G. ADE, 1976 PTB-APh 10.
T. HIBI, 1956, J.E.M. 4, 10.
T. HIBI and K. YADA, 1976, Princ. and Techn. of E.M. 6, 312.
B. LAU and G. POZZI, Optik, 51, 287.
E.N. LEITH and J. UPATNIEKS, 1967, J.O.S.A. 57, 975.
R.W. MEIER 1965, J.O.S.A. 55, 987.
A. MOLLENSTEDT and J. WAHL, 1968, Naturwiss, 55, 340.
J. MUNCH, 1975, Optik, 43, 79.
J. ROGERS, 1978, I.C.O. 11, Madrid, 235.
G. SAXON, 1972, Optik, 35, 195,359.
B.J. THOMPSON, 1978, Rep. Prog. in Phys. 41, 633.
H. TOMITA et al. 1970, Jap. J. Ap. Phys. 9, 719.
H. TOMITA et al. 1972, Jap. J. Ap. Phys. 11, 143.
A. TONOMURA et al. 1968, Jap. A. Ap. Phys. 7, 295
A. TONOMURA et al. 1979a, Jap. J. Ap. Phys. 18, 9.
A. TONOMURA et al. 1979b, Optik, 53, 147.
M. TROYON, 1977, Doctoral Thesis, Univ. of Reims.
L.H. VENEKLASEN, 1975, Optik, 44, 447.
R.H. WADE, 1974, Optik, 40, 201.
R.H. WADE, 1978, Ultram. 3, 329.
R.H. WADE, 1979 in "Computer Processing and E.M. Images" ed. P. Hawkes, Springer.
H. WAHL, 1974, Optik, 39, 585.
H. WAHL, 1975 "Bildebenenholographie mit Elektronen" Habilitationschrift, Univ. of Tübingen.

PROSPECTS OF X-RAY HOLOGRAPHY
V.V. ARISTOV

Solid State Physics Institute Academy of Sciences of the USSR,
Chernogolovka, Moscow district, 142432, USSR

The problem of X-Ray holography may be divided into three different items: the first one is the problem of using the optical holographic schemes in roentgenography; the second is the solution of the phase problem in structural analysis using holographic methods; the last one is the problem of the development of X-ray interference or holographic microscope by using X-ray optics methods. Let us analyse all these problems in succession.

a) <u>on the possibility of using optical holographic schemes in X-Ray holography</u>

At present, X-ray holograms have been obtained in Fourier's holographic schemes (fig. Ia), with beam separation by the Lloyd's mirror (Fig.Ib) and Fraunhoffer's axial holography (Fig.Ic). In these schemes, holograms of the simplest objects were obtained with characteristic radiation AlKα (λ_x = 8,34 Å), BeKα (λ_x = 114 Å), CKα (λ_x = 44,8 Å) and synchrotron radiation with λ_x = 60 Å. The image was reconstructed in laser beams (λ_l = 0,63 μm). The experiments yielded a resolution of 2 to 5 μm with objects of several tens of microns (see the list of references in paper [1]).

A comparison of these modest results of experimental investigations with the theoretical conclusions about the possibilities of developing high resolution X-ray holographic microscopy suggests that direct extension of schemes of coherent optics to roentgenography is inexpedient, as is an attempt to develop effective X-ray optics on the basis of usual optical elements. This conclusion was arrived at by estimating the maximum resolution of the holographic schemes which is limited by an insufficient coherency of X-ray radiation sources, by aberrations and low aperture ratio of holographic systems [1]. It was shown that presently the resolution of traditional schemes of X-ray holography cannot exceed the value 1 μm in the field of view of several tens of microns due to a limited power of X-ray sources. An increase in the resolution by only one order requires the source power to be increased by 10^6 times. The resolution is also restricted by wave front distortions during image reconstruction in visible light. It means that the use of the holography schemes represented in Fig. I to obtain holograms in X-rays is inexpedient, at present at least. In our opinion, in the X-ray wave lengths range, the development of coherent-optical systems is possible only by using X-ray diffraction from perfect crystals.

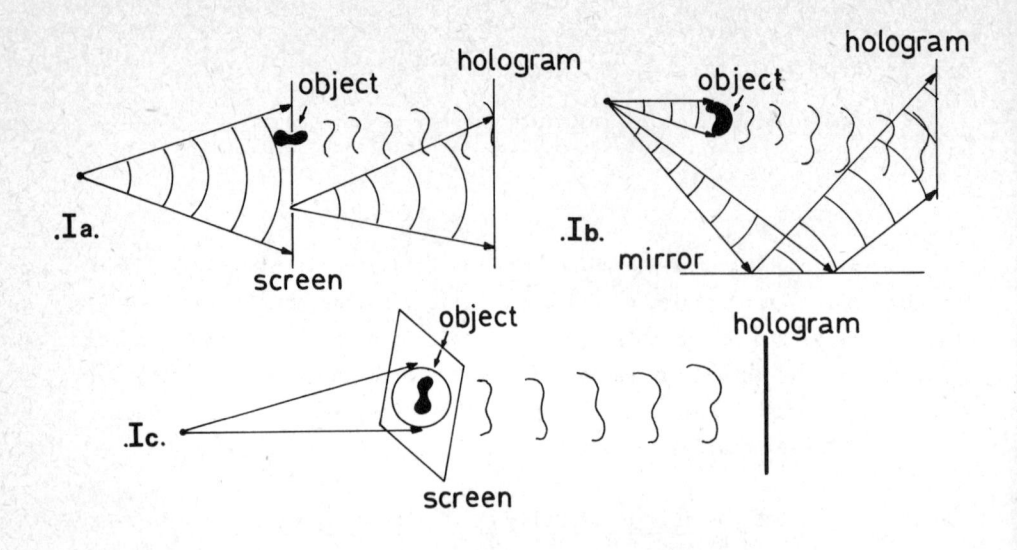

Fig. 1. Experimental schemes of hologram recording.

b) <u>on the possibility of the solution of the phase problem in structural analysis using holographic methods</u>

Let us consider a Bragg's X-ray microscope [2]. This microscope was intended for image reconstruction of the crystal lattice by light diffraction from an X-ray diffraction pattern. In a Bragg microscope the reconstruction of the atomic structure is possible only after the relative phases of the different reflections of the X-ray pattern have been determined. The methods to measure the phases are indirect, many of which, like the method of heavy atom, being similar to holographic ones. In several cases, these methods yield good results, but their application is limited to specific structures. In this respect, attempts have been made to solve the phase problem in structural analysis using direct holographic methods [3-4]. In our opinion, the possibility of utilizing holographic methods, in this case, is problematic too. The fact is, that the scattering of a monochromatic weakly diverging wave by a crystal gives rise to one diffracted wave only. To obtain another diffracted wave one has to change the geometry of the experiment. This means that apart from the usual requirement of holography to have coherent object and reference waves, one has to preserve constant the difference between the optical paths of the reference and diffracted waves for each of the diffracted waves or be able to measure it with an accuracy of fractions of the X-ray wavelength.

This requirement reduces substantially the range of possible schemes for experimental measurements of the phases. It can be easily shown that only the phases

of the beams diffracted by a large perfect crystal can be experimentally measured
using either the Kossel's method or multiwave scattering. In principle, relative
phases of different orders of diffraction of three-crystals X-ray interferometer can
be measured.

So, in the case of large highly perfect crystals not only amplitudes but
also relative phases of the diffracted waves can be measured. Unfortunately, direct
methods to measure phases are not of practical interest at present, because the
majority of crystals, including biological ones, are imperfect.

c) on the possibility of constructing an interference microscope for short-wavelength X-rays ($0,5 \text{ Å} \lesssim \lambda_x \lesssim 3 \text{ Å}$)

Lately, X-ray diffraction optics that makes use of diffraction in perfect crystals has attained a success, namely : X-ray interferometers are constructed [5], X-ray diffraction focusing is accomplished [6-7] , the possibility of utilization of polychromatic diverging beam in interference experiments is shown [8] . All what is said above suggests the idea of the construction of an X-ray high resolution interference microscope. (When obtaining interferograms of some objects by means of a traditional X-ray interferometer [9] resolution in the horizontal direction is limited by the divergence of the wave field in the crystals and is equal to the size of the Borrmann fan). One of the possible schemes for such a microscope is presented in fig. 2a). A polychromatic radiation from a point source O is diffracted in a thin crystal S. In the "mirror", the diffracted wave is diffracted again, during this process radiation for each wave-length is constricted to a narrow spot due to the effect of the diffraction focusing and the polychromatic radiation is focused in a point determined by the geometry of the experiment.

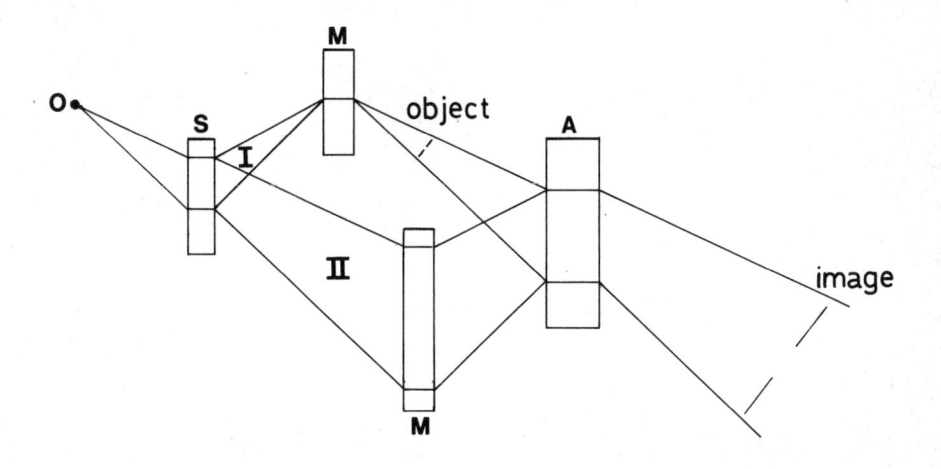

Fig. 2a. An example of a scheme of an high resolution interference microscope for short wavelength X-Rays.

This point is the source of a diverging wave with the size of the focus $2\Lambda \tan\theta/\pi$ < 10 μm (Λ is the extinction lenght, θ is the Bragg angle). If an object is placed in a diverging wave I, then, behind the crystal A one can observe its magnified image with coherent field II superposed on it. At a substantially large magnification coefficient, the distortions introduced on passing by the object wave of crystal A can be ignored. The scheme considered is the simplest example of a scheme of an holographic microscope used to record an hologram of a focused magnified image.

So, using the methods of X-ray diffraction optics one can construct an holographic microscope with resolution at least up to 1 μm (likely up to 0,1 μm [10]). Principal elements of such a microscope must be an X-ray interferometer, diffraction focusing systems and a system to form a magnified image in order to reduce the aberration effect introduced by diffraction in different crystals of the interferometer. A partial compensation of these aberrations is also possible in using optical filtration methods to process the obtained images.

d) on the possibility of developing holographic settings for long-wave X-rays ($\lambda \geqslant 30$ Å)

The development of long-wavelength X-ray high resolution microscopy [11] and construction of special mirrors, which reflect X-ray at large incidence angles [12] , suggests a setting for long-wave length X-ray holography of biological samples (fig. 2b). This is an example of the setting for recording volume X-ray holograms like that of the Denisuk holography method. In order to reconstruct the image with visible light, the surface relief of the hologram was suggested to be magnified by λ_x/λ_1 times using X-ray contact microscopy methods [12] . For thin biological samples temporal and space coherence of the usual radiation

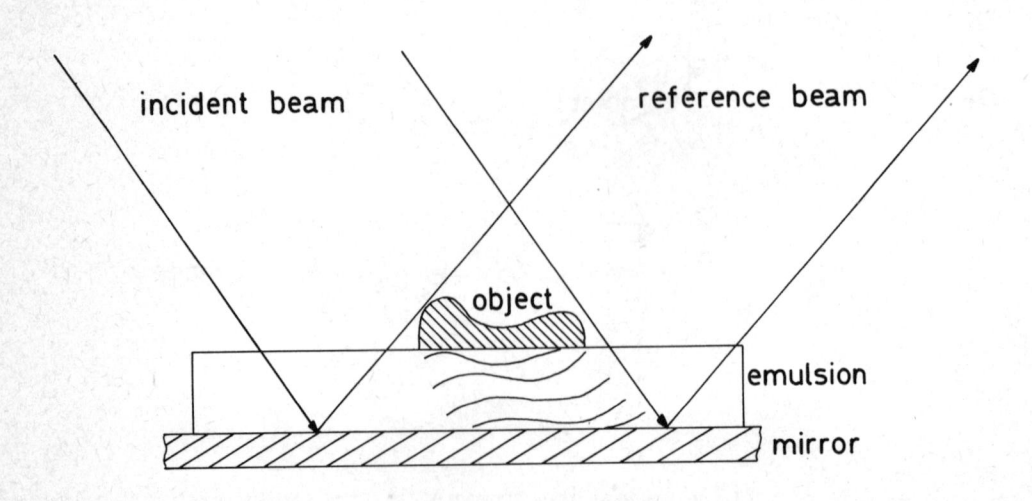

Fig. 2b. The holography schemes for long-wavelength X-rays.

source is sufficient for the hologram recording. The exposure time
is approximately the same as for contact microscopy and wave front aberrations during
image reconstruction are absent due to hologram size magnification. So, perhaps, the
scheme under consideration will be useful for the development of high resolution
(up to some tens of ängstroms) long-wavelength X-ray holographic microscopy.

REFERENCES

[1] V.V. ARISTOV, G.A. IVANOVA.(1979). J. Appl. Cryst. 12, 19.
[2] W.L. BRAGG (1942), 149, 470.
[3] L.N. KONDUROVA, A.I. SMIRNOV (1971). Zh. T.F. 41, 1043.
[4] V.I. ZAITSEV, B.K. VAINSTEIN, G.I. KOSOUROV. Krystallographiya (1968) 13, 594.
[5] W. GRAEFF (1979) see this issue.
[6] V.L. INDENBOM, E.V. SUVOROV, T.Sh. SLOBODETSKII (1976). Zh. Eksp. Teor. Fiz. 71, 359.
[7] V.V. ARISTOV, V.I. POLOVINKINA, I.M. SHMYTKO, E.V. SHULAKOV (1978). Zh.Eksp.Fis. Pisma v Red. 28,(1), 6.
[8] E.V. SHULAKOV, V.V. ARISTOV (1979) J. Acta Cryst.
[9] M. ANDO, S. HOSOYA (1972). Proceed. of the 6 International conf. on X-ray optics and microanalysis. Tokyo.
[10] P.V. PETRASCHEN, F.N. CHUKOVSKII (1976). Zh.Eksp.Fis. Pisma v Red. 23, (7) 385.
[11] D. SAYRE (1979) see this issue.
[12] E. SPILLER (1972) . Appl. Phys. Lett. 20, 365.

IMAGEING BY MEANS OF CHANNELLED PARTICLES

Yves Quéré

S.E.S.I. C.E.N.

92 Fontenay-aux-Roses, France

A positive particle is said to be channelled in a crystal when its trajectory is concentrated between two successive atomic planes (planar channelling) or along a few (three or four) neighbouring atomic rows (axial channelling). The velocity of a channelled particle is never far from being parallel to these planes (rows) so that its movement may be considered as due to a series of correlated glancing collisions on the successive atoms of a plane (row), gently repelling it to the neighbouring plane (row).

Classical mechanics apply quite well, at least for particles heavier than protons of, say, some keV's (a 1 MeV proton has a wavelength of $\simeq 2.5 \times 10^{-4}$ Å.). The repulsive potentials of planes (rows) have been calculated by Lindhard [1]. Two successive planes contribute to the formation of a potential valley along which the particles both propagate (longitudinal movement) and oscillate (transverse movement). A comprehensive review of channelling (theory, experiments and applications) has been given by Gemmel [2].

Obviously, channelling is a property of the perfect crystal. Inversely any type of lattice imperfection (phonons, defects..) is expected to dechannel particles. Defect-dechannelling was first observed for stacking faults [3] and afterwards for most types of defects like dislocations [4,5] or dislocation loops [6], interstitial atoms [7,8], gas bubbles [9], Guinier-Preston zones [11]...

Dechannelling by foreign atoms has been extensively used for locating atoms in a lattice. A recent and convincing example is the observation of octahedral and/or tetrahedral sites of hydrogen in f.c.c. metals [10].

Dechannelling by defects can also be used for the purpose of imageing. The principle is straightforward. Let us shoot channelled particles in a crystal and collect them when they emerge. The density of collected particles (i.e. the blackening of the collector) should depend, both

i/ on the local density (and nature) of defects in the crystal, giving a "defect contrast", and

ii/ on the orientation of the crystal, giving an "orientation contrast".

The simplest way to "shoot channelled particles" is just to shoot particles in a crystal thick enough so that only particles which have been channelled all the way through will emerge. This can be obtained either with sources of particles (Pu, Am, Cf ...) [11-14] or with accelerators [15]. Typical thicknesses will be $\simeq 20\mu$ (aluminium) or $\simeq 8\mu$ (gold) for 5 MeV α particles.

The "collector" is either a photographic device (generally plastic foils for α particles or protons; mica for fission fragments...) or a counter if more quantitative information is wished.

The following figures show some examples of this technique.

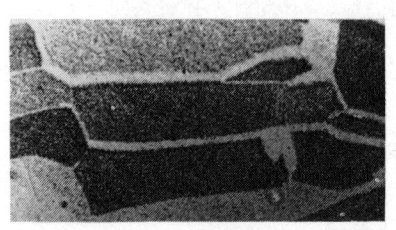

Figure 1
Grains and grain boundaries in a nickel foil. The white ribbons correspond to the projection on the plane of the foil, of boundaries which dechannel the particles. The defect-contrast (grain boundaries) and the orientation-contrast are visible. Particles are 1.9 MeV protons from a Van de Graaff [15].

Figure 2
Twins in copper, (left) and silver (right). One will observe the defect-contrast, due to dechannelling on twin-boundaries, and the orientation-contrast between twin and matrix. Particles are α's from an americium source. Thickness of the samples: 12 μ.

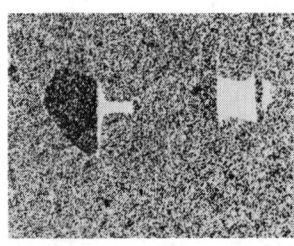

Figure 3
Insulary grains in a large single crystal of nickel showing a good example of orientation contrast. Particles are α's from an americium source.

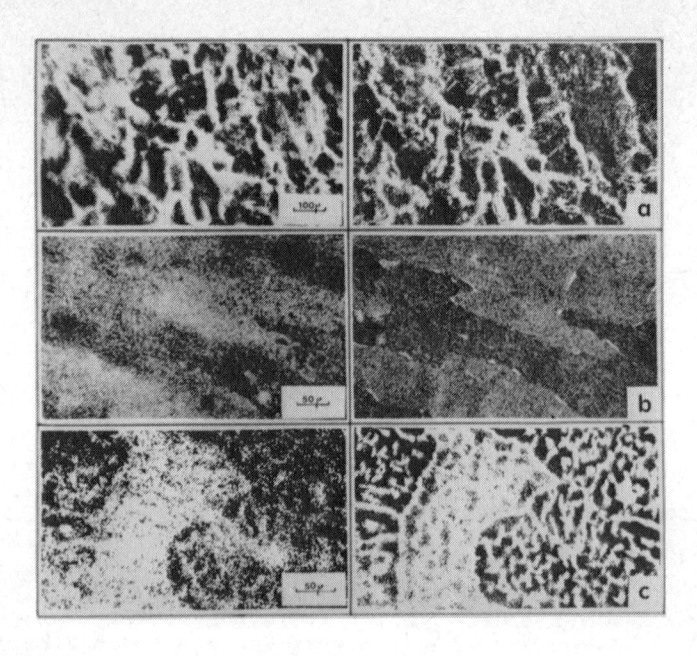

Figure 4
A comparison between contrasts obtained with two different types of particles. Left: α particles from an Am source. Right: fission fragments of uranium. The definition of the images is better for fission fragments, due to more restrictive conditions of channelling.
a : gold.   b : molybdenum.   c : tungsten. [14].

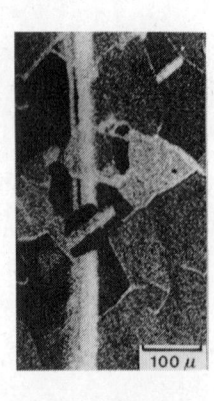

Figure 5
Dechannelling due to dislocations. Dislocations have been produced by a slight cold work in a polycrystal of nickel, along a vertical line (left). The result of a thermal anneal is visible on the right. Particles: 1.9 MeV protons [15].

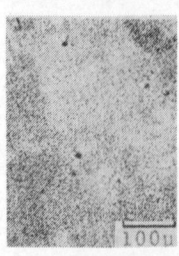

Figure 6
Dechannelling due to dislocation loops. Left: a crystal of well annealed aluminium. Right: the same crystal containing loops created by quenching and clustering of vacancies. 5 MeV α particles [12].

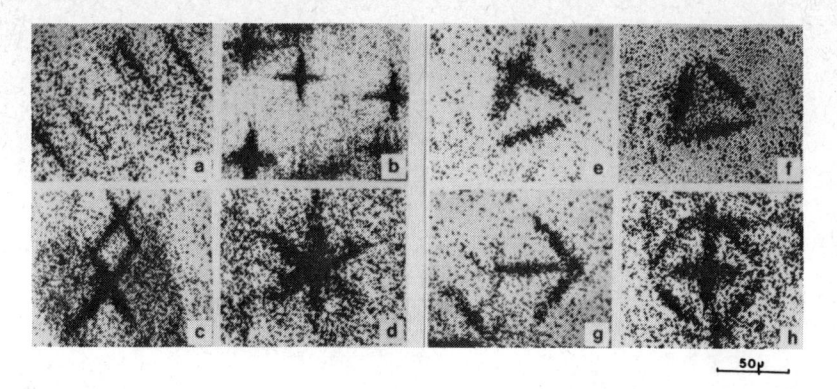

Figure 7
Some typical channelling patterns [14].
If the sources of particles are unhomogeneous, containing
"points" of activity higher than the average, geometrical
patterns appear on the collector which are images of these
point-sources through the crystal. The straight lines of
the patterns are the intersections of low-index atomic
planes with the plane of the collector. These patterns
thus allow a local orientation of the crystal. For example, the two lines of the cross in b/ indicate the $\langle 100 \rangle$
directions of a f.c.c. crystal; the three lines of the
triangle in f/ indicate the three $\langle 110 \rangle$ directions of
a $\{111\}$ plane of a f.c.c. etc. (These patterns are
obtained with α americium sources). The principles of
indexation will be found in [14].

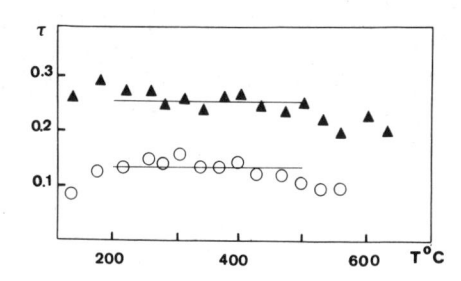

Figure 8
Evolution of gas bubbles in a solid.
If the channelled particles hit a counter instead of a collector like on figures 1-7, quantitative information on defects is easily obtained.
 Here, the sample is an aluminium foil containing He bubbles. The crystal is annealed at temperature T, which allows the bubbles to migrate
and coalesce, leading as a consequence to a large difference in size between the initial bubbles and those after anneal at $\simeq 600°C$. The transmission τ (or dechannelling $1-\tau$)
for two samples of different He concentration remains relatively constant. Here, $1-\tau$ is a measurement of the total surface of bubbles. This experiment illustrates the
"law of surfaces" stating that if bubbles at thermal equilibrium coalesce, the surfaces — not the volumes —
are additive [9].

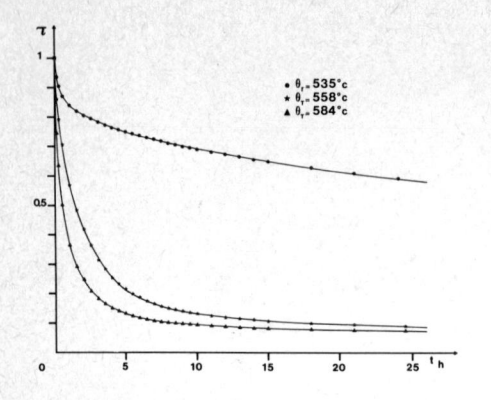

Figure 9
Isothermal ageing of an aluminium-copper alloy.
After a quench from temperature θ_T, a room-temperature "ageing" allows the Cu atoms to migrate with the help of quenched vacancies, and to coalesce into Guinier-Preston (G.P.) zones. These G.P. zones tend to dechannel particles (here α particles), thus decreasing the transmission τ (see Fig. 8) [11].

R E F E R E N C E S

1. J. Lindhard
   Math.-fys. Meddr. 34, 1, 1965
2. D. Gemmel
   Rev. Modern Phys. 46, 129, 1974
3. Y. Quéré, J.C. Resneau, J. Mory
   Comptes Rendus 262, 1528, 1966
4. J. Mory, Y. Quéré
   Rad. Effects 13, 57, 1972
5. Y. Quéré
   Phys. Stat. Sol. 30, 713, 1968
6. G. Chalant, J. Mory
   J. Physique (Lettres), sous presse, 1979
7. J.J. Quillico
   Rapport C.E.A. R-4532, 1973
8. A. Dunlop, N. Lorenzelli, J.C. Jousset
   Phys. Stat. Sol. 49, 643, 1978
9. D. Ronikier, G. Désarmot, N. Housseau, Y. Quéré
   Rad. Effects 27, 81, 1975
10. J.P. Bugeat
    Thèse, Université de Grenoble, 1979
11. G. Désarmot
    Rapport C.E.A. R-4795, 1976
12. Y. Quéré
    Ann. Physique 5, 105, 1970
13. J. Mory
    Rapport C.E.A. R-4745, 1976
14. G. Delsarte
    Rapport C.E.A. R-4027, 1970
15. Y. Quéré, E. Uggerhøj
    Phil. Mag. 34, 1197, 1976

NEUTRON OPTICS USING NON-PERFECT CRYSTALS

A.K. Freund

Institut Max von Laue-Paul Langevin,
B.P. 156, F-38042 Grenoble Cedex, France.

1. Introduction

What may be the relation between *neutron optics using non-perfect crystals* and the topic of this workshop which is *imaging processes and coherence in physics*, and what is the difference to *perfect crystal neutron optics*? The term *neutron optics* seems to be rather ill-defined because optics is dealing with light waves or photons and not with neutrons which interact in a fundamentally different way with matter. However, as has been outlined in other papers of this workshop, several principles discovered in optics can be applied also to the description of phenomena occurring at the scattering of X-rays, neutrons and electrons. The present paper deals with the application of *focusing principles* to neutron scattering techniques by the use of imperfect crystals. The aim of these methods is not to create a direct image of the microscopic structure of the sample studied like in microscopy but to concentrate the neutron beam on the sample in order to gain intensity at the expense of angular divergence. Such methods are important because neutron sources are much weaker than X-ray or laser sources and the samples studied are in most cases much smaller than typical neutron beam cross-sections. In addition to this focusing in real space the term *focusing* will be extended also to *reciprocal space* and thereby to instrumental resolution which affects directly the imaging process. The imaging process itself is *indirect* and consists in the reconstruction of atomic positions and motions by means of a Fourier transform using patterns of elastically and inelastically scattered neutron intensities. The sharper the lines in these patterns the better works the imaging process. In the following the principles of geometrical focusing will be given rather than the physics done with neutron scattering and the experimental details for which the literature cited below should be consulted. For completeness, the existence of *time focusing* in neutron scattering techniques must be mentioned; a detailed description of these methods is, however, beyond the scope of this paper.

Despite the fact that focusing principles are known from optics since a long time ago and that they have been applied to experimental methods in several fields of research such as X-ray diffraction, electron scattering and particle accelerators, it took a relatively long time before focusing techniques were proposed for neutron scattering instrumentation. The *Bragg optics* discussed by DACHS and STEHR (1962), JAGODZINSKI (1968), EGERT and DACHS (1970) and DACHS (1970) extended X-ray focusing methods (CAUCHOIS, 1932; JOHANSSON, 1933) to neutron diffraction taking into account the larger penetration depth of neutrons in curved crystals. MAIER-LEIBNITZ (1967, 1970, 1972) replaced the curved crystals by lamellar systems and proposed several new methods. Since then, much work has been done on theoretical analysis as well as on the real efficiency of focusing systems, and more recently the actual state of focusing neutron monochromators and analysers and their applications has been discussed extensively (AXMANN, 1978).

2. Some Words about Neutron Scattering Experiments

In this section a short account is given of the parameters defining a neutron experiment and of the devices used in neutron scattering instrumentation which may be chosen as focusing elements in the beam geometry. More detailed descriptions are to be found in appropriate textbooks (BACON, 1975; KOSTORZ, 1979). Fig. 1 shows a flowchart of a neutron scattering experiment taken from FREUND and FORSYTH (1979). The properties of the neutron beam emerging from a thermal neutron source (see the paper by Forsyth, this conference) are subsequently adapted by devices defining the incident beam to the experimental conditions required. Similar devices are used for the definition of the beam scattered by the sample before reaching the detector. The

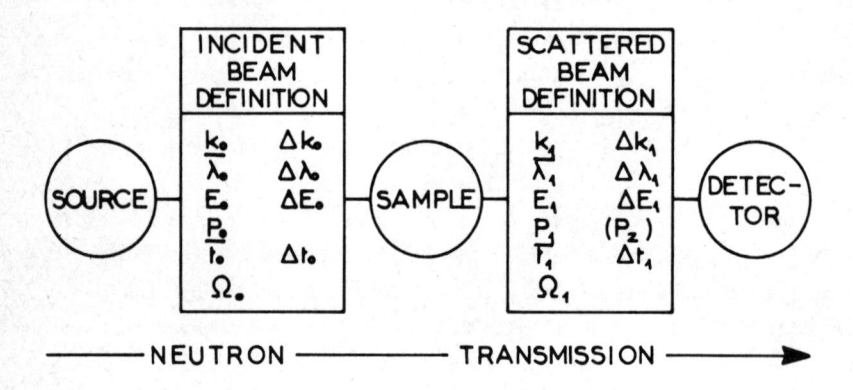

Fig. 1 Schematic diagram of a general neutron scattering experiment

monochromator/polariser determines the incident beam direction confined in a solid angle Ω_0, the polarisation $\underline{P}_0$, the neutron energy E_0 and energy spread ΔE and thereby the incident wavevector $\underline{k}_0$ ($\pm \Delta k_0$). Observation of scattering from the sample is carried out within a solid angle Ω_1 analysing the polarisation $\underline{P}_1$ and the energy E_1 ($\pm \Delta E_1$) corresponding to a wavevector $\underline{k}_1$. Of course, not all the parameters shown in Fig. 1 are independent : a neutron energy E will have an associated wavelength λ. The choice of description is largely one of convenience for the type of experiment involved, but it may also depend on the choice of the defining device. As already mentioned, this paper will not deal with the time structure of the neutron beam characterised by t_0 and t_1 but with steady state experiments in neutron diffractometry studying crystal structures by elastic scattering and in three axis spectrometry using inelastic scattering for investigating elementary excitations (phonons, magnons, etc.) in condensed matter.

A typical beam geometry of a triple axis crystal spectrometer is sketched in Fig.2. A monoenergetic beam of neutrons is obtained by Bragg reflection from the mono-

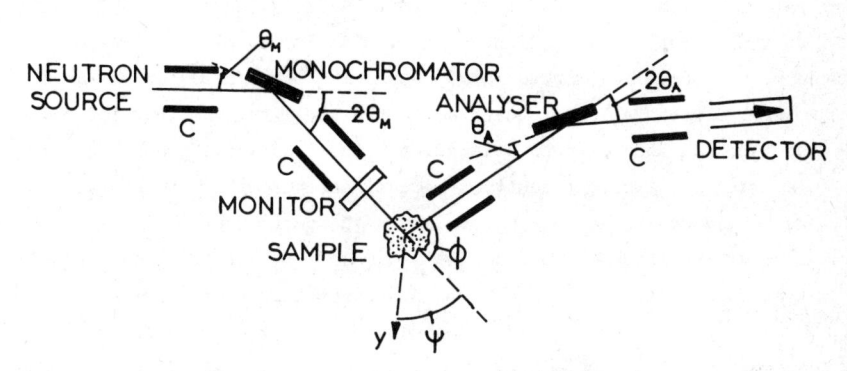

Fig. 2 Schematic illustration of a three-axis spectrometer for inelastic neutron scattering experiments

chromator crystal. Its intensity is measured by the monitor just before the sample crystal the orientation of which is specified by the angle ψ between a crystal axis y and the incident beam direction. The energy of the neutrons scattered inelastically by the sample at an angle ϕ is determined by an analyser crystal reflecting the beam in direction of the detector. Collimators, diaphragms or slits denoted by C may be inserted to limit the various neutron beams. At elastic scattering experiments on diffractometers there is in general no need for an analyser crystal except for the elimination of background intensity, and the detector is placed at the position of the analyser. In this case, Bragg scattering from the sample crystal is observed and the angle ϕ in Fig. 2 is replaced by $2\Theta_S$ whereas ψ becomes $\pi/2 - \Theta_S$ and the axis y is parallel to the diffraction vector $\underline{\tau}$ associated to the set of reflecting lattice planes in the sample. The angles Θ_M, Θ_S and Θ_A are the Bragg angles of the monochromator, the sample and the analyser crystal, respectively.

The elements defining the beam geometry in Fig. 2 which may be involved in focusing are the monochromator and the analyser crystal and to some extent also the collimators. Focusing microguides based on the principle of total reflection which may also polarise the neutron beam permit to produce a convergent beam (see FREUND and FORSYTH, 1979). Without mentioning details of these devices we focus our attention on the monochromator and the analyser which are imperfect crystals because perfect crystals have a much too narrow width of reflection compared to typical neutron beam divergences. In fact, the widths β_M and β_A of the diffraction curves characterising the degree of imperfection of these crystals should be about equal to α_0, α_1, α_2 and α_3 being the collimations of the neutron beams before and after the monochromator and before and after the analyser crystal, respectively. This is the result of an intensity-resolution optimisation (see *e.g.* MAIER-LEIBNITZ, 1967; DORNER, 1972). The crystal imperfections or dislocations leading to a broadening of the diffraction profiles are generated by plastic deformation of initially nearly perfect crystals.

The diffraction process in imperfect crystals and in particular in crystals used for monochromating and analysing purposes is described by kinematical theory including secondary extinction (ZACHARIASEN, 1945; BACON, 1975; FREUND and FORSYTH, 1979). In the frame of this theory the imperfect crystal with its complicated defect structure is replaced by the so-called *mosaic model* consisting of a composite of small crystal blocks the angular orientation of which are not exactly the same but follow a Gaussian distribution function about a mean orienation. The individual blocks in this composite similar to a brick wall scatter neutrons independently so that the scattered *intensities* will be summed up instead of *amplitudes* in the perfect crystal case. The width of the angular distribution is called the *mosaic spread* η of the imperfect crystal and gives rise to a width β of the neutron diffraction profile recorded from such a crystal. The relation between η and β depends on neutron wavelength and is such that $\beta \gtrsim \eta$.

In order to give a schematic illustration of the diffraction conditions and of the influence of beam divergence and mosaic spread on resolution, it is convenient to use the *reciprocal* or *momentum space* which is obtained by a Fourier transform of real space. The difference between reciprocal and momentum space is just a factor 2π in the length of the vectors in Fig. 3 where the diffraction conditions are shown using Ewald's construction projected onto the scattering plane. This is a geometrical construction of Bragg's law given by

$$2d \sin \Theta = \lambda \qquad (1a)$$

which can be written as

$$\underline{k}_1 - \underline{k}_0 = 2\pi\underline{\tau} \qquad (1b)$$

where d is the interplanar spacing of the reflecting set of lattice planes, Θ is the Bragg angle, λ the neutron wavelength, $\underline{k}_1$ and $\underline{k}_0$ the reflected and incident neutron wavevectors such that $|\underline{k}_1| = |\underline{k}_0| = 2\pi/\lambda$, and $\underline{\tau}$ the reciprocal lattice vector or diffraction vector.

The width β of the imperfect crystal diffraction curve is represented in Fig. 3 by a smearing of the reciprocal lattice point corresponding to an angular domain of reflection. The above construction will be used later on in section 4. An account of the *intensity* diffracted by different kinds of *mosaic crystals* has been given by FREUND (1978). For the moment, the discussion returns to real space.

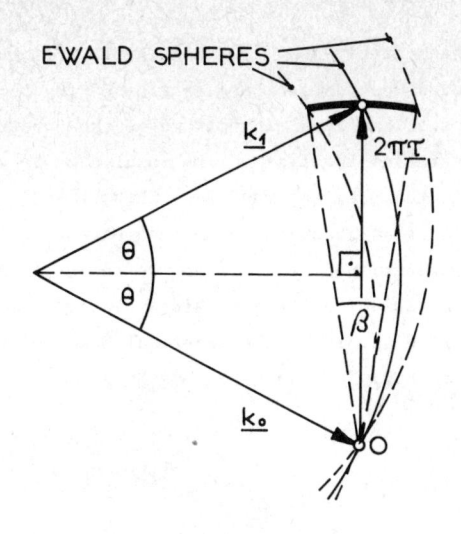

Fig. 3. Diffraction in momentum space of a parallel and monochromatic beam by a non-perfect crystal

3. Focusing in Real Space

Focusing in real space consists in concentrating a large neutron beam onto the sample and this can take place in two planes. Focusing in the scattering plane, also called *horizontal focusing* because usually the neutron beam path lies in a horizontal plane, affects directly the resolution whereas *vertical focusing* has practically no influence on the definition of neutron energies or wavevectors. In both cases, curved single crystals or composite crystal systems are used which serve simultaneously as monochromators of the incoming white neutron beam produced by the reactor. An ideal thin crystal bent to a radius R acts exactly like a mirror with the focal length

$$f_h = (R \sin\Theta)/2 \qquad (2)$$

in the case of horizontal focusing and

$$f_v = R/(2 \sin\Theta) \qquad (3)$$

in the case of vertical focusing.

3.1. Horizontal focusing

The neutron beam path in horizontal focusing is shown in Fig. 4. A crystal is bent elastically or plastically to a radius R=ON and its surface is shaped by grinding such that it has a curvature of r = R/2 = OM = MN. Any ray emerging from the source and passing through a slit S positioned at the point C on the Johansson circle J will be reflected by the crystal surface onto point C' which is also on J.

If a parallel neutron beam is to be focused, for instance when replacing the slit S by a neutron guide tube or a collimator K any ray parallel to OC is converging to the point D on the focal circle F defined by Eq. (2). As can be seen from Fig. 4, no change of the Bragg angle is involved in the first case and the focusing is monochromatic.

In the second case, however, the white and nearly parallel beam is focused non-monochromatically and the wavelength reflected depends on the direction of the secondary ray, *i.e.* after reflection from the crystal surface. In both cases the penetration depth of the neutrons in the crystal increases the size of the focal spot as well as the width of the wavelength spread reflected (see the ray CB). This effect is not very large because the crystal thickness if at least two orders of magnitude smaller than the radius of curvature. It is, however, non negligible. On the other hand, aberration effects can be neglected when replacing the Johansson monochromator by a simply bent crystal of the Johann type (JOHANN; 1931). A practical example of horizontal focusing in real space is shown in Fig. 5.

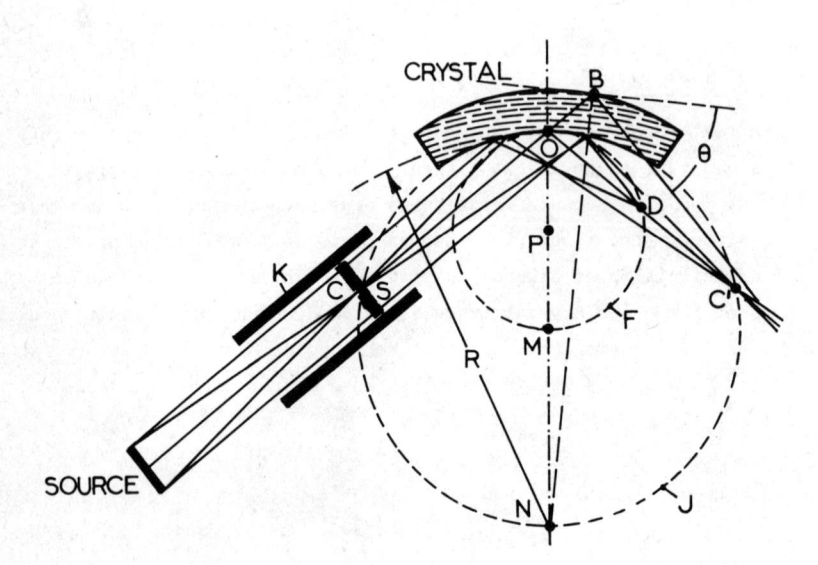

Fig. 4. Horizontal focusing in real space

3.2. Vertical focusing

The ray geometry of vertical focusing is sketched in Fig. 6 for again both a divergent and a parallel beam. A crystal is curved cylindrically about an axis lying in the scattering plane LOA. Since the linear dimensions of the crystal are usually small relative to R, the radius of curvature, aberration may be neglected and the classical formula for hollow mirrors may be used for the projection of the rays onto the plane AOB containing the optical axis OA. This leads to the following relation

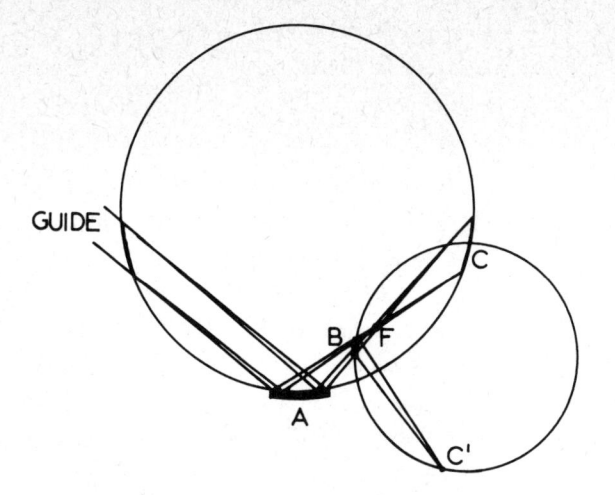

Fig. 5. Guinier geometry on a neutron guide tube. The rays reflected by crystal A meet at F, those scattered at the sample B meet at C'. (From DACHS, 1970)

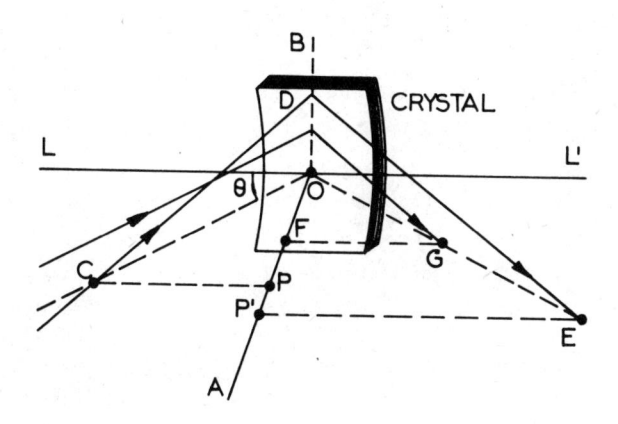

Fig. 6. Vertical focusing in real space. (After RISTE, 1970)

for a point source at C

$$1/OE + 1/OC = (2\sin\Theta)/R = 1/f_v \qquad (4)$$

and for a parallel beam (OC → ∞)

$$OG = R/(2\sin\Theta) = f_v \qquad (5)$$

Similar equations can be derived also for horizontal focusing by replacing f_v by f_h. As vertical focusing does practically not affect resolution, its application is straigthforward and has become very popular. A more detailed analysis of the gain in intensity which can be reached including crystal mosaic spread and beam divergence has been carried out by Currat (1973). In particular at neutron guide tubes, high gain factors can be reached (FREUND et al., 1978, 1979).

4. Focusing in Reciprocal Space

The possibility of focusing in reciprocal or momentum space arises because the instrumental resolution functions are ellipsoids in three-dimensional reciprocal space for elastic scattering experiments and in a four-dimensional space (energy ω and three components of wavevector transfer Q) for inelastic scattering, with typically a large ratio between the length of its major and minor areas. The orientation of this ellipsoid with respect to the reciprocal space volume defined by the sample (elastic scattering) and to the dispersion relation of the sample crystal (inelastic scattering) determines, together with the scan mode, beam divergence and monochromator mosaic spread, the degree of focusing which can be obtained on neutron diffractometers and three axis spectrometers. These focusing possibilities are rather limited when using flat crystals but may be considerably extended by means of curved monochromators and analysers in particular if the curvature can be made tunable allowing the experimentalist to optimise the experimental conditions by manipulating the resolution function.

The condition for Bragg scattering can be sketched in reciprocal space by the well-known Ewald construction which is represented by Fig. 3 in section 2 of this paper. A slightly modified version of this geometrical construction is represented in Fig. 7 showing diffraction of a divergent white neutron beam by a mosaic crystal monochromator. Shape, size and orientation of the volume in momentum space selected out of the white beam depends not only on the Bragg angle Θ_M of the monochromator but also on the beam divergence α_o and on the mosaic spread β of the imperfect monochromator crystal.

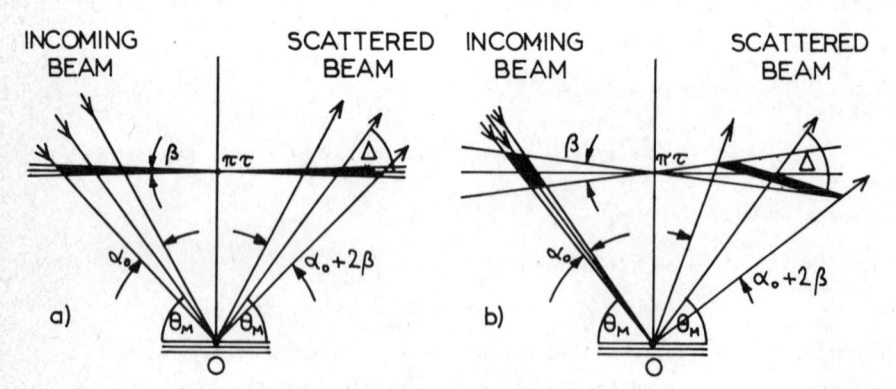

Fig. 7. Selection of momentum space volume by an imperfect crystal, a) $\alpha_o > 2\beta$; b) $\alpha_o < 2\beta$

In addition, the momentum spread $\Delta k = k_{max} - k_{min}$ may be further limited by the beam divergence α_1 after reflection. Δ gives the orientation of the volume in momentum space, $\tau_M = 2\pi/d$ is the reciprocal lattice vector and O is the origin of the reciprocal lattice. For $\alpha_0 > 2\beta$ one obtains $\Delta = \Theta_M$ whereas for $\alpha_0 < 2\beta$ the angle Δ is given by the relation $tg\Delta = 2tg\Theta_M$. The relative momentum or wavelength spreads associated $\Delta k/k = \Delta\lambda/\lambda$ are given by $\beta ctg\Theta_M$ in the first and $\alpha_0 ctg\Delta$ in the second case, respectively (KALUS and DORNER, 1973).

The volume in k-space impinging on the sample crystal is subsequently scanned by e.g. turning the sample through its Bragg position in a diffraction experiment, and the width w of the intensity profile recorded is described by the relation

$$w = \left[\alpha_0^2(1 - tg\Theta_S/tg\Theta_M)^2 + \beta^2(2 - tg\Theta_S/tg\Theta_M)^2 + \gamma^2\right]^{1/2} \quad (6)$$

where Θ_S is the Bragg angle and γ the mosaic spread of the sample crystal, respectively. The function $w(\Theta_S)$ is called the resolution curve of the diffractometer if γ is negligible. Gaussian distributions of α_0, β and γ are assumed for simplicity. For $\alpha_0 > 2\beta$ the curve $w(\Theta_S)$ has a minimum at $\Theta_S = \Theta_M$ which is called the *dispersion-free arrangement* in a double-crystal diffractometer. This minimum is shifted to the position $tg\Theta_S = 2tg\Theta_M$ for the condition $\alpha_0 < 2\beta$. By drawing the sample diffraction conditions in Figs. 7a and 7b and rotating its reciprocal lattice vector about O one can see that the minimum of the resolution curve corresponds to scanning the momentum volumes selected by the monochromator crystal in a direction perpendicular to its large axis. *This is called focusing in momentum space.*

The possibility of varying the distance L between monochromator and sample and the use of a monochromator crystal with tunable radius of curvature allows the adaptation of the experimental conditions to the requirements defined by the sample such that focusing both in real and reciprocal space may be achieved simultaneously leading to an intensity gain without loss in resolution. Fig. 8 shows diffraction from a curved crystal both in real and in momentum space. The orientation of the volume element defined by a curved monochromator crystal is described by the angle Δ and related to the beam geometry by

$$tg\Delta = tg\Theta_M/[1 - L/(Rsin\Theta_M)] \quad (7)$$

By choosing L and R it is possible to adjust Δ in a predetermined way.

The *thickness* Δk of the momentum element parallel to the vector $\underline{k}$ is determined by the crystal mosaic spread β and the crystal thickness T. By the use of nearly perfect crystals one can match Δk to the resolution required choosing the thickness according to

$$\Delta k/k = (Tctg^2\Theta_M)/R \quad (8)$$

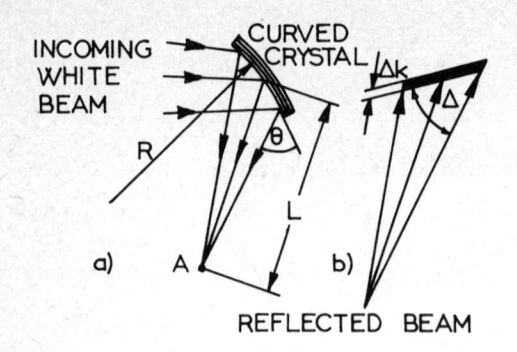

Fig. 8. Bragg scattering from a curved single crystal in (a) real space and (b) momentum space : R = radius of curvature; L = distance between single crystal and sample

This equation holds for thermally curved crystals (KALUS *et al.*, 1973) and plastically bent monochromators (EGERT, 1974; HOHLWEIN, 1975). In the case of elastically bent crystals Eq. (8) becomes

$$\Delta k/k = T(ctg^2\Theta_M - \mu)/R \tag{9}$$

where μ is Poisson's ratio, and if a sandwich of many thin crystals each of thickness t is used (KALUS, 1975; FREY, 1971, 1974, 1975)

$$\Delta k/k = nt(ctg^2\Theta_M - \mu/n)/R \tag{10}$$

where n is the number of crystals. The dimension of the momentum element perpendicular to $\underline{k}$ is given by slits or collimators defining the beam geometry.

A somewhat different approach to focusing in reciprocal and real space has been proposed by MAIER-LEIBNITZ (1967) and followed up by RUSTICHELLI (1969) and BOEUF and RUSTICHELLI (1973). A composite monochromator system (Fig. 9) is made up of a set of crystalline lamellas each of which is inclined to the following one by the width β of its diffraction pattern. If the crystal is imperfect, β corresponds to about the mosaic spread, if it is perfect β is given by the so-called Darwin width. In the general case the crystals are cut asymmetrically ("Fankuchen cut") where the degree of asymmetry is determined by the focusing requirements. Such a system as well as the bent crystals mentioned above have several advantages when compared to a classical flat mosaic crystal monochromator. The Gaussian mosaic distribution is replaced by a nearly rectangular distribution leading to a sharper definition of the wavelength and angular spread of the neutron beam impinging on the sample. The momentum distribution in the neutron beam reflected by the composite system shown in Fig. 9 (real space) is sketched in Fig. 10a (momentum space). If β is small as it is the case in typical applications, the thickness Δk of the momentum space element is determined by the divergence α_o of the primary white neutron beam according to

$$\Delta k/k = (\alpha_o/2)\operatorname{ctg}\Theta_M \qquad (11)$$

whereas its dimension parallel to $\underline{k}$ depends on the number of lamellas. The focusing is non-monochromatic and thus the wavelength of a reflected ray depends on its direction.

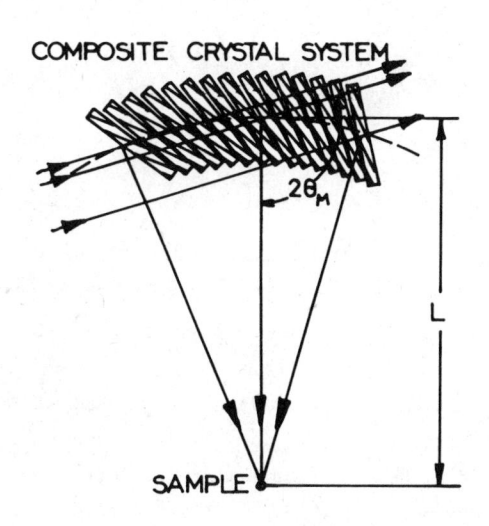

Fig. 9. Schematic respresentation of the selection of neutron energy by a neutron monochromator system consisting of plane crystalline lamellas. (After MAIER-LEIBNITZ, 1967).

There is a possibility of producing *monochromatic focusing* by a composite system when using crystals with a gradient of lattice spacing or lamellas which have just slightly different lattice parameters. Normally, this should be in the order of some percent in $\Delta a/a$. Such a system reflects a neutron beam with a momentum space distribution shown in Fig. 10b and can be considered as a *monochromatising neutron lens*. A composite system of crystals with different lattice spacings has been described recently by BOEUF et. al. (1978).

Parallel to the development of such systems also the theory for neutron diffraction by perfect crystals had to be modified for slightly distorted crystals (KLAR and RUSTICHELLI, 1973; ALBERTINI et. al., 1976 a,b). Detailed descriptions of these sophisticated systems and theories are beyond the scope of this paper despite their interest from the "optical" point of view. Comparison between experiment and model calculations have given good agreement (ALBERTINI et. al., 1977) but the application of nearly perfect crystals to neutron monochromatisation is still limited last but not least because of practical problems arising at the production of composite systems.

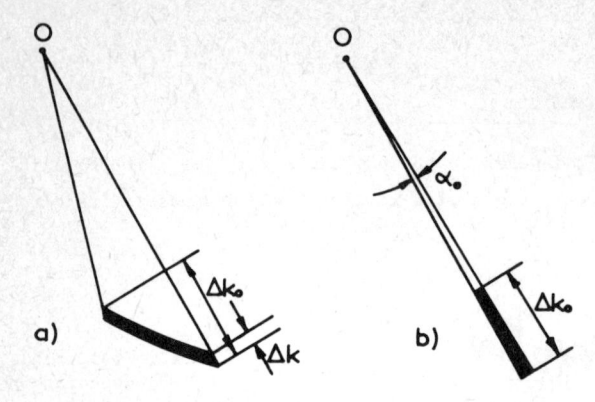

Fig. 10. Momentum space distribution in a neutron beam reflected by (a) a system without and (b) a system with a gradient in lattice spacing. (After RUSTICHELLI, 1970)

5. Some Examples of Focusing Methods in Neutron Scattering Techniques

A so-called *modified Laue method* has been proposed by MAIER-LEIBNITZ (1967) for a considerable gain of measuring time in crystal structure analysis. A bent or composite monochromator crystal reflects a large wavelength band on a stationary sample and excites a large number of reciprocal lattice points simultaneously. Whereas in the classical Laue method all higher order reflections fall in the same direction and cannot be separated the new Laue method allows to measure these reflections independently since the focusing is non-monochromatic (Fig. 11). This technique has to use a multidetector or a film (KLAR, 1973; THOMAS, 1972; HOHLWEIN, 1977).

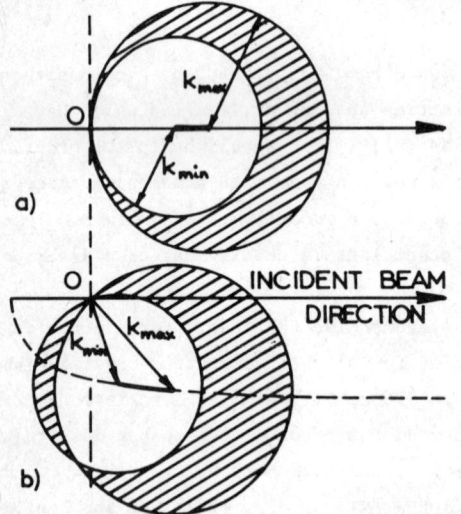

Fig. 11. Laue method in reciprocal space and Ewald spheres corresponding to k_{min} and k_{max}. (a) Classical method (b) new Laue method. All reciprocal lattice points lying in the hatched region contribute to the Laue diagram.

With increasing momentum space volume of the neutron beam impinging on the sample also the background intensity will become more important. This background which is mostly arising from inelastic scattering processes could be eliminated by a second composite system of crystalline lamellas mounted between the sample and the detector (Fig. 12).

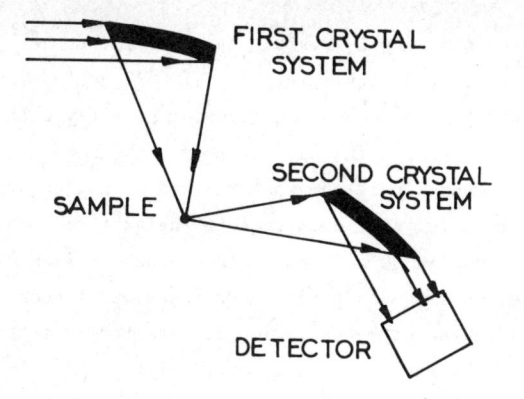

Fig. 12. Focusing method for the measurement of many reflections excluding inelastic scattering. (After MAIER-LEIBNITZ, 1967)

The geometrical considerations which are applied to neutron monochromators can also be extended to *focusing analysers* in order to gain intensity i.e. measuring time. An analyser crystal with variable radius of curvature has been developed for three-axis spectrometers by SCHERM et. al. (1977) and some applications of this device are described by SCHERM and WAGNER (1977). The focusing geometry is shown in Fig. 13 in real and reciprocal space. A curved analyser is used to turn the resolution ellipsoid in Q-ω-space by varying the radius of curvature and the analyser-detector distance L_D. To each element of the multidetector is associated a volume element in Q-ω-space designed by the numbering in Fig. 13. Three cases are presented :

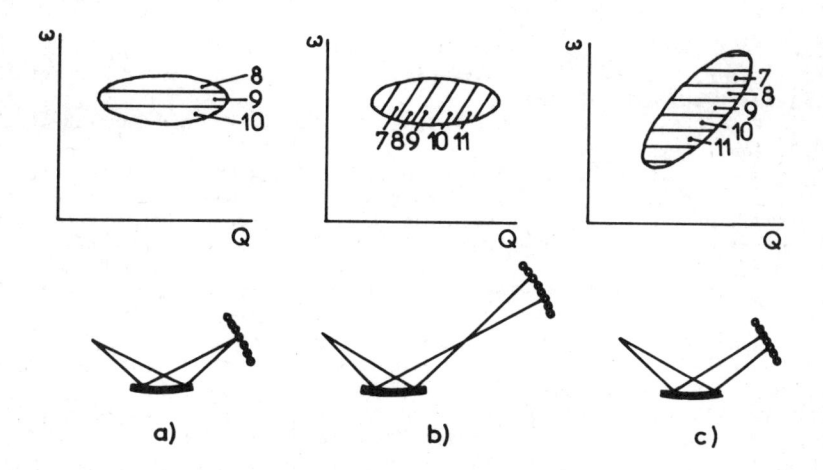

Fig. 13. Triple axis spectrometry using a curved analyser and a multidetector. (After SCHERM and WAGNER, 1977).

(a) $R = R^x$ and $L_D = 2f_h = L_A$
(b) $E = R^x$ and $L_D \gg L_A$
(c) $R > R^x$ and $L_D = L_A$

where L_A is the sample-analyser distance and $R^x = L_A \cdot d \cdot k/\pi$. The collimators between sample, analyser and detector have been removed (c.f. Fig. 2). Fig. 13 demonstrates the different kinds of possibilities arising when choosing focusing geometry with variable R and L_D. Also flat samples can be used in these methods.

The angular precision of crystal orientation is somewhat relaxed at the construction of composite perfect or nearly perfect crystal analysers in inelastic neutron spectrometers working at Bragg angles close to 90°. Several thousands of Si and CaF_2 nearly perfect crystals were glued on curved surfaces covering several square metres and serve as focusing analysers in the high-resolution backscattering instruments IN10 and IN13 at the Institut Laue-Langevin (BIRR et. al., 1971).

Some other possible applications are discussed by FREY (1978), GRIMM (1978), KALUS (1978) and AXMANN et. al. (1978). In conclusion, it can be said that focusing techniques in neutron scattering are still in the developing stage except vertical focusing. Whereas principles are clear, some problems still arise at the practical construction of high-precision composite crystal systems with variable curvature. Efforts are being made at the Institut Laue-Langevin and at other reactor stations in order to solve this problem. It is important to have the possibility of tuning the curvature of monochromators and analysers not only because the radius of curvature is energy or wavelength dependent but also for operating an instrument in the classical *flat-crystal and collimator mode* if this appears to be preferable (availability of large samples, danger of multiple scattering etc.). Further developments of focusing methods are expected in connection with the construction of pulsed neutron sources and multidetector systems.

References

Albertini, G., Boeuf, A., Cesini, G., Mazkedian, S., Melone, S., and Rustichelli, F. (1976a). Acta Cryst. A32, 863-868
Albertini, G., Boeuf, A., Mazkedian, S., Melone, S., Rozzi, V., and Rustichelli, F. (1976b). J. Appl. Cryst. 10, 118-122
Albertini, G., Boeuf, A., Klar, B., Lagomarsino, S., Mazkedian, S., Melone, S., Puliti, P., and Rustichelli, F. (1977). Phys. Stat. Sol. A44
Axmann, A. (1978). Proceedings of the workshop on Fokussierende Neutronenmonochromator und Analysatoranordnungen, Hahn-Meitner-Institut, Berlin. HMI Report B273
Axmann, A., Kasper, F., and Dachs, H. (1978). In : Axmann (1978), p. 168
Bacon, G.E. (1975). Neutron Diffraction, third edition. Clarendon Press, Oxford
Birr, M., Heidemann, A., and Alefeld B. (1971). Nucl. Instr. and Methods. 95, 435
Boeuf, A., and Rustichelli, F. (1973). Nucl. Instr. and Methods 107, 429-435
Boeuf, A., Detourbet, P., Escoffier, A., Hustache, R., Lagomarsino, S., Rennert, A., and Rustichelli, F. (1978). Nucl. Instr. and Methods 152, 415-421
Cauchois, Y. (1932). J. Phys. (7) 3, 320
Currat, R. (1973). Nucl. Instr. and Methods 107, 21-28
Dachs, H. (1970). J. Appl. Cryst. 3, 220-224
Dachs, H., and Stehr, H. (1962). Z. Kristallogr. Kristallgeom. Kristallphys. Kristallchem. 117, 135-145

Dorner, B. (1972). Acta Cryst. A28, 319-327
Egert, G. (1974). J. Appl. Cryst. 7, 564
Egert, G., and Dachs, H. (1970). J. Appl. Cryst. 3, 214-220.
Freund, A. (1978). In : Axmann (1978), p.1.
Freund, A., and Forsyth, J.B. (1979). In : Neutron Scattering in Materials Science (G. Kostorz, editor). Chapter X. Vol. 15 of the series A Treatise on Materials Science and Technology (H. Herman, editor). Academic Press, New York, in press
Freund, A., Hewat, A.W., Scherm, R., and Zeyen, C. (1977). Report 77FR115S, Institut Laue-Langevin, Grenoble
Freund, A., Hustache, R., and Zeyen, C. (1977). To be submitted to Nucl. Instr. and Methods
Frey, F. (1971). Nucl. Instr. and Methods 96, 471-473
Frey, F. (1974). Nucl. Instr. and Methods 115, 277-284
Frey, F. (1975). Nucl. Instr. and Methods 125, 9-17
Frey, F. (1978). In : Axmann (1978), p. 54
Grimm, H. (1978). In : Axmann (1978), p. 73
Hohlwein, D. (1975). J. Appl. Cryst. 8, 465-468
Hohlwein, D. (1977). Acta Cryst. A33, 649
Jagodzinski, H. (1968). Acta Cryst. B24, 19
Johann, H.H. (1931). Z. Phys. 69, 185
Johansson, T. (1933). Z. Phys. 82, 507
Kalus, J. (1975). J. Appl. Cryst. 8, 361-364
Kalus, J., and Dorner, B. (1973). Acta Cryst. A29, 526-528
Kalus, J., Gobert, G., and Schedler, E. (1973). J. Phys. E6, 488-492
Klar, B. (1973). Ph. D. Thesis, Universität Hamburg
Klar, B., and Rustichelli, F. (1973) Nuovo Cimento B13, 249-271
Kostorz, G. (1979). Neutron Scattering in Materials Science. Vol. 15 of the series A Treatise on Materials Science and Technology (H. Herman, editor). Academic Press, New York, in press
Maier-Leibnitz, H. (1967). Ann. Acad. Sci. Fenn., Phys. Ser. A6, 267, 1-17
Maier-Leibnitz, H. (1970). In : Some Lectures on Neutron Physics, Report JINR, p. 183 Dubna
Maier-Leibnitz, H. (1972). In : Inelastic Neutron Scattering, p. 681. IAEA, Vienna
Riste, T. (1970). Nucl. Instr. and Methods 86, 1-4
Rustichelli, F. (1969). Nucl. Instr. and Methods 74, 219-223
Rustichelli, F. (1970). Nucl. Instr. and Methods 83, 124-130
Scherm, R., and Wagner, V. (1977). In : Inelastic Neutron Scattering, p. 1. IAEA, Vienna
Scherm, R., Dolling, G., Ritter, R., Schedler, E., Teuchert, W., and Wagner, V. (1977) Nucl. Instr. and Methods 143, 77-85
Thomas, P. (1972). J. Appl. Cryst. 5, 78-83
Zachariasen, W.H. (1945). Theory of X-Ray Diffraction in Crystals. Wiley, New York.

ULTRASONIC REAL-TIME RECONSTRUCTION IMAGING

C. Bruneel

Université de Valenciennes 59326 VALENCIENNES France

Introduction

The results of the studies in the optical image reconstruction may not be directly generalized to the acoustical case. Two main features explain the need for further studies :

- first, specific problems appear since the ultrasonic devices dimensions may not be termed as extremely large with respect to the wavelength. So, for a given aperture, the phase rotation will be slower for the acoustical case than for the optical one.

- second, new particular abilities are given by ultrasonic waves, which allow original reconstruction schemes. The low velocity of acoustic waves leads to short wavelengths at frequencies easily generated using standard electronic equipment. It is then possible to launch pulsed ultrasonic wavetrains and to get echographic B scan images (Annexe 1). On an other hand, it is possible to use phase and amplitude sensitive receivers and so to bypass the holographic technics.

Two original schemes of image reconstruction using these specific properties, of the ultrasound will be described. The first one uses an acousto electronic lens mode with two transducers arrays, one for sampling the acoustic field reflected by the object and the other for reconstructing the image after an electronic inversion of the relative phases between the channels. In the second scheme, the reconstruction is performed optically after the transfert of acoustical informations onto a light beam via acousto-optic interaction.

These two devices have been conceived in order to allow the observation at a high image rate (up to 1000 per second) of the fast motion of the cardiac valves.

Theory

Wave propagation may be depicted by the Huyghens model of the decomposition of an arbitrary surface into elementary omnidirectionnal sources. The strength of these sources is proportionnal to their area and to the wave amplitude in their neighbourhood. The wave amplitude at some observation point P may then be computed by summing all the contributions from the elementary sources. When the phases (i.e. resulting from the summation of the relative phase of the sources and the phase lag introduced by the path between the sources and the observation point) of these terms are stationnary, constructive interference occurs and the resulting amplitude is high. In the opposite case, this amplitude remains very low. Starting from the propagation equation, Rayleigh and Sommerfeld (1, 2) have given a mathematical foundation to the Huyghens principle.

Assuming that the observation distance is much greater than the source dimensions, Fresnel has suggested a simplified description using the Fresnel's function:

$$\bar{\phi}_1(P) = \frac{1}{j\lambda z_1} \exp\left(-\frac{j\,k|P|^2}{2z_1}\right) \qquad (1)$$

The amplitude inside a plane P_1 is then expressed as a convolution integral in terms of that inside the plane P_o

$$U_1(P) = U_o(P) * \phi_1(P) \qquad (2)$$

where P stands for the two coordinates in the plane, z_1 the distance between the two planes, k and λ respectively the wave vector modulus and the wavelength.

Owing to the following property of the Fresnel's function

$$\phi_1^*(P) * \phi_{2-1}(P) = \delta(P) \text{ if } z_{2-1} = z_1 \qquad (3)$$

the information $|U_o(P)|$ may be deduced from $U_1(P)$ since :

$$U_2(P) = U_1^*(P) *{}_{2-1}(P) = U_o^*(P) * \phi_1^*(P) * \phi_{2-1}(P) = U_o^*(P) \qquad (4)$$

That means physically that if we take the complex conjugate of the information of the shadow in a plane P_1 (i.e. if we invert the relative phase), the amplitude in the plane P_2 such as $z_{2-1} = z_1$ will be the same as in this one in plane P_o (i.e. in plane P_2 we obtain the image of the scattering power of the object set in plane P_o)

We get a further step in simplication (3) and developped a model taking advantage of the linearity of the exact formulas. This allows then to consider each point source in plane P_o as radiating individually. This model gives a first approximation of the results and allows an easy interpretation of the device behaviour.

By choosing source dimensions lower than the wavelength, the amplitude may be assumed uniform, so that only the phase variations are taken into account. Moreover in the B scan (see Annex 1) echographic mode, only a single variable has to be considered. At some distance d from the source, the relative phase varies along a line perpendicular to the mean direction (see fig. 1), according to the approximate quadratic law :

$$\phi(x) = - \frac{2\pi}{\lambda} \left[\sqrt{d^2 + x^2} - d\right] \simeq - \frac{\pi x^2}{\lambda d} \qquad (5)$$

which is valid in the paraxial approximation.

The minus sign means that the lateral rays have a phase lag with respect to the central one. If after sampling (whose step will be assumed lower than an half wavelength) we reemitt these signals with the same phase distribution, the reconstructed field will be equivalent to the one sampled and the corresponding beam will seem issuing from a virtual point source (fig. 1).

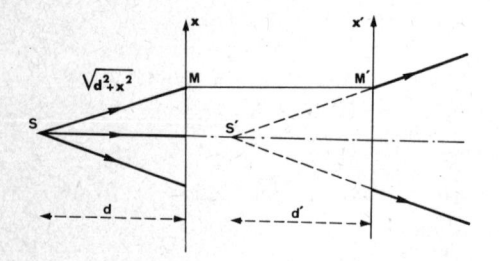

Fig. 1 : Direct connection between M and M' giving a virtual image S'

Fig. 2 : Phase inversion system (SIP) between M and M' giving a real image I'

But now if we inverted the sign of the phase distribution, the generated beam will converge toward a real image point I' (fig. 2). The relative phase lag introduced by the path M'I' is given by :

$$\phi(x') = - \frac{2\pi}{\lambda'} \left(\sqrt{d'^2 + x'^2} - d'\right) \simeq - \frac{\pi x'^2}{\lambda' d'} \qquad (6)$$

So that taking into account the relative phase of the source point M', the resulting phase of the ray issuing from M' at point I' equals :

$$\phi(M', I') = \frac{\pi x^2}{\lambda d} - \frac{\pi x'^2}{\lambda' d'} \qquad (7)$$

This phase appears stationnary versus M' if :

$$\frac{x^2}{\lambda d} = \frac{x'^2}{\lambda' d'} \qquad (8)$$

where x' represents the ordinate of the reemitting point connected to the receiving point of ordinate x and d' the distance of the real image I' to the reemitting away.

This result constitutes in fact the base of the holographic technics and we describe in the following two experimental methods performing in real time this mathematical transformation known as a Fresnel transform.

Acousto-electronic lens (4, 5)

The reflected acoustic field may be sampled by a piezoelectric transducer array and converted into an electrical form, so that the phase inversion process may be performed electronically. Several techniques are available to do so, the very first being to use electronic delays by analogy with a simple lens. However, for B. Scan echography, the distance from the object is continuously varying so that a very fast zoom effect is needed in order to keep the object well in focus. The relative phase inversion will then be performed using the non linear response of diodes in a ring modulator giving so the product of two sinusoïdal functions. Let us consider two input signals, a reference one $A \cos \Omega t$ and an other one $B \cos(\omega t + \phi)$. The output signal appears as :

$$AB \cos \Omega t \cos(\omega t + \phi) = \frac{AB}{2} \cos[(\Omega + \omega)t + \phi] + \frac{AB}{2} \cos[(\Omega - \omega)t - \phi] \qquad (9)$$

By electrical filtering we keep only the second part whose amplitude is proportionnal to B. The sign of its relative phase is inverted with respect to the input one if the reference pulsation Ω is higher than the information signal one ω. The generated wave focuses according to eq.8 at the distance d' :

$$d' = \frac{\lambda}{\lambda'} \left(\frac{a'}{a}\right)^2 d \qquad (9)$$

where a'/a stands for the period ratio of the reemetting and the receiving array. The focused signal is picked up by a third array lying in the convergence zone. By varying the reemitting wavelength λ' the distance d' may be kept constant when d is varied, so that the fast zoom effect is performed simply by changing the reference signal frequency.

The validity of this approach has been verified using a 10 channels device. The reemitted field is visualized using Schlieren's method which gives an optical image of the acoustic field distribution (fig. 3) where the converging effect appears clearly.

The system described here is an analogical computer for the Fresnel transform which is more simpler and faster than digital computers for this purpose. A sector scan apparatus which uses the same principle is now under study. Other applications may also be devised like in radioastronomy for example.

Fig. 3 : Reconstructed acoustical field distribution

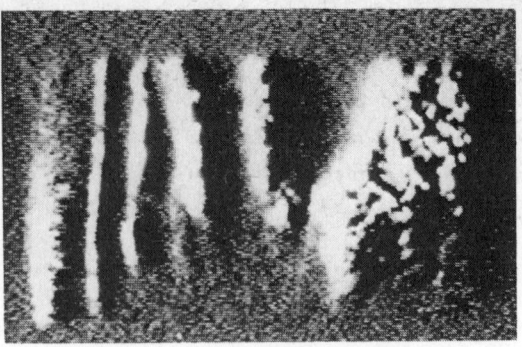

Fig. 4 : Image of the heart

Acousto-optic processing (6, 7)

Another technique is to put the useful informations onto a light beam. The acousto-optic interaction may be viewed as a nonlinear interaction phenomenon involving a reference optical signal A cos Ωt and an acoustical signal B cos $(\omega t + \phi)$. This gives diffracted optical beams such as :

$$\alpha AB \cos[(\Omega + \omega)t + \phi] \text{ for the plus one order}$$
$$\alpha AB \cos[(\Omega - \omega)t - \phi] \text{ for the minus one order}$$

where α is an efficiency factor characterizing the interaction medium.

A spatial filter allows to select either the plus one order corresponding to a virtual optical image or the minus one order giving a real optical image or the insonifying object. Cylindrical optical lens are used to form the image on a screen. A galvanometric mirror converts the temporal modulation of the light beam spatial modulation in order to restitute the second set behind this galvanometric mirror compensate for the distance variation of the successive lines in order to keep the whole image well in focus.

A complete imaging apparatus using the acousto-optic interaction principle has been built for medical applications, more specifically in the cardiologie area and an image given by this prototype is shown on fig. 4.

Conclusion

In this paper we have described two new imaging devices using the specificities given by the low frequencies used in ultrasonics. The analogic technics allow real time reconstruction schemes and in the ultrasound imaging area classical holographic technics may easily be bypassed. The two devices described appear as analogical Fresnel transform devices whose applications may be extended to other domains as the radioastronomy area. Experimental results confirm the theoretical predictions and images of in vivo structures as the heart have been obtained.

References

1) J. Goodman "Introduction to Fourier Optics", Mc Graw-Hill 1968.

2) P.M. Duffieux "L'intégrale de Fourier et ses applications à l'optique" Faculté des Sciences, Besançon 1946.

3) C. Bruneel Thèse d'Etat, Valenciennes 1978.

4) C. Bruneel, B. Nongaillard, R. Torguet, E. Bridoux and J.M. Rouvaen, Ultrasonics, Nov 1977, pp. 263-264.

5) C. Bruneel, R. Torguet, E. Bridoux, J.M. Rouvaen, B. Delannoy and M. Moriamez, IEEE Ultrasonic Symposium, Phoenix, Oct 1977.

6) R. Torguet Thèse d'Etat, Paris 1973.

7) R. Torguet, C. Bruneel, E. Bridoux, J.M. Rouvaen and B. Nongaillard, 7th International Symposium on Acoustical Imaging and Holography. Chicago 1976.

Annexe 1   B scan echographic mode

The so termed B scan echographic mode charaterizes the imaging technics commonly used in sonar and radar devices. The image is formed in the reflection mode using a short pulsed wave. The time delay (t_i) needed by the pulse to propagate from the source to the reflecting object and back to the receiver allows to separate the successive lines set at different distances (d_i).

This gives one dimension of the image. The second dimension may be obtained either by a mechanical displacement of the source and the receiver (as in the radar device) or by an electronical commutation of transducers in an array structure or by a real time, one line scheme, reconstruction device. In this last case the reconstruction scheme has to take into account the distance variation of the successive lines by using a fast zoom effect.

WHAT CAN BE DONE WITH HIGH VOLTAGE ELECTRON MICROSCOPY ?

B. JOUFFREY
Laboratoire d'Optique Electronique du C.N.R.S.,
Laboratoire propre du CNRS associé à l'Université Paul Sabatier, Toulouse.
29, rue Jeanne Marvig, F-31055 Toulouse Cedex.

Introduction

High voltage electron microscopy has been developed about twenty years ago when Dupouy and colleagues constructed the first microscope working at 1.200 kV (1). Ten years later the 3 MV microscope was built in Toulouse (2). This last microscope can work slightly above 3 MV. Now, through the world, about 52 H.V.E.M. work.

It is, in fact, difficult to define exactly what is H.V.E.M. compared to conventional electron microscopy. Most of the time it is considered that 300 kV and above concern H.V.E.M. because the relativistic effects begin and also because the technique for the high voltage generator demands new solutions.

As every method, H.V.E.M. is not a general panacea. It has advantages but also can be at the origin of difficulties. These problems are principally related to the size of the apparatus for a part but also to the fact that some energy is given to the sample by the incident electrons. Sometimes this transfer of energy can be the origin of some artefacts. We shall discuss below about advantages and disadvantages of H.V.E.M. For some more complete discussion we refer to (3).

1) - Theoretical resolution limit.

The first advantage of H.V.E.M. is the theoretical limit of resolution. The objective of the microscope has a transfer function which is depending on the spherical aberration and on the defocusing (the position of the plane which is taken as object for the lens is important). This transfer function can be written as a function of the frequency which is to be resolved within the sample. If $\vec{\nu}$ is the spatial frequency, $A(\vec{\nu})$ is the

$$T(\vec{\nu}) = \frac{1}{M} A(\vec{\nu}) \exp{-i\frac{2\pi}{\lambda}} (\frac{1}{4} C_s \lambda^4 \nu^4 - \frac{\Delta z}{2} \lambda^2 \nu^2) = \frac{1}{M} A(\vec{\nu}) \exp{i\gamma(\vec{\nu})}$$

aperture function equal to zero out of the aperture and 1 inside the aperture. $\gamma(\vec{\nu})$ which represents the phase shift due to the objective can be optimized by using the right defocusing. The behaviour of this transfer function is quite favourable to H.V.E.M. which can give information at the level of 1 Å or slightly less. More precisely from this point of view the best voltage can be depending on the level of information which is demanded. Some detailed calculations have been recently done in the laboratory for instance and confirm this point (4). This interest is related to the wavelength associated to the electron (see table 1), and to the spherical aberration ($\sim$ 1 mm at 100 keV, 4.2 mm at 1.2 MeV and 6.6 mm at 3 MeV in our laboratory) which is not varying too rapidly when increasing the energy of the incident electron.

On the other hand, some calculations by Zeitler and Thomson (5) have shown that for single atoms the gain in contrast is about 3 in bright field and 6 in dark field for carbon atoms and respectively 2 and 7 for gold atoms (without any substrate).

V	10 kV	40 kV	100 kV	500 kV	1 MV	2 MV	3 MV	6 MV
$\beta = v/c$	0.1950	0.3741	0.5482	0.8629	0.9411	0.9791	0.989	0.9969
λ (Å)	0.1220	0.0602	0.0370	0.0142	0.0087	0.0050	0.0036	0.0019
λ^{-1}(Å^{-1})	8.194	16.62	27.02	70.36	114.7	198.3	277.8	523.5
m/m_o	1.01957	1.0783	1.1957	1.9785	2.9569	4.9138	6.870	12.742

Table 1. Values of various parameters in electron microscopy, as a function of voltage.

Moreover the interpretation of contrast is simpler in H.V.E.M. because a simpler theoretical treatment can be used (weak phase approximation). More precisely at high voltage this approximation can be used for thicker samples than at lower voltages. The origin of that is due to the writting of the scattered wave, which can be expressed as :
$\Phi(t) = \exp i\chi(t)$ where t is practically the normal to the sample with $\chi = \frac{-e}{hv} \int_o^t U(t)dt$ where v is the velocity of the incident electron and U is related to the potential V by $U = \frac{2m}{\hbar^2} V$. The change in phase of the incident wave is governed by the projected potential written above.

In addition it seems it has been shown for crystalline materials that both experimentally and theroretically (6) the contrast is better and the number of informations more important in H.V.E.M.. In our laboratory single atoms fixed on organic molecules deposited on very thin boron substrats ($\sim$ 10-15 Å) have been observed (7,8) (fig.1). Some nice experiments have been also recently done by Uyeda at 500 keV (9), on molecules taking advantage of the less important energy losses as an overall for ionization

Fig.1. Single atoms of Ba observed at 3 MeV on an amorphous substrate (10-15 Å thick).

a)- localization of Ba atoms (16 Å apart).

b)-Conical illumination. Overimpression of the diffraction pattern of gold.

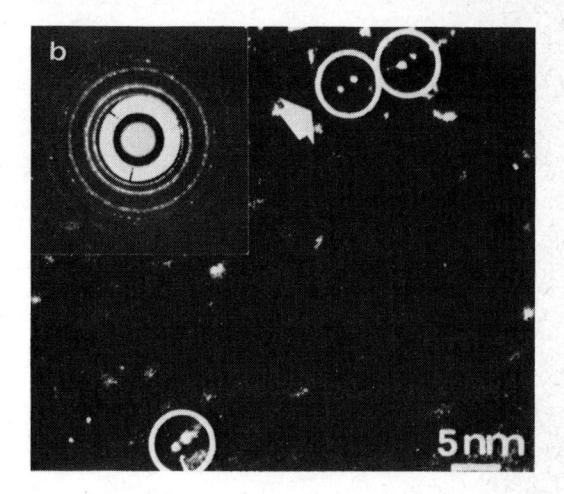

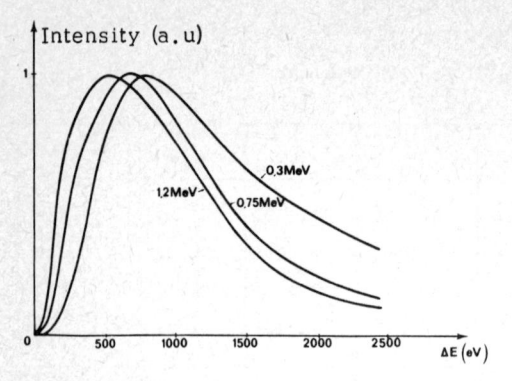

Fig.2. Electron energy loss spectra through 3 μm aluminium foils for 0.3, 0.75 and 1.2 MeV electrons (thick foils).

when the energy is increased (fig.2).

The resolution which was discussed here is only possible for thin materials even if for the reasons given above, it is possible to use thicker samples in H.V.E.M. for this purpose.

2) - Available thickness

Two types of limit of thickness can be defined : the first one is related to the range of the electrons in the sample (total penetration). It defines the thickness which gives the loss of all the kinetic energy of the incident electrons. The second one is the available thickness which gives a possibility to get informations on the sample through an image in electron microscopy.

The first penetration, the total penetration, has been studied in detail for instance by Arnal (10). An order of magnitude is in aluminum 2.1 mm, 8.1 mm respectively for electrons of 1 MeV and 3 MeV and 0.6 mm in copper at 1 MeV.

The available thickness has been studied by Humphreys (11) and Rocher et al. (12) and Thomas and Lacaze (13). The gain in penetration is an important point in favour of H.V.E.M..

The models to explain the available thickness are based on electron-phonon interactions which give a behaviour of the absorption coefficient proportional to Z^2/v^2 (scattering) where Z is the atomic number. The penetration is therefore more important in the case of light materials. However the behaviour as a function of the energy is rather pessimistic, specially for light materials, compared to the experimental results. At the present time this discrepancy is not fully understood. It could be due to the fact that the limitation in penetration is not only due to this effect but is related to the so-called top-bottom effect (14). The contrast is smoothed because of the scattering of electrons which is observed through the top-bottom effect.

The chromatic aberration is not obviously the most important factor for the penetration as it can be seen from the expression given for instance in (15, 16).

When electrons pass through a sample they have, if it is thick enough, an output most probable angle θ_p which varies as $\sqrt{t}$, where t is the thickness. The angular distribution (gaussian distribution) centered on θ_p is characterized by $\theta_G (\theta_G^2 = 2\theta_p^2)$ and the intensity collected through the aperture (see fig. 3) is given by :

$$I(\alpha) = I_o \frac{\alpha^2}{\theta_G^2} = I_o \frac{0.86 \pi^3 a_o^2 (1-\beta^2)}{N Z^2 \lambda^4 e^{2a/b} t}$$

where the detail of the expression is obtained

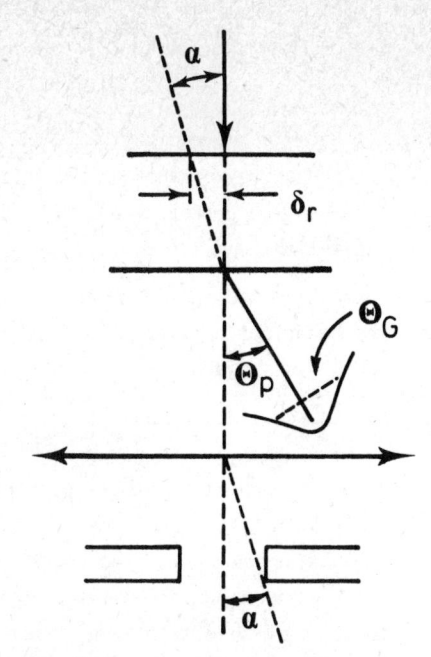

Fig.3. Drawing corresponding to the top-bottom effect.

from ref. 17. I_o is the incident intensity, α the angle of collection, a_o the Bohr radius for hydrogen, β the ratio of v with the velocity of light c, N the number of atoms by cm^3, $e^{2a/b}$ is a constant which depends on the screening of the atom. If we give for the penetration a given value of $I(\alpha)/I_o$, say 10^{-3}, we get for two voltages 1 and 2 :

$$\frac{t_1}{t_2} = (\frac{\alpha_1}{\alpha_2})^2 \frac{(1-\beta^2)_1}{(1-\beta^2)_2} (\frac{\lambda_2}{\lambda_1})^4$$

This ratio is independant of Z which means it is more available for low Z when phonon problem is not important. Moreover that model would be more useful for biological samples. It means the absorption coefficient varies with t.

3) - <u>Processus of interaction electron - sample</u>.
a)- <u>elastic interaction</u>. The basic model is well-known (18). It is based on the coulomb interaction of charged particles, with the nucleus and electrons of atoms. By using a coulomb screened potential, it is found the elastic differential cross section :

$$\sigma_e(\alpha) = 4\pi(\frac{\lambda}{a_o})^2 \frac{Z^2}{k^4} \frac{1}{(\alpha^2+\alpha_o^2)^2}$$

with α the scattering angle and α_o the screening angle $(\alpha_o) = 1/ka$, a screening radius and $k = \frac{2\pi}{\lambda}$. The total cross section is easily found and varies as $1/\beta^2$.

b) - <u>inelastic interactions</u>. These interactions are principally at small angle.
α). global treatment. The global inelastic interactions are characterized by a cross section which can be written between α_1 and π.

$$\sigma_{in}(\alpha,\pi) = \frac{\sigma_e(o,\pi)}{Z} \left(2Ln(1+\frac{\alpha_o^2}{\alpha^2}) - (\frac{1}{1+\alpha^2/\alpha_o^2}) \right)$$

The ratio of the total inelastic and elastic cross section is given by :

$$\frac{\sigma_i(0,\pi)}{\sigma_e(0,\pi)} = \frac{4}{Z} Ln(\frac{\alpha_o}{\alpha_E}) \text{ where } \alpha_E = \frac{\gamma \Delta E}{E} \frac{1}{(\gamma^2-1)m_o c^2}$$

with the ionization energy :

$\Delta E = 9.732 + 58.8 \, Z^{-0.19}$ which is available for Z above 13 about. The ratio is nearly constant with the energy of electrons.

The inelastic cross section includes different processus. Principally we can notice the excitation of electrons of the inner shells and also of the continuum.

β). inner shell excitations (19, 20, 21). The expression of the cross sections can be written :

$$\sigma_K = \frac{2\pi e^4}{m_o c^2} \frac{f_{K,L\ldots}}{E_{K,L\ldots} \beta^2} \log \frac{\alpha^2 + \alpha_E^2}{\alpha_e^2}$$ where e is the charge of the electron, f_K the oscillator strength ("proportion" of electrons participating per atom to the excitation K, L...). E_K, the energy of the edge, increases with Z, for a same type of excitation. $\alpha_E \cong \frac{E_{K,L\ldots}(E+1)}{E(E+2)}$ where E is the incident energy. The behaviour as a function of energy is shown on fig. 4.

All the excitations have a cross section which is roughly varying as $1/\beta^2$ (mean free path as β^2). If we look on fig.5, the cross section as a function of the angle ($\sigma(o \to \alpha)$) is more concentrated on the forward direction when the incident energy is increased. In this direction the background due to excitations of the continuum is minimum. Therefore it is easier to detect electrons at high voltage than at lower voltages. This is favorised by the smallness of optical aberrations at high voltage. These points explain why it has been possible to detect losses up to 12.000 eV about 1.2 MeV, instead of 2.000 eV about at 60 keV.

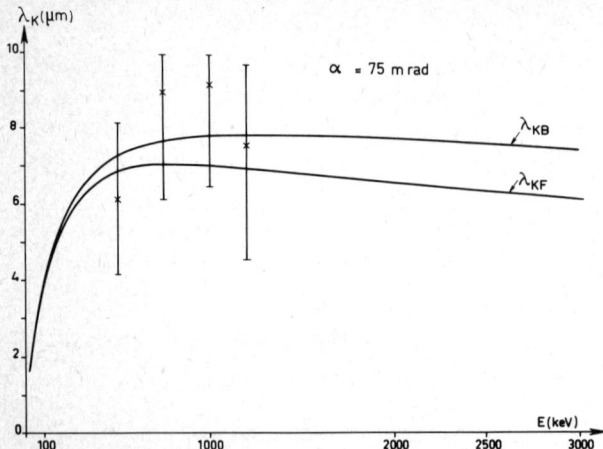

Fig.4. Inner shell excitation free path (carbon).
λ_{KB} corresponds to the Bethe Model,
λ_{KF} corresponds to the Fano treatment.
Experimental points are marked with bars of error.

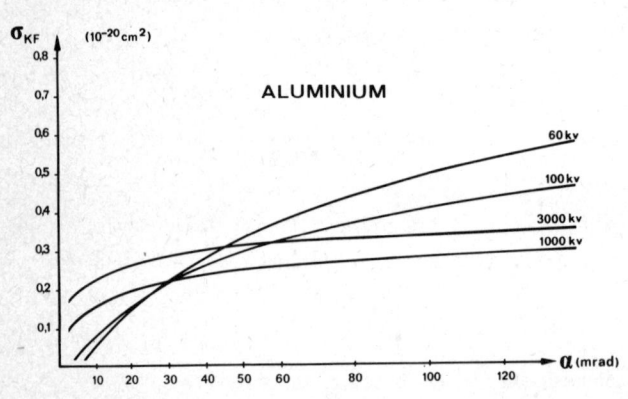

Fig. 5. Cross sections (integrated) as a function of the scattering angle for different energies in the case of aluminium.

Amongst other excitations, plasmons can play an important role. They are quite easy to observe in aluminium or carbon. Table 2 gives different order of magnitude of cross sections determined for high and low energy.

4) - Dynamical theory or classical treatment.

a)- Classical treatment. The condition for using classical treatment is $\lambdabar \ll a$ which has to be related to the uncertainty relations.

Details about the problem of semiclassical treatment, as use of WKB methods, have been

Al	σ_e	σ_i	σ_i/σ_e	σ_K	$\sigma_{plas.}$	$\sigma_{cont.}$	σ direct knock-on
100 keV	$1.288 \ 10^{-18}$ cm^2	$1.327 \ 10^{-18}$	1.03	$2.76 \ 10^{-21}$	$5.528 \ 10^{-20}$	2 to 10 σ_K	0
1 MeV	$4.37 \ 10^{-19}$	$5.23 \ 10^{-19}$	1.2	$8.73 \ 10^{-22}$	$1.675 \ 10^{-20}$	2 to 10 σ_K	$50 \ 10^{-24}$

Table 2.

given by Berry and Ozorio de Almeida (22).

The particles become more and more "classical" when increasing the energy. It means it becomes more correct to speak about trajectories.

b) - <u>Dynamical interactions</u>. Using the secular expression for dynamical theory (23, 24), it is possible to show that for two crystals and two different energies, the condition to have the same Bloch wave coefficients can be written (systematic reflexions) :

$$E_2 = \frac{g_2^2}{g_1^2} \frac{U_{g_1}}{U_{g_2}} (m_0 c^2 + E_1) - m_0 c^2 \qquad U_g = \frac{2m}{\hbar^2} V_g \ .$$ If $g_1 = g_2$ and knowing that $U_{g_1} \simeq 1.6$ MV

If $g_1 = g_2$ and knowing that $U_{g_1} \sim 3.5 \ U_{g_2}$ with reference to gold and aluminium respectively, and doing $E_1 = 0.1$ MV, we shall get the same type of contrast with $E_2 \simeq 1.6$ MV for aluminium. That has been experimentally checked by Rocher (24), Roucau and Ayroles (25) in more details taking into account the influence of temperature (Debye Waller correction on U_g).

We note the parameter g_2^2/U_{g_2} which is also important in other considerations. In a review, Humphreys and Whelan (26) discussed the condition $\frac{g_2^2}{U_{g_2}} < 1$, which is necessary to have many beam effects. Because $U_{g_2} = U_{g_{2.0}} (1 + \frac{E}{m_0 c^2})$ where $U_{g_{2.0}}$ is the term non corrected for relativity, we see that increasing the energy increases the dynamical interactions. This condition is also available for channeling.

c)- <u>Critical voltage</u>. It has been shown that for the second order bragg condition it is possible, in many cases, to find an accelerating voltage for which the intensity of this bragg spot becomes practically zero (see 3 for a review). This effect can be easily understood in a three beams treatment. The interest of this effect is due to his important sensisivity to the orientation of the sample, it means to the excitation of accidental reflexions and to the temperature also. The variation with temperature is, in the case of copper (440), about 1 kV per degree. This effect gives the possibility to test rather easily the Debye correction for instance. On the figure (6) we observe that the best Debye temperature is 325 K and also the Debye model is not correct above 600 K about (27).

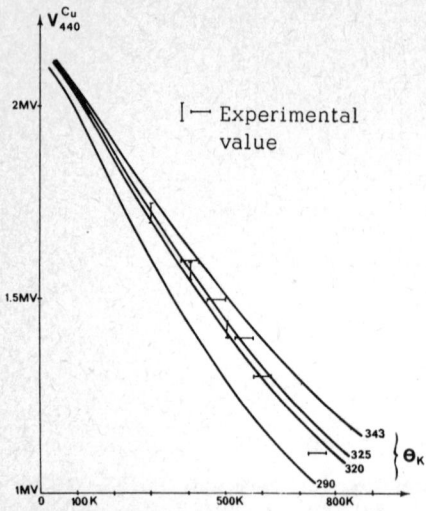

Fig.6. Critical voltage effect for the 440 reflexion in copper (27) :
- horizontal bars of determination of critical voltage by variation of temperature.
- vertical bars the temperature of the sample is constant.
Fitting is tried with different values of the Debye temperature.

This effect has been also used to obtain some information on the ordering of alloys (28).

So it is interesting to use this effect to test the structure factor or more precisely the potential which is used to deduce the scattering atomic factor curve. This point is the first interest of this relativistic effect. The determination of the atomic factor variation as a function of the scattering angle needs the determination of different points to be more precise. That is one of the interest of very high voltages ($\sim$ 3 MeV) which give the possibility to get different points on the curve (27). However this test is limited at the angles given by the lowest reflexion which is used.
($\sim 0.2 \frac{\sin \alpha}{\lambda}$ in copper). So it is quite interesting to compare the results obtained by the critical voltage and the ones given by direct scattering experiments which can give results quite close to the forward direction.

5) - <u>Applications in metallurgy and biology</u>.

a)- <u>in situ experiments.</u> The interest of H.V.E.M. in "in situ" experiments is essentially due to the available thickness compared to conventionnal electron microscopy (fig.7) (12). The spatial resolution of the method is good in comparison with the other methods which are used at the present time : X rays, Soft X rays, neutron topography (see ref. 29 for a review). The large gap between the pole pieces enables to fabricate a very "microlab" in the microscope itself.

Depending on the velocity of the phenomenon the images can be recorded on plates ($\lesssim$ 1-2 s.) or magnetic video tape recorders (exposure time $\sim$ 1/25 s. - 1/65 s.).

α) <u>plastic deformation.</u> The interest of these studies is due to the possibility of deformating a sample at low ($\sim$ 20 K), or high ($\sim$ 1100 K) temperature in the microscope itself with velocities as low as 10^{-7} up to 10^{-3}/s about. So it is possible to follow at the microscale of dislocations the behaviour of a material during the deformation.

The use of this method gives the possibility of testing some models, to understand if a mechanism is active. Some interesting results have been obtained 5 - 10 years ago by Furubayashi (30), Vesely (31), and Imura (32) for instance. The subject was concentrated on BCC materials, principally on molybdenum. The screw dislocations become quite straight at low temperature ($\gtrsim$ 110 K in Nb, room temperature in Mo) because of their

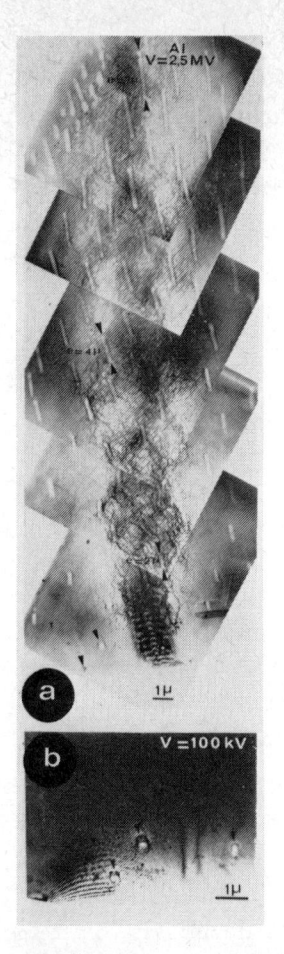

Fig. 7. Some area taken at 100 kV and 2.5 MV. The gain in penetration is about 10. Be fibres in the matrix of aluminium give an easy way to determine the penetration (ref. 12).

core structure. It was thought that these screw dislocations are responsible of stage I while stage 0 is due to the movement of edge dislocations. This point has been checked and Vesely was able to describe mechanisms of dislocations sources. It has shown two types of sources acting respectively at low and high stress (31).

On the other hand, Imura and coworkers (32) have studied the mobility of edge and screw dislocations in Fe-Si. They determined the velocity of edge dislocations in a Lüders band ($\sim 2.10^{-2}$ cm/s).

So, with the work of Louchet et al. (33, 34) concentrated on BCC materials (fig. 8), these studies, which have been done in detail, have given some quantitative informations and the behaviour of BCC materials related to the double kink mechanism (the velocity of a screw dislocation is proportional to its free length) is understood in a much better way than before. The waiting time in a Peierls valley has been determined in some cases. Some studies on softening have been also developed (35).

In FCC materials Henderson et al. (36) have studied the behaviour of dislocations. In an Al 1 % Mg alloy they have shown that the velocity of dislocations is included between 0.01 µm/s. and 1 µm/s. for stresses between 5 and 15 MN/m^2. An interesting result was also that with the same limits the density of mobile dislocations varies between 10 and 35 %.

Caillard and Martin, in aluminium, have observed different non usual gliding planes (37). Some interaction mechanisms (attractive or repulsive trees) between dislocations have been studied and creep studies are in progress (38).

We can in abstract say that the in situ experiments can be useful to study :
- geometrical informations on interaction mechanisms between dislocations on clusters and dislocations sources ; - collective behaviour ; - quantitative informations.

These two last types of informations demand a lot of experiments. Quantitative informations are obviously easier when the geometry is simple (low temperature in BCC materials).

Some discussions are around the validity of in situ plastic deformations in H.V.E.M. because of the production of point defects above the threshold. See for this point the discussion of Kubin and Martin (39) and Jouffrey (29). The best way would be to work below

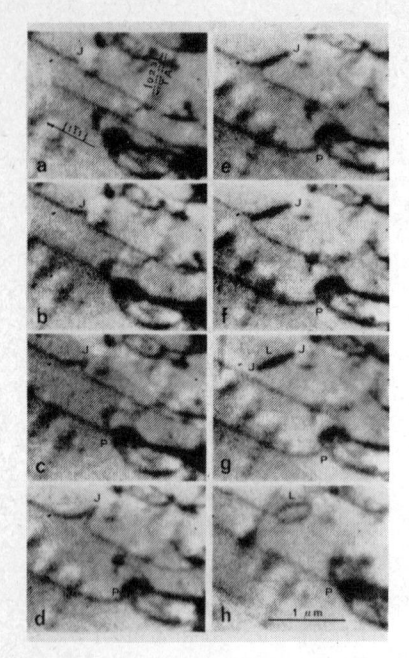

Fig. 8. Deformation of Nb at 110 K. Formation of a closed loop from a jog (F.Louchet).

the accelerating voltage corresponding to the threshold what is not always very easy but often possible. The influence of defects seems troublesome at low temperature. The diminution of the accelerating voltage limits the available thickness and the influence of surfaces can become not negligible. Therefore in many cases the experimentalist has to find an optimum for his experiments.

β)-<u>Experiments within environmental cells</u>. Since the first proposal of Marton (40) in 1935, many attempts have been done to observe samples in particular atmospheres. We refer to Swann (41) and Parsons (42) for a review respectively in metallurgy and in biology. These cells can be roughly classified in two types : - window type; the object environment is separated from the column of the microscope by electron transparent windows. So the pressure is rather well defined but their use is delicate. This type of cell can be preferable for low temperature work. - differentially pumped cells; in this attachment there is a chamber where the vacuum is intermediate between the one of the microscope and the pressure in the cell. There are four apertures to align in this case.

The applications of these cells can be :
- Low temperature application : solidified gazes (43)or liquid materials at room temperature. Radiation induced corrosions as in the case of silicon which has been reported by Swann (41).
- Room and high temperatures : it is possible to study the influence of different gazes as nitrogen, oxygen .... Swann has studied for instance the reduction of hematite to magnetite at 650° C in a 5 % H_2/Ar mixture at 30 torr (41). Other studies on water reaction with concrete have been tried.

γ)- <u>Radiation damage</u>. During the observation in the microscope, the sample can suffer different types of damage which can be troublesome in some cases. At the opposite, H.V. E.M. can be a unique tool for researchers who are interested in.

The deposited energy gives an increase in temperature which is, in most of the cases, not important at all (∼ 1 degree). It depends on the sample and its connection to the specimen holder. In defavourable cases the heating can be however very important.

The two principal effects are : formation of point defects and ionization.

- Formation of point defects. The order of magnitude of the cross section corresponding to the displacement of one atom out of its natural site is about 50 barns, what is quite small. However this processus is cumulative and so can give quite strong effects. The cross section is given by the Mac-Kinley and Feshbach (44) expression which underestimates the values for heavy elements. So it is preferable in these cases to use Oen's results (45)..

The number of pairs created by second and by atom is given by $C = \emptyset \sigma_d N_d$ with $N_d = E/2E_d$ as a first approximation when E_d is the energy of displacement and E the energy of the primary electrons. $\emptyset$ is the flux of electrons ($\sim 10^{16-20} e/cm^2/s$).

The energy transferred in a knock-on is given by : $T = \frac{2E}{M_{at}c^2}(E + 2m_o c^2) \sin^2 \frac{\alpha}{2}$ where α is the scattering angle of the incident electron through this processus.

We see that the number of pairs can be quite high after a few minutes ($C \sim 5.10^{-5}$/s. for 1 MeV).

The order of magnitude of the displacement energy is 166 keV for Al and 1.2 MeV for gold.

So it is possible to study the kinetics of formation of clusters. First, interstitial clusters are formed. Vacancy clusters can be observed later on. It is also possible to study in a very efficient way the interaction of point defects and pre-existing defects as dislocations, planar defects and defects in volume. The influence on precipitation, ordering, formation of new phases, creep, recrystallisation, the formation of voids is also interesting. In this last case many attempts have been done as was reported by Makin (46).

The simulations of swelling which can happen in nuclear reactor cores because of the neutron radiation, have been extensively studied. The conclusion is that there are some differences between the neutron and H.V.E.M. experiments but this last technique enables to get very rapid useful informations on the role of impurities, of defects, grain boundaries on the swelling rate.

See also on these problems on radiation damage the review of Urban (47).

- Ionization. These problems have ben reviewed in the inelastic events.

b) - *Chemical analysis*. It is possible to use H.V.E.M. in an efficient way to perform chemical analysis by using inner shell excitations. This method is quite useful principally for light elements but is also available for heavier materials by using L, M, .. excitations. The available thickness is superior to 2000 Å instead of 300 Å about 80 kV.

Two methods can be used (20, 21) :

α)- *an area is selected* by one aperture and the spectrum is obtained. This is shown on figure 9.

β)- *an image* is obtained by using only the electrons which have lost the energy corresponding to the inner shell excitation. So it is possible to localize the areas where is

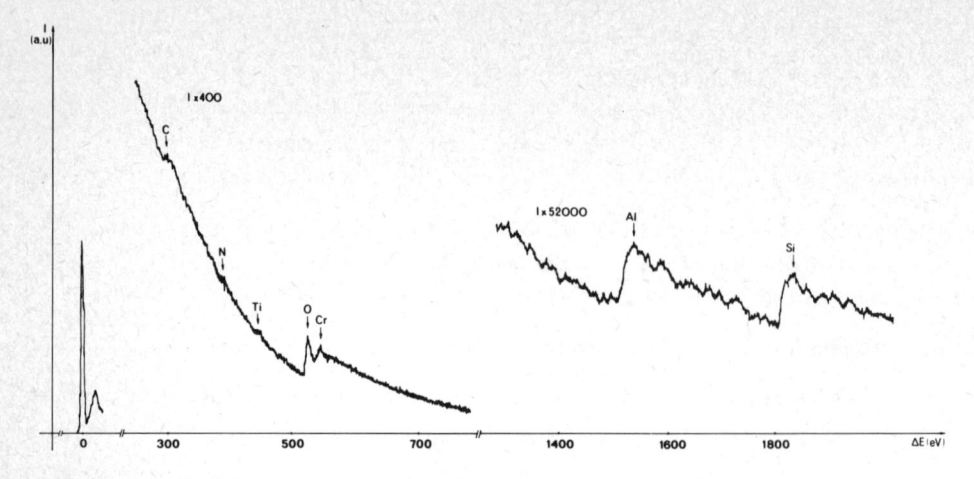

Fig. 9 - Spectrum of a thin section of a pathological lung (21).

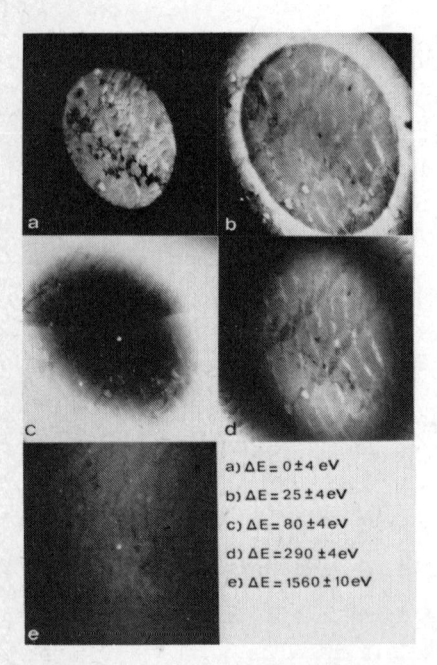

Fig. 10. Filtered images obtained by selecting a given kind of electrons. For instance, in d the electrons which have been used for the image had an energy E = 1 MeV-1560 eV $\pm$ 5 eV. In c and e the small cluster ($\sim$ 450 Å in diameter) appear white (the corresponding losses are characteristic of the L and K excitations in Al.

a) $\Delta E = 0 \pm 4$ eV
b) $\Delta E = 25 \pm 4$ eV
c) $\Delta E = 80 \pm 4$ eV
d) $\Delta E = 290 \pm 4$ eV
e) $\Delta E = 1560 \pm 10$ eV

present a given element. The figure 10 shows such application.

c) - <u>Miscellaneous applications</u>. In this part we can note : three dimensionnal reconstruction by use of quantitative stereography and also integrated stereo technique (48) concerning in the use of a system of parallel cylindrical lenses which contains as many as ninety pictures of the same area (e.g. $\pm$ 45°) ; magnetic materials (49) ; melting aspects and solidification (50) ; semiconductors (51) ; recrystallisation in aluminium (52, 53,54), in copper (55), in titanium (56) and some other materials, phase transitions in vanadium sesquioxyde (57) ; composites materials (58,59), cavitation (60) ; ceramics and minerals (61, 62, 63) ; polymers (64,65) ; ionic crystals (66) ; amino acids (67) ; superplasticity (68).

6)- Conclusion

In most of these studies the interest is related to the important available thickness which gives an aspect of the sample closer to the bulk material. For instance in precipitations it is often possible to observe big precipitates or tangles of dislocations. Sensitive materials are also interesting to be studied because of the β^2 law (see ionization).

References

1.- DUPOUY G., PERRIER F. and DURRIEU L., C.R.Ac.Sc., 1960, 251, 2836.
2.- DUPOUY G., PERRIER F. and DURRIEU L., J. Microscopie, 1970, 9, 575-592.
3.- JOUFFREY B., in Electron microscopy in materials science. Ed.Ruedl E. and Valdrè U., 1975, 979.
4.- TANAKA M. and JOUFFREY B.- to be published.
5.- ZEITLER E. and THOMSON M.G.R., Optik, 1970, 31, 258 and 359.
6.- O'KEEFE M.A., BUSEK P.R. and IIJIMA S., Nature, 1978, 274, n°5669, 322-324.
7.- DORIGNAC D. and JOUFFREY B., 4th Int. Cong. HVEM, Toulouse, Ed. Jouffrey B. et Favard P., 143.
8.- DORIGNAC D., MACLACHLAN M.E.C. and JOUFFREY B., Ultramicroscopy, 1979, 4, 85-89.
9.- UYEDA N., FUJIYOSHI Y. and KOBAYASHI T., 9th Intern. Cong. on electron microscopy, Toronto, Ed. Sturgess J., 1978, I, 242.
10- ARNAL F., Thèse d'Etat, Toulouse, 1975.
11- HUMPHREYS C.J., Proc. 25 th Anniv. Conf. of EMAG, Cambridge, Ed. Nixon W.C., 1971, 12.
12- ROCHER A., AYROLES R., MAZEL A., MORY C. and JOUFFREY B., in High voltage electron microscopy, Ed. Swann P.R., Humphreys C.J. and Goringe M.J., 1974, 436.
13- THOMAS G. and LACAZE J.C., J. Microscopy, 1973, 97, 301.
14- HASHIMOTO H., J. Appl.Phys., 1964, 35, n°2, 277-290.
15- HOWIE A., in Modern diffraction and imaging techniques in materials science. Ed. Amelinckx S., Gevers R., Remaut G. and Van Landuyt J., 1970, 295-339.
16- JOUFFREY B. 5th Int. Summer School on lattice defects in crystals, Krynica (Poland), 1976, 165.
17- SOUM G., ARNAL F., MARAIS B. and VERDIER P., C.R.Ac.sc., 1977, 284, 413-416.
18- MARTINEZ J.P., Thèse d'Etat, Toulouse, 1978.
19- JOUFFREY B., Annals of the New York Acad. of Sc., 1978, 306, 25815, 29-46.
20- JOUFFREY B., KIHN Y., PEREZ J.Ph., SEVELY J., ZANCHI G., 5th Int. Conf. on high voltage electron microscopy, Kyoto (Japan), 1977, 225.
21- JOUFFREY B., KIHN Y., PEREZ J.Ph., SEVELY J., ZANCHI G., 9th Intern. Cong. on elect. microscopy, Toronto, Ed. J.M. Sturgess, 1978, III, 292.
22- BERRY M.V. and OZORIO de ALMEIDA A.M., J. Phys. A, 1973, 6, 1451.
23- MORY C., Thèse 3ème cycle, Orsay, 1973.
24- ROCHER A., Thèse d'Etat, Orsay, 1973.
25- ROUCAU Ch., AYROLES R., Phil. Mag., 1976, 34, 4, 517-534.
26- WHELAN M.J. and HUMPHREYS C.J., 5th Int. Cong. HVEM, Kyoto, Ed. Imura T., Hashimoto H., 1977, 185.
27- ROCHER A., JOUFFREY B., 7th International Cong. on electron microscopy, Canberra, 1974, 344.
28- THOMAS L.E., SHIRLEY C.G., LALLY J.S. and FISHER R.M., 3rd International Cong.HVEM, Oxford, Ed. Swann P.R., Humphreys C.J., Goringe M.J., 1974, 38.
29- JOUFFREY B., Int. Symp. on "in situ" experiments in HVEM, Halle, 1979, to be published in Phys. Stat. Sol.
30- FURUBAYASHI E., J. Phys. Soc. Japan, 1969, 27, 130.
31- VESELY D., in High voltage electron microscopy, Ed. Swann P.R., Humphreys C.J. and Goringe M.J., 1974, 189.
32- IMURA T., in High voltage electron microscopy, Ed. Swann P.R., Humphreys C.J. and Goringe M.J., 1974, 179.
33- LOUCHET F., KUBIN L.P. and VESELY D., in 4th Intern. Conf. on the strength of metals and alloys .Ed. Laboratoire de Physique des Solides, ENSMIM, INPL, Nancy), 1976, 171.
34- LOUCHET F., Thèse d'Etat, Toulouse, 1976.
35- KUBIN L.P. and LOUCHET F., Acta Met., 1979, 27, 337-342.
36- HENDERSON BROWN M., and HALE K.F., in High voltage electron microscopy, Ed. Swann P.R., Humphreys C.J. and Goringe M.J., 1974, 206.
37- CAILLARD D. and MARTIN J.L., 4th Int. Cong. HVEM, Toulouse, 1975, Ed. Jouffrey B. et Favard P., 305.
38- CAILLARD D. and MARTIN J.L., Intern. Symp. on "in situ" experiments in HVEM, Halle, 1979, to be published in Phys. Stat. Sol.
39- MARTIN J.L. and KUBIN L.P., Ultramicroscopy, 1978, 3, 215-226.
40- MARTON L., Bull. Class. Sci.Acad. Roy. Belg., 1935, 21, 553.
41- SWANN P.R., 4th Intern. Cong. HVEM, Toulouse, 1975, Ed. Jouffrey B. et Favard P., 299.

42 - PARSONS D.F., 4th Int. Cong. HVEM, Toulouse, 1975. Ed. Jouffrey B. et Favard P. 355.
43 - VENABLES J.A. and BALL D.J., Proc. Roy. Soc., 1971, A 322, 331.
44 - Mc KINLEY W.A. and FESHBACH H., Phys. Rev., 1948, 74, 1739-1763.
45 - OEN O.S., ORNL, 1973, 4897.
46 - MAKIN M.J., 9th Int. Cong. on E.M., Toronto, Ed. J.M. Sturgess, 1978, III, 330-342.
47 - URBAN K. 4th Int. Cong. HVEM, Toulouse, 1975, Ed. Jouffrey B. et Favard P., 159.
48 - RAMBOURG A., HERMO L., MARRAUD A., and BONNET M., Biol. cell., 1977, 29, 25a.
49 - FISHER R.M., LAUY J.S. and SZIRMAE A., 4th Int. Cong. HVEM, Toulouse, 1975, Ed. Jouffrey B. et Favard P., 243.
50- FUJITA H., TABATA T. and AOKI T. 5th Int. Cong. on HVEM, Kyoto, Ed. Imura T., Hashimoto H., 1977, 439.
51 - KOLBESSEN B.O. and STRUNK H., 5th Int. Cong. HVEM, Kyoto, Ed. Imura T., Hashimoto H. 1977, 637.
52- LAGNEBORG R. and LETHINEN B., 5th Int. Cong. HVEM, Kyoto, Ed. Imura T., Hashimoto H. 1977, 381.
53 - ROBERTS W. and LETHINEN B., Phil. Mag., 1974, 29, 1431.
54 - TOYODA K. and FUJITA H., 5th Int. Cong. HVEM, Kyoto, Ed. Imura T., Hashimoto H., 1977, 451.
55 - WILBRANDT P.J. and HAASEN P. Int. Symp. on "in situ" experiments in HVEM, Halle, 1979, to be published in Cryst. and Techn.
56 - NAKA S., Thèse de Docteur Ingénieur, Orsay, 1978.
57 - ROUCAU C., AYROLES R., J. Microsc. Spectrosc. Electron., 1978, 3, 197.
58 - VALLE R., MARTIN J.L. and STOHR J.F., 4th Int. Cong. HVEM, Toulouse, 1975, Ed. Jouffrey B. et Favard P., 59.
59 - RUHLE M., SPRINGER C., CLAUSSEN N. and STRUNK H., 5th Int. Cong. HVEM, Kyoto, 1977, Ed. Imura T., Hashimoto H., 633.
60- LOVEDAY M.S. and DYSON B.F. 4th Int. Cong. HVEM, Toulouse, 1975, Ed. Jouffrey B. et Favard P., 285.
61 - THOMAS G. 5th Int. Cong. HVEM, Kyoto, 1977, Ed. Imura T., Hashimoto H., 627.
62 - HEUER A.H. and MITCHELL T.E., Proc. of the US-Japan HVEM Seminar, Honolulu, Ed. Imura T., Fisher R.M., 1977, 83.
63 - CADOZ J., CASTAING J. and PHILIBERT J., 9th Int. Cong. on electr. microscopy, Toronto, Ed. Sturgess J.M., 1978, 606.
64 - BOUDET A., private communication, 1979.
65 - KOBAYASHI K. and SAKAOKU K., Lab. Invest., 1965, 14, 1097.
66 - STRUNK H., 4th Int. Cong. HVEM, Toulouse, 1975, Ed. Jouffrey B. et Favard P., 229.
67 - GLAESER R.M., 4th Int. Cong. HVEM, Toulouse, 1975, Ed. Jouffrey B. et Favard P. 165.
68 - BRICKNELL R.H. and EDINGTON J.W., Acta Met., 1977, 25, 447-458.

TRANSFER FUNCTIONS AND ELECTRON MICROSCOPE IMAGE FORMATION

P.W. Hawkes

Laboratoire d'Optique Electronique du C.N.R.S.
B.P. 4347, F-31055 Toulouse Cedex

1. BACKGROUND

Two types of electron microscope are capable of providing high-resolution information and it is only as the limit of resolution is approached that it is useful to discuss transfer theory. These are the conventional transmission electron microscope, in which an extended area of the specimen is illuminated with electrons, which subsequently form a magnified image of this area, and the scanning transmission electron microscope, in which a very small probe is scanned over the specimen in a regular raster and the image is formed point by point, all the incident electrons contributing to each image point (pixel). In the conventional instrument, the image contrast is generated by two mechanisms which, though not independent, may be conveniently treated separately. In order to understand these, we briefly recapitulate some practical aspects of electron microscopes. (See also the chapter by Wade in this volume.)

Commercial microscopes operate with electrons accelerated through a voltage of some 100 kV, which corresponds to a wavelength of the order of 0.04 Å (4 pm). They consist of an electron source, five or six magnetic lenses and a means of observing the image (fluorescent screen) and of recording it (photographic plate). The lenses form three groups, each with a different role. The first two act as *condensers*, and direct a fine nearly parallel beam of electrons on to the specimen with very little angular spread ; the illumination is thus spatially highly coherent. The third lens, the *objective*, is immediately downstream from the specimen, which is indeed often immersed within the objective lens. This is the only lens within which the electrons travel at a (comparatively) steep angle to the axis (a few milliradians) and the resolution of the instrument is limited by the spherical and chromatic aberrations of this lens. By light-optical standards these are huge and we are therefore obliged to work at a very small numerical aperture, with the result that the theoretical limit of resolution, suitably defined, is of the order of a few ångströms, or around a hundred wavelengths. Owing to the immense chromatic aberration, the electron accelerating voltage (i.e. wavelength) and lens currents are very highly stabilized, to better than one part in 10^5, and the illumination is thus highly coherent temporally as well as spatially. We note that, in common with all electron lenses, the focal length of the objective can be varied very easily by altering the current in the windings. By this means, we can arrange that the plane conjugate to the (fixed) image plane coincides with the specimen or is at a certain distance from it, and through-focal series of images of the same specimen can readily be obtained.

Beyond the objective lens are formed the first intermediate image, in a plane conjugate to the specimen, and the diffraction pattern of the specimen, in a plane conjugate to the source. The remaining *intermediate and projector* lenses serve to magnify and image either of these onto the final "image" plane of the microscope, where the dimensions of the smallest structure recorded must be large enough to surpass any limitations due to the grain size of the recording medium.

The specimens used are essentially phase specimens, in the sense that virtually none of the incident electrons is halted - there is no absorption - but many electrons are deflected or "scattered" by the atoms of which the object is composed. Electrons scattered through large angles are intercepted by a small diaphragm, situated in the focal plane of the objective (the diffraction pattern plane), thus providing one of the two contrast mechanisms alluded to above: fewer electrons reach regions of the image corresponding to specimen areas containing heavy atoms, which are thus seen as dark patches in the image. This is known as diffraction contrast and is the dominant contrast mechanism except at high resolution. Electron deflection may be regarded as a phase shift of the incident electron wave function and we might therefore imagine that in the absence of some special accessory, analogous to a phase plate, no contrast would be generated at the image by electrons deflected in the specimen but not enough to strike the objective aperture. In fact, however, the combined effect of spherical aberration and of the defocus is similar to that of a phase plate of non-uniform thickness : a phase variation is added to that due to the specimen with the result that an interference pattern is generated at the image by unscattered and scattered electrons. This interference pattern is, for a certain type of specimen - weakly scattering objects- related simply to the atomic distribution in the latter and provides high resolution information, which cannot, however, be read off directly.

For weakly scattering specimens, we can show that the image contrast is related linearly to the phase and amplitude variations impressed on the incident electron wave by the specimen, and this linear relationship is fully characterized by a knowledge of the *phase and amplitude transfer functions*, as we show in the next section. Before turning to this, however, we must explain the presence of an amplitude term, which seems to contradict our earlier assertion that electron specimens are essentially phase specimens. The paradox is, however, easily resolved. When electrons are scattered within a specimen they may be deflected with virtually no loss of energy in which case the interaction is *elastic* or they may lose energy, in which case it is said to be *inelastic*. Transfer theory is applicable to elastically scattered electrons and any others that have been inelastically scattered are effectively lost from the theory. They therefore appear as an amplitude or absorption term.

2. TRANSFER THEORY (COHERENT ILLUMINATION)

The illumination in a conventional microscope is to a good approximation coherent, though departures from perfect coherence are extremely important as we shall see in the next section. In order to introduce the notion of transfer functions without unnecessary complication, we consider a parallel beam of monoenergetic electrons, corresponding to plane monochromatic waves, incident on a specimen. The latter is characterized by a complex transparency S (x, y), which we write :

$$S = (1 - s) \exp (i \phi) \tag{1}$$

The wave emerging from the specimen is assumed to be equal to the product of S and the incident wave. The propagation of the wave function through the electron lens system is well understood and it is easily shown that the wave function at the image plane, ψ_i, is related to that at the object plane, ψ_o, by the linear expression :

$$\psi_i (x,y) = \frac{1}{M} \iint K (x/M - x_o, y/M - y_o) \psi_o (x_o, y_o) dx_o dy_o \tag{2}$$

in which M denotes the magnification and certain quadratic phase factors have been omitted. Writing :

$$\tilde{\psi}_i (p,q) = \iint \psi_i (Mx, My) \exp \{- 2 \pi i (px + qy)\} dx dy$$

$$\tilde{\psi}_o (p,q) = \iint \psi_o (x_o, y_o) \exp \{- 2 \pi i (px_o + qy_o)\} dx_o dy_o$$

$$\tilde{K} (p,q) = \iint K (x, y) \exp \{- 2 \pi i (px + qy)\} dx dy \tag{3}$$

we see that :

$$\tilde{\psi}_i (p,q) = \tilde{K} (p,q) \tilde{\psi}_o (p,q)/M \tag{4}$$

It is not the image wave function or its transform that is observable, however, but the current density, which is proportional to $\psi_i \psi_i^*$. Nevertheless, for weakly scattering specimens, useful relations can be derived. For such objects, we assume that both S and ϕ are so small that we may write :

$$S = (1 - s) \exp (i \phi) \approx 1 - s + i\phi \tag{5}$$

whereupon we find :

$$M \tilde{\psi}_i (p,q) = \tilde{K} (p,q) \{\delta(p,q) - \tilde{s}(p,q) + i \tilde{\phi}(p,q)\} \tag{6}$$

or :

$$M^2 \psi_i (Mx, My) \psi_i^* (Mx, My)$$
$$= \left[1 - \iint \tilde{K} (\tilde{s} - i \tilde{\phi}) \exp \{2 \pi i (px + qy)\} dp dq\right]$$
$$\times \left[1 - \iint \tilde{K}^* (\tilde{s}^* + i \tilde{\phi}^*) \exp \{- 2 \pi i (px + qy)\} dp dq\right] \tag{7}$$

Since $\tilde{s}^* (-p, -q) = \tilde{s} (p, q)$ and likewise for $\tilde{\phi}$, we have :

$$M^2 \psi_i (Mx, My) \psi_i^* (Mx, My) - 1$$

$$\simeq - \iint \tilde{s} (p,q) \{\tilde{K} (p,q) + \tilde{K}^*(-p, -q)\} \exp \{2 \pi i (px + pq)\} dp\, dq$$

$$+ i \iint \tilde{\phi} (p,q) \{\tilde{K} (p,q) - \tilde{K}^*(-p, -q)\} \exp \{2 \pi i (px + qy)\} dp\, dq \qquad (8)$$

Denoting the left-hand side by C (Mx, My) and setting :

$$\tilde{C} (p, q) = \iint C (Mx, My) \exp \{- 2 \pi i (px + py)\} dx\, dy \qquad (9)$$

we see that :

$$\tilde{C} (p, q) = \tilde{s} (p, q) B_s (p, q)$$
$$+ \tilde{\phi} (p, q) B_\phi (p, q) \qquad (10)$$

where :

$$B_s (p, q) = - \{\tilde{K} (p, q) + \tilde{K}^* (p, q)\}$$
$$B_\phi (p, q) = i \{\tilde{K} (p, q) - \tilde{K}^* (p, q)\} \qquad (11)$$

Equation (10) states that the Fourier transform of the image contrast is linearly related to the transforms of the specimen amplitude (s) and phase (ϕ) distributions ; the functions B_s and B_ϕ are known as the amplitude and phase transfer functions. The transform of the point-spread function, K, is readily expressed in terms of the spherical aberration coefficient, C_s and the defocus, Δ. We find :

$$B_s = - 2 a \cos \gamma (p, q)$$
$$B_\phi = 2 a \sin \gamma (p, q) \qquad (12)$$

in which :

$$\gamma = \frac{2 \pi}{\lambda} \{ \frac{1}{4} C_s \lambda^4 (p^2 + q^2)^2 - \frac{1}{2} \Delta \lambda^2 (p^2 + q^2) \} \qquad (13)$$

and a is a function equal to unity in the opening of the objective aperture and to zero elsewhere.

The fact that the transfer functions are sinusoidal creates severe problems when we wish to synthesize a faithful image from the recorded current distribution : some contrast will be inverted but worse, much information will have been irretrievably lost in the vicinity of the zeros. Several procedures have been devised and tested, for regenerating the functions s and ϕ (or their transforms) from focal series, since the zeros will occur at different spatial frequencies as Δ is varied. The amplitude term can often be neglected, since specimens thin and light enough for the condition $\phi \ll 1$ to be satisfied will cause little inelastic scattering. We then have :

$$\tilde{C} = \tilde{\phi} B_\phi = 2 a \tilde{\phi} \sin \gamma \qquad (14)$$

This offers a convenient means of studying the transfer function of actual instruments, for by taking a very thin layer of amorphous material, typically carbon, as specimen, $\tilde{\phi}$ will be a very slowly varying function and the Fourier transform (or *spatial frequency spectrum*) of the image contrast will show sin γ directly, as a series of concentric rings. Various techniques for extracting C_s and Δ from these diffractograms have been devised.

3. PARTIALLY COHERENT ILLUMINATION

Although the illumination in an electron microscope is highly coherent, the effects of finite source size (partial spatial coherence) and of finite wavelength spread (partial temporal coherence) finally limit the resolution attainable. It has long been known that for weakly scattering objects, a transfer function formalism can again be derived and an equation identical with eq. (10) is obtained, although the transfer functions are of course different. A particularly interesting result, which has been established much more recently, however, reveals that for most practical situations, the transfer function corresponding to partially coherent illumination may be written as a product of the coherent transfer function (B_s or B_ϕ) and two modulating functions, one characterizing finite source size, the other finite wavelength spread. That each aspect of partial coherence can be separately represented by a modulating function is not very difficult to show - we refer to the bibliography (section 5) for details of this. It is, however, much less obvious that, if both aspects have to be considered simultaneously, as is the case in practice, it is legitimate to multiply the corresponding modulation functions. That this is indeed usually permissible has now been demonstrated, however, and the product representation has been used to obtain estimates of the wavelength spread and source size of various electron microscopes.

What effect do these modulating functions have on the transfer of information from specimen to image ? Broadly speaking, they attenuate the recorded signal at the image so that beyond a certain point, determined by the magnitude of the wavelength spread and source size, this signal cannot be distinguished from the natural granularity or noise. They thus create a real, practical limit to the attainable resolution, whereas the unmodulated sinusoidal transfer functions suggest, unrealistically, that information can be obtained in principle far beyond the "theoretical limit of resolution".

4. OTHER CASES

In the preceding sections, we have given an elementary account of the transfer function formalism for the common bright-field mode of operation of conventional electron microscopes. We now mention briefly some of the other modes used in electron microscopy and examine the applicability of transfer function theory to them.

Several dark-field modes are routinely available on commercial electron microscopes, but only in exceptional circumstances is a transfer function theory applicable. The reason for this is readily understood when we realise that the existence of a *linear* bright-field theory is a consequence of the fact that weak scattered wave fields interfere with a comparatively strong unscattered beam. In the darkfield modes, however, this unscattered beam is intercepted and only the scattered electrons reach the image. The possibility of establishing a linear relation between some aspect of image and object has been explored in considerable detail ; in general, no such relation exists but an exception of some interest corresponds to individual discrete specimen points such as heavy atoms bound in some characteristic geometry to a lighter molecule.

A number of other bright field modes have been attracting considerable attention recently, particularly for high resolution microscopy and for minimizing the damage caused by electron-specimen interactions. There are tilted illumination, which gives increased resolution in the tilt direction, and hollow-cone, bright-field illumination, which should give an overall increase in resolution but at the expense of reduced contrast. Since high-resolution image information is almost inevitably difficult to detect, this may be too high a price to pay, for certain types of specimen at least, but is nonetheless of considerable theoretical interest. For both types of illumination, the image-object relation can be described by a transfer function formalism but for tilted illumination, the latter is anisotropic, in the sense that the microscope response varies with the azimuth, relative to that of the tilt direction. A thorough understanding of this type of illumination is nevertheless of importance, since it should be possible to obtain three-dimensional information about easily damaged specimens with a low electron dose by using a multiple tilt scheme. For hollow-cone illumination, the transfer functions are again axially symmetric but low in absolute value which implies a diminution of contrast, as mentioned above.

To what extent can this account of the transfer theory of image formation in the conventional fixed-beam transmission microscope be transferred to the scanning instrument (STEM) ? A detailed study of the formation of the STEM image shows that the recorded current is of course affected by the properties of the probe-forming lens but that only for weak phase specimens is the relation between the signal recorded and some simple aspect of the specimen straightforward. The effect of partial coherence on STEM image formation is very different ; it can be shown that the spatial frequency spectrum of the image intensity is the same as that of a STEM with a coherently illuminated probe, modulated by the mutual intensity in the exit pupil of the probe-forming lens.

5. FURTHER READING

The texts available on image formation in the electron microscope, transfer functions and the effect of partial coherence range from non-specialist works with little mathematics to highly technical papers that pre-suppose considerable familiarity with the subject. Of the non-specialist material of a general kind, we draw attention to a book and two other accounts by the present author (Hawkes, 1972, 1975, 1978 c) and a very readable chapter by Lenz (1971). More technical discussion is to be found in a still very accessible chapter by Frank (1973 b), a long review and a book by Misell (1973, 1978) and a particularly clear book by Saxton (1978).

Transfer functions were first introduced into electron optics by Hanszen and his colleagues and the earlier work is reviewed in Hanszen (1971). The effect of temporal partial coherence was first analysed by Hanszen and Trepte (1971) and the fact that spatial partial coherence can be expressed by means of a modulating function was shown by Frank (1973 a) ; the combined effect of the two was fully investigated by Wade and Frank (1977). A long review of partial coherence in electron microscopy by Hawkes (1978 b) gives many more references in this field. Numerous papers have been devoted to tilted and conical illumination ; for a full list, we refer to a bibliography (Hawkes, 1978 a), only mentioning the papers by Hoppe et al. (1975), Wade and Jenkins (1978) and Saxton et al. (1978). The many methods of exploiting image diffractograms to obtain information about microscope operating parameters will likewise be found in the bibliography mentioned above ; we draw attention to papers by Saxton (1977), Wade (1978) and Frank et al. (1978/9), which are specifically concerned with partial coherence. The problem of filtering electron micrographs to reduce the effect of the transfer function is discussed very thoroughly by Kübler et al. (1978) and has also been explored in practice by Saxton et al.(1977).

Finally, we draw attention to the work of Burge and Dainty (1976) on STEM imagery, including the effect of partial coherence across the probe, to the very detailed analysis by Hanszen and Ade (1977) and to a paper by Spence and Cowley (1978). The question of dark-field image formation is analysed by, for example, Ade (1975).

REFERENCES

G. Ade : Nichtlinearitätsprobleme bei Hell- und Dunkelfeldabbildungen, Optik, 42, 199-215 (1975).

R.E. Burge and J.C. Dainty : Partially coherent image formation in the scanning transmission electron microscope, Optik, 46, 229-240 (1976).

J. Frank : The envelope of electron microscopic transfer functions for partially coherent illumination, Optik, 38, 519-536 (1973 a).

J. Frank : Computer processing of electron micrographs, in Advanced Techniques in Electron Microscopy, ed. by J.K. Koehler (Springer, Berlin and New York, 1973 b), pp. 215-274.

J. Frank, S.C. McFarlane and K.H. Downing : A note on the effect of illumination aperture and defocus spread in brightfield electron microscopy, Optik, 52, 49-60 (1978/9).

K.-J. Hanszen : The optical transfer theory of the electron microscope : fundamental principles and applications, Adv. Opt. Electron Micr., 4, 1-84 (1971).

K.-J. Hanszen and G. Ade : A consistent Fourier optical representation of image formation in the conventional fixed beam electron microscope, in the scanning transmission electron microscope and of holographic reconstruction, PTB-Bericht APh-11, 31 pp (1977).

K.-J. Hanszen and L. Trepte : Der Einfluss von Strom- und Spannungsschwankungen, sowie der Energiebreite der Strahlelektronen, auf Kontrastübertragung und Auflösung des Elektronenmikroskops, Optik 32, 519-538 (1971)

P.W. Hawkes : Electron Optics and Electron Microscopy (Taylor and Francis, London, 1972).

P.W. Hawkes : Computer processing of electron micrographs. Int. Rev. Cytol. 42, 103-126 (1975).

P.W. Hawkes : Electron image processing : a survey, Comp. Graph. Im. Proc. 8, 406-446 (1978 a).

P.W. Hawkes : Coherence in electron optics, Adv. Opt. Electron Micr. 7, 101-184 (1978 b).

P.W. Hawkes : Computer processing of electron micrographs, in Principles and Techniques of Electron Microscopy, ed. by M.A. Hayat (Van Nostrand-Reinhold, New York and London, 1978 c), vol. 8, pp. 262-306.

W. Hoppe, D. Köstler, D. Typke and N. Hunsmann : Kontrastübertragung für die Hellfeld-Bildrekonstruktion mit gekippter Beleuchtung in der Elektronenmikroskopie, Optik, 42, 43-56 (1975).

O. Kübler, M. Hahn and J. Seredynski : Optical and digital spatial frequency filtering of electron micrographs, Optik 51, 171-188 and 235-256 (1978).

F. Lenz : Transfer of image information in the electron microscope, in Electron Microscopy in Material Science, ed. by U. Valdrè (Academic Press, New York and London, 1971), pp. 540-569.

D.L. Misell : Image formation in the electron microscope with particular reference to the defects in electron-optical images, Adv. Electron. Electron Phys. 32, 63-191 (1973).

D.L. Misell : Image Analysis, Enhancement and Interpretation, (North Holland, Amsterdam, New York and Oxford, 1978).

W.O. Saxton : Spatial coherence in axial high resolution conventional electron microscopy, Optik 49, 51-62 (1977).

W.O. Saxton : Computer techniques for image processing in electron microscopy, Adv. Electron. Electron Phys., Suppl. 10, 289 pp (1978).

W.O. Saxton, A. Howie, A. Mistry and A. Pitt : Fact and artefact in high resolution microscopy, in Developments in Electron Microscopy and Analysis, 1977, ed. by D.L. Misell (Institute of Physics, Bristol, 1977) pp. 119-122.

W.O. Saxton, W.K. Jenkins, L.A. Freeman and D.J. Smith : TEM observations using bright field hollow cone illumination, Optik, 49, 505-510 (1978).

J.C.H. Spence and J.M. Cowley : Lattice imaging in STEM. Optik, 50, 129-142 (1978).

R.H. Wade : The phase contrast characteristics in bright field electron microscopy, Ultramicroscopy, 3, 329-334 (1978).

R.H. Wade and J. Frank : Electron microscope transfer functions for partially coherent axial illumination and chromatic defocus spread, Optik, 49, 81-92 (1977).

R.H. Wade and K.H. Jenkins : Tilted beam electron microscopy : the effective coherent aperture, Optik, 50, 1-17 (1978).

ACOUSTIC MICROSCOPY

B. NONGAILLARD

UNIVERSITY OF VALENCIENNES
59326 VALENCIENNES FRANCE

INTRODUCTION

Optical microscopy is a well established investigation technique which furnishes a satisfactory spatial resolution for many applications. However, in some areas, the ability of processing only thin slices of specimen together with the need for tedious selective coloration techniques lead to a time consuming sample preparation. The concept of using acoustic waves for microscopy is relatively new since it appeared in the three latest decades. The acoustic wave velocity in matter is very small (five orders of magnitude lower than that of light waves). The acoustic wavelength falls then in the optical spectrum at frequencies near one gigahertz where the acoustic attenuation doesn't prevent the propagation over some hundreds of wavelengths (see fig. 1).

	Optical microscopy	Acoustic microscopy
Velocity range	1.5 to 3 x 10^8 m/s	1.5 to 11 x 10^3 m/s
Frequency range	10^{14} to 10^{15} Hz	10^8 to 10^9 Hz
Range of wavelength	0.4 to 0.8 μm	1 to some hundreds of μm
Relative indexes	no more than 2	greater than 7
Attenuation in water		220 dB/mm/GHz^2

Fig. 1 : Comparison between optical and acoustic microscopy.

The most important point is that optical and acoustical microscopy are complementary techniques. Optical microscopy gives informations about the dielectric properties of the sample and acoustical microscopy gives informations on the mechanical properties of the sample

In this last case the transmitted or the reflected beam amplitude is connected to the acoustic impedance Z.
Where
$$Z = \rho v$$
ρ being the density of the material
v being the velocity of the acoustic wave.

So, we get a new information about the mechanical properties of the sample and the sample preparation may be simpler (without resorting to colorations for optically transparent samples). We can also get, with acoustic microscopy, informations in the bulk of opaque materials.

Transmission acoustic microscope.

Several acoustic microscopes have been studied and tested which rely upon different physical principles : collinear acousto optic interaction[1], radiation pressure detector[2], photopiezoelectric effect[3], optical detection of the dynamic rippling of a surface[4], and scanning acoustic microscope[5]. This last system appears to be the more suitable with respect to resolution and ease of use since only standard ultrasonic and electronic methods are implemented. As suggested by professors QUATE, the object is mechanically scanned in the common focal plane of the two lenses (fig.2).

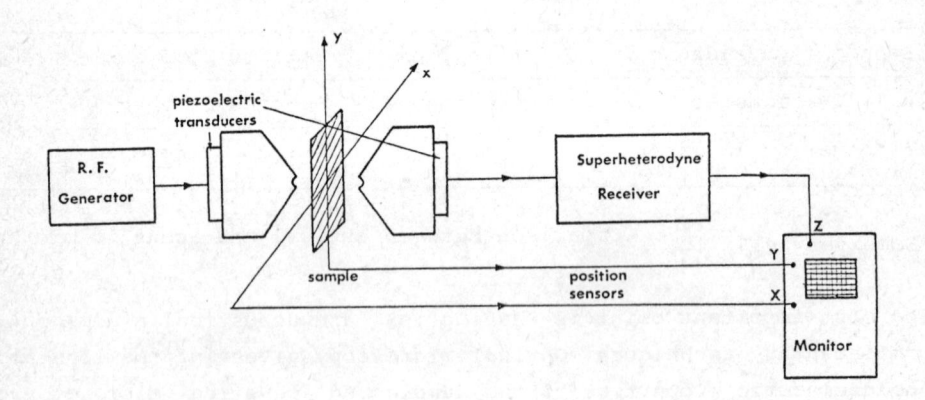

Fig.2 : Synoptic scheme of the transmission acoustic microscope.

The visualisation monitor is driven in synchronism with the object displacements and intensity modulated by an electrical signal proportionnal to the object acoustic transparency function. The fast motion of the sample is operated at about 100 Hz frequency and the slow motion at a 1Hz frequency. To observe the image of a sample we must use a remanent cathode ray-tube or we must photograph the screen of a classical cathode ray-tube : A 0.5 μm resolution has been reported at an operating frequency of three gigahertz[6]. The good performances obtained follow from the characteristics of the acoustic lenses used. The grinding of spherical sapphire-water interfaces leaves us with lenses of 7.3 relative refractive index. For such an unusual value, when thinking in optical terms, the geometrical aberrations which vary as $\frac{1}{n^2}$ (if n stands for the lens index), are very small. For an angular aperture of 120 degrees the least confusion circle is then 0.25 μm in radius. It follows that, up to gigahertz frequencies, the resolution of the confocal system remains diffraction limited. It can be shown that such lenses follows the classical relationships of paraxial optics despite the lens aperture is greater than 120 degrees. This feature arise from the fact that the dephasing on the lens input face is divided by its refractive index The most important difficulty to implement an acoustic microscope technological one. Indeed we must hollow a spherical lens of 120 μm radius to image samples at a 600 MHz operating frequency. For such a value of the lens radius the attenuation during the propagation of the acoustic wave in the liquid coupling medium is not very prohibitive, say 15 dB. But at a 2 GHz operating frequency the lens radius of curvature must be 10 μm. The technological limit attained actually is a 50 μm radius of curvature. The losses in the system increase and the image contrast becomes poorer. Actually the total losses of a transmission acoustic microscope, operating at a 600 MHz, are 80 dB and we can get a good image contrast using a heterodyne receiver which has a very high sensitivity. Some very good images have been obtained at a 700 MHz frequency in the Molecular Acoustic laboratory of Strasbourg[7] about biological samples. A very specific biological sample preparation has been worked out in this laboratory which allows to bypass the classical paraffin which sustains the thin biological specimen, and to replace this by a more homogeneous polystyren film. A test grid has been imaged with a 250 magnifying power (fig.3) and a 630 magnifying power (fig.4). The bars are 50 μm large and we can see on these photographs some very small details with a good contrast. We may observe that the contrast in these photographs is reversed. The fig.5 shows the acoustic image of a piece of thyroïd tissue with a 160 magnifying power and the fig.6 shows the details of the former image (mark b) with a 400 magnifying power. The next photograph fig.7 shows the acoustic image of breast tissue with a 160 magnifying power. All these photographs show the very high contrast characterizing conjunctiva tissue, which tissue is hardly seen with an optical microscope.

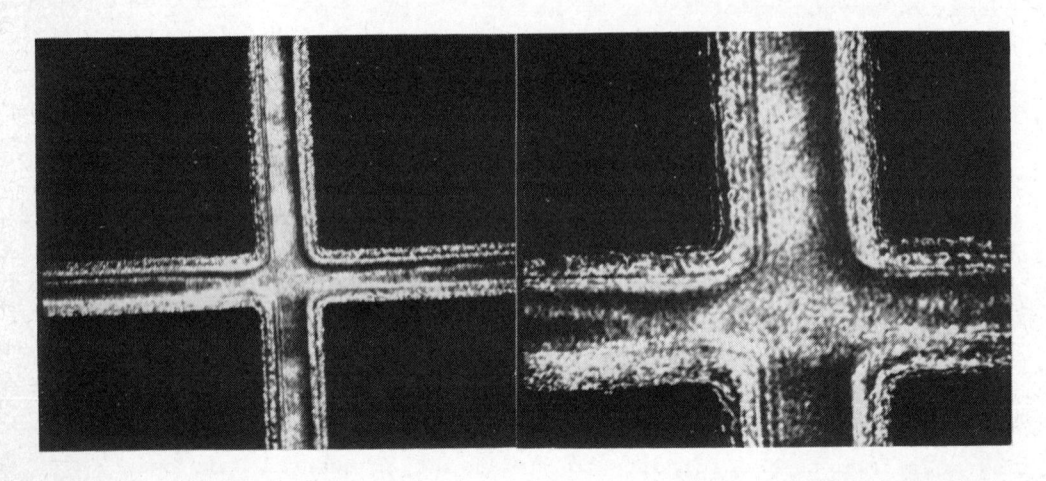

Fig.3 : acoustic image of a test grid.(x 250)

Fig.4 : acoustic image of the same test grid.(x 630).

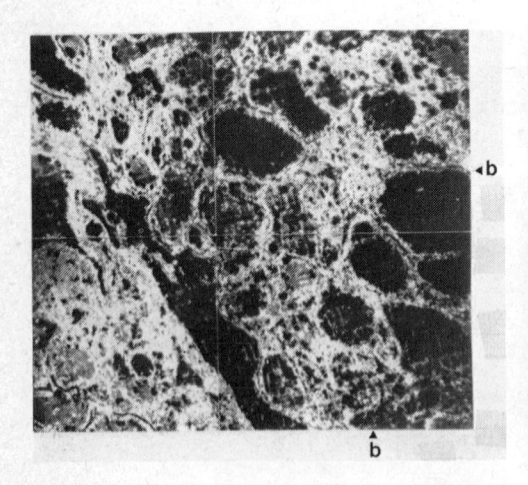

Fig. 5 : acoustic image of thyroïd tissue (x 160).

Fig. 6 : details of thyroïd tissue (x 400) in mark b)

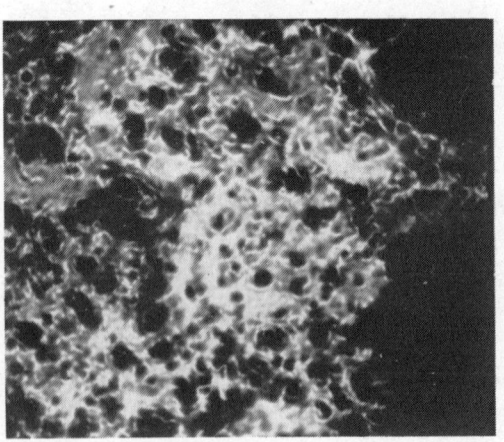

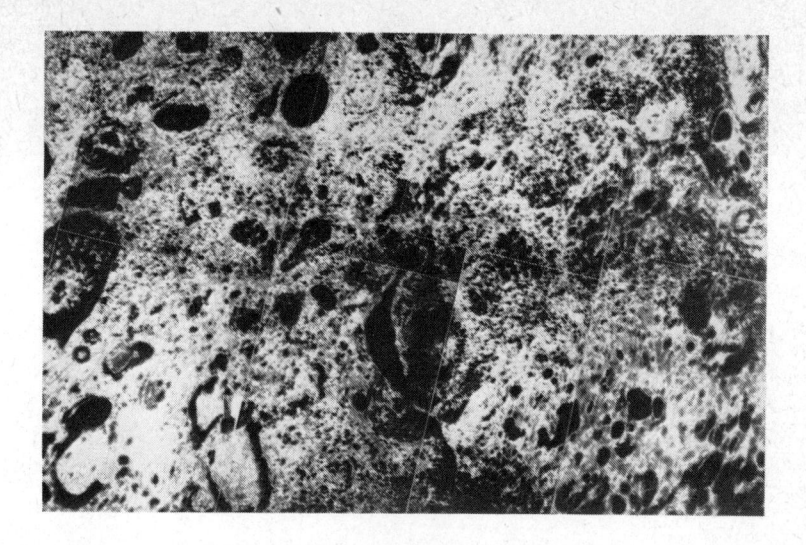

Fig.7 : acoustic image of breast tissue (x 160)

Reflection acoustic microscope

But, a transmission acoustic microscope cannot image thick specimens because the two confocal lenses distance is , at a 600 MHz frequency, 150 µm. So for thick specimens it is also suitable to use reflection acoustic microscopy in some areas like geology, metallurgy and generally non destructive testing where the thickness of the sample may be equal to several millimeters. The reflection scheme allows the visualization of internal structures lying under the specimen surface using acoustic waves. This would require careful lapping and polishing techniques if optical microscopy was to be used. In the reflection acoustic microscope there is a single transducer-delay line and a single lens grinded in this delay line. A great simplification results then in the mechanical design of the apparatus. However, the occurence of several parasitic reflected electrical signals requires a more sophisticated electronic system. One signal comes from the reflection F of the electrical energy (Fig. 8) at the piezoelectric transducer since the electrical match between the generator and the transducer is never perfect. The other signal comes from the reflection of the acoustic energy R on the highly acoustically mismatched spherical sapphire water boundary. The pulse mode of operation allows us to separate in time the usefull information E from the parasitic ones F and R. A delayed gate is used to prevent the receiver from being saturated by the parasitic signals(fig.9).

A boxcar integrator is used to get a continuous bright image.

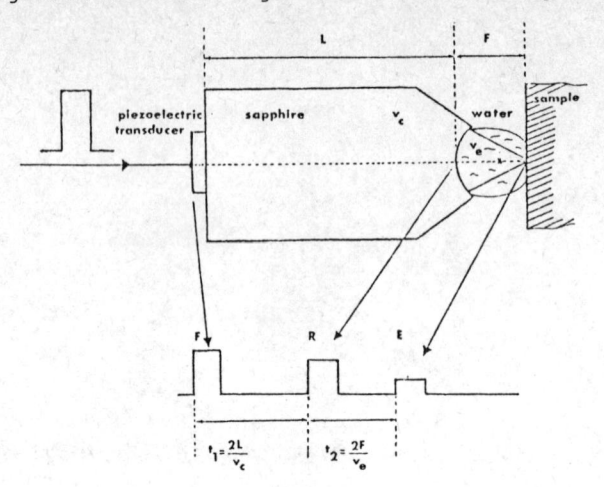

Fig. 8 : Received electrical signals

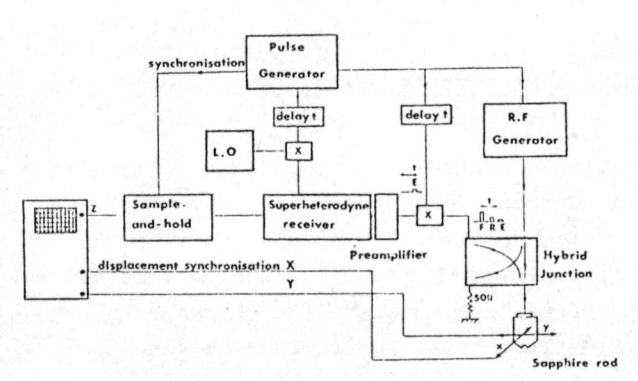

Fig. 9 : synoptic scheme of the amplifier.

Our reflection system works at a 130 MHz frequency with a spatial resolution of 10 µm inside water. This resolution has been verified using test grids with bars of 12 µm width (fig.10). The total loss in the system amounts to 50 dB and the signal to noise ratio is nearly equal to 60 dB for a typical + 20 dBm electrical power emission. So a very high quality image may be obtained.

Fig. 10 : acoustical image of a test grid.
at a 130 MHz frequency.

The practical interest of the reflection acoustic microscope for seing structures lying under the surface of an opaque material or of a thick sample has led us to the numerical calculation of the acoustic field distribution behind the surface of a plane boundary. The results of this computation allow us to determine the lens radius of curvature R together with its distance to the sample surface[8] :

$$R = \frac{n-1}{n} (e + n'Z)$$

Where n is relative index of the lens with respect to water.
n' the relative index of the sample with respect to water.
Z the exploration depth under the sample surface.

The choice of the radius of curvature and angular aperture of the lens determines the minimal distance propagated in the water coupling medium inside which the attenuation is very high. The sample, when homogeneous, has a much lower attenuation. It is the case when we search for some defects inside a sample. So the radius of curvature must be kept minimum in order to preserve the high contrast. Thin slices, 2 mm in thickness, have been analyzed and images have been taken at different depths, using a 1 mm lens radius and a 1 20° angular aperture. A typical optical image of a fossil is given in fig. 11 and the acoustical image of the sample surface is shown in fig. 12a together with that of a plane lying at nearly 150 μm depth (fig. 12b).

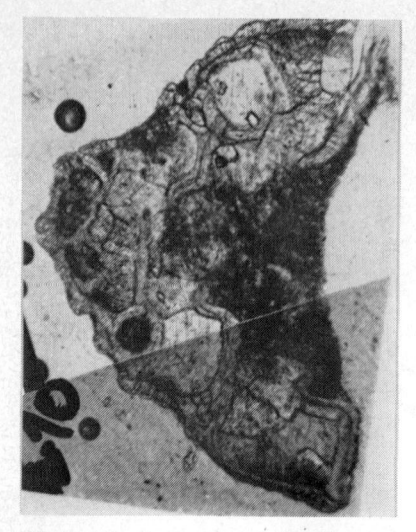

Fig. 11 : Optical image

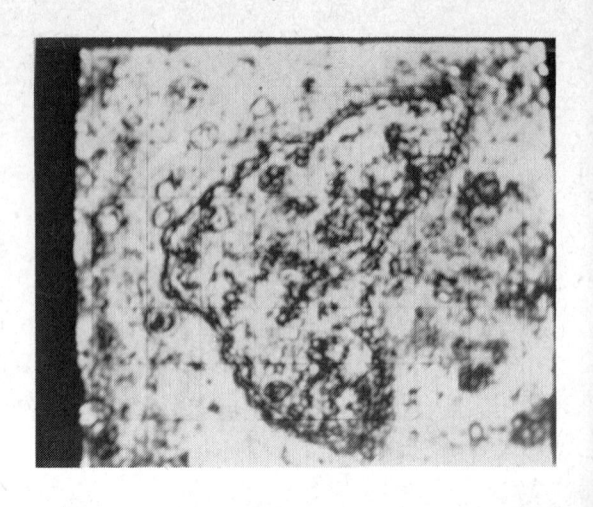

Fig. 12a : acoustical image of the sample surface.

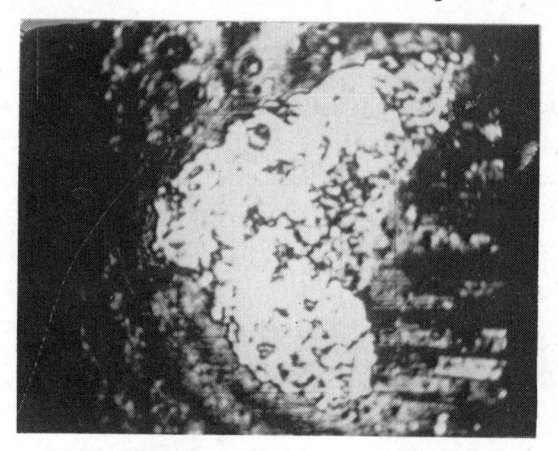

Fig. 12b : Acoustical image at 150 µm depth

To conclude, the rated resolution compares actually well with that given by the best optical microscopes and in some instances structures have been imaged acoustically which don't appear on optical micrographs. It is the case for structures lying in the bulk of opaque materials. Some images have been obtained at Standford University at a 3 GHz frequency which show some defects in integrated circuits. Some applications of acoustic microscopy to non destructive testing area are also now under study. However a particular study and a matching of the basic apparatus must be devoted to each new problem. For example, the application to the localization of very small diameter defects is now underway in our laboratory.

REFERENCES

(1) J.F. HALICE, PH. D. THESIS, Stanford University (1971)

(2) JA. CUNNINGHAM and C.F. QUATE, Stanford University

(3) B.A. AULD, RC ADDISON and D.C. WEBB, Proc. IEEE, 57, 713 (1969)

(4) LW KESSLER, PR. PALERMO and A. KORPEL, Acoustical Holography, Vol. 4, 51-71 Plenum (1972)

(5) RA. LEMONS and CF. QUATE, Applied Physics Letters, 24, 163, (1974) and 25, 251 (1974)

(6) V. JIPSON and CF. QUATE, Applied Physics Letters, 32, 789 (1978)

(7) This work was supported by the DGRST. We thank Professors R. CERF, P. LEMARECHAL and V. LIST for the good images they would to give us and we show in this paper (fig. 3 to fig. 7)

(8) B. NONGAILLARD, J.M. ROUVAEN, E. BRIDOUX, R. TORGUET and C. BRUNEEL, Applied Physics Letters (Février 1979)

USE OF CODED APERTURES IN GAMMA-IMAGING TECHNIQUES

J. Brunol
Institut d'Optique
Université de Paris XI, Bât. 503
91406 Orsay, France.

Recently some methods have been developped in the field of nuclear medicine and radiography. Transaxial tomography, multiple pinhole and coded apertures techniques are these main imaging processes which appeared in the last few years. Data processing, by numerical or analogic means, of one or several basic images is a common factor to all "these tomographic methods by reconstruction". Here we present some principles and fundamental considerations on the latest method : Coded Aperture Imaging.

I. Presentation of Coded Aperture Imaging

Coded Aperture Imaging (CAI) is a general imaging method which allows to synthetize a lens in wavelength domains where such an element cannot exist. CAI implies a two step imaging operation.
- In the first step or encoding step, a coded aperture is placed between the object and the detector. The plane containing the aperture is parallel to the detector and generally its transparency is binary. At this stage only, a shadow casting operation is performed. Each point of the object gives an image onto the detector which is the conical projection of the code from this point. The size of the projected pattern depends upon the distance ℓ_1 from the object point to the aperture plane, but also upon the distance ℓ_2 from the aperture to the detector. These different magnifications lead to a virtual tomographic capability (i.e the ability to reconstruct one object plane in focus while blurring the other emitting slices).

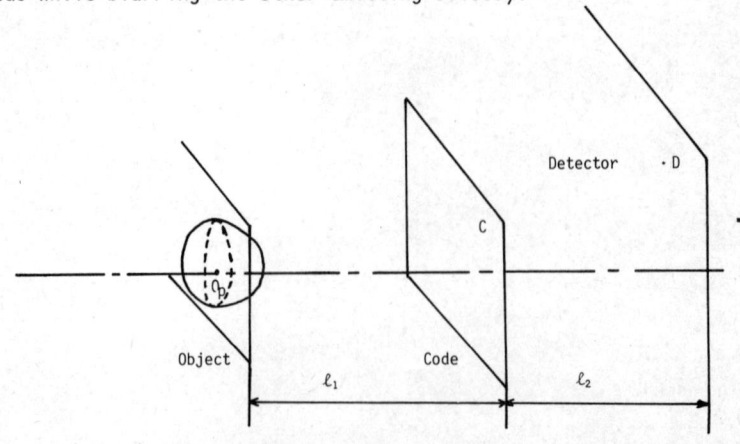

Coded aperture imaging ; encoding step

- The second step or decoding step allows to recover the informations on the three dimensional (3-D) emitting distribution from the coded pattern I(x,y). For a plane object, defined by a luminance function O(x,y), and for a code C(x,y), I(x,y) can be expressed as a convolution product defined by

$$I(x,y) = O(-x \frac{\ell_1}{\ell_2}, -y \frac{\ell_1}{\ell_2}) * C(\frac{\ell_1}{\ell_1+\ell_2} x, \frac{\ell_1}{\ell_1+\ell_2} y) \qquad (1)$$

where * denotes the bidimensionnal convolution.
In the following the magnification factors $-\frac{\ell_1}{\ell_2}$ and $\frac{\ell_1}{\ell_1+\ell_2}$ are implied because we will first study the reconstruction of a plane object and next the ability of the preceeding processing th successfully restore the 3-D information, so that we can write :

$$I(x,y) = O(x,y) * C(x,y) \qquad (2)$$

taking the bidimensionnal Fourier transform of (2), we obtain :

$$\tilde{I}(\nu,\mu) = \tilde{O}(\nu,\mu) \cdot \tilde{C}(\nu,\mu) \qquad (3)$$

The decoding of a planar object would consist of an inverse filtering (i.e obtained by dividing spectrum of the coded image by the spectrum of the code).
But such an exact deconvolution is very sensitive to noise, because, for real apertures, the functions $\tilde{C}(\nu,\mu)$ have bands of very small values. Dividing by $\tilde{C}(\nu,\mu)$ at these points will amplify mainly the noise and will damage the decoded result. For this reason we perform an approached deconvolution using a pseudo-Wiener filter.
Classical imaging can be related to this type of processing. Such an analogy gives us a very interesting approach when we will consider the overall modulation transfert function, the impulse response and the tomographic capability.

II. Lens imaging analogy

A lens imaging process (one stage) in coherent monochromatic light can also artificially be decomposed in a two step process which must be related to the preceeding technic. For that purpose we consider the complex amplitude U(x,y) just before the lens. If a(x,y) represents the complex amplitude of the object, U is given by a Fresnel transform

$$U(x,y) = a(x,y) * e^{\frac{ik(x^2+y^2)}{2d}}$$

Obtained by
- neglecting constant factor $\frac{e^{ikd}}{i\lambda d}$
- assuming $d_1 = d_2 = 2f$ (f focal length)

The second step includes the lens effect and the second Fresnel transform. It is possible to show that this second step is a deconvolution process (if we do not consider the limitation of the diffraction due to the finite extend of the pupil) because it performs a convolution between U(x,y) and $e^{-ik\frac{(x^2+y^2)}{2d}}$. When the system is diffraction limited the phase factor $e^{ik\frac{(x^2+y^2)}{2d}}$ which multiplies a(x,y), in the relation between the object and the image u(x,y).

$$u(x,y) = \int\int_{-\infty}^{+\infty} h(x-x_o, y-y_o) \, a(x_o, y_o) \, e^{ik\left(\frac{x_o^2+y_o^2}{2d}\right)} dx_o \, dy_o$$

does not change rapidly in the extend of the point spread function {1} and so that it can be neglected. In this case we can conclude that the second step performs a deconvolution with in addition a filtering by the complex amplitude $\tilde{h}(\nu,\mu)$ of the pupil of the lens.

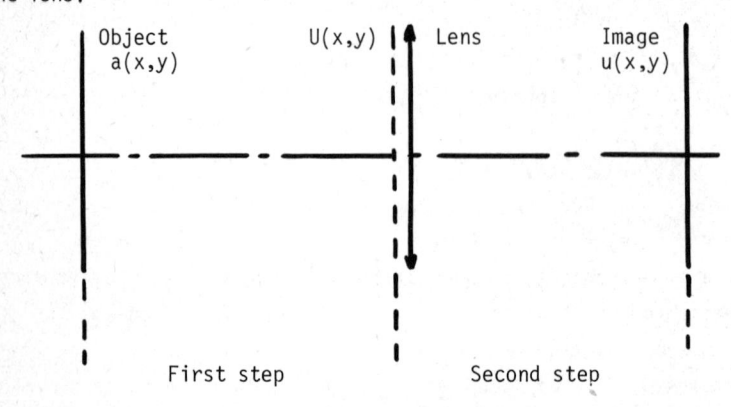

Decomposition of a classical monochromatic imaging process in a two step process

Such an analogy is very useful in practice because it gives the conditions to make images of high quality. But it is not exactly possible to take the same way in coded aperture, because we cannot obtain directly complex or real negative function, so that the synthesis of a $e^{i\alpha(x^2+y^2)}$ coding needs 3 separate encoding steps.
The loss of the dynamic capabilities of single coded images (because no motion of the object, code and detector) makes the multicoding not very used in practice. In the following we present a single recording CAI method using an annular aperture with a pseudo-Wiener deconvolution.

III. <u>Imaging problems in CAI</u>

As we explained before it is better to perform an approached inverse filtering than an exact deconvolution. This can be achieved by replacing the inverse filter $\frac{1}{\tilde{c}(\nu,\mu)}$ by :

$$\frac{\tilde{c}}{\tilde{c}^2 + \epsilon} \qquad (4)$$

The aim of this pseudo-Wiener filter (called pseudo because the conditions of application of Wiener filtering are not satisfied for instance when the fluctuations are coming from Poisson noise processes) is to perform a selective rectification of the frequencies of O(x,y) diffeently attenuated by the encoding stage. Yet,this amplification occurs only when frequencies have been encoded in an important means (in the opposite case we will only amplify noise). As we can see, it is the parameter or "noise parameter" whiwh rules the operation. The filter (4) leads to a modulation transfer function (MTF)

$$\text{MTF} = \frac{C^2}{\tilde{C}^2 + \varepsilon} \qquad (5)$$

The following figure represents the encoding transfer function for an annular aperture used in nuclear medicine.

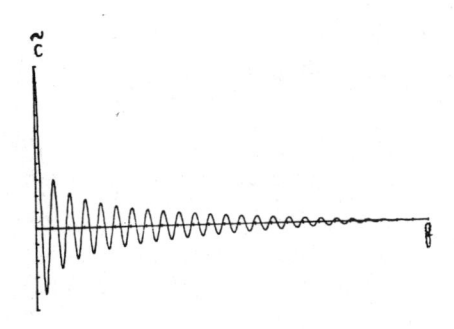

$\tilde{C}$ = encoding transfer function

Moreover, the CAI method makes it possible to reconstruct slices situated in planes parallel to the code. An analysis in the Fourier and in the object domains has shown {2} that the defocused impulse responses kept a nearly constant energy. But the extension of these functions is all the more important as defocusing increases. Such a conclusion is related to the defocusing effect on the point spread functions obtained in classical imaging.

IV. Noise problems in CAI

There exists two types of noise in CAI.
- The first is related to detector noise. This type of fluctuations is very important in infra-red domain. The use of multiplex imaging {3} allows a gain in signal to detector noise ratio which can be related to the Fellgett's advantage in Fourier Spectroscopy.
- But X and γ-ray imaging photon noise coming from Poisson process is the main fluctuation.

  Compared with pinhole imaging for instance a large gain in collection efficiency G (maybe 100 or more) is achieved. But it leads to a gain in signal-to-noise ratio only for 1 emitting point. We can understand this singular result in the following way :
  . if n is the number of photons collected in pinhole imaging and coming from one pixel, the noise is $\sqrt{n}$ and the signal to noise ratio S/B = $\sqrt{n}$.

. In coded aperture the number of photons collected from the same point is Gn, but the noise does not come only from this point. In fact each other emitting point gives, at the reconstruction step, noise in the preceeding pixel. This influence cannot be evaluated in a simple way but it implies a loss in the signal-to-noise ratio compared with $\sqrt{Gn}$ (in fact $\sqrt{Gn}$ is obtained if there is one emitting point). We have shown (4) that in nuclear medicine configurations a gain of 2 or 3 nevertheless remains. Such a factor will not be important in laboratory experiments but it allows a reduction in exposition time from 4 to 9 in "in vivo" medical studies.

V. CAI applied to nuclear medicine

In nuclear medicine the classical images consist of parallel or conical projections of the 3-D distribution of the radiotracor injected to the patient. The superimposition of the activities of the slices is one of the main limitations in the diagnosis. CAI is one way which can be used to extract a certain 3-D information.

We have mainly applied this method in heart imaging. In this field the radioisotope used is 201 Thallium which emits γ rays of 70 keV. From the interpretation of the results we obtained, one can expect that CAI, which provides a contrast enhancement, will improve the sensitivity in detecting homogeneities of the tracor distribution. Also it increases the degree of accuracy in determining 3-D site and extent of the radioactivity defects corresponding to necrotic or ischaemic tissues. On the other hand, if the ECG (electrocardiogram) signal is used to drive the gamma camera, so that photons are registered during the sytole or the diastole, it is possible to obtain tomographic images, that allows more accurate evaluation of dimensions of the carties and of wall thickness. The following pictures show the comparison between classical projections and four reconstructed slices in a patient study made in the Nuclear Medicine Division of Hospital Cochin in Paris.

Conclusion

As we say at the begining, coded aperture imaging is a general method not limited to nuclear medicine. It can be used in several domains such as neutrons, α particles imaging. We begin to develop it in microimaging for laser induced plasmas diagnosis. The first results that we obtained in this field show a resolution of 10 µm and we hope rapidly a gain of 2 to 4 with regard to this number.

References

{1} J.W. GOODMAN-"Introduction to Fourier Optics, Mc Graw-Hill (1968).
{2} J. BRUNOL, N. deBEAUCOUDREY, J. FONROGET, S. LOWENTHAL- Opt. Comm 25 (1978)163.
{3} J. APPEL, A. GIRARD - Nouv Rev. Opt. (1976)221.
{4} J. BRUNOL, N. de BEAUCOUDREY, J. FONROGET - To appear in Opt. Comm.

Mr K... Abraham Scintigraphies ^{201}Tl

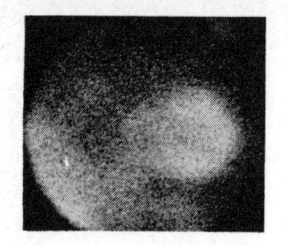

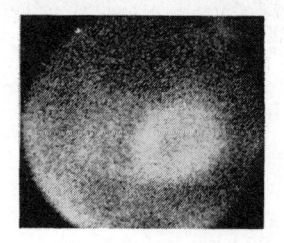

ANT VUE LAO VUE

Mr K... Abraham Coded Aperture Imaging
Reconstruction of 4 slices separated by 8mm.

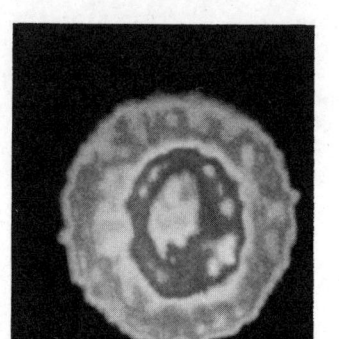

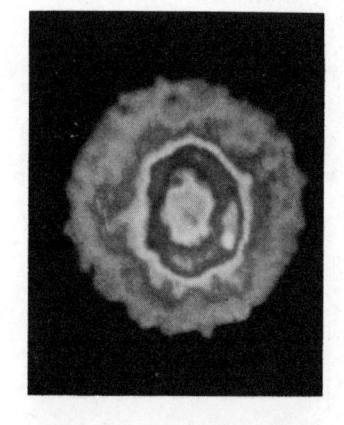

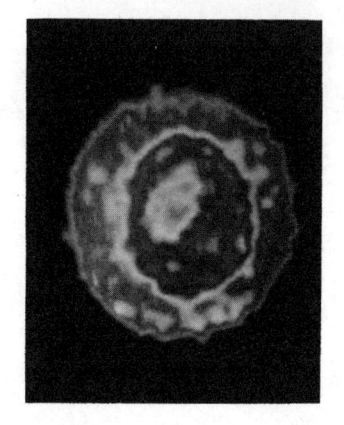

PRINCIPLES AND TECHNIQUES OF ACOUSTICAL IMAGING

Mathias FINK

Laboratoire de Mécanique Physique, Université Paris VI
2, Place de la Gare de Ceinture. 78210 St-Cyr l'Ecole.

INTRODUCTION

The ability to get good images when using ultrasound as a source of data is due to a certain number of suitable conditions which are found in biological tissues. Most of the tissues, other than bones, have a density ρ approaching that of water, and the propagation speed of the ultrasound waves c in this media is practically constant ($\simeq$ 1540 m/s), so that accurate depth ranging by echo delay time is achievable in pulse echo systems. Moreover, the relative uniformity of the acoustical impedance ($Z = \rho c$) of the tissues to be explored explains why it is possible to perform an exploration in depth. Low-level reflections occur between different soft tissues ($\Delta Z \ll \bar{Z}$), so that most of the acoustic energy is transmitted through the different interfaces and it is available for imaging deeper structures. The depth of penetration is mostly limited by the absorption of ultrasound waves which increases with frequency, and not so much by the number of interfaces which may be high. Frequency of the order of 3 MHz are generally used to achieve exploration over 20 cm.

I. THEORY OF IMAGING

Development of imaging techniques involve to choose some model for the propagation of ultrasound in a 3 D biological structure. The acoustical properties of tissues enable us to adopt as a simple model, the one of the linear acoustics in a fluid. The acoustic field is derived then from a potential Φ which obeys for a homogeneous medium, and when neglecting absorption, the wave equation :

$$\Box \Phi = 0 \quad \text{where} \quad \Box = \Delta - \frac{1}{c^2} \frac{\partial^2}{\partial t^2} \quad (1)$$

The presence of non homogeneous properties associated with the explored tissues will modify this wave equation by introducing an ensemble of source terms which are a function of the local variations of density, compressibility and attenuation and which may be considered as an ensemble of scattering centers.

$$\Box \Phi = S(\Phi, \delta\rho, \delta\chi, \delta\alpha) \quad (2)$$

The purpose of acoustical imaging is to reconstruct the map of all these scattering centers with the best accuracy. This is why we choose to submit the object to be examined, to a known acoustic beam $\Phi^°(\vec{r},t)$, and from the acoustic field $\Phi(\vec{r},t)$ that we observe, we try to extract the source terms S.

If we accept the slight disturbing characteristic of the non homogeneous properties presented to the incident field $\Phi^°$ by the tissues submitted to the beam, it is possible to simplify the equation (2) by keeping in the development in the source terms with respect to $\Phi^°$ only the term of order 0.

$$\Box \Phi = S (\Phi = \Phi^°, \delta\rho, \delta\chi, \delta\alpha) \qquad (3)$$

This classical approximation of a one diffusion process [1] allows then to completely separate the effect of different sources. The ensemble of scattering centers creates a potential Φ^1 which, added to the irradiation potential $\Phi^°$, will give the value of the observed potential $\Phi = \Phi^° + \Phi^1 \quad (\Phi^1 << \Phi^°)$

$$\Box \Phi^° = 0 \qquad (4')$$

$$\Box \Phi^1 = S (\Phi^°, \delta\rho, \delta\chi, \delta\alpha) = s(\vec{r},t) \qquad (4'')$$

Such an approximation which is equivalent to disregarding the effect of the non homogeneous properties on the potential diffracted by each source is certainly inaccurate in the areas containing some bone structure or air pockets because their impedance is extremely different from the impedance of most of the biological tissues. However, it is a very convenient hypothesis to the extent that the desired acoustic data can be acquired outside the irradiated domain in terms of the acoustic field Φ^1, and this, through a linear equation (4") which allows one to find a solution under the form of an equation of convolution because of its invariance properties (in a free space) :

$$\Phi^1 (\vec{r},t) = s(\vec{r},t) \otimes g(\vec{r},t) = \iint s(\vec{r}_o,t_o) g(\vec{r}t|\vec{r}_o t_o) d\vec{r}_o dt_o \qquad (5)$$

The effect of each point source can be made evident through the term $g(rt|r_o t_o)$: Green function associated with the equation (4") which is nothing else than the perturbation Φ^1 created by a point source located in $\vec{r}_o$ excited by an impulse $\delta(t_o)$. The theory of diffraction [1] teaches us that in a free space this function can be written under the form :

$$g(\vec{r}t|\vec{r}_o t_o) = \frac{1}{4\Pi R} \delta(t - t_o - \frac{R}{c}) \quad \text{where } R = |\vec{r} - \vec{r}_o| \qquad (6)$$

This is the classic choice of the solution of delayed potential of equation (4") which expresses the fact that each point source creates a perturbation which travels at a speed c under the form of a diverging wave front.

The seeking of the reverse convolution operation which will allow to recover the ensemble of the source terms $s(\vec{r},t)$ is going to be made easier by the two following remarks :

a) on the one hand, it is not necessary to know $\phi^1(\vec{r},t)$ in the total space : the Huygens principle and its various formulation [1] teaches us that the value of the field ϕ on a given plane π located outside the source area is sufficient to determine in all points in the half space $z > z_\pi$. Such a fact will allow us to only make a two-dimensional spatial study, limited to the exploration of a single plane π :

$$s(\vec{r},t) = \phi^1_{\vec{r} \in \pi}(\vec{r},t) \otimes g^{-1}(\vec{r},t) \tag{7}$$

We shall see later [II-III] the fundamental part played by the location of this reception plane relative to the irradiation source ϕ° (transparence imaging or echography).

b) On the other hand, the operation consisting of going back to the different source terms, starting from the field $\phi^1(\vec{r},t)$ observed on the plane π, can be seen as a true reversal of the time progression, which would bring back each of the diverging wave fronts to its initial source ; utopian operation that can be replaced rightly with a reconstruction of $s(\vec{r},t)$ in delayed time, because only the spatial relationship of the source terms is of interest to us. This operation of deconvolution which must associate to the diverging wave front coming from a source located in $\vec{r}_o$ the distribution $\delta(\vec{r}_o)$ can then be understood as a parallel processing of the data, similar to the one that is sequentially performed by scanning a plane π with a spherical ultrasonic transducer (for example, a spherical cup of piezoelectric ceramic) of "infinite" aperture and of radius $z_o - z$ (z_o and z are the depth coordinate of the source and of the plane. This type of transducer, because of its spherical geometry, delivers an infinite signal every time that a source located at the distance $z_o - z$ aligns itself on its axis. Mathematically, this spherical geometry can be considered as equivalent to a spherical correction of delay associated to a plane transducer. Such a correction of delay can be written under the form of a convolution by a distribution $g^{-1}(\vec{r},t)$, which is equal to

$$g^{-1}(\vec{r},t) \propto \delta(t - T + \frac{R}{c}) \tag{8}$$

where T is a time delay required to provide a causal characteristic to the function $g^{-1}(\vec{r},t)$.

The acoustic image of the object under study is then directly related to the spatial portion of the function $s(\vec{r}',t)$ which is equal to :

$$s(\vec{r}',t) = \iint_{\vec{r} \in \pi} \phi^1(\vec{r}, t - T + \frac{|\vec{r} - \vec{r}'|}{c}) \, d\vec{r} = \phi^1(\vec{r}',t) \otimes g^{-1}(\vec{r}',t) \tag{9}$$

A parallel processing of this kind, although being worth considering at first, once the maping of the field $\phi^1(\vec{r},t) = \phi^1(x,y,t)$ is obtained by means of numeric processes, is however excessively cumbersome. On the one hand, the amount of data to store is tremendous, that is to say : a three dimension table (two for the space and one for the time) where the number of data by dimension is larger than a hundred ; the frequency of the temporal sampling and the pitch of the spatial sampling being tied to the duration and the shape of the irradiation pulse. On the other hand, the numerical decoding of such a vast quantity of data is still an operation much too cumbersome, which could not be conducted and still maintain real time.

It is possible to consider replacing the digital processing with a purely analogic processing which would be performed in real time, by means of acoustic lenses. It is well known that a lens of this type is able to perform in real time the desired decoding operation. Its behavior is one of a time delay device able to reverse the order of arrival of the data conveyed by the various wave fronts coming from the object. It transforms, in this way, the diverging wave coming from a point source in a converging wave focusing at the point image. This stigmatism allows us to produce, in a natural way, the image of the object, plane by plane. However, this kind of processing is limited because of the difficulties encountered in the manufacturing of either quality ultrasound retinas or wide aperture acoustic lenses. The most successful lens imaging system is the Green camera [2] which uses a sophisticated acoustical lens and a system of rotative prisms, which permits a fast mechanical scanning of each plane of the object (10 frames per second) in front of a linear array of 192 ultrasonic transducers.

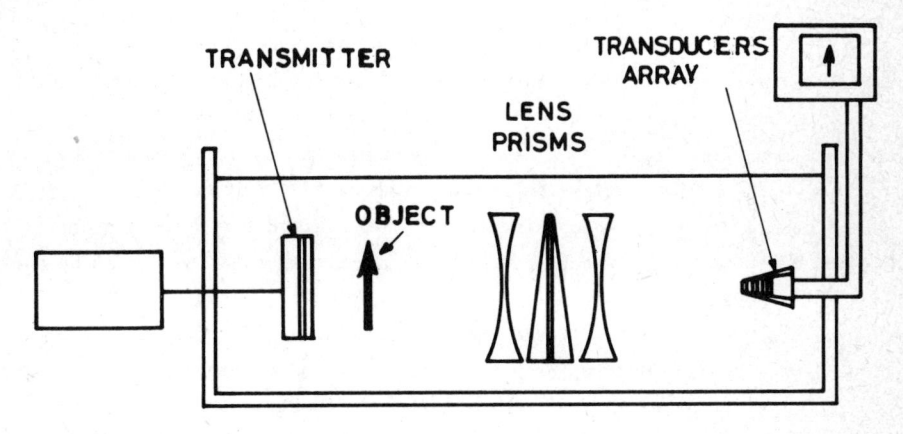

Fig. 1 - Real time acoustical imaging with a linear array and mechanical scanning of the focused image using rotative prisms

Besides these classical imaging methods, different attitudes have been retained in the field of acoustical imaging.
- The first attitude, originating from the spectacular results obtained in coherent optics, consists of illuminating the entire object with a monochromatic signal. This is the domain of Acoustical Holography.
- A second attitude consists of coming back to the use of brief signals and working

in echography, but replacing a global illumination of the object with a more selective illumination by only illuminating one plane, one line, or even only one point of the object. This increase in selectivity produces a better image quality but at the same time increases the image acquisition time.

II. ACOUSTICAL HOLOGRAPHY

The visible advantage of a monochromatic illumination is primarily found in its invariance properties which entail that in each observation point, the same monochromatic structure is found, of which the complete knowledge then requires only two parameters : the amplitude A and the phase ψ, which, because of the linearity of piezoelectric transducers, can be easily obtained.

It must be noted that in monochromatic lighting, the temporal information associated to each point source, is then replaced by a simple phase information. The spatio-temporal form $g(\vec{r}t|\vec{r}_0 t_0)$ (6) which was associated with each source in pulsed mode is reduced here to the spatial form :

$$\frac{\exp(ik|\vec{r} - \vec{r}_0|)}{4\pi|\vec{r} - \vec{r}_0|} \qquad (10)$$

The recognition of each of the point sources is then made through the signature in phase of the perturbation that it retransmits, and the deconvolution process requires now only to work in the spatial domain (which was not the case in (9)). In the case of Fresnel approximation, such a deconvolution process is nothing more than a Fresnel transform which may be realized by various methods.

Some are purely analogic, which after having transferred the phase and amplitude information of the acoustic waves under a form permitting their detection by means of an optical coherent beam, are using optical decoding processes [3,4,5,6].

Among the various solutions offered, mention must be done of the levitation effect of a liquid gaz interface due to the ultrasonic radiation pressure (Fig. 2). The optical reconstruction performed in real time with the diffraction of a laser beam by the distorded interface, realizes the required Fresnel transform. This decoding process is however degraded by some important aberrations due to the large difference between the optical wavelength and the acoustical wavelength. Some other decoding methods, purely digital, associated to the exploration of the detection plane by one or more transducers, allows us to avoid these aberrations at the expense of the data acquisition time of these images [7,8].

It must be understood that until now, even if some good images of simple test patterns have been obtained, the results obtained when observing "in vivo" organs have been disappointing, despite the use of reconstruction processes apparently ideal (large aperture, sufficient sampling, high dynamics).

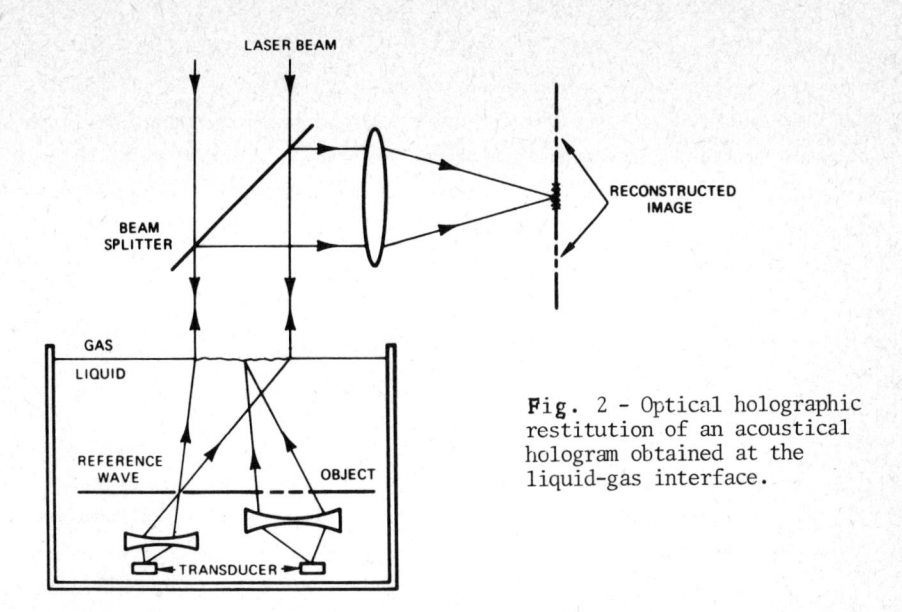

Fig. 2 - Optical holographic restitution of an acoustical hologram obtained at the liquid-gas interface.

This constatation can, in fact, be explained by noticing that in acoustical holography, the over all illumination of the object implies that the recognition of each of the point sources will be made in the presence of an important noise coming from the ensemble of the other points of the object. This interfering noise (speckle noise) decreases in practise the resolution power in a drastic way especially when the studied object contains many scattering centers. Besides the use of a monochromatic signal which can only provide locally two informations (A and ψ) no matter what the complexity of the object to be studied, is not a good choice for a source illumination when studying organs of complex structure. The decoding of the coordinates of each of the sources in the presence of speckle noise is made more efficent when the "signature" of each of the sources on the detection plane carries more information. By enlarging the detection aperture, one can hope to gain access to a larger quantity of information. However, in acoustics, the difficulties encountered to manufacture ultrasound retinas of large dimension limit, in fact, the apertures used, to some dimensions of the order of a few hundred wavelength. In the best cases, the $2N^2$ data acquired in acoustical holography, by means of a matrix retina sampled by N^2 transducers (where N as of today is no greater than 300 [9]) are not enough to reconstruct with accuracy the three dimensions of an object of reasonable size. It must be noted here that this problem is not as acute in optics where the apertures used contain a much larger number of wavelength ($\simeq 5.10^4 \times 5.10^4 \; \lambda^2$) which allows us to obtain much richer information on some objects that, it must be pointed out, are not truly three dimensions objects. In optics, in most cases, only the surface of the objects diffuses the light, and the information that it delivers and that we try to decode, as only in fact a bidimensional characteristic ; whereas in acoustics, the relative transparency of organs

to ultrasound beams of a few MHz makes them true tri-dimensional objects.
In front of these findings, the use of a temporal coding richer for illumination appears to be required in acoustics. Besides the "radar type" signals having a high compression ratio (BT≫1, B being the bandwidth and T the signal duration) which is able to provide a large quantity of data, but which requires some very cumbersome temporal decoding techniques, it has become evident that the easiest way to enrich the acoustical information was to come back to the use of brief illumination signals, while working then in echography.

III. ULTRASOUND ECHOGRAPHY

To work in the echographic mode consists in choosing for a reception plane, a plane identical to the transmission plane, while using a brief signal for illuminating the object. One gains additional information to locate the different scattering centers, taking into consideration the arrival time of each of the wave fronts diffused by each of these sources.

For an infinitely short duration signal emitted at the time $t = 0$, the signature of a given target located in $\vec{r}_o$ ($z_o, \vec{m}_o$) at the level of the reception plane ($z = 0$) varies except for an amplitude factor according to :

$$\delta (t - z_o/c - | \vec{r} - \vec{r}_o | /c) \qquad (11)$$

An expression that can be made simpler by considering it in a Fresnel paraxial approximation according to :

$$\delta (t - 2z_o/c - | \vec{m} - \vec{m}_o |^2/2z_o c) \qquad (12)$$

and which explicitely demonstrates that it is possible to go back to the coordinates of a point source not only by means of the radius and the curvature center of the wave front that is associated with it ($\rho = 2z_o c, \vec{m}_o$), but also by taking into account the arrival time of the beginning of this wave front at the level of the reception plane ($\tau_o = 2z_o/c$), time which is dependant, in an univocal way, of the depth of the target. This information is to be compared to that obtained in transmission imaging where for a fixed distance between the transmitter and the receiver equal to L, each of the sources is observed through an expression of the form

$$\delta(t - L/c - |\vec{m} - \vec{m}_o|^2 /2z_o c) \qquad (13)$$

where the arrival time L/c of the various wave fronts on the reception aperture is independant from z_o. It is an information not as rich as in echography, which only gives the distance z_o through the wave front curvature. The signals coming from the different scattering centers arrive together at the reception plane and introduce a speckle noise very undesirable for the decoding process. In echography however, the natural decoding of the z coordinate brings an important decrease of this noise, which is coming from a slice of the object and which is the thinnest when the illumination signal is the shortest. In fact, experimentally, the illumination signal i(t) is far from being similar to the infinitely short duration of the illuminating signal δ (t) and the signature of the target at the level of the reception plane must be considered as being the convolution of the distribution (12) with i(t) .

$$i (t - 2z_o/c - |\vec{m} - \vec{m_o}|^2 /2z_o c) \qquad (14)$$

where i(t) has a form and a duration directly in relation to the characteristics of the emitting source. It is possible to obtain some illumination time in the order of the microsecond, when one uses well damped transducers ; this compared to the 200 μ s required for exploring a depth of 15 cm gives an idea of the amount of data that can be obtained.

It must be well understood that the possibility of using the echographic information in order to visualize small dimension structure ($\simeq$ mm) is a specific acoustical property. It is the very slow velocity of the ultrasound, as compared to the electro-magnetic waves, that permits us to realize without difficulties the separation of echos coming from very close targets (1,3 μ s for a distance of 1 mm between two targets).

A - "Parallel" echographic processing.

The choice of an echographic exploration mode, although it appears desirable, brings in acoustics a certain number of problems. The fundamental problem is, of course, related to the decoding operation that must be performed in order to reconstruct with accuracy the mapping of the scattering centers. If one wants to take advantage of the natural decoding of the coordinate z_o, brought forth by the echography, real time decoding processors must be developed that perform a deconvolution operation at the same time the echos are coming back, by means of a time delay nucleus (8) which is adapted at all times to the cuvature of the observed wave fronts. A decoding process of this type can be perceived as a real "zoom effect" that will be accomplished for example by means of an acoustical lens with ultrafast variable focus, located on the way of the reflected beam. One would observe then, at the level of a single detection plane, the sequence of the images of each of the planes illuminated by the acoustical exploring beams ; the rapid change of focus ($\Delta f/ \Delta t \simeq 3mm/\mu s$) permitting us at all times to focus on the image of the plane of which the echos are arriving on the detection plane. If it is true that such a decoding operation is perfectly utopian when classical acoustical lenses are used, one can imagine substituting the delay effect of such a lens by the action of a delay line network connected to a transducer array which will be used as the imput face of this lens. The ensemble of the delay electrical signal then will excite a second tranducer array which, acting as the output face of the lens, would deliver in a second propagation medium, an acoustical beam which,

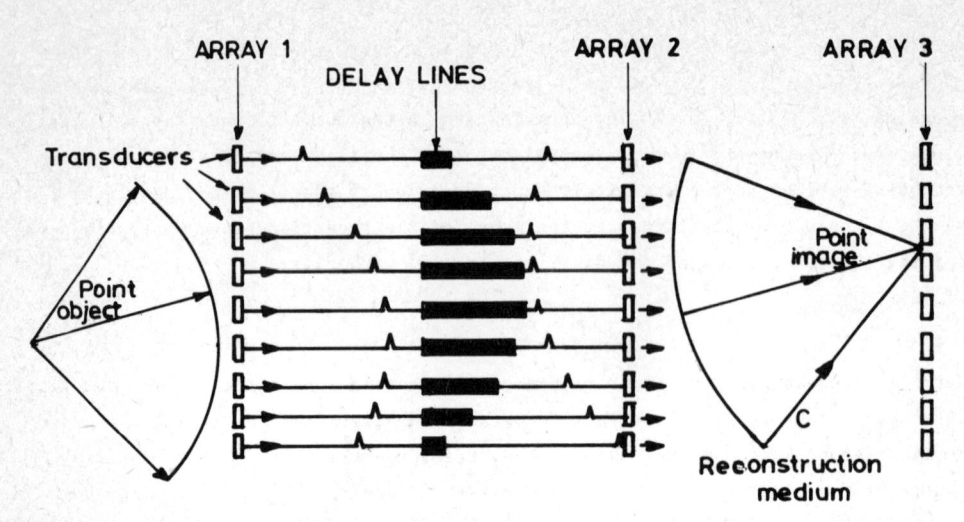

Fig. 3 - Electronic lens with delay lines and 2 transducers arrays, the third array is used to observe the acoustical image.

through the propagation effect, will create the acoustical image of the plane that is to be observed (Fig. 3). The instantaneous focus of such a system depends on the one hand on the distance z' at which one seeks to restitute the images of each plane of the object and, on the other hand on the depth $z = ct/2$ of the target observed at the time t

$$f = \frac{z\,z'}{z' - z} = \frac{ct\,z'}{2z' - ct} \qquad (15)$$

The quadratic correction of delay that must be applied by the electronic delay lines in order to simulate this focusing will be a function of the time according to an hyperbolical law

$$\Delta\tau\,(\vec{m},t) = -\frac{\vec{m}^2}{2cf} = -\frac{\vec{m}^2}{c^2}\left(\frac{1}{t}\right) + \frac{\vec{m}^2}{2cz'} \qquad (16)$$

This solution which would permit access to the tri-dimensional structure of an object during a single ultrasound "shot", no matter how attractive it looks, is, as of today, too complex technologically to be carried out. Whether it is the making of a very well damped two dimensional transducer array or the numerous use of analog controlled delay lines (for example, CCD type) while being extremely expensive, do not have a performance satisfying the requirements of an echography (high dynamic > 80 db), or also the display of tri-dimensional information, the technological problems that remain to be solved are enormous. In fact, until now, the echographic experiments have always been limited to a much more selective exploration of the objects, by illuminating either one plane or one line of the object during each ultrasonic "shot". The limitation to 2 dimensions of this type of parallel processing has been investigated by TORGUET and

BRUNEEL [10,11]. The principal interest of this parallel processes is the speed in forming an image ; each B mode echographic plane (the B plane is determined by the direction of the array and the direction of the ultrasound beam) is obtained in less than 300 µs. However, if this image rate seems interesting, one must think that in the case of a complex structure, the decoding of the coordinates of each target among the noise coming from other targets located at the same distance of the array, can deteriorate the resolution power in a significant manner.

B - Sequential echographic processing.

Until now, the sequential exploration processes of B mode echographic planes have given the results more exploitable by physicians. In these technics, one uses an even more selective illumination of the object. An ultrasound "shot" allows the formation of the image of only one line of the object and not of all the plane. To form the image of a line, after having selectively illuminated a section of the object with a relatively narrow ultrasound beam, one applies a focusing process which during the reception phase is going to select among the echos, the ones that comes from the line that is to be displayed. The image of a slice in B mode echography is then obtained sequentially, line by line, during a period that can last, in the case of the manual scanning of a transducer, some 10 seconds (Fig. 4). In the case of the electronic scanning of a transducer's array, this period can be reduced to less than a 1/50 of a second.

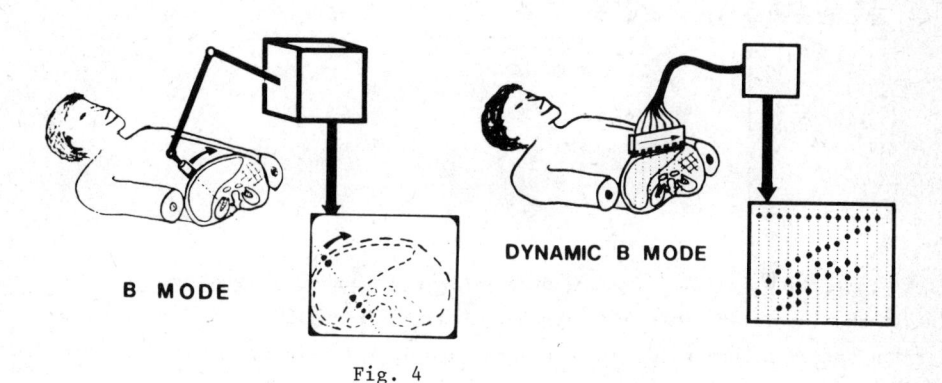

Fig. 4

In fact, practically all the echographic research has been carried out on the development of the sequential exploration systems. On the one hand, the decoding processes to set up are much simpler to make than in the parallel processing technics. On the other hand, the selective illumination by a relatively narrow beam largely reduces the spatial interference noise encountered with the other technics, in the case of the visualization of objects having a complex structure. In other words, by accepting N shots in order to form the image of a plane, one substitutes the deconvolution operation that was to be performed during a single ultrasound "shot" in the "parallel" technics, with a "deconvolution" operated in a sequential manner which reduces, in a substantial way, the problems of interference noise.

Besides the technics using a single aperture transducer, one tries to have it focus along a caustic axis long enough [12,13], but which must be moved manually or mechanically in order to get the image of the B mode echographic plane, one has tried, obviously to use transducers' arrays, associating with the electronic scanning, the synthesis of a focusing aperture. The problem to solve then is the making, behind the transducer array, of the electronic filters that are adapted to the curvature of the diverging waves coming from the target under illumination. Such a filter which simulates the action of a cylindrical transducer with a varying focus, must bring a cylindrical correction of time delay to the signals delivered by the aperture transducers, then through the summation of these delay signals, must only produce a strong signal for each of the targets being studied (Fig. 5).

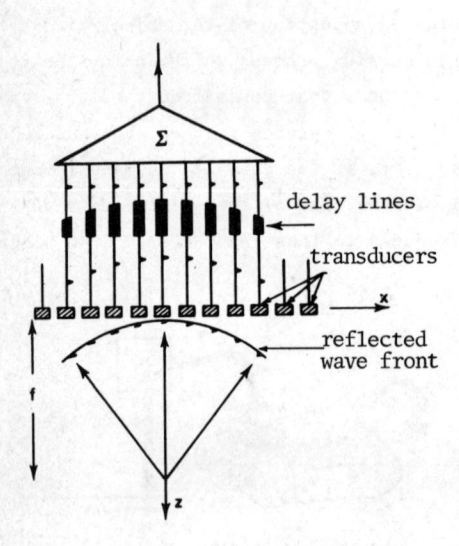

Fig. 5 - Electronic focusing during reception with delay lines.

At this time, many dynamic focusing processes have been developped that are using rather than time delay lines made of analogical shift registers (which as we know are not satisfactory enough to be used in echography), some digital time delay lines which permit one to obtain only a limited number of focal distances during an ultrasound "shot", but which, because of the relatively small aperture retained, will however give a good covering of the various focal zones. Moreover, in these systems, the use of time delay lines during the transmission can permit, when only a limited number of transducers is used, a scanning of the plane by sectors rather than linearly(Fig.6). By applying a transmit impulse slightly shifted in the time from one transducer to the other, one orientates the angle θ of the transmit beam according to a law :

$$\Delta\tau_T(x) = (tg\ \theta)\ x/c \qquad (17)$$

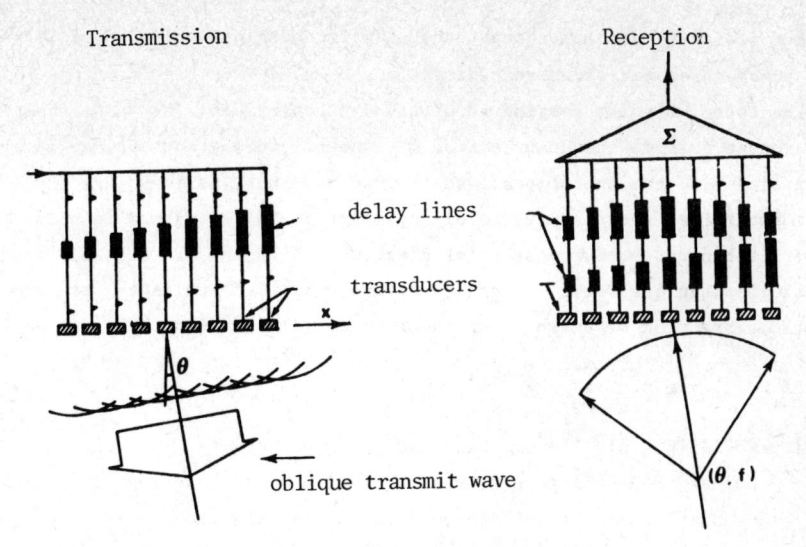

Fig. 6 - Electronic beam steering at transmission
and oblique focusing during reception.

that can be changed from one "shot" to the other. One uses then during the receiving, an oblique focusing process which combines simultaneously the linear delay corrections and the quadratic delay corrections needed to recognize the target oriented at the angle θ (Fig. 6).

$$\Delta\tau_R(x) = \frac{tg\theta}{c} x - \frac{1}{2\ cf} x^2 \qquad (18)$$

Such a device has been developped for the first time by THURSTONE [14] that uses an array of 32 transducers, which allows the focusing on 5 different areas, but requires having at its disposal a large choice of time delays. It is a device still very complex and expensive.

More recently, in order to limit the number of time delay lines used, one has become interested in the linear scanning of the arrays containing a larger number of transducers (approx. 100). By limiting the number of transducers located in the focusing aperture at less than 20, an analogic multiplexing then allows the restriction of the number of time delay lines to a reasonable value [15]. However, in this last realizations, the focusing aperture remains relatively small. In the technic quoted in reference, the use of 16 transducers at 3 MHz with a pitch of 1 mm, limits the angular aperture of reception at less than 10 degrees for targets located at 10 cm, which means that a lateral resolution of less than 3 mm at 6 db cannot be obtained.

In absence of a simple and economic way for synthetizing large focusing apertures with time delay lines, it has been thought that the focusing possibilities which are specific to the monochromatic waves and which since FRESNEL have been developped in optics, are in fact especially interesting in ultrasound echography. One knows that when considers a monochromatic wave of pulsation ω , the time delay concept can be systematically replaced with the one of phase shift. One can then, as in holography, choose to recognize the target under study, solely by means of the signature in phase of the waves that it generated. The most total electronic decoding process will consist then in substituting the quadratic correction of delay apllied behind the transducers, with a quadratic correction of phase under the form of :

$$\Delta \psi (x) = \omega \Delta \tau (x) = - \alpha x^2 \qquad (19)$$

The ability to define this law of phase modulo (2 Π) permits, in addition, the restriction of the number of dephasing networks required for the focusing. This "Fresnel" focusing mode, similar to the one obtained in optics with the kinoform lenses, appears to be extremely interesting in B mode echography where the signals used are necessarily brief. In this case, if the quadratic correction of phase made electronically is independant from the pulsation ω, that is to say that the coefficient α is a constant, the echos of the target will be decoded by such a lens on an appreciable depth of focusing. The quadratic correction of phase is in fact equivalent to a delay correction being different for each of the frequency components of the reflected signal.

$$\Delta \tau (x) = \frac{\Delta \psi (x)}{\omega} = - \frac{\alpha}{\omega} x^2 \qquad (20)$$

Therefore, one creates, by means of a single array of dephasor networks, a curvature of the delay law and a focal length that will be different for each of the echographic frequencies.

$$f = \frac{1}{2 \alpha c} \omega \quad \leftrightarrow \quad \Delta f = \frac{1}{2 \alpha c} \Delta \omega \qquad (21)$$

One reaches then, a system able to focus on a large depth, without having the need to modify the reception network in proportion to the echo return provided that the echographic signal as a wide spectrum Δ ω . It is an undeniable advantage of this focusing mode, which also can be used for the transmission in order to make a very narrow ultrasound exploring beam on a considerable depth [16]. The double focusing, at the transmission as well as at the reception, improves greatly the lateral resolution of the system. Even if this technique looks attractive, the making of frequency non-dependant dephasors is not an easy process. In fact, such dephasors can be perceived as dispersive lines, with a linear frequency dependance of the phase velocity $v_\phi = \beta \omega$. It must be noted that a "chirp" convolver (as the one used in radar technics) is able to realize this operation. However, high dynamic (80 db) convolvers of this type are difficult to obtain. Therefore, a strong phase sampling of the Fresnel law (19) is necessary ; only the phaseshift of 0 or Π can be obtained easily on wide frequency

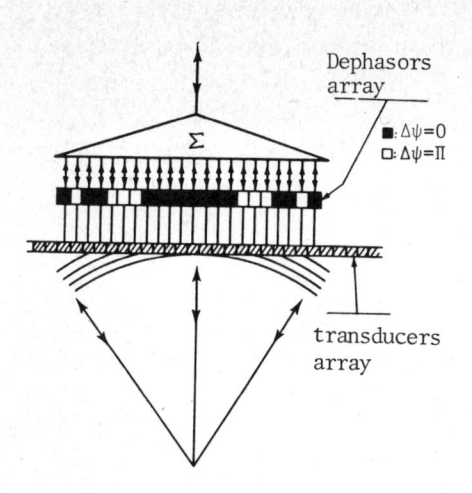

Fig. 7 - 2-state Fresnel focusing

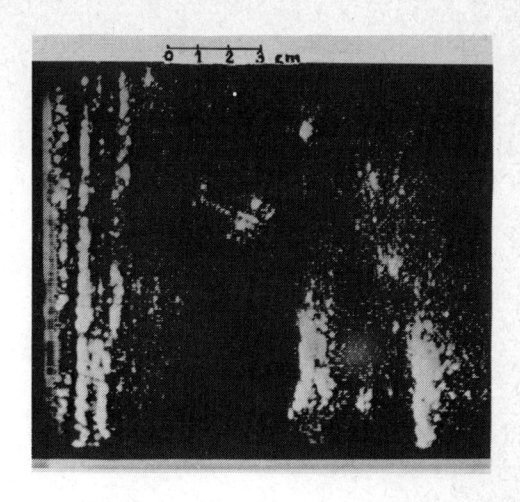

Fig. 8 - B echography of a 12 weeks fetus. Details of the vertebral column.

bandwidth. Each transducer is thus connected with a simple electronic inverter. It is a question of either keeping the signal as it has been received from the transducer, or processing it with an inverter, which will shift its phase of Π (Fig 7). The simplicity resulting from the use of electronic inverters instead of time delay lines, allows one to obtain easily very large focusing aperture sampled by an important number of transducers. We have developped such a Fresnel focusing technique from a linear array of 160 transducers [16, 17]. The focusing aperture is obtained from a group of 64 transducers. Electronic scanning of this aperture along the array allows us to obtain 25 frames of 160 lines per second with a lateral resolution of the order of 2 mm.
The clinical examination technique is simple and consists of putting the transducer array in contact with the skin by means of a coupling jelly and by changing its orientation in order to systematically explore the area of interest. When the probe is in a fixed position, the image observed on the screen is a slice of the organs in motion : for example, the motion of the liver or the kidneys when breathing, the change of caliber of the vena cava, the beats of the abdominal aorta, as well as the motions or the heart beats of the fetus. The excellent lateral resolution obtained allows us to observe some details that in practice were not easily distinguishable. On very young fetuses (8 to 12 weeks), it is possible to watch cardiac motion and to see vertebrae that are separated by less than 2 mm (Fig. 8).

In order to conclude this study on acoustical imaging, we must notice that, compared to the deceiving results of acoustical holography, echographic devices have proved their greater versatility specially after the development of the dynamic B echography.

These devices have already largely benefited from Fresnel focusing and holographic concepts to improve the quality of imaging and their lateral resolution particulary.

REFERENCES

1   MORSE, INGARD - Theoretical acoustics. 1968 , Mac Graw Hill
2   GREEN P.S., SCHAEFER L.F., JONES P.D., SUAREZ J.R. - A new high performance ultrasonic cmaera. Acoustical Holography, Plenum, 5 , 1974, 493-503
3   MUELLER R.K., SHERIDON N.K. - Sound holograms and optical reconstruction. Appl. Phys. Lett. , 9 , 1966, 328
4   MUELLER R.K. - Acoustic holography. Proc. of the I.E.E.E. 59, 1971, 1319-1335
5   MUELLER R.K., KEATING P.N. - The liquid gas interface as a recording medium for acoustical holography. Acoustical Holography, Plenum, 1, 1969, 49-56
6   SMITH R.B., BRENDEN B.B. - Refinements and variations in liquid surface and scanned ultrasound holography. I.E.E.E. Trans Sonics Ultras. SV 16, 1969, 29
7   BOYER A.L., HIRSCH P.M., JORDAN J.A., LESEM L.B., VAN ROOY D.L. - Reconstruction of ultrasonic images by backward propagation. Acoustical Holography, Plenum, 3, 1970, 333
8   KEATING P.N., KOPPELMANN R.F., MUELLER R.F., STEINBERG R.F. - Complex on axis holograms and reconstruction without conjugate images. Acoustical Holography, Plenum, 5, 1974, 515-526
9   ALAIS P. - Real time acoustical imaging with a 256 x 256 matrix of electrostatic transducers. Acoustical Holography, Plenum, 5, 1974, 671-684
10 TORGUET R., BRUNEEL C., BRIDOUX E., ROUVAEN J.M., NONGAILLARD B. - Ultrafast echotomographic system using optical processing of ultrasonic signals. Acoustical Holography, Plenum, 7, 1976, 79-85
11 BRUNEEL C., NONGAILLARD C., TORGUET R., BRIDOUX E., ROUVAEN J.M. - Reconstruction of an acoustical image using an acousto-electronic lens device. Ultrasonic, 15, n° 6, 1977, 263
12 BURCKHARDT C.B., GRANDCHAMP P.A., HOFFMANN M. - Methods for increasing the lateral resolution of B scan. Acoustical Holography, Plenum, 5, 1974, 391
13 AUPHAN M., GOUDIN R., DALE G. - Improvment of echographic diagnosis through tracking focusing process. Proc. of Biosigma 78, 122-127 ; Paris.
14 THURSTONE F.L., RAMM O.T. - A new ultrasound imaging technique employing two dimensional electronic beam steering. Acoustical Holography, Plenum, 5, 1974, 249-259
15 POURCELOT L., BERSON M., RONCIN A. - Real time high resolution ultrasonic imaging. Proc. of Biosigma 78, 116-120 ; Paris
16 ALAIS P., FINK M. - Fresnel zone focusing of linear arrays applied to B and C echography. Acoustical Holography, Plenum, 7, 1976, 509-522
17 FINK M. - Theoretical aspects of the Fresnel focusing technique. Acoustical Imaging, Plenum, 8, 1978 , 149-164

NMR IMAGING

I. L. Pykett, P. Mansfield, P. G. Morris, R. J. Ordidge and V. Bangert
Department of Physics
University of Nottingham
University Park
Nottingham NG7 2RD
England

INTRODUCTION:

The first Nuclear Magnetic Resonance (NMR) experiments to reveal internal structure in a specimen were reported independently by Lauterbur (1) and Mansfield and Grannell (2) in 1973. Since then, a number of small- and intermediate-scale NMR images have been produced by a variety of methods, and in particular those of the live human finger (3) and fore-arm (4) have given a clear indication of the medical potential of the technique. Large-scale imaging of the live human body has recently been accomplished by two groups (5,6). Most of the imaging techniques are discussed and compared in review articles (7,8).

LINE-SCAN AND PLANAR IMAGING METHODS

The principle of most of the methods relies on a fact first noted by Gabillard (9) in 1951 that the inhomogeneous broadening of the absorption spectrum produced by placing the sample in a *non*-uniform magnetic field will be dependent on the bulk shape of the sample. In practice, a linear magnetic field gradient is superimposed on the large static homogeneous field, and, providing the field gradient is large compared to the intrinsic linewidth, the shape of the broadened absorption spectrum will simply be the profile of the density of the resonant nuclei along a direction perpendicular to the field gradient. The application of this principle to our line-scan and planar imaging techniques has been described in detail elsewhere (10,11), and so is only briefly reviewed here.

In the case of line-scanning, plane selection is achieved by applying a linear magnetic field gradient perpendicular to the required plane. Spins lying along a particular line in the selected plane may be excited by applying an orthogonal gradient to the sample, and simultaneously irradiating with a tailored selective radiofrequency (RF) pulse having the required narrow frequency distribution. The Fourier transform of the ensuing free induction decay signal, recorded in a third mutually orthogonal gradient, will yield the spin density profile along the selected line. After a short delay time τ the process is repeated for signal averaging prupoases, and then the next line in the plane is irradiated. In this way, a complete image may be obtained a line at a time.

During the delay time τ between pulses, the excited spins are assumed to relax to their equilibrium state with a spatially dependent spin-lattice relaxation time $T_1(x,y,z)$. The effective mobile spin density distribution in the plane z is then given by

$$\rho(x,y,z,\tau) = \rho(x,y,z)[1 - \exp(-\tau/T_1(x,y,z))] ,$$

where $\rho(x,y,z)$ is the true mobile spin density. Unless $\tau \gg T_1(x,y,z)$, the effective density can be less than the true density, and since $T_1(x,y,z)$ can vary over the cross-section, careful choice of τ can be used to enhance picture contrast.

Although the images obtained by line-scanning methods are of high quality, the data collection time is necessarily rather long (usually greater than about 5 minutes for acceptable resolution and signal-to-noise ratio) and this can be troublesome, particularly in the medical situation. For this reason, we have turned our attention to more efficient imaging methods, and in particular, the new method of echo planar imaging in which all volume elements in the sample contribute to the observed signal, and all are uniquely defined in a single experiment.

There are two main concepts embodied in this new method: (a) the free induction decay following a non-selective RF pulse is recalled as a series of spin echoes and this, together with digital sampling, imposes a discreteness on the frequency domain projection profile; (b) all three of the orthogonal gradients are appropriately applied during the signal recording period, in such a way as to allow simultaneous resolution of *all* the discrete lattice points along a single projection axis.

In principle then, it is possible to obtain a complete three-dimensional image from a single free induction decay and echo train. However, in these early experiments, the required imaging plane is first defined by selective irradiation, and the echo-planar technique then applied to the signal arising from that specific plane. The image thus obtained may require signal averaging, and if so, the process is repeated after a cycle delay time τ, just as in the line-scan experiment.

TISSUE CHARACTERISTICS

Although NMR is applicable to any nucleus which possesses a magnetic moment, the majority of images of biological interest presented to date are proton NMR images, the signals arising from the ^{1}H nuclei contained in the water, fat and oil distributed within the sample. The rationale for proton imaging is two-fold. Firstly, the NMR signals from protons are intrinsically the strongest of those from any stable isotope, and secondly, biological tissue contains on average about 75% of water, the distribution of which is known to alter in various pathological conditions, and in trauma. In man, quite a large fraction of the water is contained in the cell cytoplasm, but there is also a significant amount of extra-cellular water in addition to be bulk body fluids. As one might expect, the largest contrast in water content is between bone with about 12%, and the soft tissues which cover a range from about 70% to 85%.

Many organs then, should be distinguishable on this basis, particularly when considered along with their geometrical location in the body. For example, grey and white nervous tissues are relatively easily discriminated in an NMR image due to their 10% difference in water content. Just as important, however, is the variation of a factor of about five between the spin-lattice relaxation times of the various tissues; in particular, it is well documented that T_1 is longer for tumourous tissue than for the normal host tissue (12).

As mentioned earlier, the effective spin density observed in the image generally includes variations in T_1. However, if two or more images are obtained using different delay times τ, it is a simple matter to separate the true spin density component $\rho(x,y,z)$ from the relaxation time component $T_1(x,y,z)$ and display these parameters separately. Other NMR parameters might also be expected to show variations among tissue types, and they could also be independently measured by suitably modifying the experimental technique.

APPARATUS

A block diagram showing the principal components of the system is given in Figure 1.

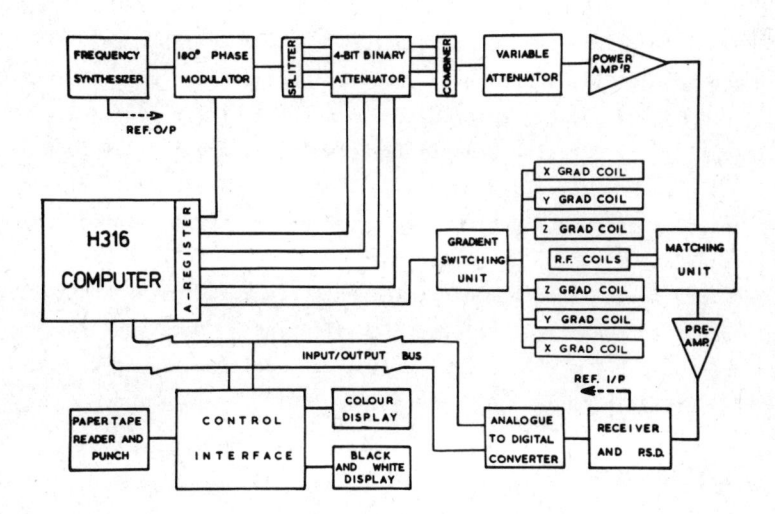

FIGURE 1: Block Diagram of the NMR Imaging System

The magnet and coil system used to obtain our first NMR images restricted the specimen size to a diameter of about 2 cm. The probe and gradient coil assembly were then situated in the 3 inch gap between the pole faces of a Varian V-3400 iron-cored electromagnet operating at 3523 Gauss, which corresponds to a proton resonance frequency f_p of 15 MHz. Many line-scan images were produced using this conventional NMR magnet, examples of which may be seen in the literature (3,13). It was also used to obtain the planar image presented below.

We have subsequently been able to scale up the system to perform whole-body line-scan imaging. In this case, the static magnetic field (939.5 Gauss, f_p = 4 MHz) is provided by an Oxford Instruments 4-coil air-cored electromagnet, with a maximum access diameter of 64.5 cm. After initial alignment, an axial homogeneity of better than two parts in 10^5 was achieved, ±12 cm from the centre of the magnet. Over a typical whole-body imaging plane, however, (36 cm x 23 cm) the homogeneity is only one part in 10^4. Much of the non-uniformity is due to the interaction of the magnetic flux with the dispersed iron in the structure of the laboratory. Future removal of the magnet to an iron-free environment, and the use of additional field correction coils, should increase the homogeneity in the imaging plane by a factor of 10.

For each of the two magnets, the designs of the three gradient coils are essentially similar. A reversed Helmholtz pair provides a gradient parallel to the main field, and each of the other two gradients are provided by four line-currents with appropriate return paths. Series-tuned, crossed-coil RF probe systems have been used.

RF pulses are provided by a mainly home-built spectrometer. Low level RF at either 15 MHz or 4 MHz is routed from the crystal controlled frequency synthesizer via a 180° phase modulator to a 4-bit binary RF attenuator which implements the required gating and amplitude modulation of the RF waveform, and from there the tailored pulse is amplified and fed into the transmitter coil.

Gating pulses for the binary attenuator, and the trigger for the 180° phase modulator, are provided by a Honeywell H316 8K minicomputer. The received signal is amplified, phase-sensitively detected and finally converted into digital form by a 10-bit analogue to digital converter which is again controlled from the computer. The computer also performs on-line signal averaging, Fourier transformation, and storage of the 8-bit image data.

The windowed image data is either displayed on a black and white monitor or is colour coded and output to a domestic colour television set. Note that for the monochrome images presented below, low or zero signal intensities are represented as black, and areas of high effective spin density are shown white.

LARGE-SCALE LINE-SCAN IMAGES

Using the simple phantom illustrated in Figure 2, we are able to assess the in-plane resolution and extent of distortion in the images obtained with the whole-body apparatus.

The phantom consists of a supportive matrix containing ¾ inch diameter test tubes, spaced at 1 inch intervals. For this demonstration, only the shaded test tubes contain water, to form the letters I P. As had been expected from the rather poor static field homogeneity, there is some geometric distortion in the NMR image of the phantom, shown in Figure 3.

The *intensity* distortions are most probably due to curvature in the 3-4 cm thick imaging plane again arising from field non-uniformity. Such a curved plane might not include the whole of each 4.5 cm long test tube. Most of the tubes in the image (which consists of 90 x 75 original picture points) are well separated, indicating an

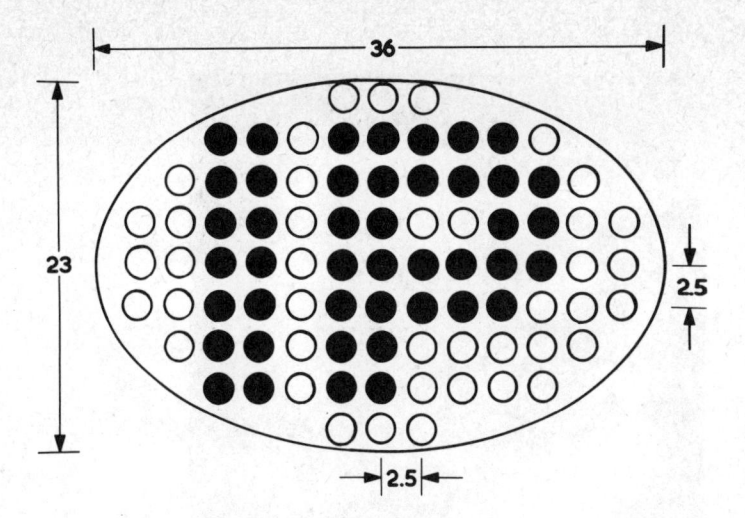

FIGURE 2: Test-Tube Matrix Phantom (Dimensions in cm)

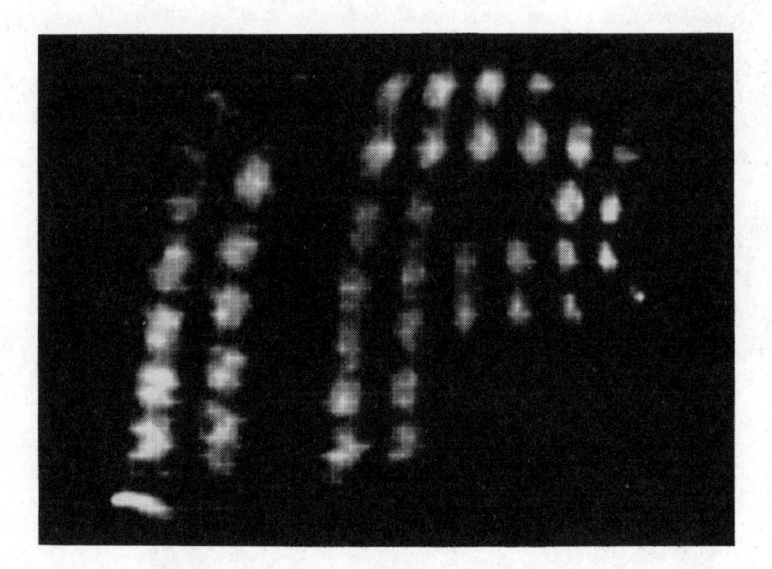

FIGURE 3: NMR Image of the Phantom Shown in Figure 2

in-plane resolution of better than ¼ inch on a Rayleigh-criterion basis.

Figure 4 shows a transverse cross-sectional image through a human cadaver leg. The sample, from the mid-thigh region, was about 6 inches thick. The scan slice is again 3-4 cm and the complete image, obtained in 27 minutes using a cycle delay time τ of 0.33 s, is an average of 128 shots. The femur and femoral vessels are clearly revealed in this image, which would indicate that line-scanning techniques could be usefully applied to limb or head imaging where movement can be easily restricted.

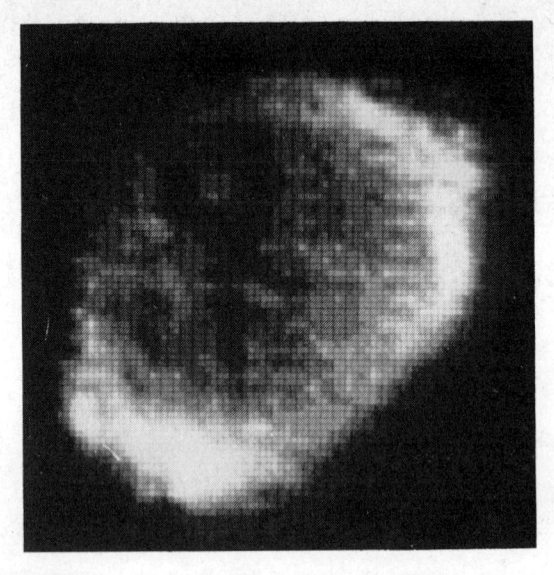

FIGURE 4: Cross-Sectional Image through a Human Cadaver Thigh

Figure 5 is the first line-scan whole-body image, and is a transverse section through the abdominal region of PM at the level of the third and lower part of the second lumbar vertebrae.

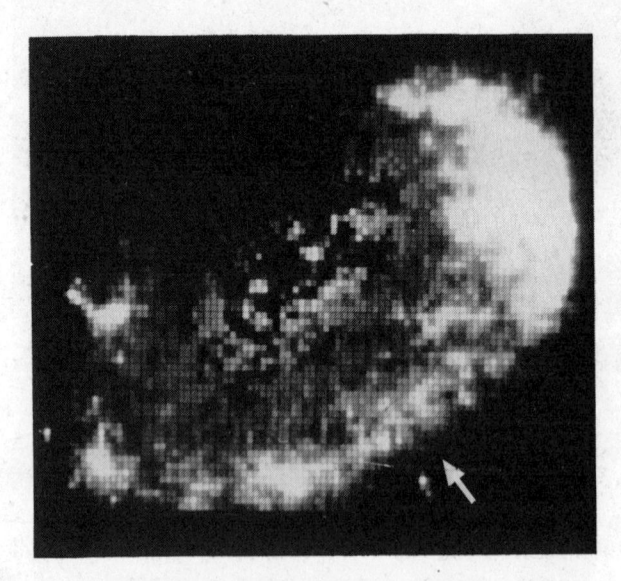

FIGURE 5: Cross-Sectional Image Through the Live Human Abdomen at L2-3. Arrow Indicates Mid-Line Posterior. Left Side Lies to the Left of the Figure.

The original 90 x 75 point image was of generally reduced intensity in the centre. This has been ascribed to lack of complete RF penetration, but intrinsic RF inhomogeneity and non-uniformity of the receiver coil response over the imaging plane may also be contributory factors. The data have been smoothed and empirically corrected for these effects. There was a progressive phase-change throughout the scan which has produced a discontinuity in the image of the anterior body wall.

Each line of the image was averaged 96 times, and the total scan time was 40 minutes. Over this period involuntary body movements, especially those due to respiration, will clearly have degraded the image quality substantially. Nevertheless, much significant detail is still visible, and a detailed anatomical assignation is given in Reference 5. In particular, the blood-saturated liver is clearly visible as a high-signal region on the right of the image. This is in direct contrast to the low signal received from the liver in cadaver scans where the blood is no longer liquid, and this highlights the danger of extrapolating results from dead to living systems. Nonetheless, it does indicate, in albeit a rather unsophisticated way, the potential of NMR to provide *functional* information, regarding the state of health of the subject.

Figure 6 illustrates the use of T_1 discrimination in breast tumour detection (14).

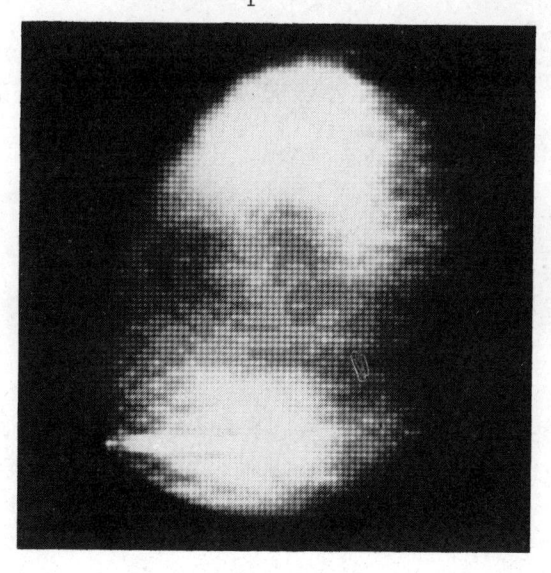

FIGURE 6: NMR Image of Excised Human Breast

The image is a coronal projection of a human breast obtained 1½ hours following simple mastectomy. The nipple is visible to the left, and a scirrhous carcinoma lies to its right. Since the slice thickness examined included the whole breast, one would expect to receive the largest signal from the central region, corresponding to the greatest tissue thickness, with a gradation of intensity downwards towards the periphery. However, in this short delay scan ($\tau = 0.15$ s), regions of long T_1 are discriminated against and the tumour appears darker than the surrounding area. When the delay time is increased to 0.3 s, the dark nipple region of low water content remains, but contrast in the tumourous region is reduced.

The tumour size and position correspond well with subsequent paraffin sections and histological findings, and the observed T_1 values are in agreement with other published data, obtained by conventional NMR techniques.

ECHO PLANAR IMAGES

Figure 7 shows one of the first images produced by the echo-planar method. The sample consisted of a thick-walled 1.4 mm bore glass capillary tube, inserted alongside a solid glass rod of diameter 4.6 mm in a 14.6 mm ID test tube of mineral oil.

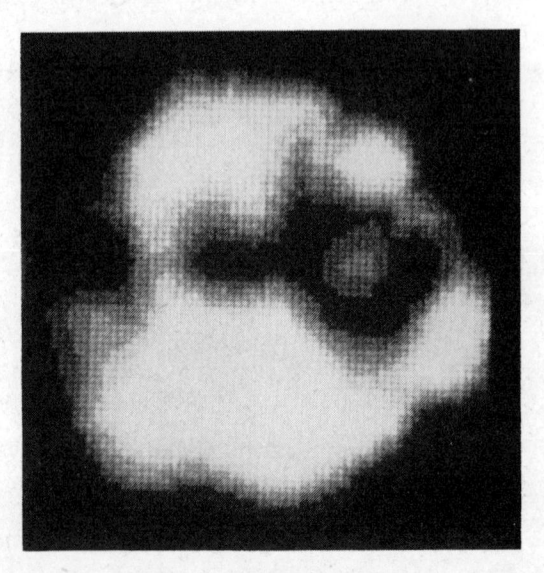

FIGURE 7: Echo Planar Image of a Capillary Tube Phantom

The original data comprised a 32 x 16 array which has been linearly interpolated here to produce an image of approximately 50^2 elements. Each full, unaveraged cross-sectional picture took only 10.28 ms to produce, and the average count in this case was 64. The distortions in the image arise from a variety of effects to be discussed elsewhere, many of which will be eliminated or corrected for on the second generation of planar images. Nevertheless, the general outlines of the phantom are apparent, and the bore of the capillary tube is resolved.

FUTURE PROSPECTS

Absolute *in-vivo* tissue characterization by NMR may or may not be possible, but even if the diagnosis of cancer and other pathological conditions were found to be somewhat equivocal, screening or adjunctive diagnosis by NMR imaging would still be of considerable value, especially since it is essentially a non-hazardous technique.

The measurement of any standard NMR parameter is in principle possible using suitably modified imaging techniques. For instance, the *in-vivo* measurement of flow and diffusion constants might provide useful diagnostic information, insofar as they are known to alter in diseased states.

Using the echo planar technique, it may be possible to perform dynamic imaging studies of the heart, gating the scan to the electrocardiograph and averaging over successive cardiac cycles if required for signal-to-noise ratio enhancement.

The imaging of other nuclear species is of course possible at a reduced sensitivity, and of particular interest would be ^{31}P studies which could in principle be used to analyze *in-vivo* the phosphate metabolite concentrations in normal and abnormal tissues.

It has already been demonstrated that NMR imaging techniques are able to reveal considerable anatomical detail, and it is envisaged that they will also be able to provide a wealth of functional information.

ACKNOWLEDGEMENTS

We are grateful to the SRC and MRC for support of the NMR imaging projects and to D. Kerr for assistance in the construction of some of the apparatus. We are especially indebted to T. Baines for the design and construction of much of the electronic apparatus.

REFERENCES

1. Lauterbur, P. C., 1973, Nature, 242, 190-191.

2. Mansfield, P. and Grannell, P. K., 1973, J. Phys. C: Solid State Phys., 6, L422-L426.

3. Mansfield, P. and Maudsley, A. A., 1977, Brit. J. Radiol., 50, 188-194.

4. Hinshaw, W. S., Andrew, E. R., Bottomley, P. A., Holland, G. N., Moore, W. S. and Worthington, B. S., 1979, Brit. J. Radiol., 52, 36-43.

5. Mansfield, P., Pykett, I. L., Morris, P. G. and Coupland, R. E., 1978, Brit. J. Radiol., 51, 921-922.

6. Damadian, R., Minkoff, L., Goldsmith, M. and Koutcher, J. A., 1978, Naturwiss., 65, 250-252.

7. Mansfield, P., 1976, Contemp. Phys., 17, 553-576.

8. Brunner, P. and Ernst, R. R., 1979, J. Mag. Res., 33, 83-106.

9. Gabillard, R., 1951, C.R. Acad. Sci. (Paris), 232, 1551.

10. Mansfield, P., Maudsley, A. A. and Baines, T., 1976, J. Phys. E: Sci. Instrum., 9, 271-278.

11. Mansfield, P. and Pykett, I. L., 1978, J. Mag. Res., $\underline{29}$, 355-373.

12. See for example: Medina, D., Hazlewood, C. F., Cleveland, G. C., Chang, D. C., Spjut, M. J. and Moyers, R., 1975, J. Natl. Cancer Inst., $\underline{54}$, 813-818.

13. Pykett, I. L. and Mansfield, P., 1978, Phys. Med. Biol., $\underline{23}$, 961-967.

14. Mansfield, P., Morris, P. G., Ordidge, R. J., Coupland, R. E., Bishop, H. and Blamey, R., 1979, Brit. J. Radiol., $\underline{52}$, 242-243.

NUCLEAR SCATTERING RADIOGRAPHY

D. Garreta

DPHN/ME - CEN Saclay

INTRODUCTION

Nuclear Scattering Radiography (NSR) consists in producing three dimensional radiographies of samples that can be large by using medium energy protons beams (E_p of the order of 1 GeV) and standard particle physics detection systems.

After the first work done at CERN (Geneva)[1,2] and SIN (Zurich)[3] this techniques has been applied at Saclay to anatomical specimen [4,5] and heavy materials [5,6]. For medical applications low radiation dose and fast data acquisition are required. A first step in this direction was done at CERN [7,8] where the solid angle of the detection was increased so that the radiography of a human head could be obtained with a radiation dose of 0.3 rad, which is satisfactory, and a faster acquisition rate but still leading to a prohibitive exposure time. Further improvement, in preparation at Saclay by the same experimental team, aims, by increasing the acquisition rate, to obtain such a radiography in 20 minutes which would allow *in vivo* applications.

PRINCIPLE OF THE METHOD (see Fig. 1)

The principle of the method consists in illuminating the sample of which one wants to have a picture with a beam of medium energy proton. Most of them will go across the sample with just a small angular deviation due to multiple coulomb scattering but some of them will experience nuclear scattering and will go out with a direction quite different from the incoming one. So if one detects proton trajectory before the sample (in counters CH1 and CH2) and after the sample (in counters CH3 and CH4) one can get the coordinates X, Y and Z of the point where the nuclear scattering took place. Therefore it is possible

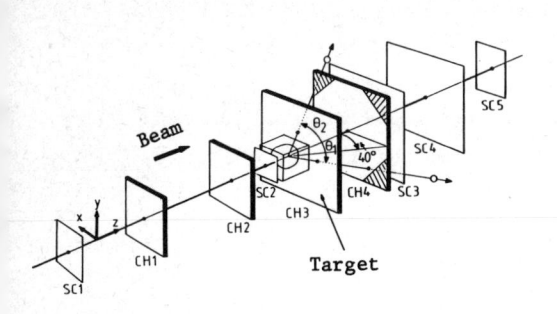

Fig. 1 : Schematic drawing of the experimental set-up of Ref 7-8, showing the set of scintillators (SC1 to 5) that trigger the position sensitive multiwire proportional chambers (CH1 to 4) used to detect the incident and scattered protons.

to count these events and store them in a three dimensional matrix n (X,Y,Z) corresponding to the countings from cells of dimensions ΔX, ΔY, ΔZ centred on the point of coordinates X, Y, Z. After corrections of possible inhomogeneity in the beam density distribution, absorptions of the incoming and outgoing protons in the sample and variation of the detection efficiency across the sample, this matrix is a direct measurement of the three dimensional *nuclear scattering density* distribution D (X,Y,Z) of the sample. The definition of D(X,Y,Z) is :

$$D(X,Y,Z) = \sum_i \frac{d_i(X,Y,Z)}{A_i} \sigma_R^i$$

where $d_i(X,Y,Z)$ is the partial density of element with atomic mass A_i, σ_R^i is the nuclear reaction cross section for giving a charged particle in the detection system.

In addition to this information it is possible to get the hydrogen density distribution in the sample. When the nuclear scattering takes place on a free proton (from a hydrogen atom) the scattered and recoil protons are coplanar with the incident one and their angles θ_1 and θ_2 are related by : $tg\theta_1 \cdot tg\theta_2 = 1/\gamma^2$ (where γ is the relativistic factor in the transformation from the CM to Lab systems). These two conditions do not hold when the two protons come from quasi-free scattering on a proton bound in a nucleus because of the Fermi momentum. Therefore the subset of events which fulfill these conditions will give a measurement of the hydrogen density distribution in the sample.

Looking at the accuracy with which the coordinates of the point where the nuclear scattering took place are determined, the resolution in the transverse direction (XY plane) is determined by the resolution of the position sensitive counters CH1 to CH4 and the multiple scattering in the sample, the resolution in the beam direction Z is equal to the transverse resolution divided by $\sin\theta$. Therefore small angle scattering has to be rejected in order to keep a reasonable resolution in the Z direction. This limits the energy of the incident protons because the higher the energy the faster is the decrease of the cross section with angle.

An interesting figure is the ratio between the number of detected events n and the incoming protons N on a sample of thickness t and density d :

$$\frac{n}{N} = 6.10^{23} \, t \cdot \frac{d}{A} \cdot \sigma_R$$

It is of the order of 1 % for t = 1 cm for organic compounds for which σ_R/A is about $1.5 \, 10^{-26}$ cm^2 with 650 MeV protons detected in an angular range of 15^0 to 40^0 [1] and drops down to about 0.5 % at 1 GeV but is so that the number of *usefull* protons is still relatively high (about 10 % for a human head at 1 GeV).

One must also keep in mind that for medical applications, the variations in nuclear scattering density that will be looked for are of the order of a few percent so that most of the problems will come from statistics, directly related to radiation dose and time of exposure.

ADVANTAGES AND LIMITATION OF THIS METHOD

- One of the greatest advantages of this method clearly appears on Fig. 2 which shows absorption coefficients of different probes as a function of the atomic mass of the sample.

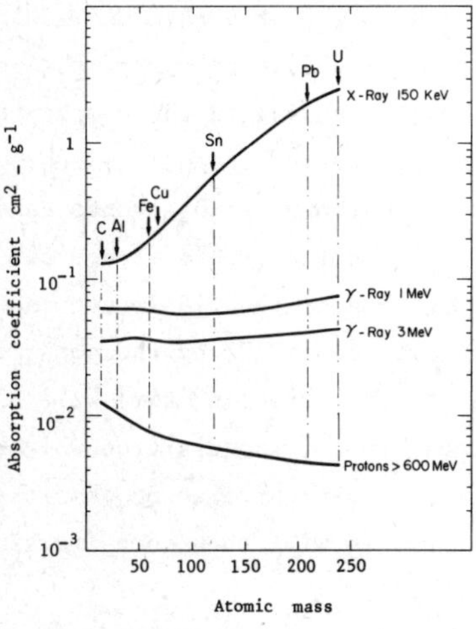

In the case of heavy elements, the transmission coefficient of 3 MeV-γ-rays through a 10 cm copper block is only 4 % when it is still more than 50 % for the protons. For a 5 cm thick Uranium block this transmission coefficient will be less than 2 % for the γ-rays and still 2/3 for the protons.

This low absorption, very useful in the case of heavy elements, is also interesting in the case of medical applications where it makes absorption corrections rather low. For instance absorption of X-rays in the bones is so large that it can produce artefacts in Computer Assisted Tomographies (CAT) of the head, making difficult the detection of tumors located near the skull, when overall corrections in the NSR of a human head were smaller than 20 % [7].

Fig. 2 : Absorption of different probes through material as a function of the atomic number A of the material.

- An other advantage of this method is the true three dimensional character of the information which allows the display of the image along any direction by simple data handling. The only anisotropy in the data is the spatial resolution which is larger along the beam direction, but there is none of the artefacts that can be generated when building a three dimensional image from independent two-dimensional cross-sections.

- The hydrogen density measurement is also an advantage but the small counting rate of such events is a serious limitation. From this point of view NMR imaging might be in much better shape although the two informations are not identical since the hydrogen density produced by NMR depends on the chemical constituents to which the hydrogen atoms belong when there is no such dependence in NSR, therefore comparison between the two informations might be interesting.

- The possibility of focussing the beam only on the useful part of the sample so that only this part and its shadow through the rest of the sample is irradiated is also interesting. In CAT for instance, a full cross section of sample has to be irradiated. Radiography of the Spine is a case where this possibility would be very useful.

- A strong limitation in the use of NSR is obviously the need of a medium energy proton beam. On existing high energy accelerators there is no problem to get such beams because they are of low intensity and do not require any special qualities so that parasite beams can be used which do not disturb the other users of the accelerator. But there are only few available in the world.

- Another limitation is the data acquisition rate. Data of Ref 4, 5, 6 were recorded at a rate of about 100 events/sec, those of Ref 7, 8 at a rate of 1 500 events/sec, the limitation comes from the on line computing time necessary to get the coordinates of the interaction vertex (350 µs to 1 ms, depending on the complexity of the event). In the next step this time should be reduced to less than 10 µs which should allow a rate of 10^5 events/sec (producing a radiography of 10^8 events, about the number of events necessary for a human head, in 20 minutes). This is not bad when one realizes that the total volume in the latter case is about 5 l and about 50 conventional CAT would be necessary to scan the same volume. Using the possibility of focussing the beam one can restrict this volume and the recording time will be decreased in the same proportion since it is the total rate of acquisition which is limited. By using a fast specialized hardware processor it might also be possible to reduce again this time by a factor of ten, but the problem will then come from the chambers which will have to stand a beam of 10^7 p/sec.

- Multiple Coulomb Scattering (MCS) is also a limitation because it spoils the spatial resolution. It is not severe in the case of medical applications because the spread in transverse position (XY plane) induced by MCS in 10 cm of water is only 0.5 mm (fwhm) for 1 GeV protons and goes up to 1.6 mm for 300 MeV protons. For heavy material it becomes more serious, 2.4 mm and 5 mm for 10 cm of copper and uranium with 1 GeV protons. This effect which goes like the power 3/2 of the thickness, can be reduced by calculating the distance between the incoming and outgoing tracks and rejecting the events for which this distance is too large, but more beam exposure will be necessary to get the same statistics. It is also possible to restore the resolution by using deconvolution techniques.

EXPERIMENTAL SET-UP AND DATA ANALYSIS

Fig. 1 is a schematic drawing of the experimental set-up used in Ref 7, 8. A set of fast counters (scintillators SC1 to SC5 in coincidence or anticoincidence with a resolution time of the order of a few nsec) detects the incident protons which have scattered with a minimum angle of about $15°$. The coincidence signal from the scintillators triggers the four multiwire proportional chambers CH1 to 4 which give the horizontal and vertical positions of the tracks with an accuracy of 1.27 mm for the incoming proton and 2 mm for the outgoing particles (these correspond to the wire spacing of the MWPC). The maximum volume that can be radiographed is determined by the size of chambers CH1 and CH2 (20 x 20 cm^2) and the room remaining free between chambers CH2 and CH3 (of the order of 40 cm). Then the coordinates X, Y, Z of the reaction point are calculated as well as the conditions which determine wether or not this scattering occured on hydrogen and the corresponding cell of the proper matrix is incremented. Along with these three dimensional matrices n (X,Y,Z) and n_H (X,Y,Z), the beam profile at the centre of the sample determined with CH1 and CH2 is recorded in a two dimensional matrix b (X,Y). The fact that the beam divergence is small is used to assume that the beam profile is independent of Z on the dimension of the sample. If it were not so, it would be necessary to record the profile in a set of XY planes at different Z positions and use interpolation for beam correction.

Then, a set of corrections is applied to the matrices :

- Beam correction : the matrix has to be corrected for beam intensity inhomogenity.

- Corrections for particle absorption, solid angle and mean scattering angle variation across the sample : since the absorption is rather low and not strongly dependent on the material, the overall correction for all these effects is assumed to be a smooth function of X, Y, Z and is determined empirically.

It is done by fitting a correction function f (X,Y,Z), chosen with such an analytical form that it can only reproduce smooth variations and not the local fluctuations of the type we are looking for, to the countings of either a *phantom* target which has approximately the shape and mean density of the sample or directly on some parts of the sample itself of which we know that they should have in average a constant density. When looking for defects in heavy material pieces which otherwise are homogeneous the latter solution is straight forward, for medical applications it is a bit more delicate and might depend on the specific case. In Ref 7, 8 where the sample was a human head, it was assumed that the mean density of soft tissues was constant over the whole sample, cells containing soft tissue were selected and used for the fit of the correction function (which was of second order in X, Y, and Z). The maximum correction was 20 % and the spread in the density distribution of soft tissues which was 30 % in the raw data was reduced to 13 % after correction (statistical fluctuation are of the order of 11 %).

- Corrections for statistical fluctuations and resolution improvement. The simplest way of decreasing statistical fluctuations is smoothing. To see the effect, we can look at a one dimentional smoothing of the type :

$$S_n = \tfrac{1}{4} O_{n+1} + \tfrac{1}{2} O_n + \tfrac{1}{4} O_{n-1}$$

(S_n and O_n correspond to channel n of the smoothed and original arrays respectively). The statistical fluctuations on S and O are related by $\sigma_s^2 = 3/8\ \sigma_o^2$. This means that the countings of O should be increased by a factor of 8/3 to decrease the relative statistical fluctuation to the level of S. But if we look at the fluctuations on the *difference* between two adjacent channels $\Delta O = O_{n+1} - O_n$ *which is a very important quantity for imaging* we see that $\sigma_{\Delta o}^2 = 2\sigma_o^2$ but $\sigma_{\Delta s}^2 = \frac{1}{4}\sigma_o^2$, therefore $\sigma_{\Delta s}^2 = 1/8\sigma_{\Delta o}^2$ which means that in order to reduce the fluctuations on the difference between two adjacent channels to the level of S the counting rate on O should be increased by a factor of 8. The trouble with smoothing is the loss in resolution.

A different way of reducing statistical fluctuation and improving at the same time the resolution is to Fourier transform the spatial density distribution and apply a filter to the frequency density distribution. This filter will have the shape of the inverse of the Fourier transform of the experimental spatial resolution in the low frequency domain in order to restore some of the spatial resolution lost in the experiment, and will cut the high frequency part of the distribution to decrease the high frequency statistical fluctuations. An alternative way of restoring spatial resolution is to reproduce the experimental density distribution by the convolution of the known experimental spatial resolution with the *true original* density distribution that is looked for. These techniques are delicate and have to be used with great care because they can easily generate artefacts.

EXPERIMENTAL RESULTS

1 - Heavy materials (Ref 4, 5)

Targets of very simple geometry were used to test the effect of MCS on spatial resolution and the capability of this method to detect defects in heavy materials.

Copper blocks shown on Fig. 3 were used to test the ability of detecting narrow defects. Fig. 4 shows the density distribution along Z axis. Fig. 4a is the raw data, on Fig, 4b where the data is corrected for particle absorption and smoothed one can easily see that the depth in the density is proportional to the width of the defect which is expected when this width is much smaller than the resolution. A preliminary test of deconvolution of the density distributions by the longitudinal resolution function is shown on Fig. 4c. From these density distribution the positions of the defects can be measured with an accuracy better than 1 mm.

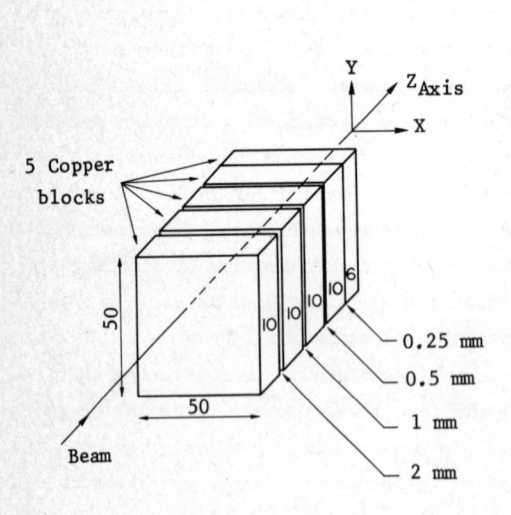

Fig. 3 : Geometry of the copper target which consists of five blocks with 50x50mm^2 cross section. Four blocks have a thickness 10 mm, the last one 6 mm. The blocks are parallel to the XY plane and separates successively by 2, 1, 0.5, 0.25 mm air intervals.

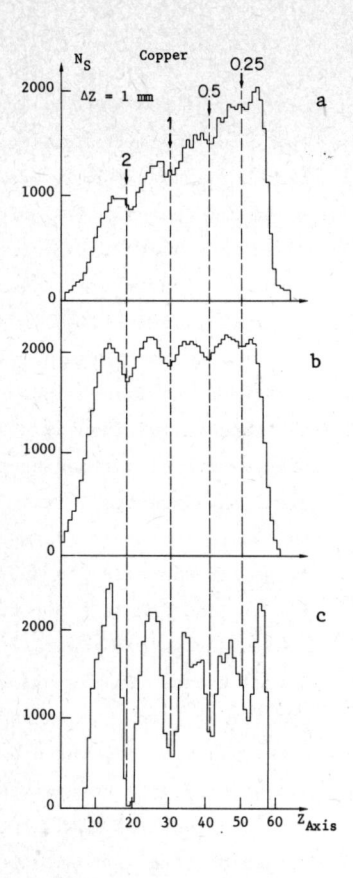

Fig. 4 : Detection and longitudinal localisation of defaults in copper.
-4a- gives the brute counting rate N_S in function of Z. On
-4b- N_S has been corrected for particle absorption ans smoothed twice to decrease statistical errors. Default thickness are indicated in mm. Channel width $\Delta Z=1$mm. The four defaults are clearly detected and precisely located.
-4c- Deconvolution of the $N_S(Z)$ spectrum of copper target shown on Fig. 4b by a gaussian function with FWHM about 6 mm corresponding to the experimental longitudinal resolution.

2 - Radiography of a human head (Ref 7,8)

The experimental set up was that described above (Fig. 1). The solid angle of the detection system was about 1.1 steradian corresponding to scattering angle ranging from $15°$ to $40°$. This large solid angle was chosen to minimize the radiation dose. The anatomical specimen, surrounded by low density expanded polystirene (.03 g/cm^3), was placed in a plexiglas box filled with formalin. Therefore the quantity of material surrounding the sample was very low so as to be as close as possible of *in vivo* radiography conditions.

A three dimensional radiography of a human head has been taken. The total number of events is $1.3 \cdot 10^8$, which corresponds to about 20 events/mm^3, they are stored in a matrix made of 124 x 124 x 96 = $1.48 \cdot 10^6$ cells of dimensions 1.4 x 1.4 x 2.8 mm^3 = 5.5 mm^3. About $6.5 \cdot 10^6$ events corresponding to scattering from hydrogen are stored in a matrix made of cells twice as large (with a volume of 44 mm^3). The radiation dose delivered during this radiography was 0.3 rad. This low radiation dose is due to the fact that 1 GeV protons are close to minimum ionisation, the cross section of the main reaction used for radiography (quasi-free scattering) is large, the information is very direct and total absorption is small. For X-rays, the information is integrated over the whole thickness of the sample and, due to strong absorption, it is necessary to deliver high radiation dose at the input of the sample in order to get a reasonable intensity at the output.

Fig. 5b and 5d represent XY cross sections, 5.6 mm thick, corresponding to the CAT presented on Fig. 5a and 5c taken from the same anatomical specimen. The hypodensisties that can be seen at the centre of left and right lobes of the brain on Fig. 5c, also show up clearly on Fig. 5d. On the other hand statistically significant density variations show up on the soft tissues of Fig. 5,

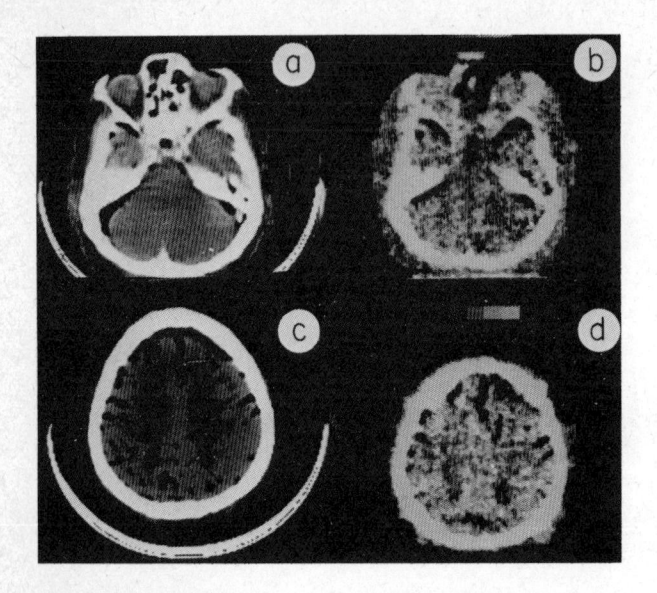

Fig. 5 : Fig. b and d : XY (horizontal) slices of the head obtained by NSR. Slice thickness is 5.6 mm and pixels are 1.4 x 1.4 mm². Each grey tone corresponds to a nuclear density variation of 4 %. Fig. a and c : XY slices of the same head approximately at the same level obtained by X-ray scanner.

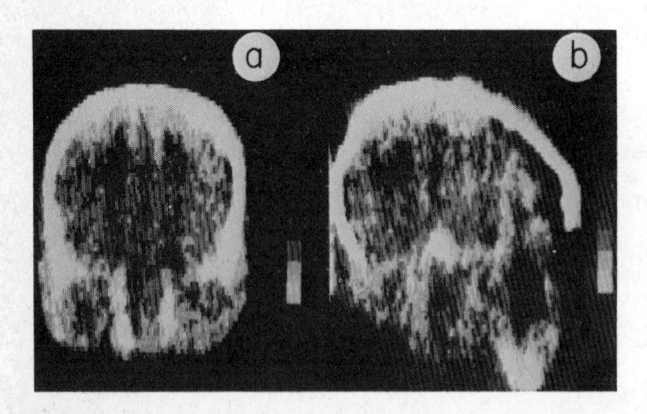

Fig. 6 : a- XZ(frontal) slice of the head obtained by NSR. b- ZY(sagittal) slice. Thickness is 5.6 mm and pixels are 2.8 x 1.4 mm². Each grey tone corresponds to a nuclear density variation of 4 %.

one level of brightness represents a variation of 4% in the density for the NSR. On Fig. 6 and 7 one level represents 4 %. Fig. 6 represents a ZX cross section (Fig. 6a), 2.8 mm thick, on which the beginning of the spine is clearly seen, and a ZY cross section (Fig. 6b) of the same thickness on which one can see the chin. These views show the capability given by NSR to provide directly cross sections along any direction. Fig. 7b and 7d represent XY cross sections for the hydrogen density, Fig. 7a and 7c are the corresponding cross sections for the normal events. On the hydrogen density pictures, bones appear dark, as expected, but one can also notice statistically significant structures with hypodensities different from the ones observed on the normal events.

CONCLUSION

NSR techniques has reached a point where it seems to have interesting applications in metallurgy. For medical applications, improvement of the data acquisition rate should make it useful for specific cases where the screening of bones makes X-rays radiography difficult, for instance for spinal cord or tumours located near the skull. But the quality of *nuclear scattering density* information has to be tested on a large number of samples.

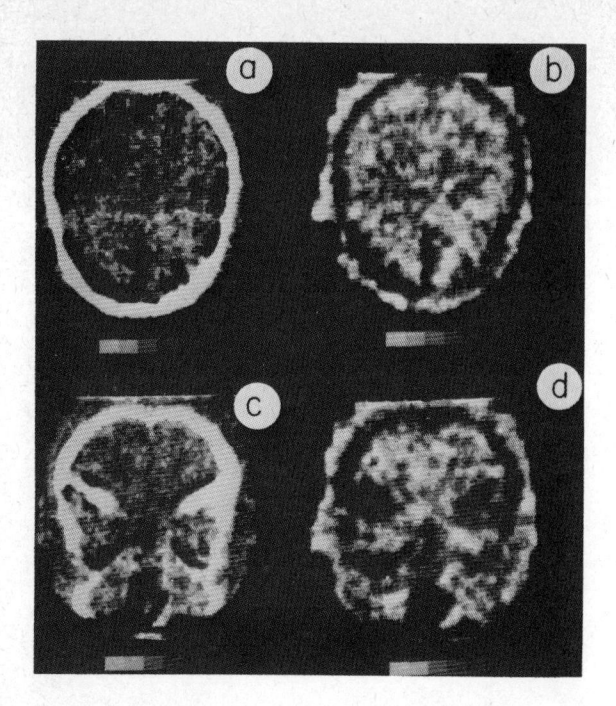

Fig. 7 : XY slices at two different levels. Thickness is 11.2 mm and pixels are 1.4 x 1.4 mm². Each grey tone corresponds to a counting variation of 4 %.
a - c - simple radiographs, b - d - hydrogen radiographs. Geometry of b and d are respectively identical to the one of a and c. On b and d bone appears black because it has a small water content and so for hydrogen.

REFERENCES

1 - J. Saudinos, G. Charpak, F. Sauli, D. Towsend, J. Vinciarelli. Nuclear scattering applied to radiography. Phys. Med. Biol. 20 : 890-905, 1975.

2 - G. Charpak, S. Majewski, Y. Perrin, J. Saudinos, F. Sauli, D. Towsend, J. Vinciarelli. Furhter results in nuclear scattering radiography. Phys. Med. Biol. 21 : 941-948, 1976.

3 - L. Dubal. Protoscopie : Annual report, Schweizerisches Institut für Nuklearforschung (S.I.N.), Villigen, Switzerland, B 14-16, 1976.

4 - J. Berger, J.C. Duchazeaubeneix, J.C. Faivre, D. Garreta, D. Legrand, M. Rouger, J. Saudinos, C. Raybaud, G. Salamon. Nuclear scattering radiography of the spine and sphenoid bone. Journal of computer assisted tomography : 2 488-498, September 1978.

5 - J. Berger, J.C. Duchazeaubeneix, J.C. Faivre, D. Garreta, D. Legrand, C. Raybaud, M. Rouger, G. Salamon, J. Saudinos. Radiographie par diffusion nucleaire. Compte rendu d'activité du département de Physique nucléaire. 1976-1977, note CEA-N-2026 p 263.

6 - J.C. Duchazeaubeneix, J.C. Faivre, D. Garreta, B. Guilleriminet, D. Legrand, M. Rouger, J. Saudinos, J. Berger. Submitted for publication in "Materials evaluation".

7 - G. Charpak, J.C. Duchazeaubeneix, J.C. Faivre, D. Garreta, B. Guilleriminet, G. Odyneic, P. Palmieri, Y. Perrin, C. Raybaud, M. Rouger, G. Salamon, J.C. Santiard, J. Saudinos, F. Sauli. Radiographie par diffusion nucleaire. Compte rendu d'activité du département de Physique nucléaire. CEA Saclay, to be published.

8 - F. Sauli. Progress in nuclear scattering radiography. International Conference on Computing tomography. Pavia. Oct. 9.10.1978.

COMPUTERIZED TOMOGRAPHY SCANNERS

R. KLAUSZ.
C. G. R.
STAINS (FRANCE)

I - INTRODUCTION -

Among the medical imaging techniques, these last years saw an important newcomer with the advent of Computerised Tomography.

The imaging process used to obtain classical radiograms had not changed in its rough lines since X-rays discovery in 1896 : a point source irradiates an object, which casts a shadow on a planar detector.

All the modern radiological tools keep these fundamentals, with improvements on the source (X-ray tubes), on the detectors (film changers, image intensifiers), or on the way to superpose such different two-dimensions projections by mechanical devices (tomography).

The CT Scanners also tend to provide radiological images ; but the major difference brought by CT is that it is an indirect and quantitative method : a large set of individual measures is obtained, then processed to finally provide a digital representation of the object on which the measures had been made.

All that concerns a "slice" of the object, as in classical tomography ; but it allows to obtain true measures of X-ray attenuation coefficients (densitometry). That's why it has also been proposed to name this technique TomoDensitoMetry (TDM).

II - THEORY -

The principle under which the CT machines work is the reconstruction of a 2-dimensionnal function from all its linear projections inside its definition plane.

A - PROJECTIONS-

The absorption of monochromatic X-ray photons by objects obey the Beer-Lambert law (fig. 1).

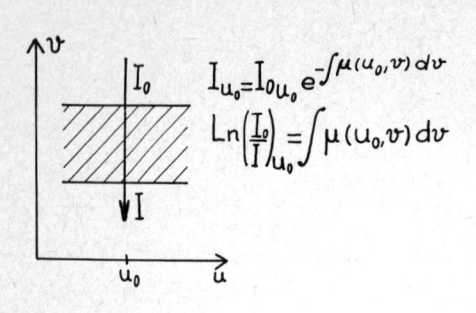

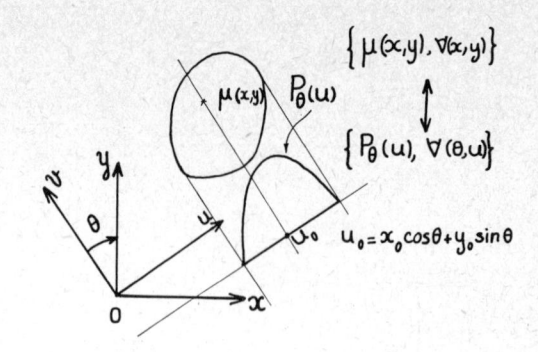

Fig. 1
X-Rays Photons Attenuation

Fig. 2
Projections

μ is the linear attenuation coefficient.

If we consider a section of a finite object, it is caracterised by $\mu(x,y)$ in any point. The set of integrals of μ along $u = u_o$ for all values of u_o constitutes the projection $P_\theta(u)$ of the object along the direction Ov defined by $(Ov, Oy) = \theta$ (see fig.2).

The relation giving all $P_\theta(u)$ from all $\mu(x,y)$ is linear ; it admits an inverse which gives $\mu(x,y)$ in any point from all projections.

B - ALGORITHMS

After some trys on iterative methods, which had proven to be valuable in similar problems with high symetries (Electron microscopy of viruses), the generally used algorithm is the filtered back-projection, deriving from the works of Radon (1), Ramachandran (2), and Shepp and Logan (3).(Application of "Central Slice" theorem).

If $\mathcal{P}_\theta(\omega)$ is the Fourier Transform of $P_\theta(u)$,

$$\mu(x,y) = \frac{1}{4\pi^2} \int_0^\pi d\theta \int_{-\infty}^{+\infty} \mathcal{P}_\theta(\omega) e^{i\omega u} |\omega| d\omega$$

with $u = x\cos\theta + y\sin\theta$

Obviously, it is possible to replace the ω integral (inverse Fourier Transform) by a direct convolution product of $P_\theta(u)$ by $\varphi(u)$, where $\varphi(u)$ is the inverse Fourier Transform of $\Phi(\omega) = |\omega|$.

From over, it is possible to decompose the algorithm in two phases :
- Computing $(P_\theta * \varphi)(u)$, directly or by Fourier Transform, $|\omega|$ multiplication, then inverse F.T.

- For a given θ (that is for one projection) adding the contribution of $(P_\theta * \varphi)(u)$ in any point (x,y) so that $x\cos\theta + y\sin\theta = u$, i.e along a line identical to the projection line.

The first operation is generally called convolution (or deconvolution or filtering), and the second : back-projection.

In real procedures, the needed precision is for now uncompatible with analog methods. It is, so, necessary to turn to digital, and, for that, to sample along projection angles (θ), along each projection (u), and inside the finite reconstruction domain.

This asks for a limitation (Shannon frequency) on the maximum ω in the physical measuring process, and drives to replace the $|\omega|$ value of Φ by a new Φ function, product of $|\omega|$ by a low pass filter (and similarly for $\varphi(u)$). Special care must be given to this filter, because it will determine the bandwidth of the system, and the "damping" of its transfer function. It will play a role similar to the apodizing functions in optics.

Another important problem comes from the determination of the value to be back projected in the (x_i, y_j) position, as in general $U_{ijm} = x_i \cos\theta_m + y_j \sin\theta_m$ is not equal to a u_n position along a projection. A compromise must be found between the precision and the computing time to interpolate $P_{\theta_m} * \varphi$ so as to know its value in U_{ijm} from the nearest positions u_n and $u_n + 1$, where $u_n \leqslant U_{ijm} \leqslant U_n + 1$.

III - BASIC STRUCTURE AND DESIGN PRINCIPLES OF A CT MACHINE

A CT-Scanner must :

- 1 - make densitometric measurements useable as projections, with :
    - X-ray producing equipments, beam conditionning devices, and X-ray detector(s), measuring the photons after object traverse.
    - A mechanical gantry, supporting the preceidings, and moving it along to allow the measure of all points of all projections.
    - An object support (the patient couch).
- 2 - ensure data management and algorithm computation, with :
    - Measuring electronics connected to the detector(s),
    - General electronics to ensure various supplies, synchronisations, interfaces, transfers, and receive orders from the operator.
    - Computing systems (including fast processors if needed).
- 3 - provide the results as interpretable pictures :
    - The reconstructed values are sent to a storage peripheral and a visualisation console (see later).

The quality of the results will be the consequence of all characteristics and performances.

Some of those limitations are pure technology, but the main ones will result from the interaction between physical limits and some severe restrictions given by the nature of the object, generally a living patient.

The theorical precision would be infinite, less the counting statistics of X-ray photons in measures. But if their number goes too high, patient exposures will also be too high.

A compromise must be found between acceptable doses and precision, for both low contrast and high frequencies resolution.

As the quantum noise on counting N events can be described by a Poisson law with a variance N, it is possible to evaluate the noise on the resulting image, for a given bandwidth. It can be established that the standard deviation is $\sigma \propto \sqrt{\omega_s^3/D}$ where D is the patient dose. This shows that for a given dose, it is important to choose the noise/bandwidth trade off.

Another important problem is the choice of photons energy. All the theory is based upon monochromatic photon sources. The optimum energy is a compromise between transmission and contrast, to achieve the best signal to noise ratio for a given dose. The nature of the detected contrasts will vary with the energy because of the different attenuation mechanisms (photoelectric absorption-elastic or inelastic scatter). The values of these contrasts ask for very high amounts of photons (typically $1.5 \cdot 10^{10}$ to 10^{12} photons per slice without object), with energies in the 60-90 keV range. The last point to determine the X-Ray source is the measuring time, which must be short for economic reasons (patient throughput) and to avoid patient moving. Typical times go from 5 minutes to 20 Seconds for head examinations, and must be less than 5 seconds (down to about 50 ms if possible for body examinations.

The only practically useable X-ray source is the vacuum X-ray tube. Its large emission spectrum is adapted by using filters, but with low selectivity to keep high efficiency. It is then necessary to correct for the polychromaticity errors by digital ways.

The detectors must too have high efficiency ; the photon rates (5.10^7 to 10^{12} per second) are too high for a counting use ; so, it is necessary to convert their analog signal to digital before use.

<u>The detectors</u> in use can be - photodetectors associated with scintillators,
- gas ionisation chambers.

In the first category, the scintillators had been NaI(Tl), then plastic scintillators or fluorite, and now ICs or, mainly, Bismuth Germanate. The points to be careful with are : high efficiency, low light absorption, and low decay time. The associated photodetectors are generally Photo Multipliers, and sometimes now photodiodes.

The second type, used only for multicells detectors are Xenon ionisation chambers used in the true ionisation mode. The high efficiency is obtained by high pressure (10 to 30 atmospheres) and thickness (10 to 100 mm). The main advantage is the simple structure giving directly an electric current. The collection time must remain low, but can be corrected (pure delay).

<u>The scanning</u> movements, combined with the detectors's structure, must allow to obtain the set of projections under a total half turn (minimum).

The number of projections and the number of samples per projection must be determined in relations with the desired precision on the reconstructed image ; current values are 128 to 512 projections by half-turn (sometimes completed on a whole turn or more), with 128 to 1024 samples per projection.

The simplest (and oldest) way to obtain a projection is to dispose one detector facing the tube, with the beam shaped by collimations. They are fixed on the same mechanical parts, and translate linearly, the beam going throuhg the field ; after each translation, the gantry rotates of an elementary angle, then a new translation is performed, and so on. The sequence of signals during the translation constitutes one projection. This way is cheap and simple, but slow :about 1 second per traverse will give a many minutes exploration time (fig. 3).

Next method on the same principle uses a number of detectors, to make the same number of projections at each translation, the elementary rotation being also multiplied by the same factor. Generally, this number is from 3 to 60, and it takes from 1 minute to 10 seconds per exploration.

Another principle is to suppress the translation, with as many detectors as samples per projection. This can be interpreted as either a conical projection in place of the parallel one, or by rearranging the samples to give back parallels projections.

These machines use 256 to 1024 detector cells, but suppress one mechanical movement. Acquisition time is down to 1 to 5 seconds (fig.4).

The last used principle is to dispose the detectors continuously on a stationnary circle ; only the X-ray tube rotates to give the different projection angles (Fig. 5). It asks for 700 to 2 000 detectors, with the same times as the preceiding.

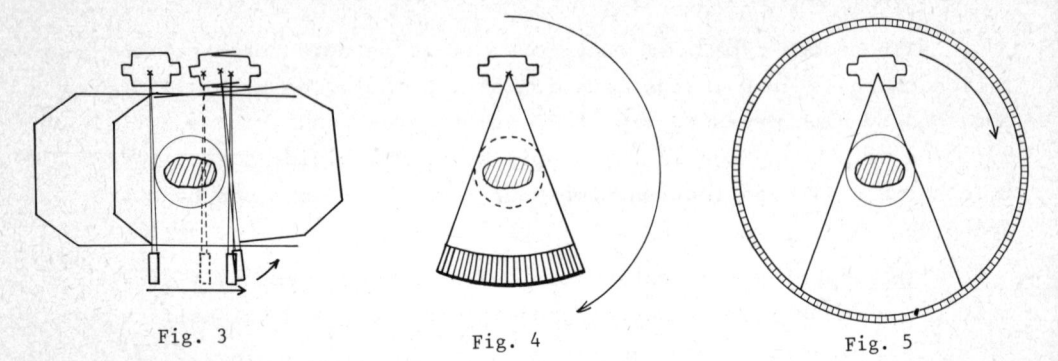

Fig. 3   Fig. 4   Fig. 5

IV - VISUALISATION

From the set of projections, the algorithm procedure allows to reconstruct the value of μ (x,y) on a sampling raster ; the number of reconstructed points is now generally between 256 × 256 and 512 × 512 (actually this number multiplied by $\pi/4$ because it has sense only in a circle).

The number of bits per point must be consistent with the precision available, which asks for 10-12 bits. The values of μ are generally expressed in a conventionnal scale called Hounsfield scale, which derives linearly from the linear attenuation values. In this scale, air is -1000, and water 0.

These values are systematically represented as continuous pictures by a zero-order interpolation (square picture elements-pixels-), analog conversion from a refresh memory and CRT viewing.

To avoid pseudo contouring effects, it is preferable to convert a minimum of 6 bits (or 8 for safety). As the values interesting the user are generally concentrated on a small interval, the digital values are treated between refresh memory and conversion by a multiplication plus offset and truncation. This is called contrast (or density) window ; the window width is the difference between the first non black value and the first white value ; the window mean is the middle of this interval. It is completely real-time adjustable to exploit the information.

Some other classical digital image treatments are also applied : smoothing, edge enhancement, enlargements ; some digital measuring is provided too : distances, angles, densities on Regions of Interest, density profiles, density histograms etc... All theses functions are called by the user on real time image consoles.

V - RESULTS - DISCUSSION

Results quality is evaluated in terms of imaging process : geometric resolution, noise, and contrast transmission ; but any performance figure has meaning only if accompanied by equivalent energy (to know the theorical values for a given object) and patient dose. The noise is described by the standard deviation on a uniform zone ; it is given in percentage of water to air difference. Standard values are between 0.1 and 1 %, with repartitions close to Gaussian.

Geometric resolution and contrast transmission are given by the Modulation Transfer function (or the Line Spread function). The cut off frequency can be found between 3 and 10 line pairs per cm, with LSF Full Width at Half Maximum between 1 and 5 mm.

These features are obtained with 60-75 keV equivalent energy and 0,1 to 20 rads maximum skin dose.

At this day, the only use of CT scanners has been for medical imagery. The first available machine (EMI 1971) was dedicated to head examination. This field has remained the best adapted, both for technical reasons (limited dynamic, fixed organs) and medical reasons (lack of other non invasive examinations). Body examinations begin to be interesting too, but the asymptotic quality in this field has not yet been reached, and other techniques are available (for example Ultra-Sound echography). The problems in this application are : large field (about 500 mm) and heavy attenuation (10^{-3} transmission), compared to 250 mm and 10^{-2} for head, and, mainly, the need for the shortest possible examination time ; the fastest human organs movements would ask for 10 to 50 ms examination time to avoid any cinetic blurring. Modern machines with 1 second time however provide a large amount of information, and prove to be quite valuable ; some improvements now on lab study will probably put them soon to a stable phase, with further developements mainly tied with new progresses in general technology.

REFERENCES

- 1 - RADON - Berichte Saechsische Akad. der Wissenschaften
        (LEIPZIG) 69, 262 (1917).

- 2 - RAMACHANDRAN & LAKSHMINARAYANAN - Proc. Nat. Acad. Sciences U.S.A.
        Vol. 68 N° 9 pp 2236.2240 (September 1971).

- 3 - SHEPP & LOGGAN - IEEE Trans. NS 21 (1974).

- 4 - CHO - General Views on 3-D image reconstr. & CTAT.
        IEEE Trans NS 21 (June 1974).

- 5 - Mc CULLOUGH & PAYNE - X-ray transmission CT.
        Medical Physics - Vol. 4 N° 2 (March/April 1977).

- 6 - HOUNSFIELD - Picture Quality of CT.
        Am J . Roentgenol. 127 : 3-9 (1976)

- 7 - JUDY - The LSF & MTF of a CT scanner.
        Medical Physics - Vol. 3 N°4 (July/August 1976).

- 8 - Mc CULLOUGH & PAYNE - Patient Dosage in CT.
        Radiology 129 : 457-463 (November 1978).

RECORDING MATERIALS AND TRANSDUCERS IN OPTICS

J.J. CLAIR
E.N.S.E.A.   CERGY

The developpment of optical processing was allowed by new recording materials for storage, and for real-time processing. Coherent optical processing has now very high performances and can be extended to incoherent light under the condition of utilizing transducers.

After a non exhaustive presentation of storage materials, non erasable and erasable, we shall describe some types of transducers.

- **RECORDING MATERIALS :**

We need to store information, as data or as repartition of phase or amplitude for filtering, processing, etc .....

To store data it may be either bit-by bit, bit oriented memories, or by holographic memories; the two general set-ups are described in figures 1 and 2.

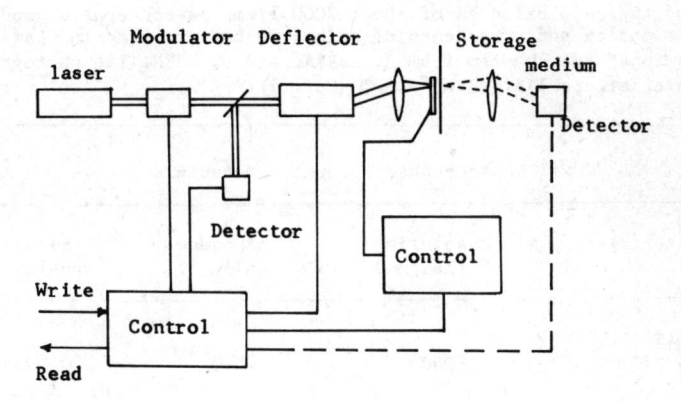

Fig.1: Bit oriented optical recording and read-out system

ref.1

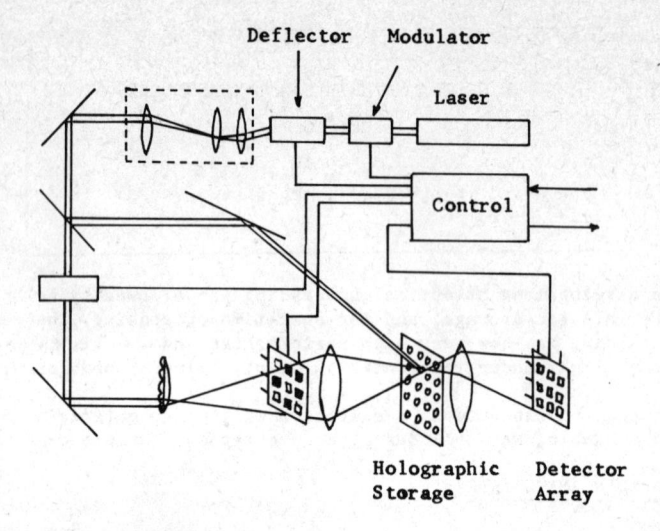

Fig.2: Holographic memory system

The emulsions used have characteristics of high resolution, sensitivity and wavelength dependance. The intensity variations may be transformed in phase variations by bleaching processes or by using directy so-called phase materials.

With classical photographic emulsions we need about $10^{-4} - 10^{-3}$ nJ/μm^2 and the resolution is of about 2000 l/mm. As concerns a good choice for optics and power considerations we have to read in "Laser Applications" the chapter from H. HASKAL and D. CHEN. The photographic materials are listed in Table 1 (Ref. 1).

Table 1: Photographic Recording Materials

Materials	Resolution (lines/mm)	Exposure (nJ/μm^2)	Sensitive wavelength (Å)
KODAK 649 F	6000	7×10^{-4}	4000-6500
AGFA-GEVAERT 8 E 75	3000	2×10^{-4}	6943
10 E 56	2800	5×10^{-5}	5145

A better efficiency is given by phase storage medium and we use polymer solutions ; the principle of storage is the photopolymerisation process in general and we give here the three principal types of the so-called photoresists. We have to note the development of phase gratings for spectroscopy, and the fabrication of videodisc systems (fig.3).

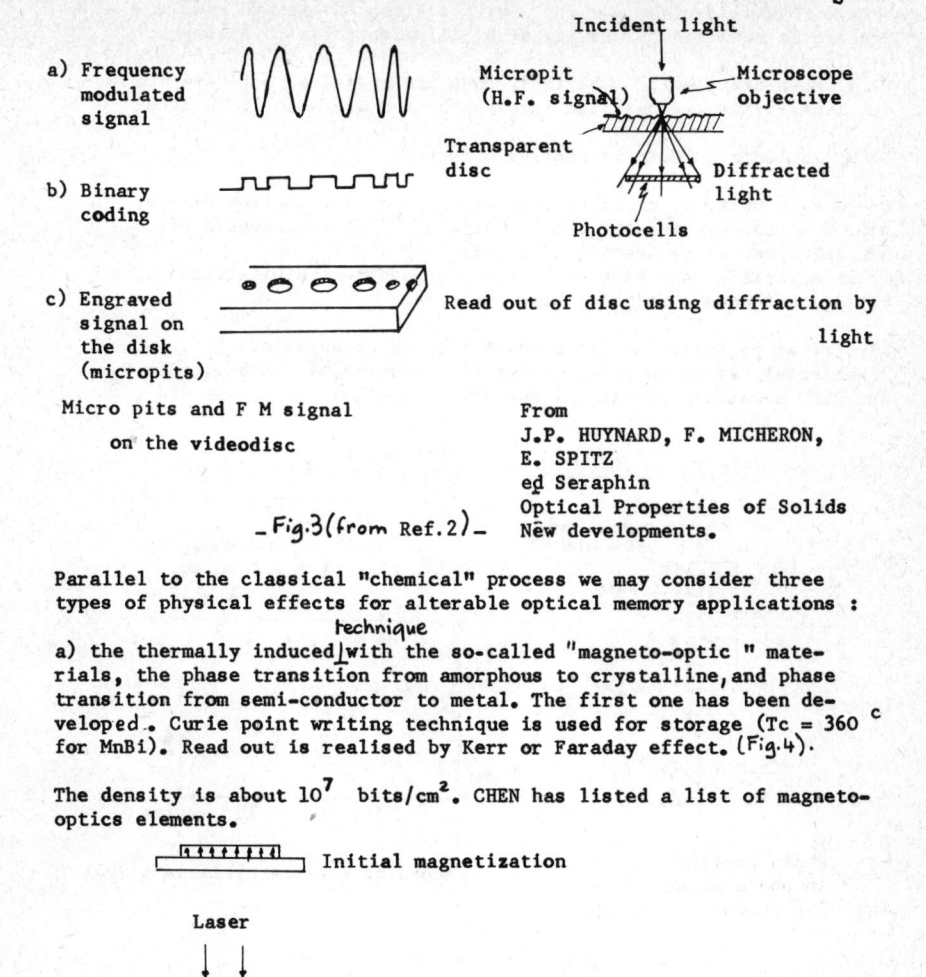

a) Frequency modulated signal

b) Binary coding

c) Engraved signal on the disk (micropits)

Micropit (H.F. signal)
Transparent disc
Incident light
Microscope objective
Diffracted light
Photocells

Read out of disc using diffraction by light

Micro pits and F M signal on the videodisc

From
J.P. HUYNARD, F. MICHERON, E. SPITZ
ed Seraphin
Optical Properties of Solids
New developments.

_ Fig.3 (from Ref.2)_

Parallel to the classical "chemical" process we may consider three types of physical effects for alterable optical memory applications :
a) the thermally induced technique with the so-called "magneto-optic" materials, the phase transition from amorphous to crystalline, and phase transition from semi-conductor to metal. The first one has been developed. Curie point writing technique is used for storage (Tc = 360 °c for MnBi). Read out is realised by Kerr or Faraday effect. (Fig.4).

The density is about 10^7 bits/cm^2. CHEN has listed a list of magneto-optics elements.

Initial magnetization

Laser

Writing "1"

from J.P HUYNARD et al.

Erasure
External field

Fig4: Write and erasure in Mn Bi

The photo-induced technique is carried out with the photochromic materials.

A change of absorption occurs with a given wavelength and erasure is realised by heating or by illuminating with a longer wavelength.
Well known are alkali halides. They are grainless but have low sensitivity and need two wavelengths and are able to fatigue.

b) Ferroelectric photo-refractive effect :

There is a drift of carriers produced by photoionization displacement of charges creating an electric field, and resulting in an electrooptic effect (Δn). Erasure is by heating or light.
These materials have high resolution, a good reversibility, no fatigue but need a relative high exposure.

Electrical polarization can also be induced by an external field in materials having no polar crystalline lattice such as $LiNbO_3$, $LiTaO_3$ and PLZT ceramics. Erasing is thermal or optical.

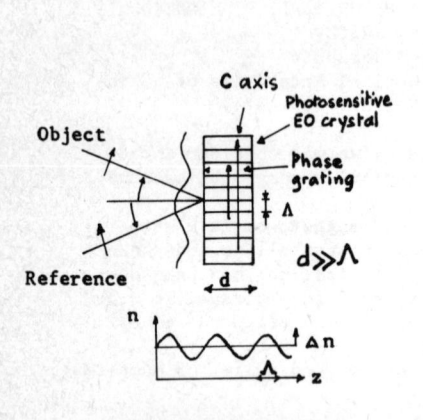

_ Fig. 4 _ (after Ref. 2)

Phase grating in photo sensitive Electrooptic crystals

Recording erasure cycle in $LiNbO_3$

c) The photon activated effect is illustrated by the thermoplastic effect.
We separate the exposure developing state. We use a sandwich technology. A layer of thermoplastic is in contact with a layer of photonconductor. A corona discharge is used and the image is transformed in a modulation of charge. Heat is applied so to deform the surface by the electrostatic charges.

The recording and erasure process is schemed below:

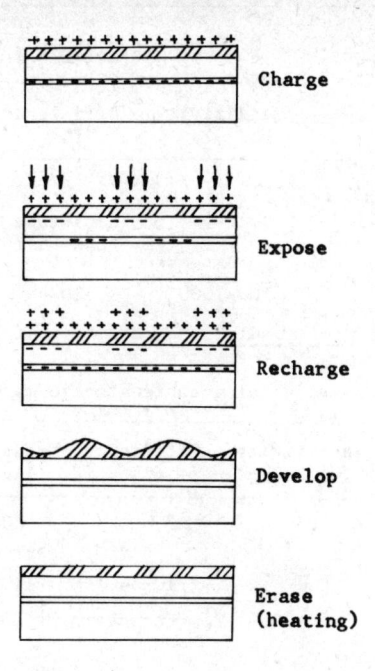

Charge

Expose

Recharge

Develop

Erase
(heating)

Recording process in thermoplastic

The two following tables give a comparison of non erasable and erasable materials with sensitivity and resolution and the properties of the photorefractive materials, from J.P. HUYNARD et al op-cité 1974.

Table 2 : Non-erasable holographic storage materials

Material	Sensitivity (mJ/cm2)	Efficiency (%)	Resolution (lines/mm)
High resolution photographic plates 649 F	0,1	Amplitude : 6 Phase : 15-30	4000
Photoresist	100-1000	20-40	3000

Erasable holographic storage materials

Material	Sensitivity (mJ/cm2)	Efficiency (%)	Resolution (lines/mm)	Erasure
Photochromic	100	3	2000	Optical heating
Thermoplastics	0,1	10-34	1000	Heating
Magnetooptic MnBi	10-100	Faraday:0,01	1000	Magnetic

Table 3 : Photosensitive electrooptic crystals

Material	Sensitivity (mJ/cm2)	Efficiency (%)	Resolution (lines/mm)	comments	Thickness (mm)
LiNbO$_3$, non doped	$10^4 - 10^5$	10	1000	Thermal or optical erasure: thermal fixing	1-10
PLZT	100	1	100-1000	Control of diffraction efficiency	1

In Fig. 5 we give the required writing energy and resolution for various types of material (H. HASKAL, D. CHEN).

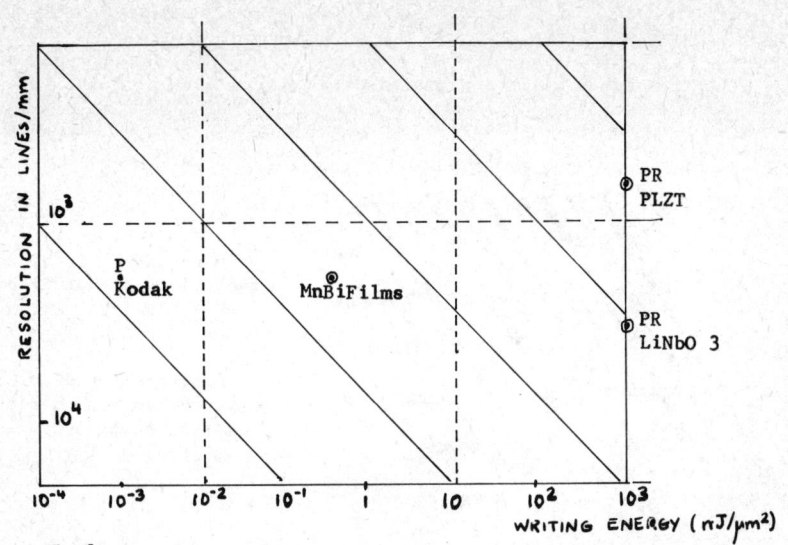

Fig 5 : Required energy/resolution

II - TRANSDUCERS :

We have to consider two types of transducers, the optically sensitive and the electronically adressed

The first ones may be adressed by scanning a laser
by an incoherent modulated signal
by a recorded pattern

In the electronically ones, we may have the scanning :

- of an electron beam or the adressing of a matrix
- we shall limit our description to two types of transducers: the liquid crystals systems and Pockels effect systems.

a) **Liquid crystals :**

The transducer will be a matrix display (Fig.6)

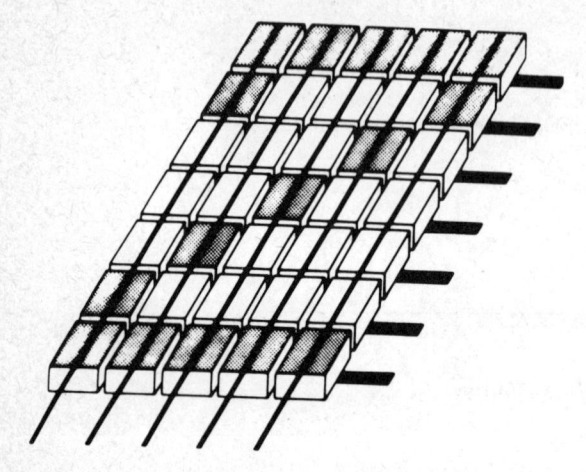

we have not to consider here the technology of matrix adressage

as concern liquid crystal we have to consider three phase, nematic, cholesteric smectics.

Fig.6 : The matrix display (Ref. 3)

The first two types are more utilised. The difference lays in the "structure" of the molecules materials which are in general biphenyl compounds or cholesteriol derivative.

we use either the dynamic scattering, of the phase charge effect, or the variable birefringence.

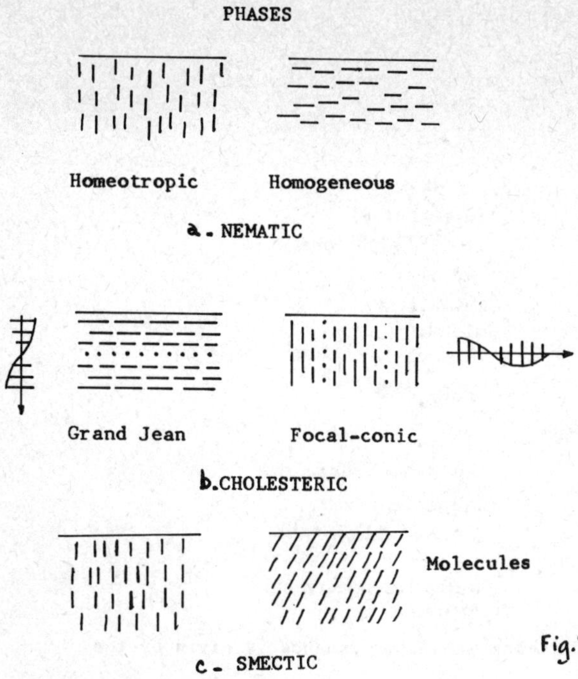

PHASES

Homeotropic   Homogeneous

a. NEMATIC

Grand Jean   Focal-conic

b. CHOLESTERIC

Molecules

c. SMECTIC

Fig. 7: MOLECULAR ARRANGEMENT OF LIQUID CRYSTAL MESOPHASES

Fig. 8: SCATTERING WITH LIQUID CRYSTAL

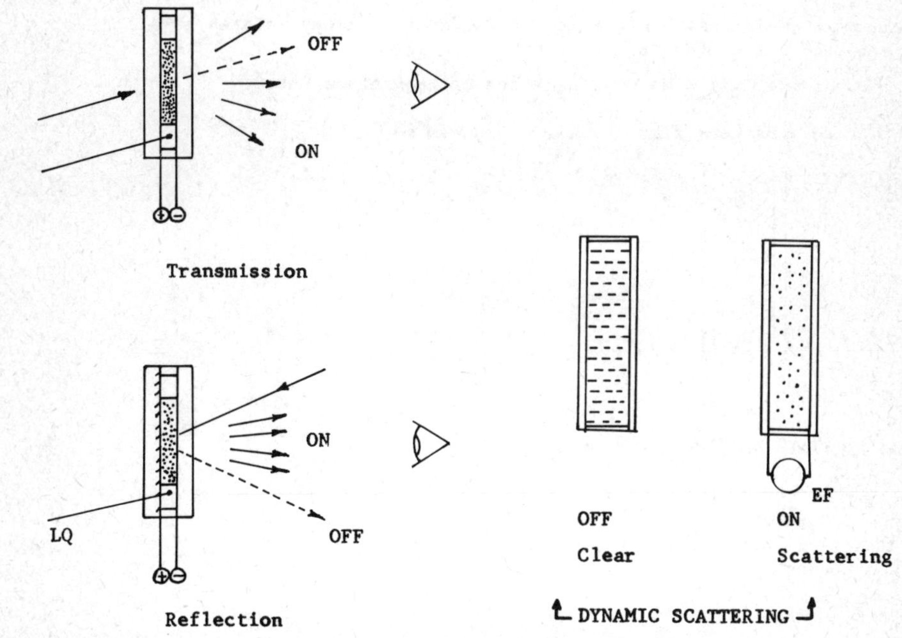

OFF
ON

Transmission

LQ
ON
OFF

Reflection

OFF   ON
Clear   Scattering
                EF

DYNAMIC SCATTERING

The disadvantage seems to be the relative slow speed, the lack of resolution and the diffusion : but liquid crystals are used for displays and screens.

a) We have to note that much work is devoted to non emissive displays. The characteristics have to be as follows:

1. Type of Construction — Flatness
   Ruggedness
   Versatility of format
   Simplicity

2. Legibility — Brightness
   Colour   (from J. KIRTON
   Contrast over wide   Ref 4)
   Viewing sector

3. Life
4. Temperature Range
5. Drive Cost — Operating voltage
   Operating power
   Discrimination ratio
   Speed
   Refreshed performance
   Memory

We may consider electrochromism whose performance is given by the basic equation :

$$Mn + C- + A+ = A+ Mn-1$$
non colored   colored

where M is an ion which can exist in different valency states ; "A" is a mobile cation (H+ ic)

$$WO_3 + (H+ +x') = H+ WO_3$$ formation of tungstene bronze.

The devices are very simple and are schemed in Fig.9 :

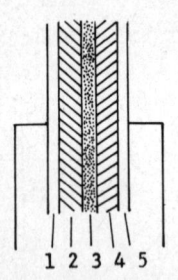

Fig.9: SOLID STATE ELECTROCHROMIC CELL   (from H.R. ZELLER
                                                Ref 4)

1-5 electrodes
2 E C material
3 solid electrolyte
4 conductor

There are also electrophoretic displays whose mode operation is given here :

Fig. 10 : Electrophoretic device

Ref 4

There is a suspension of pigment particles in a densely colored liquid pigments form a lyophobic sel.
FINALLY well-known ferroelectric displays are (PLZT ceramics)
(Pb, La) (Zr,Ti) O3)
With PLZT we may control birefringence
light scattering
surface deformation

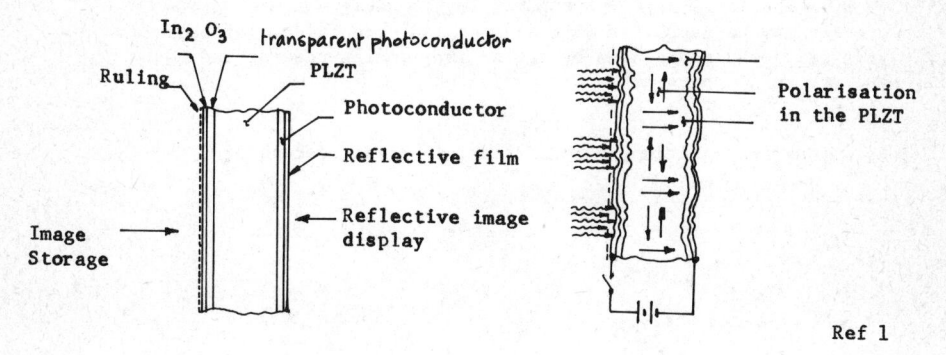

Fig.11 : FERICON STRUCTURE - SURFACE DEFORMATION

Ref 1

b. Pockels Devices :

Two types of light valves are developed, one from ITEK (Prom) and the Phototitus from LEP.

The sequence is the following one :

We apply a half wave voltage (2000 v); excited photoelectons are displaced by the applied field and stored on the insulator film. A charge field cancels the applied field. By short circuiting the electrodes the field is reversed. The image to be recorded is projected on the PROM (blue light). In the bright parts field is cancelled, and is inchanged in the dark one. Reading is done by reflecting red light.

Contrast conversion is achieved by reapplying to the PROM the initial voltage : 300 l/mm is the resolution, sensitivity is 10 nJ/cm^2, storage is few minutes in dark.

PHOTOTITUS :

We use a crystal (KD2 PO4) justabove the Curie temperature. Pockels effect is a linear electro-optic effect occuring in certain crystals where a linearly polarized light is made elliptical by electrically induced birefringence.
Modulation may be obtained with an electric field; the phase shift is proportional to the field E so the intensity transmitted is of the form of "sin 2 kv".

In the phototitus, an image is projected on the selenium layer, while a field is applied across the device, this results in the transport of charges to the K D P crystal, positive or negative charges may be deposited on the adressing side (signal of applied voltage), reading is done by reflection; during reading and evasure, electrodes are short circuited. (see Fig. 12).

Applications are addition, substraction, spectral processing .

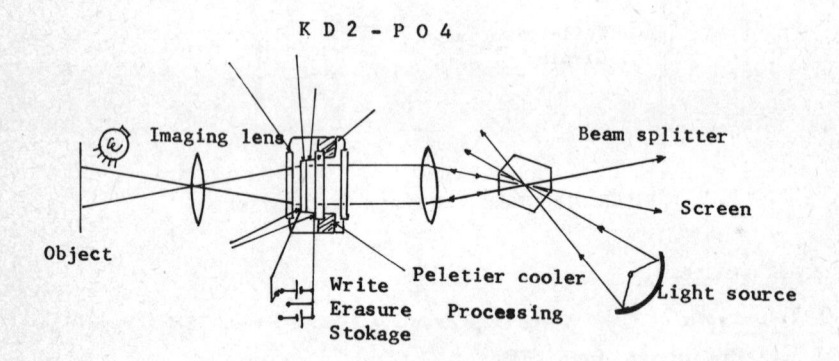

Fig.12: PHOTOTITUS OPERATION

Ref 5

In conclusion we give a brief review of transducers and materials. Information density is expected to be up to 10^8 bit/cm2 (10^{10} in volume) Continuing research effort is desirable in new physical phenomena.

For devices the liquid crystals ,PROM and KDP are promis ing.
Storage appears with PROM and DKDP but it requires large voltage
and difficulties in fabrication. Thermoplastic seems interessant
for non long lifetime;the search for new devices of course will
stimulate research in many fields. (Ref 6).

REFERENCES :

Non emissive electrooptic displays

1   Laser Applications volume 3
    ed. MONTE ROSS SCAD PRESS 1977
    article from D. CASASENT, H. HASKAL and D. CHEN.

2   Optical properties of solid, new developments
    by O. SERAPHIN 1976 NORTH HOLLAND

3   Progress in Electro-optics
    CAMATINI Nato adv. study institute series Plenum 1973.

4   by A.R. K METZ, F.K. VON WILLISEN
    Plenum Press 1975.

5   Photonics by BALKANSKI, P. LALLEMAND, GAUTHIER VILLARS 1973.

6   Progress in optics
    volume 16 1978 NORTH HOLLAND
    article CLAIR et al.

SOME APPLICATIONS IN HYBRID IMAGE PROCESSING

D. CHARRAUT

Laboratoire de Physique Générale et Optique
(associé au CNRS LA 214 : *Holographie et Traitement Optique des Signaux*)
Université de Franche-Comté, 25030 Besançon Cedex, France

The natural adaptation of optical methods to record light distributions as they occur to an observer, and the realization, in real time, of simple mathematical operations, accords a priority to the optical techniques in the image processing. The development of these methods is closely connected to the utilization of coherent light sources. These whole operations concern the usual Fourier transform analysis of 1-D or 2-D signals.

Facing the advantages of the analogic optical processing, one may object the limited range of operations available in this way. Automatic classification and pattern recognition need statistical operations which can not be performed by pure optical processing. Moreover, the possibilities to obtain large series of data and there analysis place the computers in a privileged position.

Among the activities of our laboratory, we have developed a hybrid - optical/digital - image processing in which the image analysis yields optical data, that are the start of point of a statistical digital computing expansion.

Applications to handwriting analysis are presented, starting from series of complete pages, as long as the extraction of cadastral structures from aerial photographs.

I- COHERENT OPTICAL PROCESSING - SOME APPLICATIONS

Optically, the information can be defined as the amplitude distribution of light which emerges out of a given photographic image. Then the optical systems act simultaneously on the entire spatial information so coded. For example, in the focal plane of a simple convergent lens, the light repartition can represent the Fourier transform of a signal placed in its object focal plane. This mathematical operation is achieved from a number of points in order of 20 millions which represents an usual photograph. The direct access to the Fourier spectrum permits the transposition of the well-known electrical filtering techniques in the spatial frequency domain (1), (2).

1) - *Amplitude filtering*

An illustration of the spatial filtering is given on *Fig. 1a*. The spectrum bandwidth of a printed text is gradualy enlarged to the high frequencies : one observes the successive restitution of lines, words, letters and details of the graphism. The corresponding frequency bands of these coding elements can be defined in this way (3).

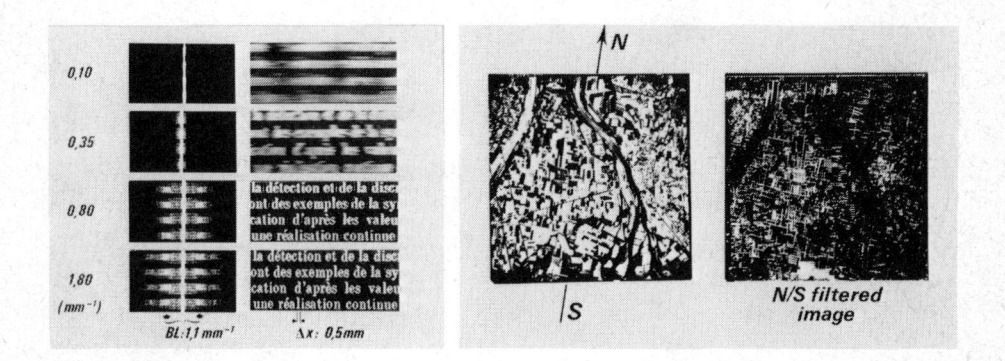

Fig. 1a

Fig. 1b

A Fourier transform property expresses a correspondence between image and spectrum geometrical directions. Thus, an oriented structure along a given direction can be selected by means of a sector filter which acts as a small aperture (4). *Fig. 1b* shows a kind of directional filtering to underline remains of Roman cadastral structures from an aerial survey. The N/S axis to be know as a well direction, the enhancement of corresponding elements is visible in the filtered image, since the angular position of the filter is close to this direction.

2) - *Complex filtering*

In optics, the holographic devices are suitable to resolve identification and location problems (5). A typical correlator is presented on Fig. 2.

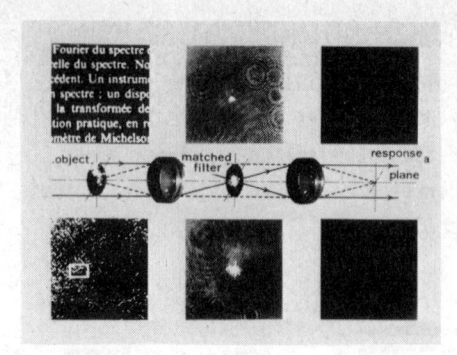

Fig. 2

The signal is positionned in the input plane of the correlator ; its spectrum is displayed in the image focal plane of the first lens. A complex filtering is obtained by introducing in this plane a Fourier spectrum hologram of a memorised signal - here a word or a piece of a radar map-. In the image focal plane of the second lens, a spot points the correlation result of the two signals. The identification study of hebraïc letters in a text page gives successful applications defining a ressemblance degree criterium based on the mesure of correlation function peak (6).

3 - Limitations of optical processing

Facing the general problem of variations in shape and size in classes of images, correlation-peak criteria only gives a gross approximation : that is the case of cursive handwriting letters. One other approach consists to evaluate statistical invariants by means of data analysis. A numeric processing is suitable to realize this analysis so that the images are digitalized.

In fact, others problems arise if we adopt this approach. The set of data can attain a large number of significative parameters : 2000 samples are necessary for one text line analysis with a resolution egal to the drawing size. In this case, a classical covariance matrix analysis needs the establishment of a 2000 x 2000 rank matrix. This pure numeric process needs large size memories as well as a long computing time which increase the analysis cost.

The two examples present here the possibilities to reduce the space-bandwidth product by modeling the spectral content of images. Optical Fourier transform takes a dominant part as a preprocessing step and compresses the information into a few number of samples to be digitalized.

II - MODELISATION OF THE INFORMATION IN FOURIER SPECTRA

1) - Handwriting modeling : the writer modulation transfer function

According to the regularity of the draw ing, we suppose that a writer gives at time t a degraded versus of a model letter he learned at time to : he acts as a linear imaging system without aberrations. The definition of a corresponding impulse response - the form factor - yields to represent the handwriting by a convolution (7). An experimental verification of this modelisation is given on Fig. 3.

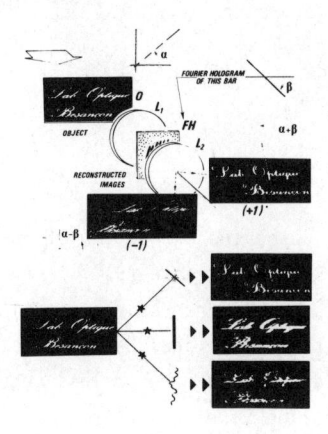

Fig. 3

Three form factors are stored in a Fourier hologram that is put in the filtering plane of an optical correlator. The model being in the input plane, three different writings are displayed in the output plane. Under this assumption, the Fourier spectrum of a handwritten page is the product of the writer's modulation transfer function by the resultant of the models, to be constant as the number of letters is sufficient.

Preliminary studies have shown that the most important part of the variations of the transfer function occurs in the low frequencies. Then, defining a smaller space - bandwidth product, the number of samples to be digitalized is reduced by a factor 100 (8).

Moreover it is possible to separate two variables controlling a page : the first locating each letter and the second being the form factor. On *Fig. 4.* a sphero-cylindrical lens removes the location parameters by spatial filtering and delivers the signal which is sampled by a photodiode.

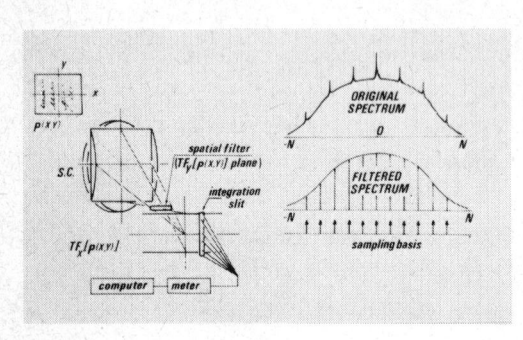

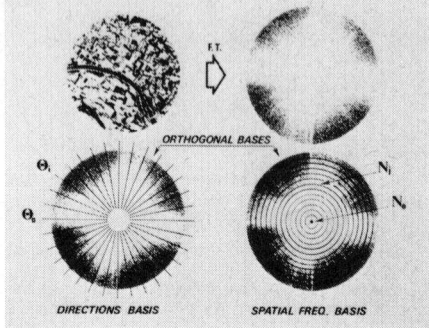

Fig. 4   *Fig. 5*

2 - *Aerial photograph modelling*

The second example concerns the extraction of Roman cadastral structures from aerial photographs of Southern France (9).

Each photograph must be considered as representing a mixture of two structures : the Roman and the actual one. Each is a statistical distribution of generally rectangular fields, with more or less regular dimensions, orientated along a given direction. The Fourier spectrum of such a distribution presents two characteristics :

- it is a grating - like spectrum,
- and this grating is stretched along an axis that gives general orientation of the cadastral structure.

Two kinds of criteria are available in order to determine a structure :

- orientation criteria if the two structures exhibit sufficient angular separation, and
- spectral content criteria if the directions are closely displayed.

Then the information must be digitalized in two spatial sampling : a directional and a spatial frequency one (Fig. 5).

The first consists of 36 directions separated from 5° to 5°, taking account of the knowledge of the aera, and the second case is composed of 44 spatial frequencies which are multiples of the fondamental frequency of a typical Roman cadastral structure.

III - STATISTICAL ANALYSIS

Different writers -that is different classes of transfer function, or different cadastral structures - that is different classes of grating spectra, are to be recognized starting from digitalized samples of Fourier spectra. Although a Fourier transform gives a dimensionnaly reduced description of data, it is not sensitive to statistical variations characterizing class properties. Then a proper mathematical tool, that can take into account relative variations over all the spectra, will permit this classification.

A transformation of coordinates, the Karhunen-Loeve transform, constitutes a typically well-adapted tool. It consists of computing the covariance matrix of data in the starting space (10).

The diagonalisation gives a set of eigenvectors which supports another equivalent space. Ordering eigenvectors by decreasing eigenvalues rank and selecting those associated to the largest, one definies an optimal reduced space -optimal in point view of minimization error truncation and information loss_.

1) - Writer recognition by the Karhunen-Loeve transform

In the case of writer recognition (11), the data to be analysed consist of 16 pages written by the German poet H. Heine (denoted H_i) and 6 pages written by his secretaries (noted S_j). The covariance matrix is computed and diagonalized : the sum of the first two eigenvalues represents more than 90% of the total dispersion. Consequently the corresponding eigenvectors Kl_1 and Kl_2 provide an optimal description for the considered pages. Fig. 6a shows a mapping in the (Kl_1, Kl_2) plane and the two domains associated with the two kinds of data. These classes are quasi-lineary separable, except the slight overlapping of points S_1, H_2 and H_9. An alternative to strong distance criteria is in the use of similarity measurement able to describe the tendency for points to be representative of a same classe. The shared near-neighbours rule (12) appears versatile enough to the evaluation of writing similarities by separation of ambiguous points in a classification tree (Fig. 6b).

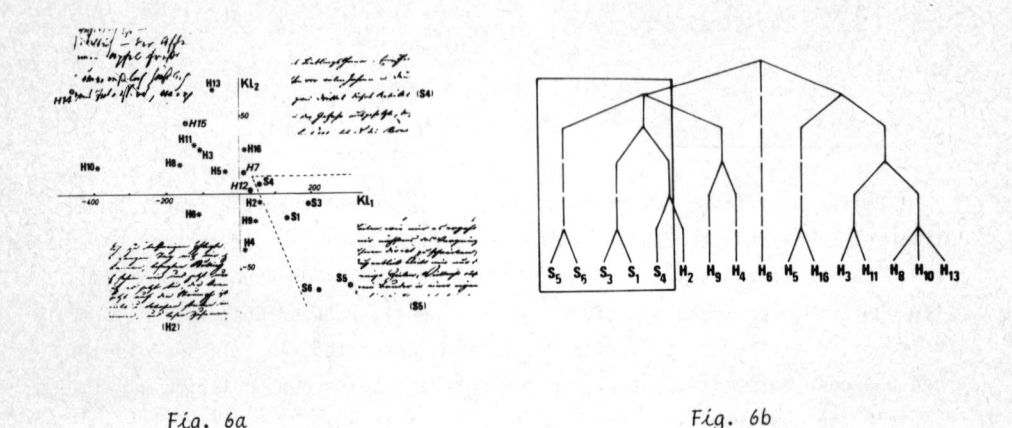

Fig. 6a   Fig. 6b

2) - Enhancement of cadastral structures

Facing the interpretation of Fourier spectra in the case of scattered cadastral structures, the K.L. transform appears as a well-adapted process (13).
An accurate analysis of the spatial frequencies of each direction is capable of distinguishing close directions associated with different interwoven cadastral structures. Conversely, spectral information is definable by searching for statistical variations applied to spatial frequencies as they are considered as functions of directions.
In both cases, covariance matrix are computed, the dominant eigenvectors of which are expected to facilitate the class separation. Fortunately the classification is directly achieved by the Karhunen-Loeve transform (Fig. 7).
The analysis of directions in the spatial frequencies bases shows three classes, separated from 5° corresponding to three cadastral structures sustained by three roads which define an angle of 35°. The analysis of spatial frequencies in the directions bases emphasizes are cluster of medium frequencies which seems to be significant. Another cluster located in high frequencies is associated with a common noise ; the low frequencies class is to be related to the continuous background.

It is convenient to visualize an optimal image, from the point of view of class separation, by using the dominant eigenvectors as Fourier filters *Fig. 8* shows on the one hand pieces of a cadastral structure extracted by statistical analysis of spatial frequencies and on the other hand three groups of axes extracted by statistical analysis of directions.

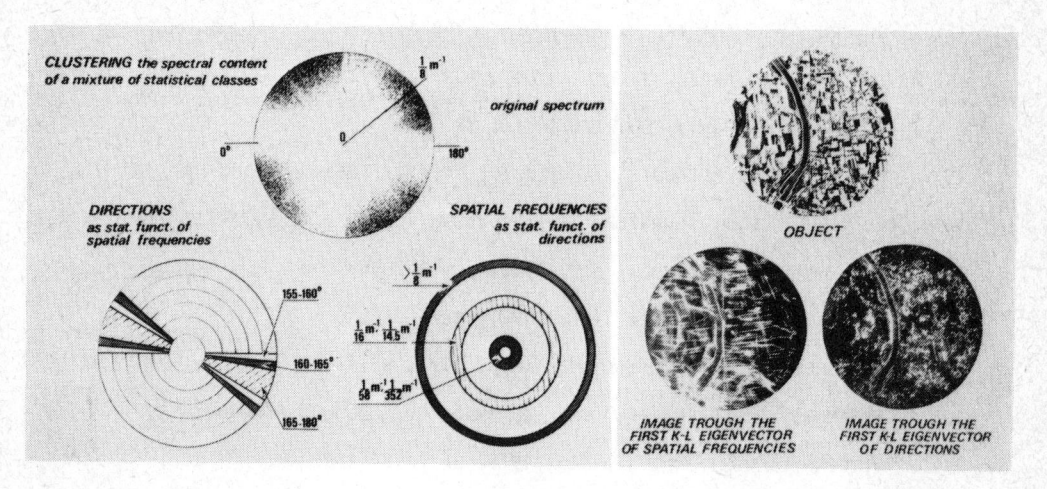

Fig. 7

Fig. 8

IV - CONCLUSION

Optical informations processing is adapted to the statistical analysis of classes of geometrical elements in images. The dimensionality of the problem is reduced by modeling the information and by computing its intrinsic dimensionality. Estimation procedures avoid the computational cost of a direct analysis of the spectral content of images. Dominant eigenvectors play the role of spatial filters and provide principal images that carry a given statistical information on differents observations.

Statistical spatial filtering not only concerns the analysis of single photographs but also series of images. In this case, eigenvectors are extracted after learning on a set of training signals ; they can be widely used to process images to be classified. It is the case of medicinal images to obtain automatic classification or diagnostic (14).

REFERENCES

(1) A. MARECHAL, P. CROCE : *C.R. Acad. Sc.*, __237__ B, 1953, pp. 607-609

(2) J.W. GOODMANN : *Introduction à l'optique de Fourier*, Masson et Cie, Paris, 1972.

(3) J. DUVERNOY, D. CHARRAUT : *Opt. Comm.*, __14__, 1, mai 1975, pp. 56-60.

(4) A. FONTANEL, G. GRUAU, J. LAURENT, L. MONTADERT : *Actes du IIe Symposium International de Photo-Interprétation*, septembre 1966, Paris, pp. III.14-III.32.

(5) J.Ch. VIENOT, J. BULABOIS : *Rev. Opt.*, __44__, 11, 1965, pp. 588-592.

(6) J.Ch. VIENOT, J.M. FOURNIER : *Israel Jl. of Techn.*, __9__, 3, 1971, pp.282-287.

(7) J. DUVERNOY : *Opt. Comm.*, __2__, 7, 1973, pp. 142-145.

(8) D. CHARRAUT : Thèse de 3e Cycle, 1977, Besançon.

(9) P.Y. BAURES : Thèse Docteur-Ingénieur, 1977, Besançon.

(10) S. WATANABE : *Conf. Inform. Theory*, 1965, Prague, pp. 635-660.

(11) J. DUVERNOY : *Appl. Opt.*, __6__, 15, 1976, pp. 1584-1590.

(12) R.A. JARVIS, E.A. PATRICK : *IEEE Trans. Comp.*, C22, 1973.

(13) P.Y. BAURES, J. DUVERNOY : *Appl. Opt.*, __17__, 1, novembre 1978, pp. 3395-3401.

(14) J. DUVERNOY : *Opt. Comm.*, __27__, 3, décembre 1978, pp. 333-338.

X-RAY IMAGE DETECTORS

J. MILTAT

Laboratoire de Physique des Solides
BAT.510, Université Paris-Sud
91405 ORSAY, France

INTRODUCTION

Image detectors used in X-Ray Topography are either photographic emulsions, or electro-optical systems aimed at providing a direct view of the diffracting crystal.

The resolution requirements are in conventional experiments determined by beam divergence and/or wavelength spread. A resolving power of 1 to 2 µm is usually requested for high quality topographic work. Since no optics in the classical sense is available for X-Rays, the basic picture element has an area typically of 1 to 4 µm^2.

Better resolution may even be requested in cases where extremely low divergence and wavelength spread beams are obtained via complex monochromators [1].

Amongst factors determining image potential quality, spatial resolution measured with an appropriate technique, dynamic range and linearity appear to be the most important. Before examining the potentialities of the two types of image detectors mentioned above, typical figures of photon fluxes available in X-Ray Topography experiments need to be considered because they impose intrinsic limitations to image quality.

I.- X-RAY PHOTON FLUXES

The most intense X-Ray sources are nowadays synchrotron radiation sources. For instance, the synchrotron radiation facility at LURE, Orsay (F) delivers, when operated at 1,72 GeV, 200 mA a continuous bremstrahlung with flux, at 20 m from the source, ranging from 0,4 to $1,6.10^{12}$ photons/sec/mm^2/Å according to wavelength. In a white beam topography experiment, a nearly perfect crystal will accept a wavelength spread $\frac{\Delta\lambda}{\lambda}$ of the order of 10^{-4} and, at $\lambda = 1$ Å, the useful photon flux impinging on the crystal will be about 10^8 ph/sec/mm^2.

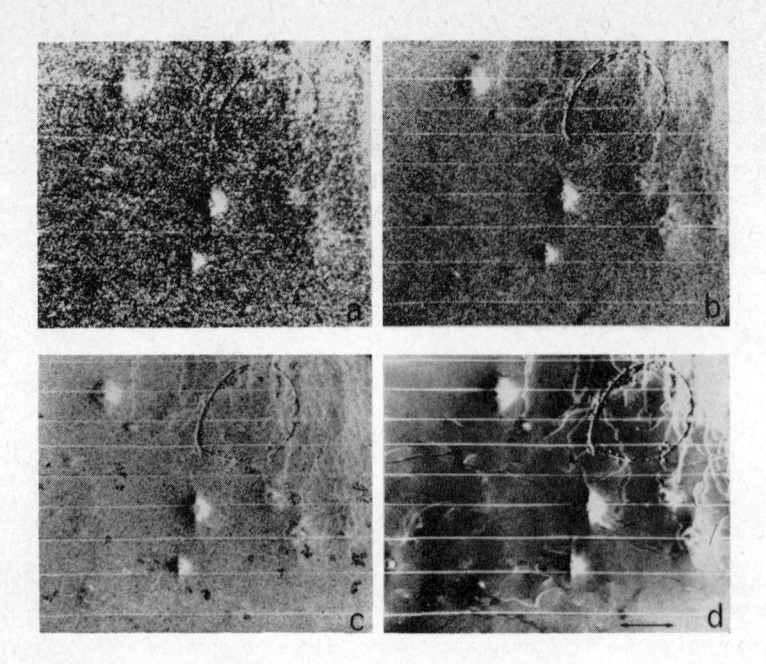

Fig. 1
Topograph of a Fe-Si single crystal recorded with decreasing speed emulsions : a) Kodak Kodirex, b) Kodak Industrex A, c) Kodak Indutrex M X-Ray films, d) Ilford L4 Nuclear emulsion 50 µm thick - MoKα radiation (18 keV). Scale mark : 200 µm.

This flux may suffer through diffraction an attenuation ranging from a few units to several orders of magnitude according to experimental conditions.

A 1,6 kW Elliott rotating anode X-Ray generator emitting MoKα characteristic radiation has a "useful" out put flux of about $\frac{1}{80}$ of the above value. By "useful", it is meant that the beam is sufficiently collimated in order to prevent superposition of the Kα_2 and Kα_1 images. The "useful" photon flux is therefore of the order of $2,5.10^6$ ph/mm^2/sec. For the sake of comparison, it may be shown that a low power Na lamp with luminance $1,5.10^4$ nits delivers at 0,5 m from the source a flux of 10^8 ph/mm^2/sec in a light tube of section 1 mm^2, assuming an emission surface of 1 mm^2. Therefore visible light fluxes are in most cases (television, photography,...) much larger than fluxes obtained from the most powerful X-Ray sources.

II.- PHOTOGRAPHIC EMULSIONS

X-Ray photographic emulsions are characterized by a high silver halide content (AgBr mostly) in order to optimize absorption. Various emulsions differ through their respective thicknesses and grain sizes before and after development. The larger the grain size is, the quicker the emulsion is, but the poorer the resolving power becomes. Fig. 1. shows the same topographic image of a FeSi crystal recorded on four different emulsions. It is clear that only the slowest one (finest grain size) is a really satisfactroy recording medium. Faster emulsions may however be used for quick inspection and in the limited scope of this paper, only a fast, coarse grain emulsion, Kodak "No Screen" X-Ray film and a slow, fine grain emulsion, Ilford L4 "Nuclear" plates will be compared (see table 1).

a) Dynamic range - Linearity

The properties of photographic emulsions are commonly expressed in terms of density curves [2] :

$$D = \log_{10} \frac{i_1}{i_0} = f(\log_{10} It)$$

TABLE I

	"No Screen" Kodak X-Ray film	Ilford L4 Nuclear Plates
Grain size (µm) before development	2,5 [3]	0,14 [5]
Grain size (µm) after development	6,5 [3]	0,25 [7]
Emulsion thickness (µm)	2 x 28 [3]	25-50-100 [5]
Indicative relative speeds at 15 keV	1	1/18
Photon lumination (ph/mm^2) for D = 2 at 15 keV	$2,7.10^6$ [3]	$\sim 50 \cdot 10^6$
Minimum observable contrast over area $\epsilon \times \epsilon$ (µm^2) C	0,15 0,03	0,3 0,06 0,03 0,006
ϵ	10 50	1 5 10 50

where
- i_0 is the intensity of light impinging on the developed emulsion
- i_1 is the transmitted light intensity
- It is the X-Ray lumination
- D is the optical density.

In opposition, however, to visible light photons, only one absorbed hard X-Ray photon is necessary to render one grain developable. A linear relation between density and lumination may therefore be expected. One absorbed photon however emits photo-electrons (photoionization), the average number of which is a function of energy. Photo-electrons render an average additional number n_1 of grains developable.

Despite of this difficulty, it is still desirable to plot D as a function of the lumination It. Figure 2 shows such plots for Kodak No Screen according to [3] and for L4 Nuclear plates according to [4]. The density appears to be a linear function of It up to a density close to 1. Above this value, deviation occurs, leading to saturation. In X-Ray topographs, a background density between 1 and 2 is a common value. Great care must therefore be taken when relating measured densities to impinging X-Ray fluxes.

The dynamic range of X-Ray emulsions is large as may be seen from fig. 2.

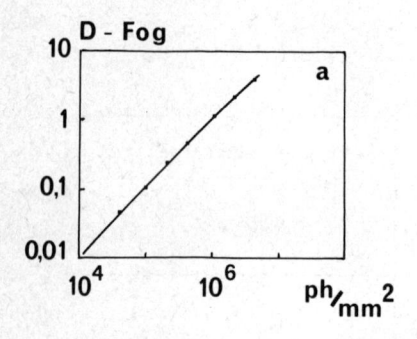

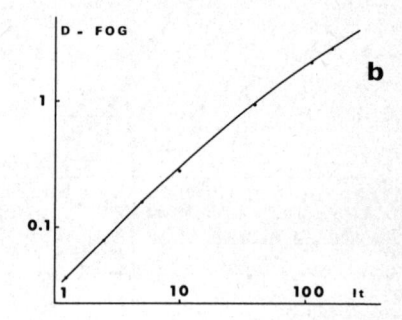

Fig. 2

Density minus Fog. versus lumination for a) Kodak No Screen X-Ray film at 8 keV according to [3] ; b) Ilford L4 Nuclear emulsion, 100 μm thick, (50 kV), Cu target, Al filtration, according to [4].

b) Spectral response and sensitivity :

The behaviour of emulsions in the energy range 1 to 100 keV may be qualitatively understood by considering that photo-electron emission increases as energy increases whereas absorption decreases. Maximum sensitivity is therefore expected at intermediate energy. Absorption moreover varies abruptly at absorption edges

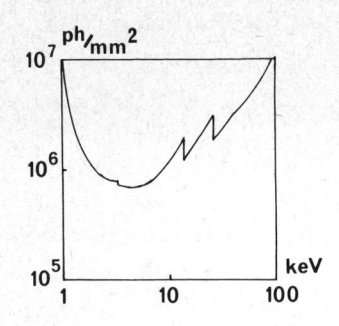

(0,486 Å [25,5 keV] and 0,920 Å [13,5 keV] for AgBr). Figure 3, extracted from ref [3], indicates the number of photons per mm^2 necessary to reach density 1 above fog as a function of energy. A similar behaviour is anticipated for other emulsions.

c) Resolving power

Fig. 3
Exposure to obtain diffuse density 1 above fog for Kodak No Screen X-Ray film as a function of energy according to [3].

Modulation Transfer Functions (MTF) [2] of X-Ray emulsions are not available and only general considerations do lead to acceptable figures for resolving power.

Photographic emulsions resolution is limited by grain size, photo ionization spread and statistical fluctuations.

According to the theoretical model proposed by Brown et al. [3], the number of developable grains per absorbed photon appears to be of the order of 15 for L4 nuclear emulsions and for energy 15 keV. This figure is close to the maximum value measured by Herz [6] in this energy range.

Photo electron tracks may be randomly distributed and only average experimental photo ionization spread may be evaluated. Lang [7] gives the following practical attainable resolutions for L4 plates as a function of X-Ray wavelength :

1 µm   at λ = 1,54 Å (8 keV)
1,5 to 2 µm at λ = 0,71 Å (18 keV)
2 to 3 µm   at λ = 0,56 Å (25 keV).

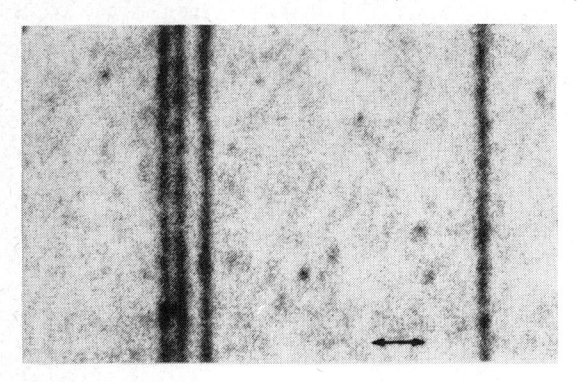

Fig. 4
Closely spaced dislocations in an epitaxial layer of $Ga_{0,7}Al_{0,3}As_{1-y}P_y$ on GaAs (Ilford L4 emulsion 50 µm thick). Weak beam technique. λ = 1,2378 Å. Scale Mark : 10 µm. Courtesy of J.F. Petroff, P. Riglet, M. Sauvage.

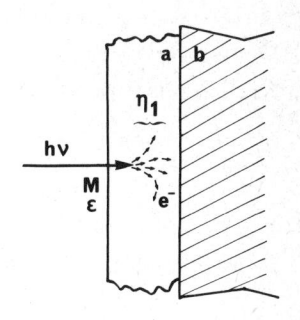

Fig. 5
Photographic emulsion (a) on substrate (b). Photoelectrons which render η_1 grains developable are emitted by the absorption of one X-Ray photon.

A published figure relative to "No-Screen" X-Ray film indicates a resolution of 50 line pairs/mm, that is 10 μm [8]. These figures are well confirmed by practice and Fig. 4 shows very closely spaced dislocations which are clearly resolved on a L4 Nuclear plates.

The last point to be considered now is statistical fluctuations. Let us consider a photographic plate (fig. 5) absorbing X-Ray photons with absorption efficiency α. Each absorbed photon is supposed to render n_1 grains developable. The signal around point M of the emulsion is :

$$S = \alpha \, n_1 \, \nu \, t \, \epsilon^2$$

where ν is the photon flux
t the exposure time
ϵ^2 is the surface element around M.

The photon noise may be defined as the standard deviation from S, which, may, if Poisson distributions are assumed for $\alpha \, \nu$ and n_1, be expressed as [9,10] :

$$N = [\alpha \, n_1 \, \nu \, t \, \epsilon^2 (1 + n_1)]^{1/2}$$

Let us consider the image of a defect of area ϵ^2 with contrast $C = |S_d - S_p|/S_p$ where S_p and S_d are the signals from the perfect and deformed regions, respectively. A defect is detectable if $R = |S_d - S_p|/N_p$ is larger than a threshold value $k > 1$.

R is equal to :

$$R = \epsilon \, C \, [\alpha \, \nu_p t \, \frac{n_1}{1 + n_1}]^{1/2}$$

where ν_p is the photon flux from the perfect region. This relation enables one to estimate the minimum contrast over a given area which may be detected. The determination of k is however delicate as it depends on the entire chain of instruments used in the analysis of the image. If a small value of k (k = 1,5) is choosen, Table I indicates the values of C obtained for α = 0,5 , n_1 = 15 and $\nu_p t$ such that density 2 is obtained.

III.- ELECTRO OPTICAL IMAGE DETECTORS

Two recent reviews [11, 12] have been dedicated to electro-optical systems used in X-Ray Topography. Two broad categories of such detectors may be distinguished: image intensifiers and video cameras, these two types of equipment being possibly associated.

In an image intensifier, X-Ray photons strike a screen which converts the photons into electrons. The electrons are accelerated and focused via electron optics on a cathodoluminescent screen. The image is therefore built instantaneously.

Quite a large amount of various types of electronic tubes may be used in video cameras. The most common is the so-called vidicon. In a vidicon, photons hit a target made of a photoconductive material (Pb O, Se, Pb S, semiconductors, diode arrays...). This photoconductive layer is charged on the photon entrance side by an applied positive potential and on the back side by a fine electron beam which scans the target. If light hits the target, electron-hole pairs are created which allow for a local discharge of the capacitor. The current of the electron beam necessary for recharging the back side of the target may be monitored. It provides the video signal which then is amplified. In such systems, the image is built sequentially, with generally the standard TV repetition rate of 1/25 second.

Two types of video systems may be differentiated, direct and indirect systems. In direct systems, the photo conductive layer is sensitive to X-Rays. In indirect systems, the topographic image is first converted into a visible light image (which may eventually be enlarged) by means of a fluorescent screen. This image may now be read by a visible light video camera.

In addition to these detectors, channel plates have also been used [13]. Resolution depends directly on channel diameter and the only system so far reported had a quite poor resolution.

TABLE II

		Ref.	Resolution	Additional references
Image Intensifier (+ Vidicon)		14	35 μm	15, 16, 17, 18
Direct systems	P b O Vidicon	9,19	15 μm	
	Diode Array	20	15 μm	
	Se Vidicon	21		
Indirect systems	Image Orthicon [Zn S + (ZnCd)S] (Ag)	22, 23	300 μm / 90 μm	25
	Screen	24	25 μm*	
	EIC Vidicon + mag. lens + Zn_2SiO_4 or $Gd_2O_2S(Tb)$ Screen	12,28	10 μm	
	Nocticon + Cs I (Tl) Screen	29		

* Due to specific experimental arrangement, not intrinsic quality of the system.

Video cameras and image intensifiers have fair dynamic ranges and linearity but have a resolution which is never much better than about 15 μm. Resolution is basically limited by reading electron beam diameter and the target thickness necessary to produce a reasonable signal (or by cell dimensions in discrete targets such as diode arrays).

Quoted resolutions of various detectors as well as a fairly comprehensive list of references are given in table II.

Resolutions quoted in Table II have not been measured in a consistant way and there certainly is a strong case for defining a standard procedure for resolution determination. One may suggest that, since spatially modulated X-Ray intensities may be obtained (Pendellösung fringes, Kato fringes...) M T F curves should be more widely used.

Since resolutions no really better than 15 μm may be achieved with electronic tubes, the only solution to improve resolution consists in converting the topographic image into a visible image and treat that image afterwards. Let us for such a system (Fig. 6) call α the absorption efficiency of the fluorescent screen, n_1 and n_2 the quantum conversion efficiencies of the fluorescent screen and of the photoconductive layer of the tube, respectively. Let us further assume that an objective (magnification M, numerical aperture N A) allows for magnification of the optical image.

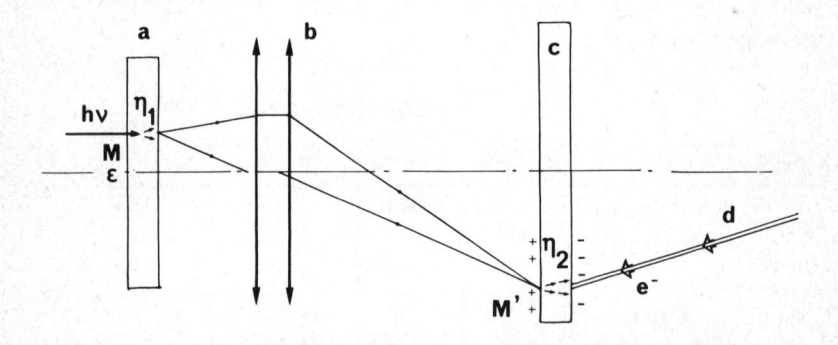

General video camera detector a) fluorescent screen, b) objective, c) camera target, d) reading electron beam

Fig. 6

Then, similarly to the equations derived for photographic plates, one gets :

$$S = \alpha\, n_1\, n_2\, vt\, \varepsilon^2 \left(\frac{NA}{M}\right)^2$$

Assuming the amplifier noise to be zero, one gets :

$$R = \varepsilon C \left[\frac{\alpha \nu_p t}{1 + \frac{1}{\eta_1} + \frac{1}{\eta_1 (\frac{NA}{M})^2} + \frac{1}{\eta_1 (\frac{NA}{M})^2 \eta_2}} \right]^{1/2}$$

which, for $\eta_1, \eta_2 \gg 1$, reduces to :

$$R \sim \varepsilon C \left[\alpha \nu_p t \frac{\eta_1 (\frac{NA}{M})^2}{\eta_1 + \eta_1 (\frac{NA}{M})^2} \right]^{1/2}$$

If the following values are considered : t = 1/25 sec : standard TV frame period, M = 6,3 ; NA = 0,16 ; α = 0,8 (excellent screen) ; C = $\frac{1}{2}$; ε = 5 µm, one finds that ν_p is to be greater than

$$\nu_p > 4 \cdot 10^7 \text{ ph/mm}^2/\text{sec. if R is to exceed 1,5.}$$

This simple calculation shows that only synchrotron radiation sources approach necessary values of photon fluxes for high resolution direct viewing topographic work. Moreover, the manufacturing of fine grain, homogeneous and thin fluorescent screens with reasonable absorption efficiencies is delicate. A brief review of the materials used up to now has been given by Stevels [30] .

As a conclusion to this section, the choice of an electro-optical system appears to be primarily governed by the magnitude of the available photon-flux. Once the average photon flux is determined, which sets the potential resolution of the system and its necessary gain , a clever compromise has to be found in order that the various elements all contribute to the same average image quality with the highest efficiency.

IV.- FUTURE PROSPECTS AND CONCLUSION

Three lines for future development appear important :

- First, more powerful synchrotron sources will presumably be built in the future. Their large power may give birth to new electro-optical systems with improved resolution.
- Second, high resolution experiments may require detectors with resolutions better than L4 nuclear plates. The development of photo sensitive resins in the energy range 6-25 keV may provide such high resolution image detectors.
- Third, image treatment from discretized images (images cut into picture elements) is certainly a powerful tool for future applications. This may be performed either : by digitalizing the video signal of a camera (such a system is tested in the present days by Hartmann [31] or by using two dimensional charge

- coupled devices (CCD) which may possibly be built in the future with small picture elements. Such systems have already been tested in the 1 to 14 keV energy range [32 33]. Their spatial resolution is at present time poorer than "No Screen" X-Ray films and they suffer radiation damage. Therefore much progress still needs being done.

REFERENCES

[1] J.F. PETROFF, M. SAUVAGE : Private communication.
[2] P. GLAFKIDES in "Chimie et Physique Photographiques", P. Montel - Paris 1967
[3] D.B. BROWN, J.W. CRISS, L.S. BIRKS : J. Appl. Phys. 47 (1976) 3722
    C.M. DOZIER, D.B. BROWN, L.S. BIRSKS, P.B. LYONS, R.F. BENJAMIN : J. Appl. Phys. 47 (1976) 3732.
[4] Y. EPELBOIN : J. Appl. Cryst. 9 (1976) 355.
[5] Ilford Technical leaflet.
[6] R.H. HERZ : Phys. Rev. 75 (1949) 478.
[7] A.R. LANG in "Modern Diffraction and Imaging Techniques in Material Science" North Holland, 1970, p. 407.
[8] K. ROSSMANN, B. LUBERTS : Radiology 86 (1966) 235.
[9] E. ZEITLER : this volume.
[10] J.I. CHIKAWA : J. Crystal Growth 24/25 (1974) 61.
[11] R.E. GREEN in "Advances in X-Ray Analysis" vol. 20 - Plenum Press (1977) p. 221.
[12] W. HARTMANN in "X-Ray Optics" Topics in Applied Physics - vol. 22, Springer Verlag, Berlin 1977.
[13] D.Y. PARPIA, B.K. TANNER : Phys. Stat. Sol.(a) 6 (1971) 689.
[14] A.R. LANG : J. Phys. E. Sci. Instruments 4 (1971) 921.
[15] A.R. LANG, K. REIFSNIDER : Appl. Phys. Lett. 15 (1969) 258.
[16] K. REIFSNIDER : J. Soc. Motion Picture and Television Engineers 80 (1971) 18.
[17] R.E. GREEN Jr., E.N. FARABAUGH, J.M. CRISSMAN : J. Appl. Phys. 46 (1975) 4173.
[18] T. VREELAND Jr., S.S. LAU : Rev. Sci. Instruments 46 (1975) 41.
[19] J.I. CHIKAWA, I. FUJIMOTO : Appl. Phys. Lett. 13 (1968) 387.
[20] G.A. ROZGONYI, S.E. HASZKO, J.L. STATILE : Appl. Phys. Lett. 16 (1970) 443.
[21] W. ECKERS, H. OPPOLZER : Siemens Forsch. u. Entwicklungs Ber. 6 (1977) 47.
[22] H. HASHIZUME, K. KOHRA, T. YAMAGUCHI, K. KINOSHITA : Appl. Phys. Lett.18 (1971)213.
[23] H. HASHIZUME, K. KOHRA, T. YAMAGUCHI, K. KINOSHITA : Proccedings of the Sixth International Conference on X-Ray Optics and Microanalysis - Univ. of Tokyo Press (1972) p. 695.
[24] S. KOZAKI, H. HASHIZUME, K. KOHRA : Jap. J. Appl. Phys. 11 (1972) 1514.
[25] H. HASHIZUME, H. ISHIDA, K. KOHRA : Jap. J. Appl. Phys. 10 (1971) 514.
[26] E.S. MEIERAN, J.K. LANDRE, S.O'HARA : Appl. Phys. Lett. 14 (1969) 368.
[27] E.S. MEIERAN : J. Electrochem. Soc. 118 (1971) 619.
[28] W. HARTMANN, G. MARKIEWITZ, U. RETTEN MAIER, H.J. QUEISSER : Appl. Phys. Lett. 27 (1975) 308.
[29] M. SAUVAGE : Nuclear Instruments and Methods : 152 (1978) 313.
[30] A.L.N. STEVELS : Journal of Luminescence 12/13 (1976) 97.
[31] W. HARTMANN : Private communication.
[32] M.C. PECKERAR, W.D. BAKER, D.J. NAGEL : J. Appl. Phys. 48 (1977) 2565.
[33] L.N. KOPPEL : Rev. Sci., Instruments 48 (1977) 669.

ELECTRON DETECTORS

Elmar Zeitler
Teilinstitut für Elektronenmikroskopie am Fritz-Haber-Institut
der Max-Planck-Gesellschaft
Faradayweg 4-6, 1000 Berlin (West) 33

1. Introduction

It is assumed that the reader's interest is in setting up an experiment which requires detection of electrons in a quantitative way. He knows about his experiment but needs some guidance into the vast field of particle detection. He wants to appreciate the problems, understand their causes and perhaps even develop a new idea for reducing them. Only under this premise can these pages--limited to a few--be a valid contribution, since the whole story requires and warrants an encyclopedia [1].

The energy of the electrons to be detected plays an important role in determining the suitable strategy. Such questions as "should the detector be compatible with UHV conditions" or "can your experiment afford magnetic fields" will be the decisive factors. Since our world is created neutral, electrons can be created only together with an equivalent positive charge. In the case of electron detectors, the positive charges are either ions or holes in the valence band of semiconductors. The detection of an electron relies totally on its electromagnetic interaction with matter. The energy transferred from the primary electron to the matter is used to produce photons, an equal amount of positive and negative charge, or heat. Therefore photon detectors, Coulombmeters or calorimeters should be applicable. A photon detector itself, however, is based on the photoelectric effect; that is, on the liberation of an electron from an atom by the photon. Hence if photons are involved in the detection of the electron, the detector must consist of one or several converters which transform electrons into photons and photons into electrons. An example of such a chain for detecting and counting charged particles is the fluorescent screen and the retina in Rutherford's famous experiment.

The separation of a positive and negative elementary charge requires a given amount of energy, and hence the primary electron can produce a given number of pairs. The principle of quantitative detection is to measure only the negative charge; i.e., keep this so-called recombination under a tolerable amount.

In those sections of the detector system in which the signal is carried by electrons, amplification can be effected. The electron can be readily accelerated, and this energy input leads to the production of more electrons. The various stages of an electron multiplier is an example. The additional energy put into the detector reappears in the higher energy of the output signal. The chemical energy of a developer transforms the silver specks initially produced by an electron on its passage through the photographic emulsion into large silver grains. The output signal, namely the electric energy in the registering photometer, will then also be increased and detectable, whereas the speck's signal, for example, might have been drowned in the noise of the recorder.

This introduction shows that electron detection in general involves conversion of one particle type into another (electron into photon and vice versa), multiplication of the particle numbers, and increment in energy of the charged secondaries. All those events are subject to statistical fluctuations. The various stages of the detection system must be chosen such that their intrinsic noise contribution does not submerge the signal. Of course this contribution is determined by the physical events in that stage or, conversely, the suitability of a certain stage depends on the physical parameters of the electron to be detected. It seems advantageous first to formulate how the noise contributions of the stages degrade the detection, and then apply this result to the special cases.

2. DQE

The complete characterization of a converter-multiplier stage is encompassed by the set of probabilities P_n which indicate the likelihood that n particles appear at the output of the device when one particle (of proper nature) impinges on the input. Although this overwhelming knowledge can be readily handled with so-called generating functions [2], for our purposes we shall reduce the characterization to the mean gain g, or the average number of output-particles per one input-particle, and to the variance $\sigma^2 = \overline{(n-g)^2}$ (σ, standard deviations, rms), inherent in the stage.

It is obvious that f independent input particles yield f·g more output particles; their variance will be the sum of every single particle variance; that is, $f \cdot \sigma^2$; if those f particles came already from the output of a statistical multiplier stage, the above result holds, with f being the average gain of that "preamplifier." What happens to the standard deviation τ^2 of the first amplifier? Since the second stage amplifies each particle independently, the standard deviation τ will be multiplied by the same factor as well. As a result of cascading two stages (capital symbols) we obtain at the output of the second stage the sum of the amplified variances (lowercase symbols)

$$G_2 = f \cdot g; \quad \Sigma_2^2 = g^2\tau^2 + f\sigma^2.$$

This expression can readily be generalized for k stages by assuming that f and τ^2 stem already from k-1 stages and hence are written as G_{k-1} and Σ_{k-1}^2; therefore

$$(\Sigma/G)_k^2 = (\Sigma/G)_{k-1}^2 + (G_{k-1})^{-1} (\sigma/g)_k^2 \tag{1}$$

or, in words, the relative variance of a k-stage amplification is that of the k-1 previous stages augmented by the relative variance of the additional stage, however reduced by the over-all gain of the k-1 stages. The fact that not the relative variances are added but the one weighted by the amplification is most significant for design of actual detectors. The additional variance or noise becomes less influential the higher the gain in the previous stages, which in its logical consequence requires that the first stage has the highest gain.

Although rms and variance are familiar concepts, the signal-to-noise ratio seems more appropriate in appraisal of detectors. Since noise is a fundamental and ubiquitous phenomenon in any event which exhibits the quantum nature of physics, the detector quality is best judged by the reduction of the signal-to-noise ratio it causes. The optimum would be that no noise is added during detection; i.e., the signal-to-noise ratio at the output coincides with the one at the input (their ratio is unity). Since for large numbers of individuals the statistical variance grows with the square root, one prefers to deal with the square of the signal-to-noise ratios.

Denoting $(\Sigma/G)^{-1}$ as the signal-to-noise ratio (S/N), we obtain from Eq. (1)

$$\frac{(S/N)_k^2}{(S/N)_{k-1}^2} = \left\{ 1 + \frac{1}{G_{k-1}} (S/N)_{k-1}^2 (\sigma/g)_k^2 \right\}^{-1}. \quad (2)$$

This ratio can be unity if the preamplification is very high; then the noise contribution of the kth stage is insignificant, or σ^2 is negligible. Because of the factor $(S/N)_{k-1}^2/G_{k-1}$ this ratio depends not only on the kth member but also on the previous members of the chain. For a Poisson distribution describing the statistical arrival of particles and photons or quanta in general, this factor is just unity; therefore

$$\frac{(S/N)_k^2}{(S/N)_{k-1}^2} = \frac{1}{1 + (\sigma/g)_k^2} \quad (3)$$

is aptly called the Detective Quantum Efficiency (DQE) pertaining to the kth stage (provided the input obeys a Poisson distribution) [3].

Obviously a gain smaller than unity, that is, an actual loss of particles, diminishes the DQE. Only then can the output remain a Poisson distribution because Eq. (1) requires that $(\sigma/g)^2 = 1/g-1$. That can be fulfilled only if $g < 1$ and $\sigma^2 = g(1-g)$, which holds for the binomial distribution. This case prevails when from the impinging electrons (1-g)% are backscattered from the input surface of the detector and its DQE is reduced to g%.

3. Physical phenomena

When electrons impinge on the detector two basic processes occur, namely, scattering and energy loss [4,5].

(1) The primary electrons undergo angular deviations by interacting with the Coulomb field of the detector atoms. This scattering may enable the primary electron to escape the detector with more or less reduced energy, as 'backscattered' electrons.

(2) The primary electron loses its energy gradually, in contrast to photons, by exciting and ionizing the detector molecules or atoms; that is, by interacting with electrons. These effects free charge and/or produce light and are the base for the primary electron detection. Some of the freed electrons may receive enough energy to leave the detector as so-called secondary electrons.

There is, however, a second mechanism of energy loss. The primary electron can produce Bremsstrahlung which may or may not be absorbed by the detector. This again is an interaction with the nuclear Coulomb field.

The occurrence and the relative importance of these effects depends on the physical conditions of the experiment--the kinetic energy of the primary electron, the atomic number of the detector atoms, the mass thickness and shape of the target. A successful detection scheme must be designed so that under the given experimental conditions the DQE is as high as possible, requiring the loss of primary events to be as low as possible. Therefore the necessary facts are compiled here.

3.1. Backscattering

The phenomenon is a volume effect. The effective depth from which backscattered electrons escape is about one-half the range of the electrons in the material [6]. The yield η, that is, the number of backscattered electrons per incident electron, is dependent on the atomic number Z of the material and the angle α of incidence, yet nearly independent of the primary energy. It is well described by the empirical expression [7]

$$\log \eta = -\sqrt{Z_1/Z} \log (1 + \cos \alpha); \qquad Z_1 = 81.$$

The energy spectrum of the backscattered electrons is broad, but exhibits for higher atomic numbers a maximum more pronounced and closer to the primary energy E_o [8]. This is reflected in the empirical expression for the mean energy $\bar{E}$ [9] carried away from the target by the backscattering

$$\bar{E} = (0.45 + 2 \times 10^{-3} \, Z) \, E_o.$$

The energy $E_o - \bar{E}$ may produce pulses which are too small to be detected, in other words, noise. Backscattered electrons can also produce secondary electrons [10].

3.2. Energy loss and range

The rate of energy loss by excitation and ionization depends on the number of electrons per unit volume $\rho L Z/A$ of the stopping material which is almost independent of Z, and on its average ionization potential. The primary electron enters only with the kinematical parameter β, its velocity in units of c [4]. The range R in units of mass thickness is just a function of the primary energy, empirically [11]

$$(R/R_o) = (E_o/\text{MeV})^s; \qquad R_o = .412 \text{ g/cm}^2, \quad s = 1.265 - 0.0954 \ln (E_o/\text{MeV}).$$

In the case of energy determination, the sensitive thickness of the detector must be larger or at least comparable to the electron range, and insensitive covers or windows of the detector must be negligibly thin compared to R. For counting, these conditions can be relaxed.

The average energy ε required for the formation of an ion pair (a name carried over from ionization of air molecules in ionization chambers) or an electron-hole pair is not necessarily the ionization energy I of the atoms or the band gap of a semiconductor, because parts of the energy lost are "wasted" for optical excitation. This distribution of the primary energy into charged and uncharged events influences also the statistical fluctuation or variance of the number of ions formed by a given amount of primary energy [12]. The average energies ε for various materials are:

Air (NST)	34 eV	Ge	2.95 eV
Si	3.62 eV	AgBr	7.6 eV

3.3. Secondary electrons

The physics of energy loss favor the transfer of small amounts to the atomic electrons [4]. Therefore many secondaries with energies probably around 1-5 eV are produced along the track of the primary (δ rays). Eighty percent have energies less than 20 eV and hence make secondaries readily distinguishable from backscattered electrons, provided they escape [10]. Their yield δ, that is, the number of secondaries per incoming primary, is difficult to measure since backscattered electrons also produce secondaries. From the energy loss formula above it is obvious that δ only depends on the number of atomic electrons and hence only slightly on Z. The yield falls with increasing energy of the primary electron, empirically as [13]

$$\delta \cdot E_o^{0.8} = \text{const.}$$

Because of the low energy of the secondaries, they stem only from the surface layers of the target, 10-100 Å. By the same token, secondary emission, the base for electron multipliers, is very sensitive to contamination or other surface changes.

3.4. Radiation loss

The second loss is the production of Bremsstrahlung. It is proportional to Z^2 and therefore relatively more important for heavy than for light elements. Furthermore it increases roughly in proportion to the total energy of the primary electron; i.e., it stays fairly constant up to energies of mc^2 (511 keV) but then increases proportionally [4]. These facts must be observed in choosing scintillators.

3.5. Scintillation

Organic scintillators convert 1-5% and inorganic about 20% of the absorbed electron energy into photon energy. The output is proportional to the electron energy loss [1]. The mechanism is rather involved, and much energy is lost in radiationless transitions. The single photons have energies of ∼ 3 eV, corresponding to the visible region. This low energy per photon compensates for the very low conversion efficiency, so that photomultipliers can be used to amplify the signal, although their photocathode efficiency for converting photons into electrons might also be 20%.

Organic and inorganic scintillators differ in the following respects: The pulse duration is for organic substances typically around 1-2 nanosec, and their average atomic number is low ($Z \sim 6$). They are radiation-sensitive and show aging after a dose of 10^{12} e/cm^2. Backscattering and X-ray production, both detrimental to high DQE, are negligible. Anthrazene is the standard substance to which all other organic scintillators are compared.

Inorganic crystals are mainly used because of their high atomic number (X-ray converters) and their high efficiency. The light pulses are typically milliseconds long, because the electrons must migrate through the lattice to excite an impurity center. Zinc sulfide is the classical substance for electron detection, and sodium iodide for X-rays. They are more radiation-resistant than organic materials.

Note that the scintillator must still be coupled to a photomultiplier, so that again losses can occur.

4. Devices

Now the actual detectors are described and ranked with increasing complexity. Using the results of previous sections for guidance, it should be easy to delineate the area of possible applications and, for given physical parameters, devise systems with the maximum DQE.

4.1. Faraday cup

The so-called Faraday cup provides the simplest arrangement. All primary electrons fall into a container whose aperture subtends a small solid angle when viewed from the inside walls. Secondary electrons can be kept inside by a low reversed bias. Radiation loss and a high backscattering yield is avoided by cladding the inside of a high-Z cup with low-Z material.

Combined with today's current amplifiers which show a noise output equivalent to 10^{-14} to 10^{-15} Amps at the input with integration values of 1-10 sec, currents of 10^{-15} Amps and charges of 10^{-12} Coulombs can be measured. This is far away from counting single electrons. But such detectors, properly designed, can cover a large range of electron energies and may be made UHV-compatible.

4.2. Channel electron multiplier

In this device the secondary electrons produced by the primary are accelerated and multiplied in many generations. Instead of a cup, a tube is used whose inside is covered with material of high resistivity and high secondary yield. With a few kV along the tube, gains of 10^8 can be attained. The high resistance of 10^9 Ohms limits the pulse rates to less than 10^5/sec, which is very high in comparison to the accidental noise rate of 0.1 pulses per second. For typical capacitors of 10^{-11} F a pulse of 10^{-12} C produces a voltage pulse of 0.5 V.

Since the first step in this continuous multiplier is the production of secondaries, the DQE can be adequate only for relatively low primary electron energies (10-1000 eV) [14]. The multiplier tolerates baking to 400°C; i.e., it is UHV-compatible. Magnetic fields have a slightly detrimental effect by influencing the amplification.

4.3. Solid state detector

These detectors are the solid state version of ionization chambers [15,16]. The energy per electron-hole pair is much lower and the utilization of primary energy much higher than in gases. The sensitive volume, however, is also small. The obvious problem is to collect the produced charges before recombination sets in. Only the p-n junction of semiconductors permits applications of a voltage for a quick removal of the electron-hole pairs produced by the passage of the primary electron in the otherwise charge-depleted sensitive volume without introducing a current that would mask the desired pulse.

The reversed bias voltage determines the width of the depletion region, which in turn must match the range of the primaries. Typical values are 1-1000 V. A limit is set by the field strength ($\sim 10^4$ V/cm) tolerable across the depletion zone and thus also an upper bound for the primary energies. The output signals are small, depending on the capacitance, i.e., on the width and the area of the depletion zone. Therefore one quotes the specific capacitance, which is about 1 pF per mm^2. In a typical detector (200 mm^2) a charge of 10^{-12} C produces a pulse of only 5 mV. Since the capacitance depends on the voltage as well, charge-sensitive preamplifiers must be used for further processing. The pulse rise times range from 1-10 nanosec.

A thin gold layer of 50 µg/cm^2 assures a sound frontal electrode and causes a lower bound for the energy of the primary. The backing is most often an aluminum plate. Although this gold layer is high-Z material, the bulk of silicon or germanium is not, so that backscattered electrons do not lower the DOE. Solid state detectors are suitable for electron detection in the energy range between 10 keV and 3 MeV. They are, however, also subject to radiation damage; they lose sensitivity after a dose of 10^{13} e/cm^2, through an increase of crystal imperfections which behave effectively as impurities. These detectors operate at ambient temperatures and are UHV-compatible.

4.4. Scintillation counters

These devices consist of a scintillator or phosphor that converts in part the electron energy into visible photons, and a photomultiplier whose cathode converts the photons into photoelectrons which, in turn, are accelerated and multiplied in number by secondary emission at the dynodes of a photomultiplier [1]. Here the DQE is determined by the matching of energy and scintillator and that of the spectral distribution of the photons to the spectral response of the cathode.

The average gain of modern photo cathodes is of the order 10 e/photon, and the gain in the electronic multiplication can be as high as 10^6, depending on the number of dynodes and the voltage between them. The bottleneck again is the first step; i.e., the effective conversion of electron energy into photons, as discussed previously.

The shot noise of the cathode $\overline{\Delta i^2} = 2\,ei\Delta f$, where i is the cathode current and Δf the bandwidth. The Johnson noise in the load resistor R is $\overline{\Delta i^2} = 4\,kT\Delta f/R$, so that a pure photo diode produces a rms voltage $\overline{\Delta V^2} = eR\Delta f\,(2\,Ri + 4\,kT/e)$. In the multiplier the input shot noise is amplified G times and thus the output shows voltage variations

$$\overline{\Delta V^2}_{out} = eR^2\Delta f\,(2i_{out}/G + 4kT/e).$$

Proper choice of gain and load resistor is indicated.

Another noise contribution comes from the dark current, which is virtually independent of the operating voltage.

These scintillator combinations are readily applicable to electron energies from, say, 10 keV to a few tens of MeV. For lower energy electrons, an acceleration towards the scintillator is standard procedure to increase the DQE [17]. For very high energies only plastic scintillators avoid a predominance of Bremsstrahlung production [18]. By careful design single-electron counting is possible.

4.5. Image intensifiers, TV-solid state devices

These economically important devices are cascades of two-dimensional arrays of channel multipliers, so-called channelplates and transparent photocathodes. The amplification can be increased above that of ordinary scintillator detectors, but the precautions for achieving a high DQE are the same. Single electron counting can be achieved so that the picture can be recorded as a two-dimensional array of electron counts [19]. Because of commercial and military interest, this field has grown into a vast specialty requiring more know-how than is normally available in a physics laboratory. The DQE can be as high as that of the photographic plate, ≈ 0.7.

4.6. Photographic emulsion

This device responds in a linear fashion to the charge density of the impinging electrons [20]. An exposure of 10^{-11} C/cm^2 or about one electron per μ^2 produces a suitable response; i.e., a sufficient amount of developed silver per unit area. The response is measured as optical density [21].

The sensitive thickness of an emission is in most cases larger than the range of 2-3 MeV electrons. But there is also a protective layer on top, requiring 5-10 keV electrons to reach the sensitive volume. The laws of response become universal when distances are measured in terms of the electron range (neglecting the effect of the dead layer). The number of grains developed is proportional to the optical density D. The variance of this number is composed of the variance of the number of impinging electrons N and the variance in the number of grains produced by one electron. This multiplication g can amount to 2-10 grains (depending on the primary electron energy) [22]. The total relative variance of this converter-multiplier follows, from the formulae derived earlier, to

$$(\sigma/D)^2_{out} = (\sigma/N)^2_{in} + 1/N \, (\sigma/g)^2 = 1/N \, (1 + r/g),$$

assuming Poisson distributions.

With an empirical value r of about 2, the corresponding DQE values are of the order of 0.6 to 0.7. Since the average Z of a photographic emulsion is higher than that of an organic scintillator, backscattering can degrade the DQE. The relatively high DQE together with the integrating property makes the photographic plate a convenient and indispensable electron detector. In addition, the spatial resolution, or the size of separable picture elements, is of the order of 10-20 μ [23]. Therefore a plate of 50 cm^2 can accommodate $\sim 10^8$ elements, each capable of storing and displaying ten or so grey steps.

Most of the manufacturers provide complete technical descriptions of their products and excellent compilations of the physical effects involved, which may serve interested readers as guides into the field.

Acknowledgement

Fruitful discussion with Dr. K.-H. Herrmann, an expert, is acknowledged with thanks.

References

1. W. E. Mott, R. B. Sutton, Encyclopedia of Physics, vol. 45, ed. S. Flugge (Berlin: Springer-Verlag).
2. T. Jorgensen, Amer. J. Phys. 6: 285 (1948).
3. R. Clarke Jones, Adv. El. and Elec. Phys. 11: 87 (1959).
4. W. Heitler, Quantum theory of radiation (Oxford: Oxford University Press).
5. N. F. Mott, H. S. W. Massey, The theory of atomic collisions (Oxford: Oxford University Press).

6. H. Niedrig, Scanning Electron Microscopy 1978, ed. O. Johari, vol. 1, p. 841.
7. F. Arnal, P. Verdier, P. D. Vincensini, C. R. Acad. Sci. B268: 1526 (1969).
8. H. Kulenkampff, W. Spyra, Z. Phys. 137: 416 (1954).
9. E. J. Sternglass, Phys. Rev. 95: 345 (1954).
10. E.g., H. Seiler, Z. Angew. Phys. 22: 249 (1967).
11. L. Katz, A. S. Penfold, Rev. Mod. Phys. 24: 28 (1952).
12. U. Fano, Phys. Rev. 72: 26 (1947).
13. L. Reimer, Optik 27: 86 (1968).
14. F. Bordini, Nucl. Inst. Meth. 97: 405 (1970).
15. J. M. Taylor, Semiconductor particle detectors (London: Butterworths, 1963).
16. G. Bertolini, A. Coche, Semiconductor detectors (Amsterdam: North-Holland, 1968).
17. T. E. Everhart, R. F. M. Thornley, J. Sci. Inst. 37: 246 (1960).
18. H. Feist, D. Harder, R. Metzner, Nucl. Inst. and Meth. 58: 236 (1968).
19. K.-H. Herrmann, D. Krahl, H.-P. Rust, Ultramicroscopy 3: 227 (1978).
20. E. Zeitler, J. R. Hayes, Lab. Invest. 14: 1324 (1965).
21. C. E. K. Mees, The theory of the photographic process (New York: Macmillan, 1942).
22. R. C. Valentine, Lab. Invest. 14: 1334 (1965).
23. J. F. Hamilton, J. C. Marchant, J.O.S.A. 57: 232 (1967).

IMAGE PROCESSING OF REGULAR BIOLOGICAL OBJECTS

Jan E. Mellema
Department of Biochemistry,
State University of Leiden,
64, Wassenaarseweg,
Leiden, The Netherlands.

1. INTRODUCTION

The main techniques for biological structure determination at the molecular level are X-ray diffraction and electron microscopy. The first method has proven its power by solving the 3D structure of large viruses, assemblies composed of a large number of protein molecules and an amount of genetic material with a total molecular weight of several millions of daltons (1, 2). A necessary condition for this type of research is the availability of single crystals of the material to be investigated. Information to a resolution of about 3 Ångstroms becomes available about the crystal structure.

Electron microscopy does not require crystalline material for imaging, but up to now the technique is not so powerful for biological structure determination in terms of resolving power. This may seem paradoxal as modern electron microscopes possess a resolving power of about 2 Ångstroms. The limitations mainly arise from the way in which the biological material is imaged in the instrument. The preparation of the specimens and the detoriating effects of the radiation damage during imaging play a much bigger role in this type of research than they do in crystallographic studies. A number of these drawbacks may be overcome by utilizing techniques that exploit the spatial frequency decomposition of the digitised electron images obtained by calculation of the Fourier transform (3, 4). These techniques are analogous to those employed by X-ray crystallographers in solving crystal structures. Particularly when the biological specimens are regular, application of these so called image processing techniques is very rewarding. Fortunately a large number of interesting biological structures are isolated in an ordered form or can be induced to associate in a regular way.

Two kinds of analysis can be performed, namely 2-dimensional spatial filtering and 3-dimensional image reconstruction. The first technique provides an averaged 2-dimensional image of the projection of the asymmetric unit. A 3-dimensional image may be obtained by combining several views of the object and the method also provides the structure of the averaged repeating unit. In those cases where the object possesses helical symmetry the different views are easily obtained, in other cases the required projections must be collected by tilting the specimen in the microscope or by a search procedure from a field of randomly oriented particles. Many of these methods have been used to derive structural information of biological assemblies to a resolution of about 20 Ångstroms. A number of initial applications are described in the book edited by Huxley and Klug (5), a more recent survey is written by Crowther and Klug (6) and by Mellema (7).

In the following the application of the 3D reconstruction technique will be illustrated. The object of study is the contractile protein sheath of the bacterial virus PBS-Z. The sheath structure is constructed from about 500 identical protein molecules of molecular weight of 50.000 daltons which are arranged in a regular way around a tube or core made of protein of unknown composition (8, 9). The results not only show what kind of structural information can be derived, but also illustrate the effects of the preparation procedures and radiation damage in this type of research. By choosing this example most aspects of digital image processing of regular objects will be discussed, so that one hopefully gets a good impression about the course of such an analysis. A full account of the work presented here will be published elsewhere (10).

2. THE BIOLOGICAL OBJECT IMAGED IN THE ELECTRON MICROSCOPE

The electron microscope has a large depth of focus and for structures not exceeding a few hundreds of Ångstrom units a projection of the scattering density of the object will be recorded. Because of the low inherent scattering power of the biological material for electrons, the specimens must be treated with electron dense material. For isolated particles or ordered aggregates (crystals) the negative staining technique is commonly employed. In this technique a heavy metal salt solution is mixed with the object and allowed to dry on the specimen support film. The object will be enveloped and penetrated by the metal ions which not only enhance its scattering power but also provide a mechanical support. This last factor is very important as the specimen must stand against the drying forces when brought in the vacuum of the electron microscope. Under the usual conditions of imaging the stain distribution within the object will be recorded as an intensity distribution on the photographic material and the intensity on the photographic plate will be linearly related to the stain distribution in the object.

The biological object may exhibit different types of symmetry. The only symmetry operators which occur in biological systems are translational and rotational displacements and these elements can be combined to form the so called line, plane and space groups. If rotational axes only describe the symmetry in the point groups, the biological structure has a finite size contrary to the previous mentioned groups which may have a infinite extent. In nature many biological structures are assembled in a regular fashion and these assemblies may be isolated as such. For example in muscle and in membranes the constituting components are frequently organized in a regular way as is the case with most if not all of the simple virus structures. The macromolecular complex which will be analyzed in the following paragraphs exhibits a line group symmetry, the relative displacements of the asymmetric unit within the sheath structure are described by a rotation and a translation. The rotational symmetry of the structure is 6-fold.

3. THE ANALYSIS OF ELECTRON IMAGES BY FOURIER METHODS

The most convenient technique to assess the information in electron images of regular biological objects is to analyze the Fourier transform of the image. In general this may be done in two ways, by making use of a coherent optical system or by calculating the Fourier components after digitizing the image. The Fourier transform yields information about the state of preservation and the type of symmetry present in the image. For example, the spatial distribution of the amplitudes of the Fourier components allows one to determine the unit cell dimensions in the case of a regular 2-dimensional object and the highest spatial frequencies yield information on the state of preservation or the resolution attainable, as in the case of X-ray diffraction. Filtered images may be obtained in which only the information present at and around the reciprocal lattice positions is used in a Fourier synthesis. Again, this analysis can be performed optically as well as by a digital Fourier summation. In many cases the averaged unit cell obtained after the experiment allows one to draw better conclusions on the cell content. Especially when a number of superimposed lattices have been recorded and the translational vectors do not coincide, the separation of the individual layers may be feasible (11). A spectacular example of such an analysis has been carried out by Unwin and Henderson (12), which made it possible to obtain a density map of a membrane by averaging over about 10^4 unit cells.

In the case of biological structures with helical symmetry the analysis takes another course. The most convenient way of decomposing the images is no longer in terms of plane wave components since structures exhibiting helical or cylindrical symmetry can be best analyzed by a decomposition in cylindrical waves. The Fourier transform of a projected structure with helical symmetry has a very characteristic appearance in which the values of the Fourier components are sampled on a set of so called layerlines. The layerlines are equally spaced and the transform value on a given layerline can be expressed as :

$$F(R,\Phi,l/c) = \sum_n G_{n,l}(R) \exp[in(\Phi + \pi/2)].$$

In the expression l designates the layerline number, c the repeat distance and n the order of the cylindrical harmonics $G_{n,l}(R)$. The coordinates R, Φ and $Z = l/c$ are cylindrical polar coordinates in transform space, whereas r, ϕ and z describe the structure in real space. The finite extension of the object may be used to limit the number of n for which $G_{n,l}(R)$ is effectively zero for any given R. The selection rule links the layerline number l with n the order of the cylindical density wave. The selection rule has the following form :

$$l = tn + um ,$$

in which t represents the number of turns of the basic helix per repeat and u expresses the number of units per repeat. The allowed values of m are any integer which fulfills the equation. The graphical representation of the selection rule, the n - ℓ plot (see Figure 2), can be considered as the reciprocal lattice of the radial projection of the structure. From this plot the various harmonics at a given layerline may be deduced.

At low resolution the inner part of each Z-plane has only one G_n contributing so that it will be possible to get a 3D reconstruction at low resolution from one single view. At higher resolution different G_n terms may overlap so that there is more than one term contributing to each annulus. The reconstructed density $\rho(r,\phi,z)$ is obtained in the following way :

$$\rho(r,\phi,z) = \sum_n \sum_l g_{n,l}(r) \exp[in\phi] \exp[-2\pi i l z/c],$$

$$g_{n,l}(r) = \int G_{n,l}(R) J_n(2\pi Rr) 2\pi R dR.$$

Generally the Fourier transform of the image will yield two estimates for G_n so that two individual reconstructions are obtained. If there are overlapping functions G within the working resolution, more projections will be required and the problem of determining the individual G_n terms can be expressed by solving a series of linear equations. Because of the nature of the data it will be advantageous to have more projections than unknown functions G_n (4, 14). As there exists a high level of background noise due to irregularities in the stain and from the support film it will be advantageous to average out these effects by combining the data from a number of projections, in order to improve the statistical weight of the reconstructed density. This can be achieved by a correlation procedure in which the relative orientation of the different projections can be determined (15). These orientations are expressed in terms of z, the relative shift along the helical axis and ϕ the difference in azimuth between two particles. A residual can be calculated which measures the degree of correlation between particles as a function of these variables. A useful residual is for example given by :

$$R = \sqrt{\sum_i |F_i| \cdot |\Delta\theta|^2 \Big/ \sum_i |F|} \quad ,$$

in which $|F_i|$ is the mean amplitude of the two corresponding Fourier components. A number of other expressions which may be useful in correlating transforms have also been described by Amos and Klug (15). The expected phase difference ($\Delta\theta$) between corresponding points in reciprocal space along a layerline of meridional spacing Z is given by :

$$\Delta\theta = -n\Delta\phi + 2\pi Z \Delta z.$$

The number of n is the order of the Bessel function on that layerline (13). After having determined the set of values ϕ and z for each particle against a reference all the data are brought to the same orientation and the Fourier components are averaged. Also one is able to estimate the radial scaling of the transforms in this procedure as there are always differences in projected size due to shrinking and stretching forces during preparation or slight changes in the magnification if particles are collected from different micrographs. The scaling can easily be achieved by expanding or compressing the transform values according to the best correlation value. Many structures are polar and therefore a search is required in which the particle can be placed upside down; this is equivalent to changing ϕ to $-\phi$ and z to $-z$.

Apart from this type of 3-dimensional analysis, filtering experiments can be carried out by computer, so that the 2 sides of the helical particle (in the terminology of Klug and Berger (16)), can be recovered separately. This analysis is very useful if one knows that there is almost no radial dependence of the density or in other words if the protein wall of the aggregate is very thin.

4. THE QUANTITATIVE ANALYSIS OF STRUCTURAL CHANGES IN CONTRACTILE SHEATH AND THE INFLUENCE OF THE ELECTRON DOSE AND PREPARATION METHOD ON THIS STRUCTURE

The contractile sheath of a bacteriophage is an illustrative example of an ordered helical assembly of protein molecules, which is able to change its structure during contraction. This structural change can be correlated with the injection process of the genetic material in the host cell. The sheath structure discussed here is derived from phage PBSZ, which possesses an abnormal long sheath structure of about 2560 Å in the extended (E) form and 1270 Å in its contracted (C) form (9). The changes in structure have been monitored as a function of electron dose and of the type of negative stain used in the preparation. The technique used to analyze the images has been the 3D reconstruction technique. Two electron doses were assayed in this way; the first micrograph was taken at about 5 electrons/Å^2 and a picture was taken of the same object at a value of about 250 electrons/Å^2. The same structures were analyzed in pairs. In the first instance optical diffraction patterns were recorded to select a number of well preserved particles from the micrographs. Radiation effects can be detected by inspection of these diffraction patterns . Two types of negative stain were used in order to prepare the specimens : uracyl (U) and phosphotungstate (P) ion containing solutions. The ions have the formula UO_2^{2+} and $PW_{12}O_{40}^{3-}$ respectively. From the optical diffraction patterns the packing parameters of the protein subunits in the structures can easily be derived. There is not much difference in the parameters for the E structure with the different stains. This can be seen in Figure 1 where a number of optical diffraction patterns is presented.

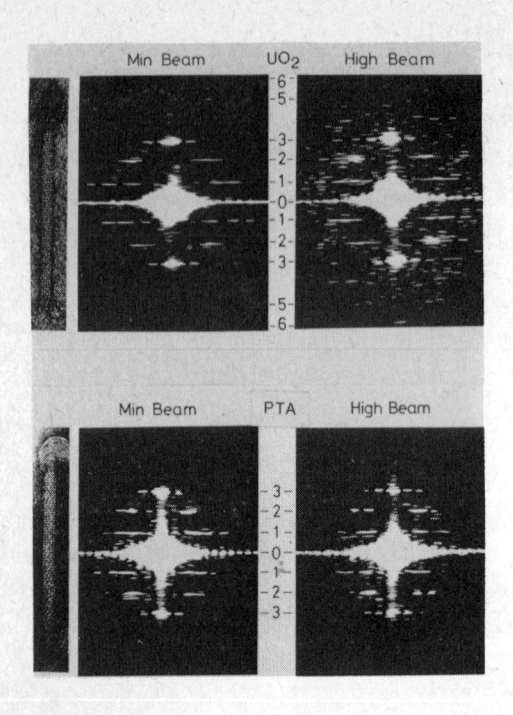

Figure 1.

Optical diffraction patterns of E-type structures, taken at different doses and which illustrate the effects of radiation damage. The intensities can be indexed on a set of approximately equally spaced layerlines. The approximate repeat of the structure is about 120 Å. The higher dose images show more noise particularly in the U-stained case. The corresponding sheath structure is shown left.

The two types of surface lattices for the E and C structure are displayed in Figure 2, in which also a graphical representation of the selection rule (the n,ℓ plot) is shown. The symmetry of both structures is 6-fold. The repeat of the E structure is about 120 Å and for the C structure a value of 320 Å or about two times higher was derived depending on the particular stain. These repeat values do not play an important role in the result of the 3D reconstruction. The surface lattices are derived from the optical diffraction patterns by drawing the unit cell at one particular radius of the structure. In this way the packing parameters of the protein subunits associated with the unit cell content, are displayed in a graphical way.

In order to monitor the effects of the radiation damage digital transforms were calculated by the procedures outlined by DeRosier and Moore (17). From these transforms the layerline information was extracted for the E and C structures stained by U and P and imaged at low (L) and high (H) dose. For corresponding images the relative orientation parameters were determined and if necessary also the polarity of the structure. For the E-U-L, E-U-H and the C-U-L and C-U-H particles Table 1 lists the results.

It can be seen from these data and the data of the analysis of the P stained structures (not shown) that the correlation between data sets from the same particle agree reasonably well. All the projections at one particular dose and obtained with a particular stain were averaged and a number of these averaged layerline data are shown in Figure 3.

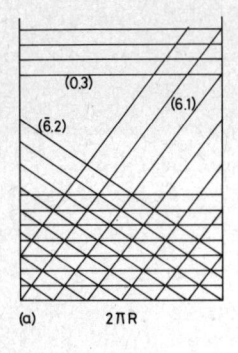

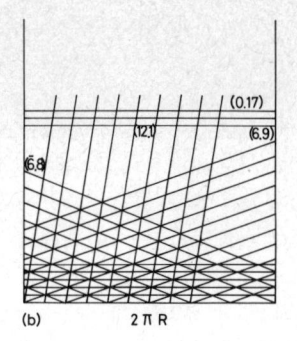

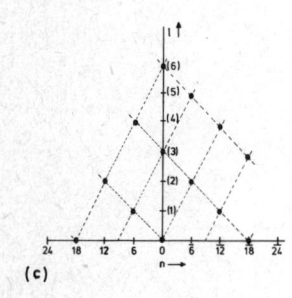

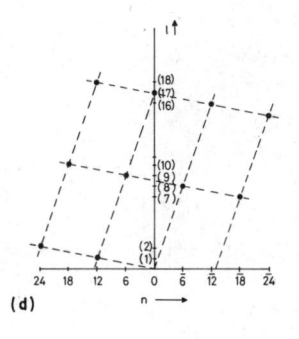

Figure 2.

The surface lattices of E- and C-type structures are respectively shown in (a) and (b). The helical families are drawn in the lattices. The hand of both structures, that is whether the helical family(6,1) is right or left handed and consequently the family (6,2) left or right handed is not known. Similarly for the C.lattice the absolute configuration is unknown, but from analogies with other structures the one postulated is very likely. The corresponding (n-ℓ) plots are shown underneath.

Table I. Survey of the correlation procedure of the data derived from the E- and C- structures, stained with U and imaged at a dose of about 5 electrons/Å^2 (L) and about 250 electrons/Å^2 (H).

Projection	L Residual	H	L ϕ	H ϕ	L z	H z	L Rscale	H Rscale
E-U-1	–	–	0.0°	0.0°	0.0 Å	0.0 Å	1.0	1.0
E-U-2	25°	18°	1.0	1.0	-1.5	0.0	.96	.96
E-U-3	22	42	25.0	27.0	-1.0	2.0	1.02	.96
E-U-4	36	45	23.	28.	-0.5	1.0	.98	.94
E-U-5	27	41	58.	62.	34.	36.	1.15	1.13
E-U-6	22	44	58.	63.	34.	36.	1.10	1.12
C-U-1	–	–	0.0	0.0	0.0	0.0	1.0	1.0
C-U-2	54	60	2.0	-3.0	-2.0	-4.0	1.06	.98
C-U-3	57/77+)	44	37.	38.	34.	21.	.96	.83
C-U-4	51/78	69	43.	41.	35.5	22.	1.06	1.01
C-U-5	47/90	63	26.	29.	57.5	51.	.94	1.00
C-U-6	59/79	48	29.	29.	56.	50.	1.04	1.04

+) The second value listed in this column represents the value in the case of reversed polarity. In the case of the E-structures the attachment of the sheath to the phage head made the determination of the polarity unnecessary.

It is concluded that the amplitude parts of the curves generally move towards a lower value of R under the influence of the electron beam. This is clearly visible in the E-U data, the corresponding phases in the two curves remain more or less equal, a sign that the dose effects are not taking place in a different way over different parts of the specimens. The changes in the curves can be explained by a movement of the negative stain. In the case of the C particles no clear rationale can be derived although the radiation effects are evident. In this case it appears as if there is not only migration of stain but also a change in the structure because form of the graph of the complex amplitudes along the layerlines also changes upon irradiation (10).

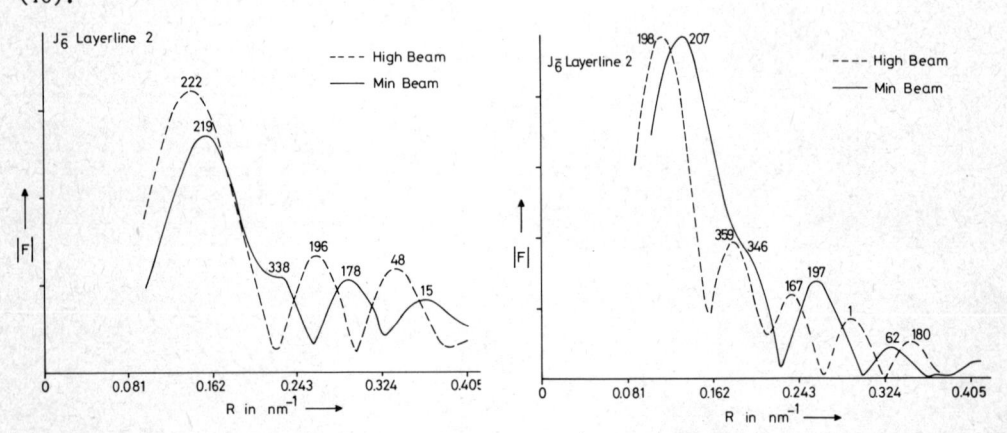

Figure 3. Graphs of the Fourier components of averaged projections obtained from the E-structures in U and P stain at different doses. Left the uranyl stained data at layerline 2 are shown and right the corresponding data in P-stain.

Also no striking difference appears in this behaviour as far as different stains are concerned. The translation of these effects in real space can be visualized by calculating a 3D density distribution by a Fourier-Bessel synthesis. A number of sections perpendicular the helical axis obtained in this way are presented in Figure 4 and the results show that in a number of these sections the density of the protein (white) changes drastically. Therefore only the interpretation of the L-data seem justified when it comes to monitoring the changes during contraction.

The resulting structures obtained by the two different stains and under low dose are in reasonable agreement as can be seen in Figure 5, which summarizes the main conclusions. Although it is not possible to determine the exact shape of the protein subunit from this type of work, it appears that in the E- as well as in the C-form of the sheath the protein subunit has an elongated shape. The long subunit axis lies in a plane perpendicular to the helical sheath axis. In going from the E- to C-form the subunits can be thought to rotate over about 35 degrees in this plane, which finally will result in a larger sheath diameter and a change of the packing parameters.

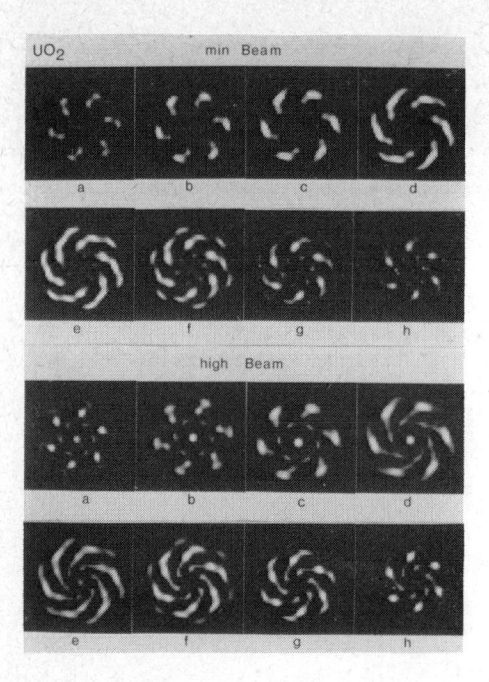

Figure 4.

A gallery of sections from the 3D density map of the reconstructed E-structure. The sections are perpendicular to the horizontal axis and spaced by about 5 Å. The sections are marked a, b, c etc. and corresponding sections through the helical axis in the two cases (minimum and maximum beam respectively L and H) have the same letter underneath. White is protein.

Whether or not this movement parallels a change of shape of the protein subunits is hard to demonstrate.

Figure 5.

A survey of the data obtained from the 3D reconstructions. In each of the 4 drawings contourlines are superimposed from three adjacent sections of the 3D Fourier. The 3 sections are 10 Å apart and the contours drawn have the same density value. The lines are drawn at .... 0 Å, ——— 10 Å, and ---- 20 Å. It can be seen that the orientation of the long subunit axis in the E-reconstruction and P-reconstruction is similar. This also applies for the C-data obtained with different stains.

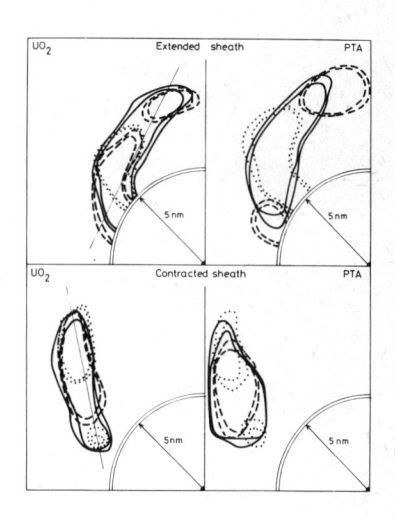

This example has served to illustrate the power of digital image processing of regular biological structures. No doubt, these methods will be applied to new structures in future. However, the breakthrough towards a higher resolution will not depend on these techniques themselves, but rather will depend on the developments in imaging the biological object without stain under minimum dose conditions, or on finding methods of preserving the objects under high electron doses.

5. LITERATURE

1. J.T.Finch (1972), Contemp.Phys.13, 1-21.
2. S.C.Harrison, A.J.Olson, C.E.Schutt, F.K.Winkler & G.Bricogne (1978), Nature 276, 368-373.
3. A.Klug (1971), Phil.Trans,Roy.Soc.London B 261, 173-179.
4. R.A.Crowther, D.J.DeRosier & A.Klug (1970), Proc.Roy.Soc.London A 317, 319-340.
5. H.E.Huxley & A.Klug (1971), New Developments in Electron Microscopy, London, The Royal Society.
6. R.A.Crowther & A.Klug (1975), Ann.Rev.Biochem.44, 161-182.
7. J.E.Mellema (1979) In Computer Processing of Electron Micrographs, Edited by P.W.Hawkes, Springer, Heidelberg.
8. H.E.Hemphill & H.R.Whiteley (1975), Bacteriol.Rev.39, 257-315.
9. A.F.M.Cremers, H.Y.Steensma & J.E.Mellema (1978), Eur.J.Biochem.89, 389-395.
10. A.F.M.Cremers & J.E.Mellema (1979), To be published.
11. U.K.Laemmli, L.A.Amos & A.Klug (1976), Cell 17, 191-203.
12. P.N.T.Unwin & R.Henderson (1975), J.Mol.Biol.94, 425-440.
13. A.Klug, F.H.C.Crick & H.W.Wyckoff (1958), Acta Cryst.11, 199-213.
14. A.F.M.Cremers, J.C.Fischer & J.E.Mellema (1979), Ultramicr.4, 91-96.
15. L.A.Amos & A.Klug (1975), J.Mol.Biol.99, 51-73.
16. A.Klug & J.E.Berger (1964), J.Mol.Biol.10, 565-569.
17. D.J.DeRosier & P.Moore (1970), J.Mol.Biol.52, 355-369.

CATHODOLUMINESCENCE TOPOGRAPHY

A. R. Lang

H. H. Wills Physics Laboratory

University of Bristol, England

I. INTRODUCTION
II. ELECTRON PENETRATION IN SOLIDS
III. THE GENERATION OF CATHODOLUMINESCENCE
    3.1 Review of radiative processes
    3.2 Donor-acceptor pair recombination
    3.3 Vibronic spectra
    3.4 Dislocation luminescence
IV. EXPERIMENTAL TECHNIQUES
V. APPLICATIONS
    REFERENCES

I. INTRODUCTION

Many luminescence phenomena bear description names which indicate the mode of excitation: photo-, chemi-, tribo-, electro-, bio-, radio-luminescence, etc. Cathodoluminescence (hereafter abbreviated to CL) refers to excitation by bombardment with fast-moving electrons. The optical processes of CL are closely related to those in other methods of stimulation with exciting energies greater than band gap energies (as in photo-excitation with short UV or X-rays), or in methods where minority charge carrier injection engenders electron-hole recombination, as in the light-emitting diode. Thus features peculiar to CL are associated mainly with the processes of conversion of the primary energy. Therefore, in the following discussion, the energy-dissipation processes and range of fast electrons in solids will be discussed first, and the lower-energy, optical emission processes subsequently. The description of techniques deals only with image-forming experiments (by photographic or electronic means); and the brief account of applications likewise restricts itself to CL topography (but hopes to illustrate the value of the topographic approach).

Sir William Crookes, father of CL, described observations on various solids in 1879[1]. Later, concerning diamond, he picturesquely wrote [2]: "It has been ascertained that the cause of phosphorescence is in some way connected with the hammering of the electrons, violently driven from the negative pole on to the surface of the body under examination ...". This epitomises an important attribute of CL, the ability to excite continuously with high energy density. Although this advantage relative to photo-excitation has been diminished in some respects by the advent of high-power CW lasers available at many frequencies (permitting excitation of emission systems selectively, as against the blunt non-selectivity of CL stimulation), and lasers can inject energy per unit area comparable with that carried by the electron beam, it must be remembered that the latter will dump its energy within $\sim 10\mu$m of the specimen surface under typical experimental conditions, whereas similar volume localisation in the photo-excitation case would require an extinction coefficient of the exciting radiation greater than 10^3cm^{-1}. For best topographic resolution, CL must win, because an electron probe can be made much smaller than an optical wavelength. Indeed, in the rapidly-developing field of CL recording in the Scanning Transmission Electron Microscope (STEM), resolutions down to nanometre dimensions are possible.

II. ELECTRON PENETRATION IN SOLIDS

Four parameters are of interest in CL studies: (i) the variation of light output(L) with electron voltage(V) and, (ii) with current density(σ); (iii) the efficiency of light production, and (iv), (highly important in topographic studies) the variation of light output with depth below the specimen surface. Study of (i) goes back to the work of Lenard[3-6] who found $L \propto (V-V_0)^q$, a relation which applies well to polycrystalline phosphors, with the index q lying between 1 and 3. The so-called 'dead voltage' V_0 can be reduced to a low value by avoiding surface contamination. Regarding (ii), plots of L versus σ generally show a negative curvature, corresponding to saturation of optical centres. Differences in saturability of various centres have proved to be an important experimental observable in CL topography; but care is needed to distinguish between effects due to specimen heating and to excitation density per se. Variations (i) and (ii) are texture and structure sensitive. One component of (iii) is not structure dependent, but is calculable, well-investigated, and of great importance in scanning electron microscopy (SEM). This is the back-scattering coefficient, η_B. It is not sensitive to V, but is roughly proportional to atomic number, Z, of the specimen up to Z=30, when $\eta_B = 0.3$, then increases less rapidly, reaching about 0.5 when Z = 90[7]. For a compound, use the weighted mean Z to estimate η_B. The overall efficiency of cathodoluminescence, given by the ratio (luminous energy emitted)/(electron energy absorbed), can be as high as 23% [5].

Turning to (iv), consider first the energy-loss processes involving the primary electrons. These can be (a) loss by Bremsstrahlung (not significant except with high V and Z), (b) atomic displacement and (c) inelastic collision with electrons, producing ionisation and excitation of atoms in the specimen. The maximum kinetic energy, T, that an electron, rest mass m_0, given accelerating voltage V, can transfer to a static nucleus, mass M, is

$$T = (2eV/Mc^2)(eV + 2m_0c^2). \tag{1}$$

The energy needed to displace an atom is typically about 25 eV[8]. With 30 kV electrons, this value is exceeded only in the case of hydrogen, which can receive 65 eV 'knock-on' energy. With heavier elements, the minimum electron energies to produce displacements can sometimes be found experimentally, e.g. 118 kV for S and 285 kV for Cd in CdS[9]. These figures provide a general guide. Note, however, that in alkali halides displacements can be effected by radiationless conversion of excitation energy (of ~5 eV only) to kinetic energy, with consequent F-centre production[10]. Thus, the dominant mode of energy loss is electron-electron interaction, with production of many secondary electrons in a cascade process which terminates when no electron has sufficient energy to produse further ionisation. The energy-loss rate of electrons, on the Bloch-

Bethe theory, with non-relativistic energies, is (Segrè[11])

$$-dE/dx = (2\pi N\ e^4/E)\ln[(E/E_i) - \tfrac{1}{2}\ln 2 + \tfrac{1}{2}], \qquad (2)$$

or, more simply, by (Garlick[12])

$$-dE/dx = (2\pi N\ e^4/E)\ln(E/E_i). \qquad (3)$$

N is the number of electrons per unit volume, E_i is an averaged ionisation energy for all atomic electrons. The assumption of a constant E_i, and in particular, the use of (3), become inadequate as E approaches E_i. Integration of $\frac{dE}{dx}$ along the electron trajectory down to $E = E_i$ gives the Bethe range, R_B, which significantly exceeds the actual linear penetration depth because of multiple scattering. Bethe ranges, expressed as mg cm^{-2}, have been graphed for various values of Z from 6 to 79, and for $4\,kV < V < 50\,kV$ [13]. Over the range of voltages used in CL topography, one can take $R_B \propto V^n$, with n decreasing from 1.8 for Z=6 to 1.5 for Z=79. Multiple scattering of electrons expands the ionized region into a teardrop shape, pendant from the point of electron beam impact. In CL topography one wishes to know the maximum depth at which significant light production occurs and the depth where light production is greatest. These data have been ingeniously determined in some phosphors by microphotography of 'glow lobes' [14]. In CL topography of diamond one can get information from line or sheet-like features that make known angles with the specimen surface, e.g. luminescent dislocations and 'giant' platelets [15]. In diamond, observing with an optical depth resolution of about $2\mu m$, it becomes apparent that maximum light generation is below the surface when $V \geqslant 20\,kV$. At $50\,kV$ maximum light is produced $4-5\,\mu m$ below the surface, and only a small fraction from below $10\,\mu m$. The domain of light production will appear larger than that of ionisation if appreciable fluorescence or outward diffusion of charge carriers before recombination occurs.

III. THE GENERATION OF CATHODOLUMINESCENCE
3.1. Review of radiative processes

CL studies are performed chiefly on covalent and ionic crystals, especially on inorganic semiconductors; and the following discussion relates particularly to the last-named class. The end product of the cascade process described in section II is a high density of electrons in the normally empty conduction band (CB) and a high density of holes in the normally filled valence band (VB). The mean energy for production of an electron-hole pair in the cascade process has been estimated by Shockley[16]: one may take this as typically about three times the bandgap energy, E_g. Electrons in the CB rapidly 'thermalise' to within $\sim kT$ of the lowest point in the CB (via intraband transitions such as 1 in fig.1), and holes likewise to

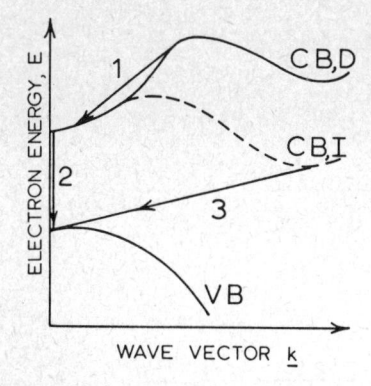

Fig. 1. Band structure in a semiconductor. VB: valence band edge. CB, D: conduction band edge for direct gap. CB, I: conduction band edge for indirect gap.

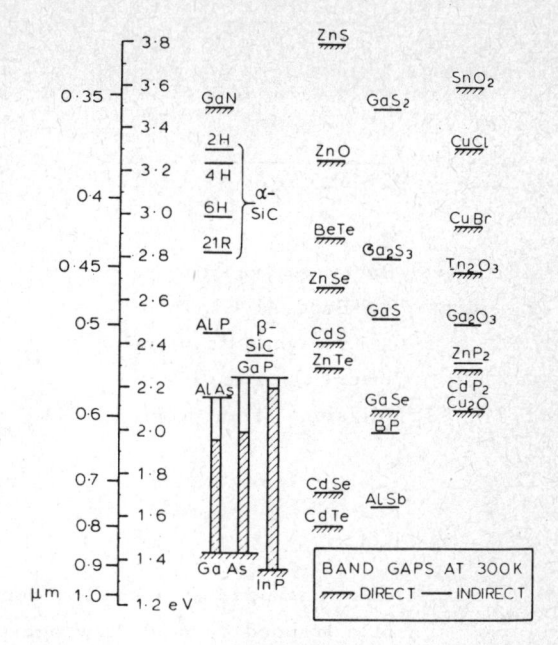

Fig. 2. Band gap energies of semiconductors in near UV to near IR range.

the top of the VB. In <u>direct gap</u> semiconductors the absolute minimum of the CB occurs at the same k-vector as the absolute maximum of the VB. Electron-hole recombination occurs with conservation of energy and momentum, in a transition (2) which is vertical on the scale of fig. 1 because $\underline{k}$ (photon) is negligible compared with $\underline{k}$ (electron). In the <u>indirect gap</u> semiconductor, the transition (3) occurs via a virtual energy state involving emission or absorption of a phonon in order to conserve momentum. Fig. 2(after [17]) displays some band gaps in the energy range covered by CL topography. To this list about a dozen ternary compound semiconductors could be added: the number of substances which are candidates for study by CL topography increases continually. Diamond has an indirect band gap of about 5.3 eV. The large differences in E_g for SiC polytypes shown on fig. 2 may appear remarkable, but it must be remembered that indirect band gaps are highly structure sensitive: the direct band gap of SiC varies little with stacking sequence. Fig. 3 schematically represents possible radiative transitions in crystals both perfect and imperfect. Transition (a), the band-to-band recombination discussed above and (b), free exciton recombination [18], occur in pure materials; but (b) can equally well represent recombination of excitons bound to impurities. The dissociation energy of an exciton is of order 0.1 eV. Recombination of a hole in

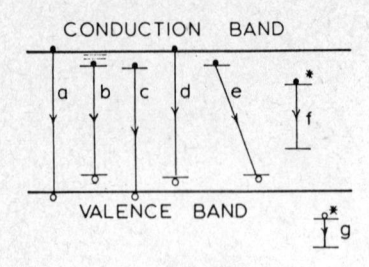

Fig. 3 (above) Radiative transitions. Fig. 4 (upper R) 'Band A' CL from natural Type Ia and synthetic diamond. Fig. 5 (lower R) The 'H3' and '3H' CL emissions from diamond.

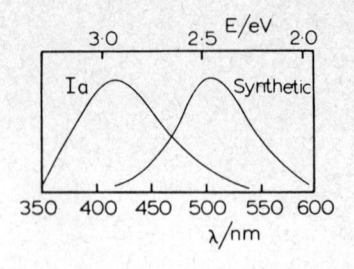

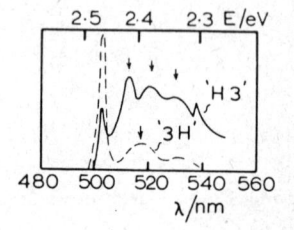

VB with an electron trapped at a shallow donor level, and of an electron in CB with a hole trapped on a shallow acceptor level, are shown by (c) and (d) respectively. The donor-acceptor pair recombination (e), and the vibronic system represented by (f) are of sufficient importance in CL to merit separate discussion below. The excitation of core levels (g) can occur in rare-earth and transition elements: the resulting emission spectra are characteristically sharp.

3.2 Donor-acceptor pair recombination

Consider a crystal containing comparable concentrations of donors and acceptors, the donor ionisation energy being E_D measured from the CB, and the acceptor ionisation energy E_A from the VB. In the case of 'shallow' centres $E_D + E_A$ will be small compared with E_g. The starting point for the process represented by (e) in fig. 3 is an ionised donor and ionised acceptor separated by distance R. These centres then capture an electron and a hole, respectively. Electron and hole subsequently combine leaving the donor and acceptor again in the ionised state and releasing energy

$$h\nu = E_G - (E_A + E_D) + e^2/\epsilon R, \tag{4}$$

(neglecting a polarisation term significant only at small R). If donor and acceptor lie on well-defined lattice sites, then possible values of R are defined and denumerable, leading to line spectra with relative line intensities proportional, inter alia, to the number of inter-site vectors in each 'shell' of radius R. Some semiconductors e.g. GaP, display such line spectra remarkably well at low temperatures[19]. Now the CL spectrum of both the common natural Type Ia diamond and and synthetic diamond consists of a broad, structureless emission (blue for natural, green for synthetic crystals) known as 'Band A' emission, upon which vibronic spectra (such as discussed below) are superimposed[20-25]. Fig. 4 sketches the shape of Band A spectra. These emissions are generally regarded as due to donor-acceptor pair recombination, but there are difficulties in this explanation, not least being the absence of line structure; for with smaller lattice parameter and smaller dielectric constant, ϵ, than in other semiconductors one would expect (from formula (4)) lines much more widely spaced than in, say, GaP. In favour of the proposed explanation is the observation that increased excitation shifts the centroid of emission to shorter wavelengths: close pairs (giving higher $h\nu$) have faster recombination rates, and saturate less easily. CL topography[15] makes it vividly apparent that there are several types of centre responsible for Band A-type spectra; and the strong polarisation of the green emission from cube growth sectors in synthetic diamond ($\underline{E}$ vector parallel to local growth facet) provides a further unexplained feature[26].

3.3. Vibronic spectra

Fig. 5 represents the density patterns on spectrophotographs of two vibronic CL emission systems found in different topographic contexts in natural diamonds. A vibronic system consists of a relatively sharp 'zero-phonon' line (corresponding to electron de-excitation represented by transition (f) in fig. 3) and, in emission, a band on the low-energy side in which several phonon peaks (such as those arrowed in fig. 5) can be recognised. The zero-phonon line can be sufficiently narrow (a few meV) for studies of the symmetry of the optical centre concerned to be studied by Zeeman, Stark or uniaxial stress splitting, or to reveal (through variations in breadth) presence of local internal stresses in the specimen. Vibronic systems occur in both emission and absorption; the absorption spectrum is a reflection of the emission spectrum about the zero-phonon line, transferring the phonon sideband to the high-energy side. Mirror symmetry is usually not exact. Note also that the 'H3' CL emission spectrum shown in fig. 5 has a small sharp peak at 538 nm, evidently an electronic transition. This does not occur in the absorption spectrum. Increase of temperature increases the area under the phonon sideband relative to that under the zero-phonon peak. Three features characterise a vibronic system: 1, the zero-phonon energy, 2, the associated phonon energies and 3, the strength of coupling of the electronic transition to lattice vibrations. This last is denoted by S, the Huang-Rhys factor[27, 28]. Fig. 5 represents an intriguing case of two vibronic systems, doubtfully distinguishable as regards zero-phonon energy, but quite distinguishable by the strong coupling (relatively large S) to phonons of energy 44 meV in the 'H3' system and weak coupling to phonons of 66 meV energy in the '3H' case. Topographic studies[15] reveal that some of the chief sources of 'H3' emission are slip bands and occasionally individual dislocations, whereas the '3H' system is emitted by the α-radiation-damaged 'rind' associated with some natural diamond surfaces.

3. 4. Dislocation luminescence

Dislocations introduce local energy states into the lattice[29, 30]: it is likely that these promote electron-hole recombination. It does not follow that they lead to a net increase of photon energy emitted, for they may promote radiationless more effectively than radiative recombination. One might suppose that generally in a direct-gap material there will be a net loss of luminescence, and the reverse in an indirect-gap material. In either case, the spectral distribution of dislocation luminescence will differ from that of dislocation-free matrix, and herein lies the possibility of 'seeing' dislocations by CL topography, even pristine, undecorated dislocations. The 'dark lines' seen in infrared luminescence micrographs of GaAs LEDs[31] are doubtless dislocations, and most convincing SEM CL images of individual (but decorated) dislocations in doped GaAs have been recorded using a photomultiplier with S1 response (giving sensitivity in the near IR, down to 1.1 eV)[32]. SEM observations of visible luminescence at indentations in MgO, attributed to dislocations, has also been recently reported[33]. Without doubt the most striking dislocation CL is seen in diamonds[15, 26, 34]. It has been possible to establish a one-to-one correspondence between CL and X-ray topographic images of individual dislocations[35]; and a strong polarisation of the blue CL emission from dislocations has been demonstrated, the $\underline{E}$ vector being parallel to the dislocation line.

IV. EXPERIMENTAL TECHNIQUES

The advantages of visual observation of the specimen cannot be too strongly emphasized. This facilitated the discovery of polarised CL from diamond, for example. It also helps in avoiding the mistaking of trapped light escaping from surface steps, indentations, cracks and scratches for genuine CL! A simple apparatus that has been used highly profitably[36] consists of a continuously-pumped sample chamber placed on a microscope stage. An inclined gas discharge tube provides a specimen current density of about 8 μA cm^{-2} at 6 kV. The chamber is reversible so that samples can be bombarded from above or below; and it has top and bottom windows

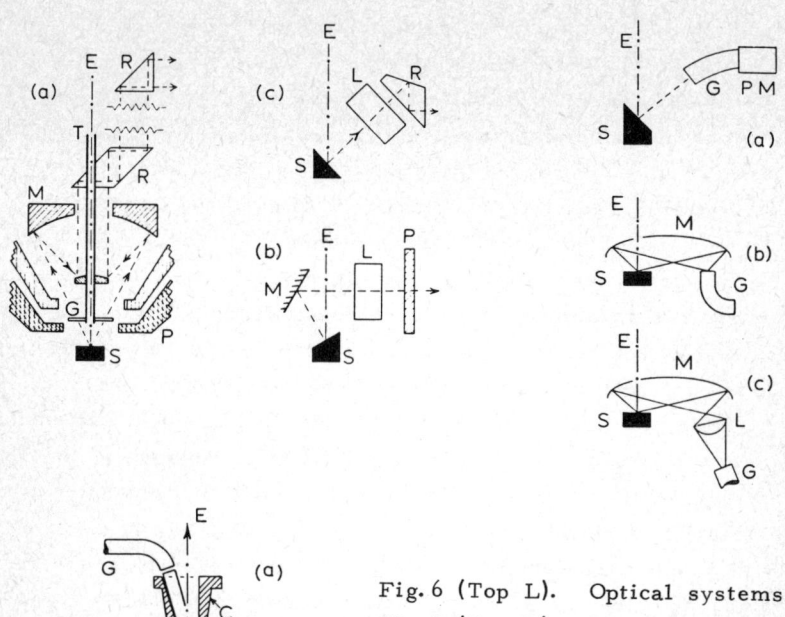

Fig. 6 (Top L).   Optical systems
Fig. 7 (Top R)   CL recording by SEM
Fig. 8 (Bottom L) CL recording by STEM

so that the sample can be examined and photographed by transmitted or reflected illumination as well as by CL, with or without crossed polarisers.

For work requiring controlled variations of kV, illuminated area and current density, more elaborate electron sources are needed. Fig. 6(a) shows the optical system in an X-ray microprobe. The electron beam E passes down tube T on the axis of the reflecting objective M, to hit the specimen S. P are the polepieces of the final electron lens. A thin glass G protects the reflecting surface of M from bombardment by backscattered electrons. Reflecting prisms R direct the light beam out of the apparatus. M can have a numerical aperture of ∼0.4 and give X35 magnification of a field 0.6 mm in diameter. Fig. 6(b) shows an optical system usable in an SEM chamber. The electron beam E falls on the specimen S and the mirror M reflects CL into the lens L placed in the chamber, from which light emerges via the glass port P. With appropriate lens and beam defocus, illuminated areas up to 20 mm diameter can be photographed. Shorter working-length lenses can be used in the set-up shown in fig. 6(c). The system shown in fig. 6(b) was used in IR CL topographic experiments involving ciné-recording of the TV monitor output of an IR-sensitive vidicon and the visual, photographic and TV observation of the output of an IR image converter used alone and also with further stages of electronic image intensification[37].

Fig. 7(a) shows arrangements for indirect imaging in the SEM via scanning electron probe, light guide G and photomultiplier PM. Use of 3 such detection systems, each with appropriate colour separation filter, or a single system with rotating filters, offer possibilities for colour TV presentation of CL topographs. To improve light-gathering, an ellipsoidal mirror M can be used together with a high-aperture fibre-optic light guide (fig. 7(b)). If the collected light is to be transmitted to a monochromator for CL spectroscopy, a collimating lens L (fig. 7(c)) is advantageous for reducing the beam aperture so that upon emergence from G it does not exceed the monochromator aperture[38, 39].

Light-collecting arrangements in the STEM are severely constricted by the presence of pole-pieces, cooled anti-contamination shields and other components. Fig. 8(a) shows a simple arrangement for use in the VG HB5 STEM[40]. A light guide G looks into a thin-walled silver tube T which sits obliquely in the specimen cartridge C and collects light from the specimen S. The tube T must have a slot to pass the electron beam E. Fig. 8(b) shows arrangements appropriate for the JEM 200B STEM[41]. The separation of pole pieces P allows room for an ellipsoidal mirror M which has a hole to allow passage of electrons E to the specimen S. A pair of mirrors R, backed by lead shields, are interposed between M and the light guide G to prevent X-rays generated in the specimen chamber from passing out into the light-detection system.

V. APPLICATIONS

The extensive lists of gem and other minerals which show colourful CL patterns indicate a profitable field for further, more analytical investigation[42, 43]. A fascinating application of CL topography is the study of lunar minerals, particularly plagioclases[44, 45]. Work on diamond and LED materials has already been noted. Particularly interesting is the combination of CL topography with the Castaing-Slodzian ion analyser in the correlation of luminescence with impurities in cassiterite[46]. Noteworthy work is being done on II-VI compounds at high beam currents and temperatures down to 4 K, in which it is hoped to observe lasing action[47].

REFERENCES

1) W. CROOKES, Phil. Trans. Roy. Soc. Lond. 170 (1879) 641-662.
2) W. CROOKES, Diamond (1909) London: Harper.
3) P. LENARD, Ann. Phys. 12 (1903) 449-490.
4) G. F. J. GARLICK, Luminescent Materials (1949) Oxford: Clarendon Press.
5) D. Curie, Luminescence in Crystals (translated by G. F. J. Garlick) (1963) London: Methuen.
6) G. F. J. GARLICK, in Luminescence of Inorganic Solids (ed. P. Goldberg). (1966) New York: Academic Press.
7) C. W. OATLEY, The Scanning Electron Microscope. Part I. The Instrument (1972) Cambridge: University Press.
8) L. T. CHADDERTON, Radiation Damage in Crystals (1965) London: Methuen.
9) F. J. BRYANT & C. J. RADFORD, J. Phys. C3 (1970) 1264-1274.
10) D. POOLEY, Proc. Phys. Soc. Lond. 87 (1966) 245-262.
11) E. SEGRÈ, Nuclei and Particles (2nd printing) (1965) New York: Benjamin.
12) G. F. J. GARLICK, Brit. J. Appl. Phys. 13 (1962) 541-547.
13) H. E. BISHOP & D. M. POOLE, J. Phys. D6 (1973) 1142-1158.
14) W. EHRENBERG & D. E. N. KING, Proc. Phys. Soc. Lond. 81 (1963) 751-766.
15) P. L. HANLEY, I. KIFLAWI & A. R. LANG, Phil. Trans. Roy. Soc. Lond. 284 (1977) 329-368.
16) W. SHOCKLEY, Czech. J. Phys. B11 (1961) 81-121.
17) A. A. BERGH & P. J. DEAN, Light-emitting Diodes (1976) Oxford: Clarendon Press.
18) D. L. DEXTER & R. S. KNOX, Excitons (1965) New York: Interscience-Wiley.
19) D. G. THOMAS, M. GERSHENZON & F. A. TRUMBORE, Phys. Rev. 133 (1964) A269-279.
20) P. J. DEAN, P. J. KENNEDY & J. E. RALPH, Proc. Phys. Soc. Lond. 76 (1960) 670-687.
21) A. T. COLLINS, Industr. Diamond Rev. 34 (1974) 131-137.
22) G. DAVIES, in Diamond Research 1975 (1975) London: Industr. Diamond Information Bureau.
23) G. DAVIES, in The Chemistry & Physics of Carbon, Vol. 13 (eds. P. L. Walker, Jr. & D. A. Thrower) (1977) New York: Marcel Dekker.
24) G. DAVIES, in The Properties of Diamond (ed. J. Field) (1979) New York & London: Academic Press.
25) P. J. DEAN, Phys. Rev. 139 (1965) A588-602.
26) G. S. WOODS & A. R. LANG, J. Cryst. Growth 28 (1975) 215-226.
27) K. HUANG & A. RHYS, Proc. Roy. Soc. Lond. A204 (1950) 406-423.
28) D. B. FITCHEN, in Physics of Color Centres (ed. W. Beall Fowler) (1968) New York & London: Academic Press.

29) R. LABUSCH & W. SCHRÖTER, Lattice Defects in Semiconductors 1974, I.O.P. Conference Series No. 23 (1975) 56-72.

30) R. WAGNER & P. HAASEN, ibid, 387-397.

31) K. H. ZSCHAUER, Solid State Commun. $\underline{7}$ (1969) 335-337.

32) C. SCHILLER & M. BOULOU. Philips Tech. Rev. $\underline{35}$ (1975) 239-246.

33) S. J. PENNYCOOK & L. M. BROWN, J. Luminescence $\underline{18}/\underline{19}$ (1979) 905-909.

34) I. KIFLAWI & A. R. LANG, Phil. Mag. $\underline{30}$ (1974) 219-223.

35) I. KIFLAWI & A. R. LANG, Phil. Mag. $\underline{33}$ (1976) 697-701.

36) J. E. GEAKE, G. WALKER & A. A. MILLS, in 'The Moon' (eds. H. Urey & S. K. Runcorn) (1972) Dordrecht: Reidel.

37) A. R. LANG & A. P. W. MAKEPEACE, J. Phys. E$\underline{10}$ (1977) 1292-1296.

38) J. B. STEYN, P. GILES & D. B. HOLT, J. Microscopy, $\underline{107}$(1976) 107-128.

39) L. CARLSSON & C. G. VAN ESSEN, J. Phys. E$\underline{7}$ (1974) 98-100.

40) S. J. PENNYCOOK, A. J. CRAVEN & L. M. BROWN, Developments in Electron Microscopy and Analysis 1977, I. O. P. Conf. Ser. No. 36 (1977) 69-72.

41) P. M. PETROFF, D. V. LANG, J. L. STRUDEL & R. A. LOGAN, Scanning Electron Microscopy/1978, Vol. 1 325-332. (O'Hare, Il., USA: SEM Inc.)

42) R. A. P. GAAL, Gems & Gemology, (Winter 1976/1977) 238-244.

43) E. NICKEL, Minerals Sci. Engng. $\underline{10}$ (1978) 73-100.

44) J. E. GEAKE, G. WALKER, A. A. MILLS & G. F. J. GARLICK, Proc. 3rd Lunar Sci. Conf., Geochim. cosmochim. Acta Suppl. 3, $\underline{3}$ (1972) 2971-2979.

45) J. E. GEAKE, G. WALKER, D. J. TELFER & A. A. MILLS, Phil. Trans. Roy. Soc. Lond. A$\underline{285}$ (1977) 403-408.

46) G. REMOND, Bull. soc. fr. Minéral. Cristallogr. $\underline{96}$ (1973) 183-198.

47) A. D. YOFFE, K. J. HOWLETT AND P. M. WILLIAMS, Scanning Electron Microscopy/1973 (Part II), 301-308 (Chicago: IIT Research Institute).

IMAGE PROCESSING IN ELECTRON MICROSCOPY : NON-PERIODIC OBJECTS

O. Kübler

Institut für Zellbiologie der ETH-Z, CH-8093 Zürich

1. Introduction and Summary

Processing of bright field phase contrast electron micrographs of aperiodic specimens has the aim of achieving atomic resolution by correcting "a posteriori" for the severe imaging defects of electron optical lenses. Image processing of non-periodic and periodic objects may be distinguished by regarding them as exercises in image restoration and in signal extraction by crystallographic methods respectively. The periodic image component of crystalline biological specimens rarely extends into a resolution range where imaging defects become important ; even in such cases the distinction between image restoration and crystallographic methods remains valid as both techniques have to be combined.

Image restoration in the modern sense began with the optical spatial filtering experiments of Maréchal [1] and has since witnessed rapid growth largely due to communication theory approaches and the versatilitiy and power of modern digital computers. Attempts have been made to use non-linear restoring methods, to include a priori imaging constraints, and to incorporate physiological aspects. Most of the work has been devoted to the restoration of degraded photographs taken in incoherent illumination and books summarizing the state of the art have recently become available [2, 3].

In the restoration of electron micrographs not all of the modern techniques seem to be helpful. The situation is simple on the one hand, since the image recording is essentially linear and physiological aspects appear to be of minor importance ; on the other hand high resolution electron micrographs are extremely noisy coherent images whose restoration leads to technical problems which may for example exclude using a priori imaging constraints.

We present results of restoring electron micrographs with a limiting resolution of 2.5 Å. The presentation starts with a discussion of optimal imaging conditions for minimizing radiation exposure of the object whilst allowing image restoration. Implications of phase and Wiener filtering in particular are discussed, the results depend critically on the exact determination of the imaging parameters but uncritically on the object-to-noise ratio above a detectability threshold. The improvement in image quality and the accuracy of digital processing are demonstrated by visual display of the original and filtered images, difference images and diffractograms. In the presentation a number of parallels are drawn to the case of restoring blurred photographs taken in incoherent illumination and extensive use is made of results previously published in greater detail [4].

2. Phase Contrast Imaging

Image formation in the electron microscope may be modelled as a linear space-invariant process in the presence of additive noise. The action of the imaging system is then completely specified by its impulse response $h(Q)$ or, in optical terminology, point spread function (psf). The image plane intensity distribution $i(Q)$ due to a phase object $\varphi(Q)$ then includes a constant background term, the convolution product of the psf and the object, and the noise $n(Q)$

$$i(Q) = i_o \{1-2[\varphi \otimes h](Q)\} + n(Q) \qquad (1a)$$

where i_o stands for the intensity of the illuminating beam. Distances Q are normalized as usual according to Hanszen [5] rendering the intensity expression (1a) independent of instrumental parameters (wavelength of the illuminating electrons, spherical aberration constant of the lens). Many aspects of imaging are easily visualized by giving the behavior of the psf h(Q). The mathematical treatment of lens aberrations, defocus, and the effects of partial coherence, however, is commonly made by going from direct space to the spatial frequency domain, i.e. by Fourier transforming the intensity expression which yields

$$I(R) = I_o \{\delta(R) - 2\emptyset(R)H(R)\} + N(R) \tag{1b}$$

Fourier transforms are represented by capital letters, the symbol $\underline{\delta}$ stands for the Dirac delta function and spatial frequencies are denoted by R (R = 1/Q). The Fourier transform of the psf h(Q) is called the transfer function (tf) H(R). It is in good approximation given by

$$H(R) = -E(R)H_c(R) \tag{2a}$$

$$H_c(R) = \sin \pi \{R^4/2 - R^2\Delta\} \tag{2b}$$

where E(R) is a slowly varying attenuating function expressing the partial coherence of the illuminating electrons, $H_c(R)$ stands for the tf valid for perfect coherence, and normalized defocus is denoted by Δ.

More detailed accounts of electron microscopical imaging and transfer functions may be found in these Proceedings in the contributions of R. Wade and P. Hawkes.

3. Atomic Resolution and Quantum Noise

The electron microscope image to be restored is usually given in the form of a photographic recording. Photographic emulsions show a response to electrons which differs markedly from that to photons in that :

(a) the optical density (OD) is approximately linearly related to the exposure for electrons [6] while it follows the well known H-D curve (logarithmic response) for photons

(b) the detective quantum efficiency (DQE) is close to one for electrons while it amounts to no more than a few percent for photons.

Since a linear relationship holds between the intensity distribution (1a) and the spatial OD distribution of the electron micrograph, linear restoration schemes appear sufficiently accurate for electron microscope image processing while more sophisticated models may become important for normal photographs. Because of the nearly ideal DQE the noise properties of electron micrographs are essentially determined by the quantum noise of the electrons. No further distinction between the recorded OD distribution and the intensity expression (1a) will be made.

Specimens in the range of atomic resolution represent very weak phase objects which will give low contrast images implying a fundamental problem in high resolution electron microscopy. Small low contrast objects can only be perceived at very low noise levels which means that high illuminance levels (imaging doses) are required. Such doses will lead to severe radiation damage of most specimens. An estimate of attainable noise levels is provided by modelling the quantum behavior of the electrons by a Poisson process.. The condition for discriminating different contrast levels in contiguous cells of area F in the presence of Poisson noise may be stated following early considerations of Rose [7] in the field of video engineering

$$i_o 2 \{\int_a [\varphi \otimes h] dQ - \int_b [\varphi \otimes h] dQ\} \geq SNR \; \sigma_n = SNR \; (i_o F)^{1/2} \tag{3}$$

Expression (3) simply means that the average image modulation must exceed the standard deviation σ_n of the noise n(Q) by a certain amount. The image modulations $[\varphi \otimes h]$ are integrated over nonoverlapping cells F_a and F_b, SNR is an empirical constant (SNR = 5 is usually quoted to guarantee detectability), and the Poisson standard deviation is given by $(i_o F)^{1/2}$.

The relationship between the detectability of a small object like a single atom and the instrumental resolution at a given illuminance, i.e. noise, level may be studied by considering an ideal diffraction limited coherent imaging system. Such a system is characterized by a tf $H_{id} = 1$ which is constant to some cut-off radius R_c imposed by the objective aperture. The psf of the diffraction limited system is given by

$$h_{id}(Q) = K_o(R_c) \{2J_1(2\pi QR_c)/2\pi QR_c\} \tag{4}$$

where $K_o(R_c)$ denotes central contrast ($K_o = \pi R_c^2$) and J_1 is the first order Bessel function. In the electron microscope only small objective apertures are practical meaning that an atom behaves almost like a point scatterer $\varphi_o \delta(Q)$. Contrast consideration may consequently be reduced to studying the behavior of the psf alone.

A reasonable estimate for the detectability of an isolated atom is derived by comparing the average intensity inside the central lobe of the psf to the background illuminance. The central lobe is defined by the smallest radius Q_o at which the psf changes sign ; at the same time, this provides a measure of resolution ($Q_o = .61R_c$). An isolated point scatterer (atom) $\varphi_o \delta(Q)$ will be detectable when Rose's inequality (3) is fulfilled

$$2.8\, \varphi_o \geq SNR\{Q_o\,(\pi/i_o)^{1/2}\} \tag{3'}$$

In the derivation of condition (3') the remarkable fact was used that the integral $V(Q) = 1 - J_o(2\pi QR_c)$ of h_{id} over a circle of radius Q yields the constant value $V(Q_o) = 1.4$ (Q_o is the radius of the central lobe) independent of the resolution. The detectability (3') of a single atom imaged in a diffraction limited system is determined by the resolution Q_o and the inverse square root of the imaging dose i_o. At modest resolutions the problem of radiation damage becomes less severe than at high resolutions since a linear increase in Q_o allows a quadratic decrease of the specimen irradiation for a fixed SNR index.

4. Image Restoration : Phase and Wiener Filtering

Electron optical lenses suffer from enormous spherical aberration as compared to light optical instruments, the electron microscope will therefore behave differently to an ideal diffraction limited system. Spherical aberration can be compensated to a certain extent by controlled underfocussing. In 1949 Scherzer [8] determined operating conditions ($\Delta = 1$, $R_c = \sqrt{2}$) for which the tf is of uniform sign and nearly constant in a wide annular spatial frequency region. Unfortunately even for modern instruments the Scherzer conditions are not sufficient to achieve a resolution in the range of chemical bond lengths ($\lesssim 3$ Å) and about 1000 electrons/Å^2 have to be used before an isolated heavy atom becomes detectable in the noisy micrograph.

Resolution and detectability may only be improved beyond the Scherzer limit by letting higher spatial frequencies contribute to image formation. Due to the oscillatory nature of the tf (2) such contributions will be transferred with phase shifts of π and the ensuing contrast reversals will render the micrographs illegible. Resolution and legibility may be reconciled following the proposal of Gabor [9] by regarding the micrograph as an in-line hologram to be decoded by spatial frequency filtering. In early applications of spatial filtering to light-optical photographs [1] sign reversals of the tf were corrected but spatial frequency attenuation was left uncompensated and noise was disregarded; this procedure will be called phase filtering (PF).

High resolution electron micrographs differ considerably from normal photographs by their high noise content which a refined restoration procedure must take into account. A simple and well controlled method of incorporating noise suppression into the restoration of one-dimensional time dependent signals has been given by Wiener [10] and a generation later his method was adapted to image restoration by Helstrom [11]. In Wiener filtering (WF) the analytical expression for the filter

f_w is defined by the condition that the mean squared error $\varepsilon = E\{|\varphi - i \otimes f_w|^2\}$ between object φ and restored image $i' = i \otimes f_w$ be minimal (E denotes an expectation value). The analytical expression is normally derived in the spatial frequency domain where Parseval's relation is used to express the error $\varepsilon = E\{|\emptyset - IF_w|^2\}$; applying the variational method yields

$$F_w(R) = H^*(R,\Delta)/\{H(R,\Delta)H^*(R,\Delta) + N(R)N^*(R)/\emptyset(R)\emptyset^*(R)\} \qquad (5)$$

where the asterisk denotes the complex conjugate and the tf H becomes real for pure phase contrast imaging. The square root of the ratio of object and noise power spectra $\emptyset\emptyset^*$ and NN^* will for clarity be called object-to-noise ratio (ONR = $(\emptyset\emptyset^*/NN^*)^{1/2}$) and should not be confused with the signal-to-noise ratio (SNR) in the degraded image. The restorability of degraded images is controlled by the magnitudes of H and ONR. The image becomes perfectly restorable for ONR $\to \infty$ ($F_w \to 1/H$) and, conversely no restoration is possible for $H \to 0$ ($F_w \to 0$).

A minimal amount of a priori information on the statistics of object and noise has tacitly been used to derive the Wiener filter (5) which makes it appear more suited for the restoration of electron micrographs than for normal photographs. Non-periodic high resolution objects consist of atoms (point scatterers) which have a nearly stationary random spatial distribution and the quantum noise of the electrons may be assumed to an even better approximation to be stationary. Normal optical scenes usually have a non-stationary irradiance distribution, stationarity of object and noise, however, is a crucial assumption in deriving the Wiener filter (5). Finally, minimizing the squared error is not the criterion that the human visual system normally employs. This physiological aspect, however, has little relevance for electron micrographs since high resolution objects bear no resemblance to everyday scenes and special ways of "seeing" need training. A more detailed account of the statistical and physiological aspects of Wiener filtering and its ramifications may be found in the book by Andrews and Hunt [2].

The performance of the electron microscope with and without image restoration has been assessed for the underfocus range $0 \leq \Delta \leq 4$ by comparing it to a diffraction limited system. The volume and the radius of the unfiltered (NF) and corrected (PF and WF) psfs were calculated. The side lobe behavior of the different psfs was characterized by computing the ratio of the side lobe and total variances; the ideal diffraction limited system has a constant value of 0.16 independent of the resolution. The results are given in Fig. 1.

An oscillatory behavior for the three criteria of comparison is common to all psfs with optimal conditions recurring close to the defocus values $\Delta(N)=(1.5+2N)^{1/2}$, N=0,1,2,3 etc.. The NF central lobe volume has positive or negative sign depending on the amount of spatial frequencies transferred with incorrect phase. The NF volume shows jumps from vanishing to pathologically large values due to a psf which has a completely corrupted shape at the corresponding defoci. The WF-psf comes closest to the psf of the diffraction limited system, the improvement over PF or even NF at the Scherzer conditions, however, is rather limited due to the limited coherence of the illuminating beam. More details on the coherence conditions and the calculations may be found in [4], the Wiener filter results are based on a frequency independent ONR = 4.5.

In the calculations of detectability, resolution, and side lobe oscillations the parameters entering into the restoration procedure (defocus, partial coherence, ONR) were assumed to be perfectly known. In an actual experiment the imaging conditions can only be set with limited precision. We tried [4] to assess how accurately the imaging parameters have to be known to obtain valid results and to lay down operational procedures to derive the parameters to within the admissible tolerances. Whereas the partial coherence conditions can be measured by independent methods and may therefore by regarded as known, defocus and ONR have to be determined a posteriori from the micrographs themselves. Beyond a detectability threshold (ONR = 2.3) the WF-psf depends only weakly on ONR and beyond ONR = 4.5 saturation is observed. Below the

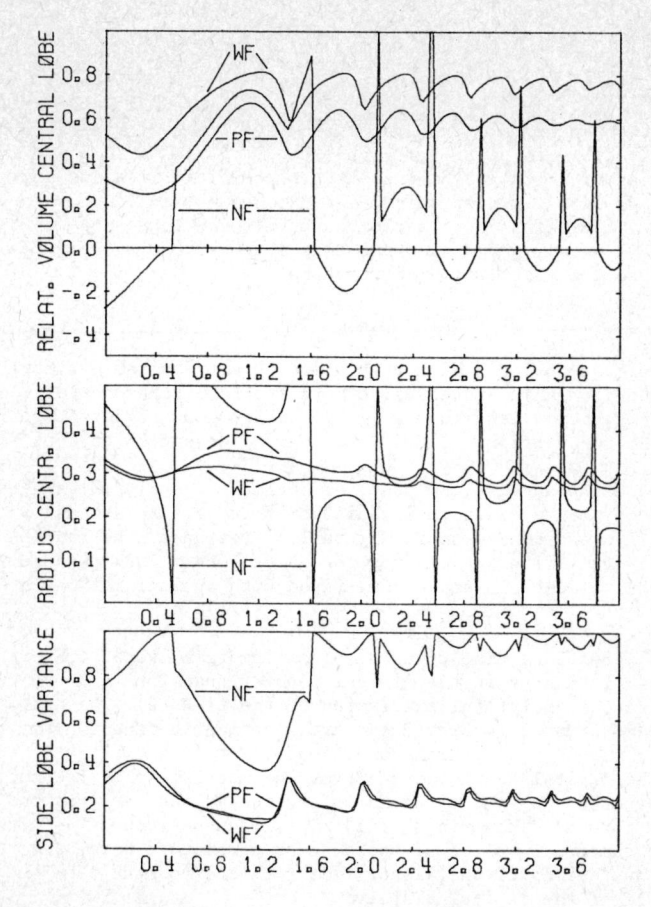

Fig. 1 : Defocus dependence of the volume of the central lobe, radius of the central lobe and the side lobe variance of the image of a point scatterer (NF = not filtered, PF = phase filtered, WF = Wiener filtered). The volume of the central lobe is normalized to the value of a focused diffraction limited system. The radius is given in normalized coordinates. The side lobe variance is normalized to the total variance.

detectability threshold noise is so predominant that restoration becomes meaningless. The best compromise between radiation damage and restorability is obtained for an imaging dose which will yield ONR ∼ 4.5. In that dose range the actual value of the ONR may be derived with sufficient precision from the variances measured on micrograph areas with and without specimen images.

For realistic ONR values the WF restoration results depend critically on defocus because the attenuating regions ($F_w < 1$ or $H < ONR^{-2}$) of the Wiener filter where noise is important are then very small, every where else the shape of F_w is essentially determined by the tf H. As may be seen from the factorization of H into an oscillating part H_c and a slowly varying envelope function E, attenuation only occurs close to the bandlimit imposed by E and for narrow annular zones centered on the zeros of H_c. The zero positions of H_c are so susceptible to defocus changes that an error of 3 % may shift a zero by more than the width of the attenuating zone. Numerical assessment of WF with incorrect defocus values [4] has also given tolerances of 3%. Axial astigmatism, meaning simply that different azimuthal directions in the micrograph are imaged with different defocus, is normally not corrected to 3 % of mean defocus. In contrast to WF of defocused normal photographs WF of electron micrographs must therefore be able to treat axial astigmatism. Fortunately defocus may be derived very accurately from the Fourier intensity of the micrograph $II^* = 4I_0^2H^2\emptyset\emptyset^* + NN^*$, $(R \neq 0)$ since the object power spectrum $\emptyset\emptyset^*$ is continuous and lines of minimal intensity reflecting the zeros of H can clearly recognized. Finally WF must use fairly large image fields since an object point "leaks out" into a whole area due to the poor psf and only the ONR imposes a limit on which dissipated contributions can be collected in the restoration. It remains to be explored whether the long range of the psf will allow more involved computational methods like maximum entropy restoration [12] to be applied to high resolution micrographs.

5. Experimental Results

The choice of non-periodic high resolution objects suitable for image restoration is limited to inorganic specimens. Due to the poor coherence of the illuminating electrons WF achieves less than a five-fold increase in detectability as compared to Scherzer's conditions so that the imaging dose must be kept substantially larger than the critical dose for biological material (1 electron/$Å^2$ typically causes severe radiation damage). We used vermiculite crystals which were randomized to a certain extent by heavy electron irradiation to have both stochastic structures of higher spatial frequencies and small crystalline residues. The results presented were obtained by digital filtering of image areas of 1024 x 1024 picture elements (pixels), more technical details and a comparison with optical methods are given in [4].

The defocus parameters (including axial astigmatism) were derived by locating the lines of minimal intensity in the digital Fourier transform and performing a least squares fit to the analytical expression of the tf H (2). The quality of defocus determination is demonstrated in Fig. 2 comparing optical diffractograms of vermiculite micrographs taken close to the optimal conditions $\Delta(N = 1)$ and $\Delta(N = 3)$ to a gray level display of the digital Fourier transform. The modulation of the continuous object power spectrum $\emptyset\emptyset^*$ by the tf H and in particular the zero regions of H are apparent as the pattern of "Thon-rings" [13]. The frequency cut-off of $\emptyset\emptyset^*$ occurs for $(2.5 Å)^{-1}$. The numerical values of the defocus parameters were used to calculate the extremely narrow regions of the Wiener filter F_w and to mark them out by a light OD-value in one half of the digital displays.

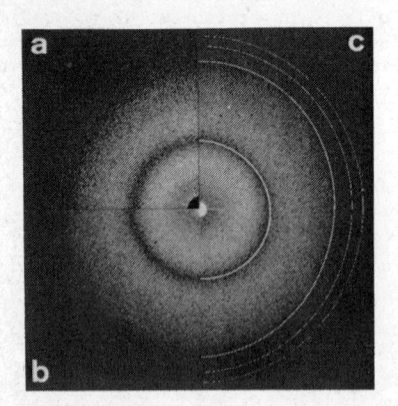

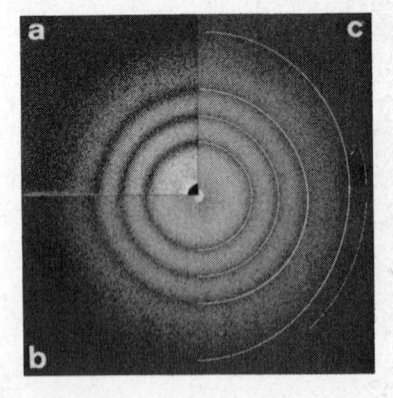

Fig. 2 : Comparison of image spectra derived by optical (a) and digital means (b,c). Demonstration of quality of digital determination of defocus for micrographs taken near $\Delta(N = 1) = 1.83$ (left half) and $\Delta(N = 3) = 2.73$ (right half). Attenuating regions of the Wiener filter are shown as white lines (c).

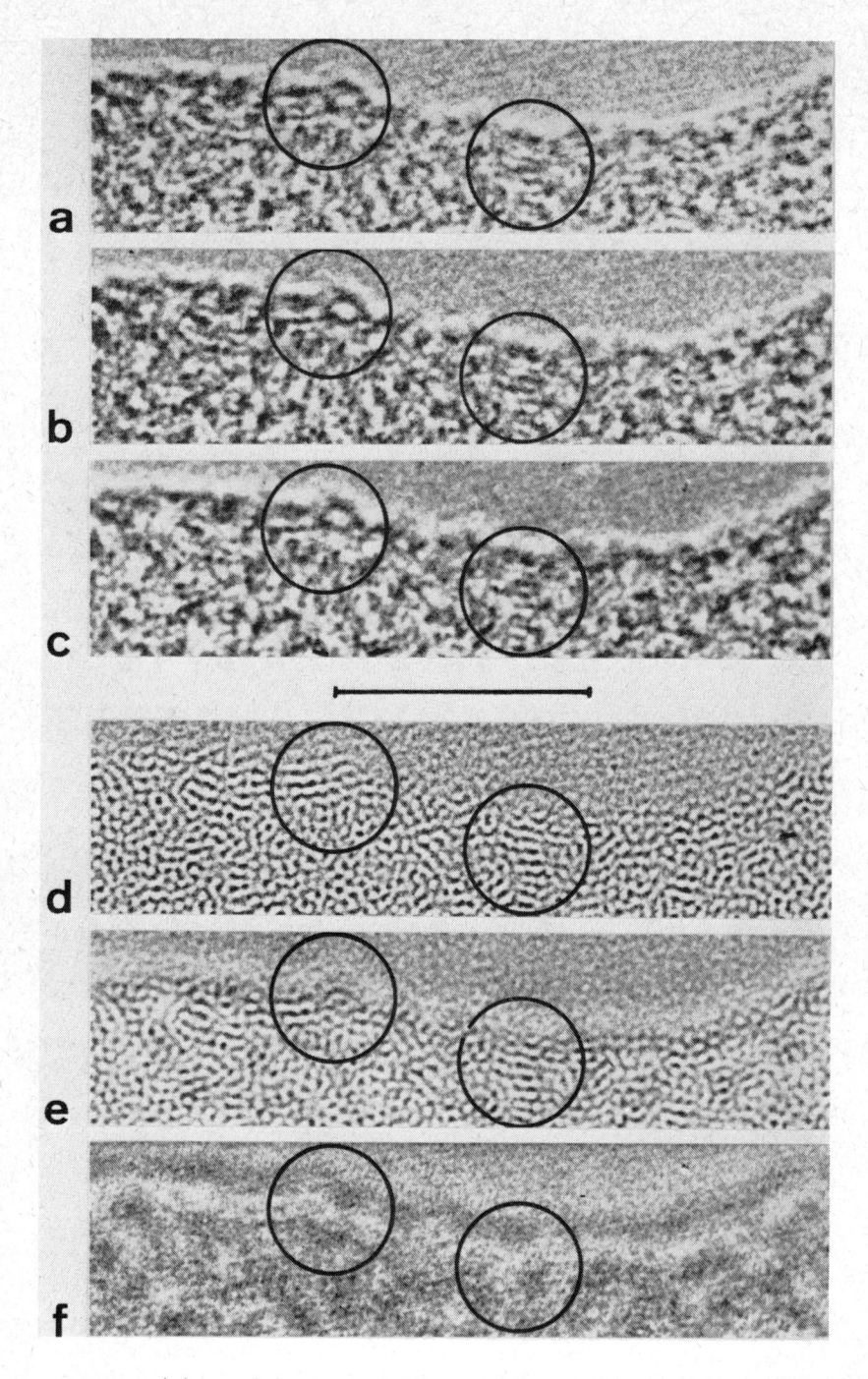

Fig. 3 : Original (a), PF (b) and WF (c) restorations, and difference images (d,e,f) of vermiculite micrograph taken at $\Delta = 1.78$ ($\Delta(N=1) = 1.83$). Changes due to restoration are prominent in difference images (d : original-PF, e : original-WF, f : PF-WF). Random and lattice like areas are encircled. The bar denotes 100 Å.

The PF and WF restorations of the vermiculite micrograph taken near $\Delta(N = 1)$ are given in Fig. 3 which shows an edge of the specimen, the gray areas at the top show the electron noise. The essentially random nature of the object is apparent making it difficult to detect the changes due to PF or WF. No conspicuous deblurring occurs as in the restoration of defocussed normal photographs since the micrographs are in-line holograms which always show sharp interference fringes. The improvement due to spatial filtering is more easily demonstrated on difference images of the original and its filtered versions. If no change occurred, the difference images would show a homogeneous appearance devoid of modulations. Contrast reversals will show up as dark and light OD-values.

The difference of the original and PF images looks similar to images having only one transferred frequency zone [14]. In the encircled areas prominent lattice like features can be recognized proving that light and dark lines have indeed exchanged position. Contrast reversals are abundant as is demonstrated by the finely grained contrast fluctuations in the rest of the image. In the difference of the original and WF images an additional low frequency modulation is seen which is separately displayed in the difference of PF and WF images. These coarse grained contrast fluctuations are due to the different properties of the phase and Wiener filter for very low frequencies and to the contamination of low frequencies by non phase contrast contributions, a more detailed discussion of this effect is found in [4].

The changes due to the different kinds of spatial filtering as well as their quality can be assessed in a global way by generating the Fourier transforms of the differently processed micrographs. The digital restorations and difference images were used as synthetic holograms in a light-optical diffractometer ; the diffractograms are shown in Fig. 4. The diffractogram for pure PF is not displayed since it is identical to the Fourier intensity of the original (Fig. 2). WF (a) is seen to produce the desired narrowing of the frequency gaps and an amplification of higher spatial frequencies where the envelope function E (2a) is strongly attenuating. The diffractograms of the various difference images show which spatial frequencies are most affected by the processing steps. The spectrum of the difference of original and PF image (b) shows a complete lack of spatial frequencies out to the first zero of the tf since no sign reversals have to be compensated in this region. Apart from the very low frequencies hidden by the beam stop, WF is seen to affect mostly the high frequencies (c), in particular those close to the bandlimit. Finally the difference (d) between PF and WF essentially consists in enhancing spatial frequencies close to the zeros of the tf. The effect of astigmatic filtering is reflected most clearly by the anisotropy of the Fourier intensity.

At higher defoci $\Delta(N>2)$ PF and WF become increasingly difficult as an increasing number of sign reversals of the tf H have to be compensated. The digital methods developed were successfully tested on micrographs taken at $\Delta(N=3)$ and should also work for larger defocus values. Attemps to restore micrographs taken beyond $\Delta(N = 3)$ appear unreasonable at present because detectability and resolution will not increase unless the coherence of the illuminating electrons can be improved considerably.

6. Conclusions

Restoration of phase contrast micrographs is an experimentally viable procedure which allows the instrumental resolution to be increased to the order of chemical bond lengths. The detectability of small objects, however, remains marginal due to the quantum noise of the imaging electrons. Sufficient signal-to-noise ratios are only obtained with high imaging doses which limits the choice of eligible objects to inorganic specimens. Biological material will not tolerate doses which are of the order of magnitude required for atomic imaging.

The restoration of micrographs poses a problem which is conceptually simpler than processing normal photographs taken in incoherent illumination. Basic concepts of estimation theory like Wiener filtering (WF) appear adequate for treating the situation encountered in electron microscope imaging. On a technical level, WF

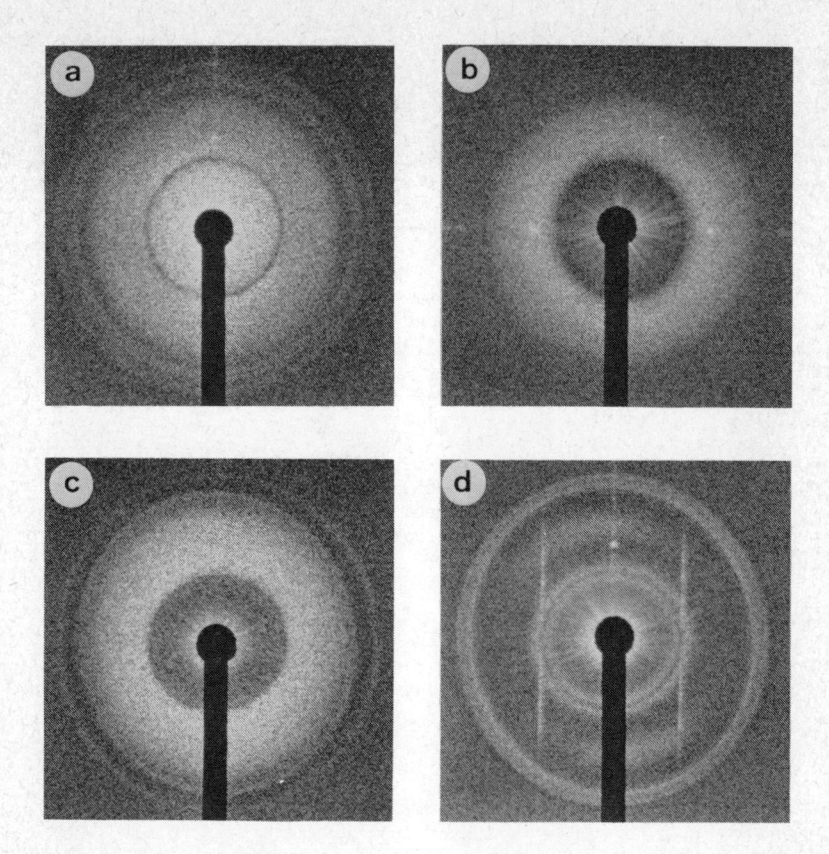

Fig. 4 : Optical diffractograms of WF restoration (a) and difference images (b,c,d) for global assessment of spatial frequency filtering. Spectra of difference images are given for original-PF (b), original-WF (c), and PF-WF restorations.

of electron micrographs is more demanding than WF of normal photographs due to the strongly oscillating behavior of the transfer function, due to the precision required in determining the electron optical imaging parameters, and due to the high noise level of the micrographs. Processing micrographs has a more subtle effect than "deblurring" defocussed photographs since phase contrast images are in-line holograms which always appear sharp, and non-periodic objects in the atomic resolution range provide quite unfamiliar scenes for visual appreciation.

The problems of high resolution electron microscopy cannot be regarded as solved by combining, in the sense of Gabor [9], electron holography and image restoration. Additional stratagems have to be used to overcome the problem of radiation damage either by working at considerably lower signal-to-noise ratios or by finding means to increase contrast. Correlation methods to superimpose many images of like particles [15] may lead the way to exploiting low signal-to-noise ratios, new image processing methods which make better use of a priori information [12] may possibly provide a means to increase contrast. The parochial viewpoint has to be abandoned since high resolution microscopy of non-periodic objects needs all the help it can get.

References

[1] A. Maréchal, P. Croce, and K. Dietzel, Opt. Acta, $\underline{5}$ (1958) 256
[2] H.C. Andrews and B.R. Hunt, Digital Image Restoration (Prentice-Hall, Inc., Englewood Cliffs, N.J. 1977)
[3] W.K. Pratt, Digital Image Processing (J. Wiley & Sons, Inc., New York, N.Y. 1978)
[4] O. Kübler, M. Hahn, and J. Seredynski, Optik $\underline{51}$ (1978) 171 and Optik $\underline{51}$ (1978) 235
[5] K.J. Hanszen, Z. Angew. Phys., $\underline{19}$ (1966) 427
[6] G.C. Farnell and R.B. Flint, J. Microsc., $\underline{97}$ (1973) 271
[7] A. Rose, J. Soc. Motion Picture Engrs., $\underline{47}$ (1946) 273
[8] O. Scherzer, J. Appl. Phys., $\underline{20}$ (1949) 20
[9] D. Gabor, Nature, $\underline{161}$ (1948) 777
[10] N. Wiener, Extrapolation, Interpolation, and Smoothing of Time Series (MIT Press, Cambridge, Mass. 1942)
[11] C.W. Helstrom, J. Opt. Soc. Amer., $\underline{58}$ (1967) 297
[12] B.R. Frieden, J. Opt. Soc. Amer., $\underline{62}$ (1972) 511
[13] F. Thon, Z. Naturforschg. $\underline{20a}$ (1965) 154
[14] F. Thon, and B.M. Siegel, Ber. Buns. Bes. Phys. Chem., $\underline{74}$ (1970) 1116
[15] W.O. Saxton and J. Frank, Ultramicroscopy, $\underline{2}$ (1977) 219

INTERPRETATION OF X-RAY TOPOGRAPHY

Y. EPELBOIN

Laboratoire de Minéralogie-Cristallographie, associé au CNRS,
Université P. et M. Curie 75230 PARIS CEDEX 05, France

INTRODUCTION

X-Ray topography is an optical means to visualize defects in crystalline solids. The experimental result is an image which support is usually a photographic film or, more seldom, a television camera and the associated monitor. The experiment is appreciated through the eye of the observer and the interpretation is dependant from its subjectivity. X-Ray topography is a diffraction experiment, in the general meaning used in optics, thus it is theoretically possible to calculate the repartition of light sources which diffract inside the crystal from the knowledge of the amplitude and phase of the light at any point on the exit surface of the crystal. The photography provides only the modulus of the amplitudes and the phase factor remains unknown. As it is impossible to rebuild this information there is no way to directly come back to the perturbation which creates the image from the study of its contrast. This unknown perturbation is the local deformation of the crystal ; X-Rays are sensible to slight deformations and a local distortion of less than 1/10" is enough to perturbate the wave-fields and give rise to a contrast. Thus it is necessary to build a model for the studied deformation, to treat it with the diffraction equations of the dynamical theory and to compare the final result i.e. the calculated repartition of intensity to the density of grey of the photographic film.

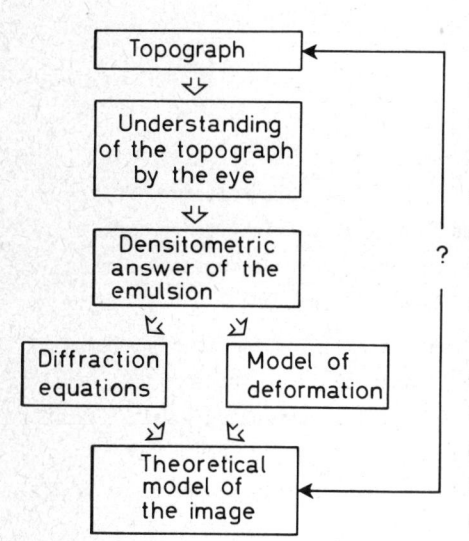

The interpretation of X-Ray topograph is to find an answer to the following questions :
1) How the eye is sensible to the different grey levels of the film and how does it understand them ?

Fig. 1

2) What is the relationship between the density of grey and the intensity of X-Rays ?
3) Which form of the diffraction equations is valid to understand the contrast ?
4) What kind of deformation is at the origin of the studied contrast ?

We will only briefly mention the first point since it is not exactly the subject of this paper. The second one has been treated by J. Miltat. The third one may be found in papers by A. Authier and F. Balibar. It is enough to say that the level of approximations of the Takagi-Taupin (1962, 1964, 1969) equations which are now used in X-Ray dynamical theory has never yet been the reason why a contrast could not be understood. Moreover in a lot of different cases the simplified formalism developped by Kato (1961) leads to excellent explanations.

The main difficulty comes from the fourth question : we will show how it is delicate to find a good model for the deformation of the crystal ; when it is known there is a very good agreement between theroy and experiment.

All this purpose is summarized in fig. 1.

I.- THE ANALYSIS OF THE IMAGE : EYE AND DETECTOR

1.- Sensibility of the eye

It is well known that the eye is not a reliable receptor. Physiological and psychological reasons modify its answer and it would be tricky to analyze these points ! Anyway attention must be payed to the fact that the eye does not present the same sensibility when analysing a wide range of contrast or a small one (Fig.2). For a given mean illumination, the eye will be able to distinguish much more details in a short range of densities than in a wide one. If the brightness of the image varies tremendously in a small surface the eye will automatically adapt its sensitivity to the mean brightness and details will not be seen. Experimenters may pass hours looking at photographs with a microscope to find the fine structure of the image defects, focusing their eye on small areas and this is of the greatest importance since these fine details, visible only to the trained experimenter, will often made the difference between various kinds of defects.

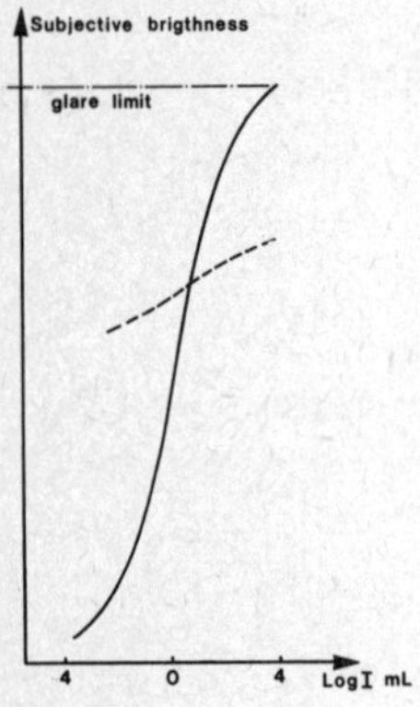

Fig. 2 Photopic vision

2.- Densitometric answer of the film

When using photographic films a saturation phenomenum occurs for the higher illuminations (Morimoto and Uyeda 1963, Epelboin, Jeanne-Michaud and Zarka 1969). Thus it is difficult to distinguish details in the darkest areas of a topograph and the best sensibility is achieved for a range of densities between 0.5 and 2. A main difficulty is that both contrast and shape of the image depend upon the range of exposure choosen for the experiment. This is a well known fact to all photographers which can easely be solved, in most cases, by comparison between the image and the model. In X-Ray experiments there is no model directly visible to the eye and the photography is not only a means to record an image but mainly a method to visualize it. Fig. 3 presents the same image of a dislocation in a section topograph for various times of exposure, all other experimental conditions being the sames. All these photographs look different and they all are true representations of the same defect. The contrast of a defect may vary from on experiment to another ; however it is surprising to find that a mean standard exists in most laboratories.

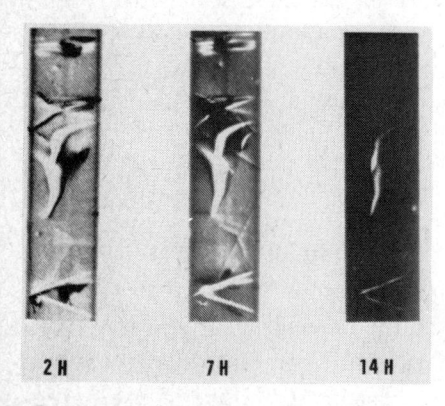

Fig. 3
Section topograph of a dislocation for various exposure times
Epelboin, Lifchitz, 1974)

The definition of a good topograph is more a qualitative criterion than an objective one : a given image may be a good one to those who want to study a density of dislocations and not good enough to understand the fine structure of these defects. A qualitative or semiquantitative interpretation will be sufficient in the first case ; in the second case it may be necessary to use the most sophisticated tools of the dynamical theory.

The method of interpretation depends upon the aim of the study.

II.- QUALITATIVE INTERPRETATION OF TOPOGRAPHS

1.- Translation and section topography

A great variety of topographic methods exit. A brief review may be found in Tanner (1976). In this paper we will only present results from the most widely used : the Lang method (1959).

As it has been explained by C. Malgrange the image of a defect may be divided in three parts : the kinematical image which is a black contrast mostly due to a kinematical diffraction by the most disturbed areas of the defect, when it crosses the refracted beam, the dynamical image which is the white shadow of the defect and the intermediary image which appears as a complicate network of fringes, rather difficult to analyse and which contains most of the information on the fine structure of the defect ; all these images are visible for defect D4 presented in the paper by C. Malgrange (fig. 5). When both film and crystal are translated together successive section topographs are integrated giving an image of a whole volume of the material. In a translation topograph most of the fine details of the section topograph are lost and the main contrast is due to the integration of the kinematical image. Thus the experimenter has two choices : either to look in a single experiment at large volumes but without any knowledge of the fine structure of the defects or to record only a section topograph ; in this case it is theoretically possible to analyze the structure of the defects but this remains difficult in most cases. Moreover only a small area is seen in the experiment. Both methods are complementary as will be shown.

2.- Contrast of a dislocation

Line D4 in fig. 5 (C. Malgrange) is the image of a dislocation in a traverse topograph. The same defect may be seen in the section topograph of fig. 5a (C. Malgrange). The white shadow and the fringes of the intermediary image indicate that this dislocation is inclined in the crystal. The line emerges on the bottom surface of the crystal. The translation topograph gives the general shape of the dislocation and the section topograph permits to calculate exactly its inclination. Let us consider now dislocations D1 and D2 ; the section topograph shows that they lie parallel to the surface of the crystal which may not be known from the translation topograph. Generally speaking translation topographs give a qualitative analysis of the defects, section topographs add quantitative measurements.

The determination of the Burgers vectors of a dislocation needs many experiments. It may be shown that the contrast of a screw dislocation vanishes when $\vec{g}.\vec{b} = 0$, the one of an edge dislocation when $(\vec{g}, \vec{b}, \vec{u}) = 0$ and $\vec{g}.\vec{b} = 0$, where $\vec{u}$ is a unit vector lying along the line, $\vec{g}$ the reciprocal lattice vector of the reflection and $\vec{b}$ the Burgers vector of the dislocation. A mixed dislocation will never be completely invisible. However its contrast may be so faint that the approximate criterion $\vec{g}.\vec{b} = 0$ is used in most cases. Thus it is necessary to find three extinctions to determine $\vec{b}$. Its sense may not be found except in simulations of the whole image or by surface effects (Dunia, Malgrange and Petroff, 1979, Lang 1965, Hart 1963).

Unfortunately in anisotropic materials extinction may not be found (see for example Barnes and al. 1978, Epelboin and Patel 1979).

We must mention that it is possible to see a translation topograph in 3D. looking with stereo glasses at two different topographs. A first method is to used $\pm \vec{g}$ reflections and a better one is to record two images with a slight rotation of 10° about $\vec{g}$ (Haruta 1965).

Dislocations are a good example of the possibilities and limitations of a qualitative approach. It is possible to determine the most important parameters of the defect and this explains why this method has become so popular in the study of isolated dislocations. In anisotropic materials this is not possible and a quantitative approach becomes necessary.

III.- SEMI-QUANTITATIVE INTERPRETATION

We may classify in this part all studies which need some understanding of the origin of the contrast and some calculations. A good example is the understanding of the contrast of dislocations through their direct image.

Authier (1967) has suggested that the most disturbed areas around the core of a dislocation line may be considered as a mosaic crystal and diffract kinematicaly. The direct image may be explained as the projection of these areas along the diffracted directions. This permits to explain single or double contrast of dislocations (Authier 1967). A detailed verification of this theory has been made by Miltat and Bowen (1975) who found that the real size of this image may vary from one half to twice the predicted one. This theory has permitted to estimate the composition of a bunch of dislocations (Epelboin and Ribet 1974) ; it has been possible to understand the contrast of dislocations loops (G'Sell and Epelboin 1979).

The understanding of the contrast of the direct image is typical of the interpretation of X-Ray topographs described in the scheme of fig. 1. The observation of the image does not provide any hypothesis for the characteristics of the defect. As shown before the defect is already known and the object of the study is to identify it precisely. For example, it is not obvious to distinguish a dislocation from a channel by topography only. It is easy to say that it is a linear defect but only some peculiar comportment of its contrast in various diffractions permit to suspect the presence of a channel. Moreover it may be difficult to distinguish a channel from a decorated dislocation. Thus the study starts by a reasonable choice for the model of deformation. Then characteristics of the image of the defect are established through the use of the dynamical theory and the model is compared to the experimental image. There is no way to assume that the solution is unique and that a different model could not explain the contrast. Finer the details of the topograph closer to the reality will be the model, at the level of approximation of the diffraction theory. For instance the study of the direct image of a dislocation can only give an estimate of its characteristics since this theory is an approximation. It is not the case of simulations when the image is computed using the most sophisticated theories for both diffraction and   deformation of the crystal.

IV.- QUANTITATIVE INTERPRETATIONS : THE SIMULATIONS

Simulation is a very familiar method to all scientists working in TEM (Head 1967). Although the first simulation in X-Ray topography has been done by Balibar and Authier the same year, the complexity of the diffraction problem and the slowness of the computers did not permit to use this method before 1974 (Epelboin 1974). Simpler simulations were done by Kato and Patel in 1968; using a simplification of the diffraction theory (Kato 1963, Penning and Polder 1961), they were able to compute the intensity of diffracted X-Rays for a silicon wafer strained by an oxyde layer. The contrast of a stacking fault was analytically solved in 1968 by Authier and this was well verified in various experiments (Authier and Patel 1976, Patel and Authier 1976).

In the most general case, simulations of section topographs must be performed by numerical integration of the Takagi-Taupin equations. They are integrated step by step along a network of characteristics parallel to both refracted and reflected directions. It may be shown (fig. 4) that the knowledge of the amplitudes of the reflected and refracted waves at a given point A may be established from their knowledge at points B and C in the network. This calculation is not obvious since the Takagi-Taupin equations are partial derivative hyperbolic equations ; the boundary conditions and step of integration have to be choosen carefully to avoid numerical errors (Epelboin 1977, Epelboin and Riglet 1979). The most widely used algorithm is based on the principle of the half step derivative as explained by Authier, Malgrange and Tournarie (1968) , the step of integration is fixed all along the calculation and anisotropic elasticity may be taken into account for the perturbation due to the defect (Epelboin 1974). More recently a variable step algorithm has been suggested by Petrashen (1977) which permits to simulate translation topographs (Petrashen and Chukovskii 1978).

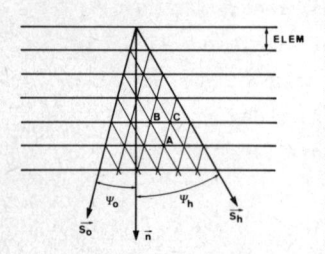

Fig. 4
Network of integration of Takagi-Taupin equations

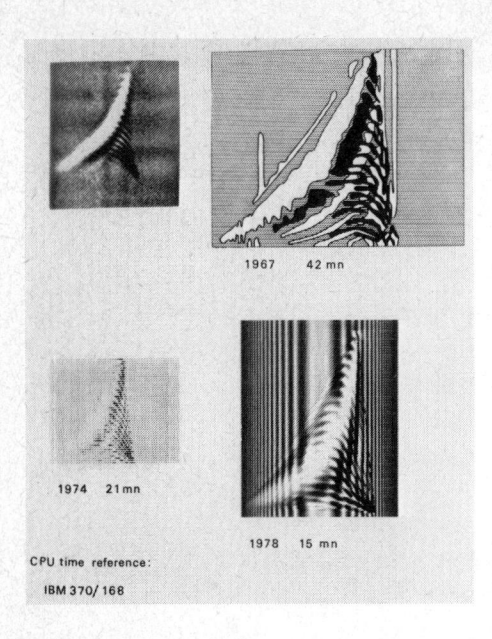

Fig. 5
Enhancement of the same simulation
Balibar and Authier 1967,
Epelboin 1974)

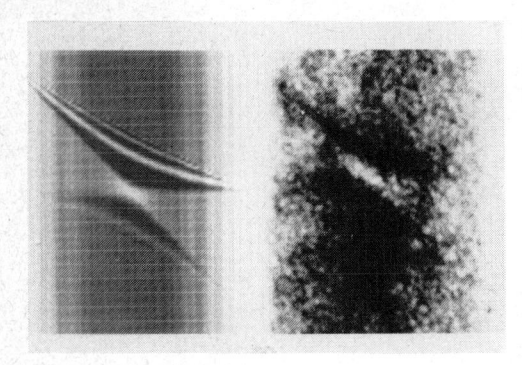

Fig. 6
Comparison of a theoretical model
to the image of a ferroelectric
wall (topography courtesy
C. Malgrange)

Simulations are presented in the form of a photography and the limitations due to this part of the simulation are of the greatest importance. Fig. 5 presents recent progresses in the simulation of defect D4 (see Malgrange) by a change in the output device and algorithm of simulation of the photographic film. Similar improvments have been obtained by Wonscievicz and Patel (1976).

Simulation is the only tool which permits a complete identification of a defect : the sense of the Burgers vector of a dislocation and its magnitude may be found, Burgers vectors may be identified although no extinction could be seen in any topography (Epelboin and Patel 1979). Moreover simulation may be used as the experimental part of a theoretical study : diffraction effects in disturbed crystals could be studied (Balibar, Epelboin and Malgrange 1975, Epelboin 1979).

The main difficulty remains, as already mentionned, the choice of a model for the deformation of the crystal. Simulations have been successful in the understanding of the contrast of dislocations (Epelboin and Patel 1979) and of ferromagnetic walls (Nourtier and al. 1977) ; they have permitted to establish the rules for the contrast of stacking faults in the Laue-Bragg case (Epelboin 1979) and this seems to be in agreement with an actual study on the contrast of ferroelectric walls in $Gd_2(MoO_3)_4$ (Capelle, Epelboin and Malgrange 1979) (fig. 6).

CONCLUSION

X-Ray topography is a powerfull method to study isolated defects in macroscopic crystals. The contrast of the most common defects is rather well-known and their study may be done without a good knowledge of the dynamical theory. Very simples rules permit to everybody the use of this method. An accurate model for the deformation due to a defect may be verified by comparison of experiments and simulations. The difficulty remains the choice of this model but modelization assisted by computer will become more and more an ordinary tool in the interpretation of X-Ray topographs.

REFERENCES

AUTHIER A. (1967), Adv. in X-Ray Analysis, 10, 9.

AUTHIER A. (1968), Phys. Stat. Sol., 27, 77.

AUTHIER A., MALGRANGE C., TOURNARIE M. (1968), Acta Cryst., A24, 126.

AUTHIER A., PATEL J.R. (1975), Phys. Stat. Sol. (a), 27, 213.

BALIBAR F., AUTHIER A. (1967), Phys. Stat. Sol., 21, 413.

BALIBAR F., EPELBOIN Y., MALGRANGE C. (1975), Acta Cryst., A31, 836.

BARNS R.L., FREELAND P.E., KOLB E.D., LAUDISE R.A., PATEL J.R. (1978), Acta Cryst. Growth, 43, 676.

CAPELLE B., EPELBOIN Y., MALGRANGE C. (1979), to be published.

DUNIA E., MALGRANGE C., PETROFF J.F. (1979), Phil. Mag., to be published.

EPELBOIN Y. (1974), J. Appl. Cryst., 7, 372.

EPELBOIN Y. (1977), Acta Cryst., A33, 758.

EPELBOIN Y., JEANNE-MICHAUD A., ZARKA A. (1979), J. Appl. Cryst., 12, 201

EPELBOIN Y., LIFCHITZ A. (1974), J. Appl. Cryst., 7, 377.

EPELBOIN Y., PATEL J.R. (1979), to be published.

EPELBOIN Y., RIBET M. (1974), Phys. Stat. Sol.(a), 25, 507.

EPELBOIN Y., RIGLET P. (1979), Phys. Stat. Sol.(a), to be published.

HART M. (1963), PhD. Thesis, University of Bristol.

HARUTA K. (1965), J. Appl. Phys., 36, 1789.

G'SELL C., EPELBOIN Y. (1979), J. Appl. Cryst., 12, 110.

HEAD A.K. (1967), Aust. J. Phys., 20, 557.

KATO N. (1961), Acta Cryst., 14, 526.

KATO N. (1963), J. Phys. Soc. Japan, 18, 12.

KATO N., PATEL J.R. (1968), Appl. Phys. Letters, 13, 42.

LANG A.R. (1959), Acta Cryst., 12, 249.

LANG A.R. (1965), Z. Naturfor., 20a, 636.

MILTAT J., BOWEN D.K. (1975), J. Appl. Cryst., 8, 657.

MORIMOTO H., UYEDA R. (1963), Acta Cryst., 16, 1107.

NOURTIER C., TAUPIN D., KLEMAN M., LABRUNE M., MILTAT J. (1977), J. Appl. Cryst., 10, 328.

PATEL J.R., AUTHIER A. (1975), J. Appl. Phys., 46, 118.

PENNING P., POLDER D. (1961), Philips RES. Rep. 16, 419.

PETRASHEN I.V. (1976), Fiz. Tverda Tela, 18, 3729.

PETRASHEN P.V., CHUKOVSKII F.N. (1978), Fiz. Tverda Tela, 20, 1104.

TAKAGI S. (1962), Acta Cryst., 15, 1311.

TANNER B.K. (1976), X-Ray diffraction Topography, Pergamon Press N.Y.

TAUPIN D. (1964), Bull. Soc. Fr. Min. Crist., 87, 469.

WONSIEWICZ B.C., PATEL J.R. (1975), J. Appl. Cryst., 8, 67.

"COHERENCE IS BEAUTIFUL"

J. Joffrin

Institut Laue-Langevin, 156 X, 38042 Grenoble Cedex

I - INTRODUCTION

The roots of coherence lie in the dual nature of all elementary particles : they can be treated either as mechanical particles with a mass, a momentum and a kinetic energy or they can be treated as a wave which oscillates with a frequency and a wave vector. This is true for all of those which are of interest for this assembly : photons, phonons, spins, bosons in general, plasmons, neutrons, électrons. Of course, much attention has been paid to the coherence of light ; this brings us back to the end of the 19th century ; consequently, the more experienced people in the field of coherence, at least those who have practicised that concept for the longest time, are the opticians. The decomposition of white light has been observed as early as the 16th century in Europe ; very early, too, the interference phenomena using a non-completely coherent source of light (the sun) has been used to prove many aspects of physics ; it was even an argument in the controversy about the description of the outside world. But until the famous paper of Glauber in 1963 /1/, the statistical properties of light beams had only been discussed exclusively in terms of classical or semi-classical theories. In fact, the wave particle duality which should be central to any convenient treatment of statistics does not survive the transition to the classical limit ; it was the aim of that paper to reconcile the two points of views. The same could be said about the acoustical beams which have a very similar status.

A second starting point to look at the coherence consists in considering the interference phenomena which are obtained with neutrons, electrons or X-rays. All these particles in general have a very small wavelength in comparison with phonons or photons ; in principle, it changes nothing for our point of view except that it is much more difficult to perform experiments to make them interfere (if we forget diffraction experiments) ; it was only recently (a few ten years ago) that it was possible to have fringes interference patterns or interferometers with these particles. You all know the beautiful results which have been obtained with X-rays to study the defects in a nearly perfect crystal /2/. The same is true for neutrons ; but the gravity field of these last particles add more fun, if necessary, than in the previous case /3/.

A third point of view to approach the coherence has to pay royalties to those who work with "spin assemblies". Opticians and acousticians, I may say, are very clever people ; all what they teach to us by making holograms or by inventing devices for the treatment of the signal, is fine art ; but I must say that most of the experiments which have been performed in the field of magnetic resonance, that is working with a spin assembly (electronic or nuclear) is, as well, fine art. The concepts which are behind the different experiments done with X-rays, phonons or neutrons are not less sophisticated, and most of them have been borrowed from the concept of coherence of a spin population. You have already had a glance at that aspect in the particular case of neutrons because these particles have not only a mass but also a spin ; in that respect a population of neutrons can be treated as a spin assembly /4/.

The fourth point of view, the last but maybe not the least, comes from a completely different horizon. It manifests itself essentially in condensed matter : I would like to speak of superfluidity or superconductivity ; in all these cases, there is a Bose-Einstein condensation in one mode, and it is similar to specify that there is some off-diagonal-long-range-order which appears in the low temperature phase of the system ; that macroscopic "condensation" induces necessarily the existence of a "phase" which can be measured by experiments on a macroscopic scale.

My point of view is that these four types of approaches are not very far from each other. My purpose is to bring them all together.

II - MACROSCOPIC OCCUPATION NUMBER AND PHASE OPERATOR

a) Let us consider the simplest picture to describe a field (electromagnetic field, acoustic field). These fields are boson fields and they can be reduced to the simple hamiltonian.

$$H_o = \hbar\omega \, a^+ a \qquad (1)$$

In that formula, ω is the frequency of the field and a^+ and a the boson field operators. This expression is the source of a very fruitful analogy between the mode amplitude of the field and the coordinates of an assembly of one dimensional harmonic oscillators. The eigen state of (1) is wellknown and the commutation relations of the operators a^+ and a induce the following properties :

$$a^+a \mid N> = N \mid N> \qquad (2)$$
$$a^+ \mid N> = (NH)^{1/2} \mid NH>$$

N is the eigen value of the hamiltonian H_o and the state $\mid N>$ is an eigen state. The mean value of the operator $(a + a^+)$ that is of the amplitude of the field is always 0 on any of these states $\mid N>$. It would tend to prove that the mean value of the "position" is 0. Let us suppose now that to H_o we add the hamiltonian H_1 which is the result of a linear coupling of the oscillator with an external source :

$$H_1 = \hbar\omega \ (\alpha a^+ + \alpha^* a) \qquad (3)$$

Let us raise the question : is it possible to discover the eigen states of the operators a or a^+, that means states which obey

$$a \mid \alpha > = \alpha \mid \alpha > \qquad (4)$$

This is in principle possible and we can try to develop the $\mid \alpha>$ states on the basis of the proper states $\mid N>$ of H_o.

After some tedious calculations, we end up with the following formula /1/ :

$$\mid \alpha> = \exp(-\frac{1}{2}\mid\alpha\mid^2) \sum_N \frac{\alpha^N}{(N!)^{1/2}} \mid N> \qquad (5)$$

By adding H_1 to H_o we are able to generate states for which the mean value of the displacement is non zero ; physically, this is equivalent to say that H_o corresponds to noise vibration ; H_1 on the contrary drives coherently the oscillator.

b) The proper state $\mid\alpha>$ is called a "coherent state" ; it may help in understanding what we have obtained if we discuss the form they take in coordinate space or in momentum space. As you probably remember, quantically, the position of the harmonic oscillator or the amplitude of the movement is given by the formula

$$x(t) = \left[a \ \exp(-i\omega t) + a^+ \exp(i\omega t)\right] \left(\frac{\hbar}{2m\omega}\right)^{1/2} \qquad (6)$$

Classically the answer would be that the amplitude of the field is given by
$$x(t) = 2 x_o \cos(\phi - \omega t) \qquad (7)$$

If we try to estimate the mean value of $x(t)$ given by (6) on any coherent state we find

$$<\alpha\mid x(t)\mid\alpha> = 2\mid\alpha\mid \cos(\phi - \omega t) \left(\frac{\hbar}{2m\omega}\right)^{1/2} \qquad (8)$$

provided that α is a complex quantity which we have written

$$\alpha = |\alpha| \exp(i\phi) \tag{9}$$

There is now a strong similarity between (7) and (8) ; this is due to the fact that we have taken the mean value on the coherent state by contrast to the $|N\rangle$ states. If we remember that the energy of the harmonic oscillator is $\frac{1}{2} m\omega^2 (2 x_0)^2 = N\hbar\omega$ and thus $x_0 = N^{1/2} \frac{\hbar}{2m\omega}$ a comparison between the equations (6) and (8) suggests that there may exist a representation of the creation and annihilation operators which could be written in the following form

$$a = N^{1/2} \exp(i\tilde{\phi}) \tag{10}$$

where in that formula N is the occupation number quoted in formula (2) and $\tilde{\phi}$ is a phase operator. Of course (10) is a formula that one guesses only from a comparison or by analogy. In fact, the change of variables introduced in (10) was given in the original paper of Glauber on the quantization of the electromagnetic field and what remains to be given are the commutation relations of $\tilde{\phi}$ with N.

c) Let us assume that (10) is true (it is not a rigorous result) and let us check what are the commutation relations which are induced for N and $\tilde{\phi}$ when we know the commutation relation for a and a^+

$$[a, a^+] = 1 \tag{11}$$

It is simple to prove that we have the formula /5/

$$[N, \tilde{\phi}] = i \tag{12}$$

As usual that commutation relation induces an uncertainty relation which in that case is called the number-phase uncertainty relation (In fact for those of you who are seriously interested in mathematical problems, N and $\tilde{\phi}$ are not well defined hermitian operators : N for instance has a spectrum limited to positive values and then the relation (12) is not correct). Nevertheless, the consequences of (12) is a relation of uncertainty for the energy which is in the wave and which is measured by N and the phase of the wave : we have necessarily :

$$\Delta N \, \Delta\tilde{\phi} \geq \frac{1}{2} \tag{13}$$

III - <u>THEN COMES THE QUESTION OF SUPERFLUIDITY</u>

We will discuss it more or less on a semi-microscopic basis and we will particularly show the dynamical consequences of the definition of the order parameter.

To start with, the definition of that state is given by the amplitude of the particle field operator $\psi(\bar{r},t)$ which has a microscopic mean value like

$$< \psi(\bar{r},t)> = f(rt)\ \exp\left[i\ \phi(\bar{r}t)\right] \qquad (14)$$

In that equation, we allow only slow variations in space and time and we suppose that the mean value of ψ has a thermodynamic meaning. It is very important to recognise that ψ is a complex quantity as well as a thermodynamic quantity. ψ has both an amplitude f and a phase ϕ ; as we will sketch rapidly, it is the behaviour of ϕ which is responsible for the various specific superfluid properties. On the other hand, ϕ is not only a thermodynamic but also a dynamical variable : it comes from its being the dynamically canonically conjugate variable of N where N is a total number of particles in the system. From (12) we have at our disposal two representations, one in $\tilde{\phi}$ where we write for N

$$N \rightarrow -i\frac{\partial}{\partial \tilde{\phi}} \qquad (15)$$

or the conjugated relation

$$\tilde{\phi} \rightarrow i\frac{\partial}{\partial N} \qquad (16)$$

These two last equations are almost correct as long as N and $\tilde{\phi}$ can be considered as continuous variables ; this is true for a thermodynamical state where N is large or equivalently when

$$\frac{\Delta \phi}{\phi} << 1 \qquad \frac{\Delta N}{N} << 1 \qquad (17)$$

are satisfied. In any case, the equation of motion for N and ϕ are the simple hamiltonian relations

$$i\hbar \dot{N} = i\frac{\partial H}{\partial \phi}$$
$$i\hbar \frac{\partial \phi}{\partial t} = \frac{\partial H}{\partial N} \qquad (18)$$

These two equations are the two fundamental equations of superfluidity /6/. The first one is a London equation for the current and the second one is a Josephson-frequency equation which gives the acceleration equation for the superfluid flow ; it is central for the description of the superfluid state. It can be read either

$$\hbar \frac{\partial \phi}{\partial t} = \frac{\partial E}{\partial N} = \mu \qquad (19)$$

These equations are very important to understand what happens : for ϕ to be constant in time, this requires that μ is a constant everywhere in the superfluid phase : no potential difference is allowed. The second point is that if any potential difference exists between two points, the difference in phase between them must change in time or equivalently

$$\frac{d}{dt}(\hbar \overline{\nabla\phi}) = \overline{\nabla\mu} = \overline{F} \quad \text{with} \quad \hbar\overline{\nabla\phi} = m\overline{V}_s \tag{20}$$

where $\overline{V}_s$ is the superfluid velocity.

It tells that the acceleration of the superfluid takes place without any viscosity. The other consequence of (19) is that ϕ being a phase, it needs not be single valued in a multiply connected system ; this is the idea of quantization of angular momentum which can be written in the following form :

$$\oint \overline{\nabla\phi} \cdot \overline{dl} = 2\pi n \tag{21}$$

where n is an interger.

The important point is that the experimental consequences of these equations have been checked experimentally ; in particular very beautiful experiments on interference have been done to check these ideas ; the phase has been proved to have an existence on a macroscopic scale /7/.

IV - COHERENCE OF A POPULATION OF OSCILLATORS

The two paragraphs above looked at the properties of a single mode ; in case of electromagnetic waves or acoustical waves, these modes had a well defined frequency and wave vector. In case of the superfluidity, there is also only one mode which is concerned : the condensation of particles occur in a single mode well defined in the momentum space ; and for that very reason it is possible to establish commutation relations between the number of particles condensate and the phase of the condensate.

Here we would like to enlarge the coherence phenomena by considering many modes together trying to establish what is the coherence among these different modes. Of course, this supposes at the very beginning that each of these modes can have a well defined phase ; the problem that we face is that, in general, if we excite a population of modes, they have, independently of their natural linewidth, a distribution of frequencies ; even if you excite all of them by a single external source at time t = o they tend to dephase afterwards because they have slightly different frequencies.

a) Consider the figure 1 : This is the distribution in frequency of the different modes. Each of them has a natural linewidth $\delta\omega$ and the resonance frequencies are widely distributed : this is an inhomogeneous linewidth ; it is associated with a density of states.

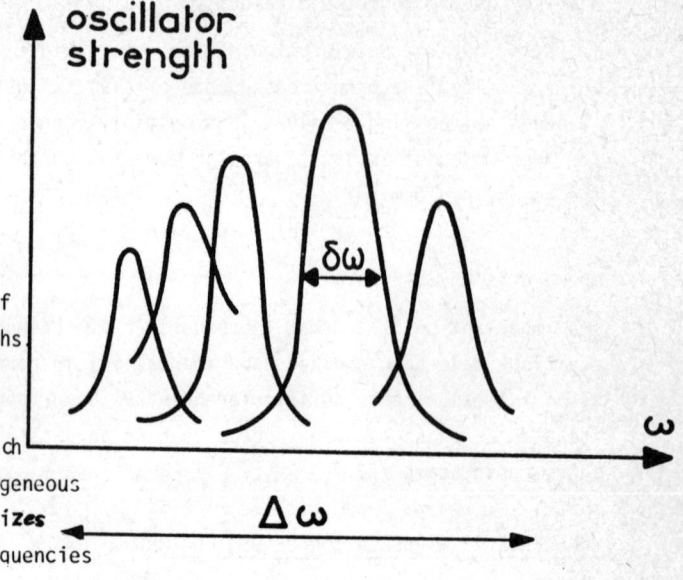

Fig. 1 : Distribution of the oscillators strengths. As a function of frequency, $\delta\omega$ is the natural linewidth of each mode ; $\Delta\omega$ is the inhomogeneous linewidth. It characterizes the distribution of frequencies of the oscillators.

Fig. 1

The signal that we can detect in an experiment is a macroscopic amplitude which is the result of the contribution of all these modes ; that sum is given by

$$\text{signal}(t) = \int f(\omega) \cos\left[\omega t - \phi(\omega)\right] d\omega \qquad (22)$$

In this formula $f(\omega)$ includes the density of states, the coupling of each mode to the source, and the force of each oscillator ; $\phi(\omega)$ it the phase of the modes of frequency ω. The formula (22) shows that the signal at any time is the Fourier transform of the distribution of the frequencies of the population of modes ; if we call $\Delta\omega$ the width of the distribution of $f(\omega)$, you observe no signal at all at a time $t > T_2^*$ where T_2^* is roughly of the order of $\frac{1}{\Delta\omega}$.

$$T_2^* \simeq \frac{1}{\Delta\omega} \qquad (23)$$

T_2^* is called the inhomogeneous relaxation time. In terms of Fresnel vectors, all the contributions tend to be orientated at random in the complex plane and for a time $t > T_2^*$ the signal is almost 0 : the amplitudes add destructively ; you observe nothing.

Let us suppose that for some reason which we will explain later, we have the possibility to "reverse the time" : at time τ, there is a process where each Fresnel vector (figure 2) which was turning anticlockwise in the complex plane sees suddenly its angular velocity exactly reversed. We ask the question : what happens later on ? It is very easy to see that at the time t = 2τ all the Fresnel vectors tend to be aligned along the original direction in the complex plane ; at that particular moment all the amplitudes of the different modes add coherently. You measure a macroscopic quantity which is the amplitude of vibration of the population of modes. In such a process, you have reestablished the original amplitude ; at time t = 2τ you have produced what is called an "echo" ; of course, at a time later (t > 2τ) the signal disappears once more.

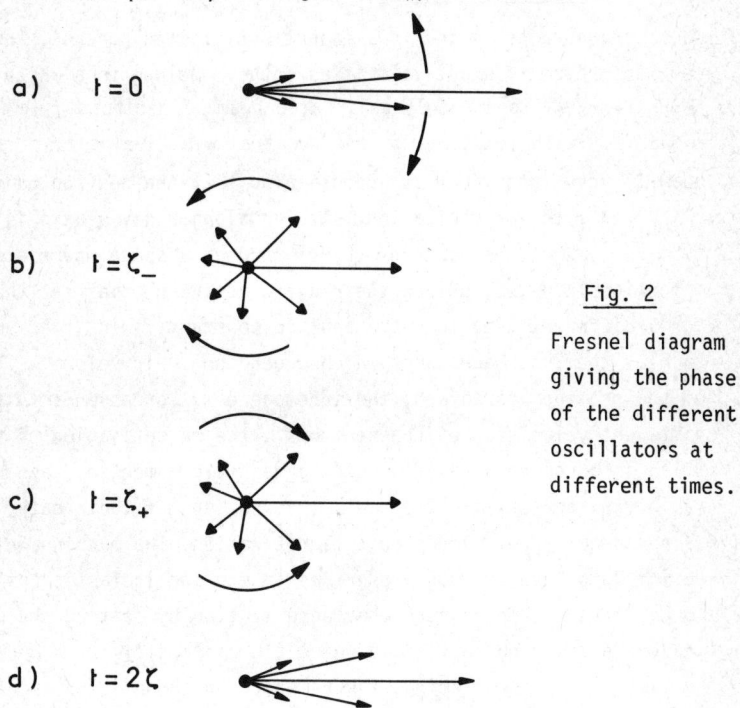

a)   t = 0

b)   t = τ_

c)   t = τ_+

d)   t = 2τ

Fig. 2

Fresnel diagram giving the phase of the different oscillators at different times.

b) It is important, I believe, to specify what are the different ingredients which characterize the phenomena of echo. First, in general, you have to start with a distribution of modes ; the second point, in order to produce an echo which is a typical coherent effect, consists to outline that a non-linear process is required to operate time reversal. In that respect, the history of the individual modes which is dependent of processes which are not time reversable (we call them in general irreversible processes) will contribute to decrease the observed echo ; (random processes, noisy perturbation of the oscillators due to the surrounding medium). That irreversible decay of the echo as a function of τ is characterized by an homogeneous relaxation time T_2. On the contrary, the inhomogeneous broadening of that assembly of modes is quite reversible and what you gain by performing an echo is that you can escape from the distribution in frequency. Practically, this is a big advantage because in general $T_2^* \ll T_2$. The interesting physical effects are also contained into T_2. You have had at least one example of such a thing during this session ; I would like to quote that example in the words I have just used.

c) <u>Spin echoes with neutrons</u>

Consider neutrons placed in a magnetic field. In general, the neutrons have different velocities ; they belong to a white beam of neutrons or to a poorly calibrated beam of neutrons ; in comparison with light beams, the spectral width velocity of a neutron beam is very large. Let us suppose also that the neutron beam is polarized : the magnetization measured along a given axis is different from zero. When the neutrons travel through a space where a constant magnetic field is applied, their different velocities let them stay for a different time into the applied constant field ; the spin of each of them has made turns which depend on their velocity. In terms of the previous paragraph, the inhomogeneity corresponds to the different velocities of the neutrons ; the magnetisation of the beam which is a coherent addition of the magnetic moment of each neutron rapidly disappears in the field ; at the end of their path, there is no longer a net macroscopic magnetization. The machine which was designed to have so good properties to explore inelastic scattering is conceived so as to restore the magnetization by letting the neutrons travel in a second half-space where the applied field is opposite ; for any elastic scattering, the velocity of the neutron is not changed. Thus for each of them whatever their velocity is, there is an exact compensation of the rotation of their spin in the two half spaces (figure 3). This is in fact a time reversal operation. It can be proved that when this machine is used in the inelastic mode of operation, its resolution is much better than any other /4/.

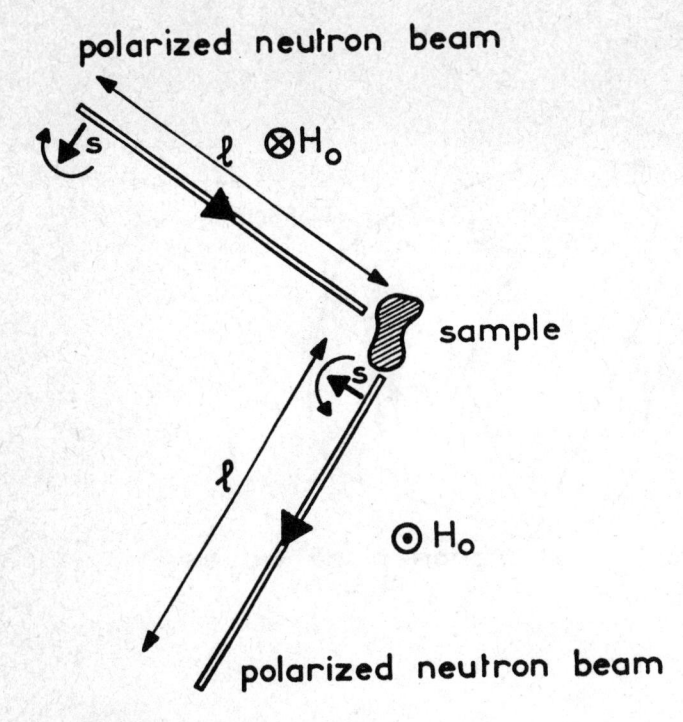

Fig. 3

Schematic diagramme of the spin echo spectrometer for neutrons.

d) <u>Phonon echoes</u>

The last example to which I would like to draw your attention corresponds to the phenomena which is called "phonon echoes" ; it is less common than "spin echoes" and for that reason, it requires maybe some more explanations, but all the principles involved are similar.

Consider a body and suppose that at time t = 0 you excite in a particular point of the surface phonons of the same frequency ; they have in fact different wave vectors and they can propagate in many directions. Very rapidly, the resulting amplitude of vibration in any point of the body decreases to zero because it is the result of many contributions of phonons which have made quite different paths into the crystal and because they dephase progressively (inhomogeneous broadening). But suppose that there is a microscopic process which is such at a given time τ you can reverse the wave vector of these different elastic modes, for each of them, one by one : if you wait long enough, at a time 2τ, all these modes will focus onto the same point where you have excited your body and you will observe a macroscopic amplitude of vibration (figure 4). What you have done and obtained is an "elastic echo" or a "phonon echo" /8/.

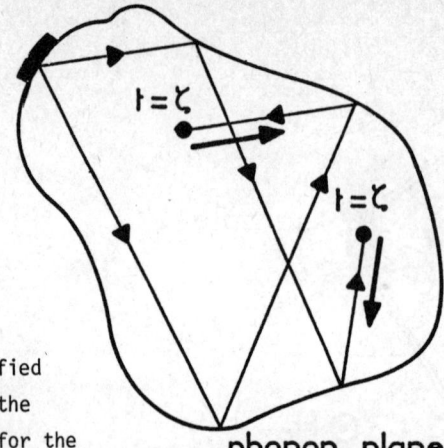

Fig. 4 : Simplified description of the different path for the different phonons modes excited at the emitter.

In such an example, the inhomogeneity corresponds to the different velocities of the modes and to their different wave vector simultaneously with their density of state ; the non-linear process, in fact, depends on the symmetry of the crystal ; that effect is only possible in a piezoelectric crystal or more generally in a non-centro-symmetric crystal if you apply for example a uniform electric field at the frequency 2ω to operate the time reversal process ; for such a symmetry, there is always a constant which couple two phonons of opposite wave vectors with the electric field. It is thus a parametric process which excites the forward phonon and which stimulates the creation of the backward phonon. In such a case, as well as in the case of neutron spin echoes, after the initial time t = 0, the coherence of the different modes has not disappeared ; it was only hidden because of the inhomogeneous broadening of the system. To be manifested once more it needed that magic operation of time reversal.

V - HOLOGRAPHY

As we have learnt, holography is a wonderful technique to observe defects, to memorize information. What I would like to suggest in that last paragraph is a device or a point of view which has something to do with both "echo phenomena" and holography technique. Let us suppose that we consider a body whose surfaces are perfect mirrors for light beams (Fig. 5) ;

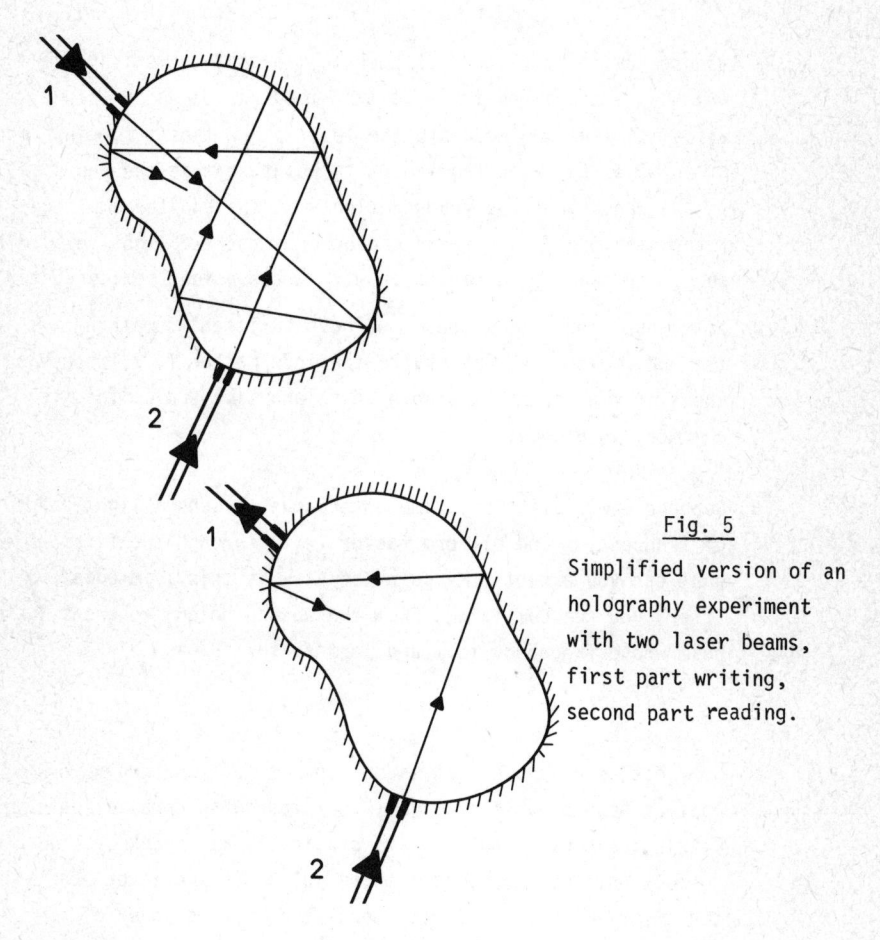

Fig. 5

Simplified version of an holography experiment with two laser beams, first part writing, second part reading.

suppose now that in the surface of that closed body we have two holes 1 and 2 through which we can send coherent light beams of the same frequency. Consider now the successive operations :

1) Through the first hole we send a light beam of frequency ω with a defined form factor $f_1(t)$; the light occupies the whole body in a very small interval of time : the light turns and turns within the body.

2) Suppose now that through the second hole, we send another light beam of the same frequency, with a form factor $f_2(t)$; in each point of the body, there is a mixing of two light beams which have the same frequency but which can be decomposed in many plane waves of different k-vectors.

The result of these two combined beams is that there is a non-uniform static electric field which is established into the body.

3) Suppose now that the body is such that a static electric field which is established for a certain time (duration of the optical pulses) can modulate the local properties ; for instance, it can move the traps from point to point so that the repartition and the density of the traps reflects at the end the amplitude of the static electric field produced by the two light beams. This can be done easily for instance with a convenient semiconductor.

4) After the operation 1 and 2, we have "written" something within the body which keeps memory of the information $f_1 * f_2$ provided the system of traps is stable on a long time scale. This is a memory in volume.

5) Suppose now that at any time afterwards we send a light beam of frequency ω and of form factor $f_3(t)$ through the first hole. What can you expect ? My point of view is this : immediately after, you see coming out from the second hole a coherent light beam whose frequency is ω and form factor /9/ is

$$f(t)_{\text{signal}} = f_1(t) * f_2(t) * f_3(t) \qquad (24)$$

6) How this is possible : it is due to the fact that when you apply a second pulse of light, you produce an array of defects within the body : you modulate statically its index of refraction. When you send a third pulse of light, you read the information : the static modulation of the index of refraction diffracts your pulse : it delivers a coherent light beam through the second hole. This operation is completely analogous to a diffraction pattern obtained from a periodic structure of atoms.

This device is very similar to an holographic device ; it could very well work with optical beam to store information on a surface or in a volume ; I must say that in the case of acoustical beams, it has already been tried and has even been considered to make memories with three-dimensional capabilities.

As a conclusion, I would like to say that there is much fun to play with coherence phenomena ; it is not only a game, but it is a very fruitful point of view as was illustrated from different fields in physics and I am convinced that in 10 years from now, when we meet again on the same subject, we will have many more examples of beautiful physics and wonderful and/or useful devices to report on.

REFERENCES

/1/ R.J. Glauber, Phys. Rev. 131, 2766 (1963)

/2/ A.R. Lang, in Diffraction and Imaging Techniques in Material Science, edited by S. Amelinckx, R. Gevers and J. Van Landuyt, North Holland, Amsterdam, 1978, p. 623 (1978).

/3/ U. Bonse, W. Graeff in X-Ray Optics, edited by H.J. Queisser, Springer-Verlag Berlin, p. 93 (1977)

/4/ F. Mezei, Zeitsch. für Physik, 255, 146 (1972)

/5/ J.R. Johnston, Am. Jl. of Physics, 38, 516 (1970)
P. Carruthers, M.M. Nieto, Rev. of Modern Physics, 40, 411 (1968)

/6/ F.W. Cummings, J.R. Johnston, Phys. Rev. 151, 105 (1966)
P.W. Anderson, Rev. of Modern Physics, 38, 298 (1966)

/7/ G. Garusta, Phys. Rev. Letters, 33, 1428 (1974)

/8/ A. Billmann, C. Fresnois, J. Joffrin, A. Levelut, S. Ziolkiewicz
Jl. de Physique, 34, 453 (1973)

/9/ J. Joffrin, A. Levelut, Cours à l'Ecole Internationale de Physique,
E. Fermi - Varena (1974) p. 291.

T. Kohonen
Content-Addressable Memories
1980. 123 figures, 35 tables. Approx. 400 pages
(Springer Series in Information Sciences, Volume 1)
ISBN 3-540-09823-2

Contents: Associative Memory, Content Addressing, and Associative Recall. – Content Addressing by Software. – Logic Principles of Content-Addressable Memories. – CAM Hardware. – The CAM as a System Part. – Content-Addressable Processors. – References. – Subject Index.

Laser Speckle and Related Phenomena
Editor: J. C. Dainty
1975. 133 figures. XIII, 286 pages
(Topics in Applied Physics, Volume 9)
ISBN 3-540-07498-8

Contents: *J. C. Dainty:* Introduction. – *J. W. Goodman:* Statistical Properties of Laser Speckle Patterns. – *G. Parry:* Speckle Patterns in Partially Coherent Light. – *T. S. McKechnie:* Speckle Reduction. – *M. Françon:* Information Processing Using Speckle Patterns. – *A. E. Ennos:* Speckle Interferometry. – *J. C. Dainty:* Stellar Speckle Interferometry. – Additional References with Titles. – Subject Index.

T. Pavlidis
Structural Pattern Recognition
1977. 173 figures, 13 tables. XII, 302 pages
(Springer Series in Electrophysics, Volume 1)
ISBN 3-540-08463-0

Contents: Mathematical Techniques for Curve Fitting. – Graphs and Grids. – Fundamentals of Picture Segmentation. – Advanced Segmentation Techniques. – Scene Analysis. – Analytical Description of Region Boundaries. – Syntactic Analysis of Region Boundaries and Other Curves. – Shape Description by Region Analysis. – Classification, Description and Syntactic Analysis.

Picture Processing and Digital Filtering
Editor: T. S. Huang
2nd corrected and updated edition. 1979. 113 figures, 7 tables. XIII, 297 pages.
(Topics in Applied Physics, Volume 6)
ISBN 3-540-09339-7

Contents: *T. S. Huang:* Introduction. – *H. C. Andrews:* Two-Dimensional Transforms. – *J. G. Fiasconaro:* Two-Dimensional Nonrecursive Filters. – *R. R. Read, J. L. Shanks, S. Treitel:* Two-Dimensional Recursive Filtering. – *B. R. Frieden:* Image Enhancement and Restoration. – *F. C. Billingsley:* Noise Considerations in Digital Image Processing Hardware. – *T. S. Huang:* Recent Advances in Picture Processing and Digital Filtering. – Subject Index.

Springer-Verlag
Berlin
Heidelberg
New York

Selected Issues from
Lecture Notes in Mathematics

Vol. 594: Singular Perturbations and Boundary Layer Theory, Lyon 1976. Edited by C. M. Brauner, B. Gay, and J. Mathieu. VIII, 539 pages. 1977.

Vol. 596: K. Deimling, Ordinary Differential Equations in Banach Spaces. VI, 137 pages. 1977.

Vol. 605: Sario et al., Classification Theory of Riemannian Manifolds. XX, 498 pages. 1977.

Vol. 606: Mathematical Aspects of Finite Element Methods. Proceedings 1975. Edited by I. Galligani and E. Magenes. VI, 362 pages. 1977.

Vol. 607: M. Métivier, Reelle und Vektorwertige Quasimartingale und die Theorie der Stochastischen Integration. X, 310 Seiten. 1977.

Vol. 615: Turbulence Seminar, Proceedings 1976/77. Edited by P. Bernard and T. Ratiu. VI, 155 pages. 1977.

Vol. 618: I. I. Hirschman, Jr. and D. E. Hughes, Extreme Eigen Values of Toeplitz Operators. VI, 145 pages. 1977.

Vol. 623: I. Erdelyi and R. Lange, Spectral Decompositions on Banach Spaces. VIII, 122 pages. 1977.

Vol. 628: H. J. Baues, Obstruction Theory on the Homotopy Classification of Maps. XII, 387 pages. 1977.

Vol. 629: W. A. Coppel, Dichotomies in Stability Theory. VI, 98 pages. 1978.

Vol. 630: Numerical Analysis, Proceedings, Biennial Conference, Dundee 1977. Edited by G. A. Watson. XII, 199 pages. 1978.

Vol. 636: Journées de Statistique des Processus Stochastiques, Grenoble 1977, Proceedings. Edité par Didier Dacunha-Castelle et Bernard Van Cutsem. VII, 202 pages. 1978.

Vol. 638: P. Shanahan, The Atiyah-Singer Index Theorem, An Introduction. V, 224 pages. 1978.

Vol. 648: Nonlinear Partial Differential Equations and Applications, Proceedings, Indiana 1976-1977. Edited by J. M. Chadam. VI, 206 pages. 1978.

Vol. 650: C*-Algebras and Applications to Physics. Proceedings 1977. Edited by R. V. Kadison. V, 192 pages. 1978.

Vol. 656: Probability Theory on Vector Spaces. Proceedings, 1977. Edited by A. Weron. VIII, 274 pages. 1978.

Vol. 662: Akin, The Metric Theory of Banach Manifolds. XIX, 306 pages. 1978.

Vol. 665: Journées d'Analyse Non Linéaire. Proceedings, 1977. Edité par P. Bénilan et J. Robert. VIII, 256 pages. 1978.

Vol. 667: J. Gilewicz, Approximants de Padé. XIV, 511 pages. 1978.

Vol. 668: The Structure of Attractors in Dynamical Systems. Proceedings, 1977. Edited by J. C. Martin, N. G. Markley and W. Perrizo. VI, 264 pages. 1978.

Vol. 675: J. Galambos and S. Kotz, Characterizations of Probability Distributions. VIII, 169 pages. 1978.

Vol. 676: Differential Geometrical Methods in Mathematical Physics II, Proceedings, 1977. Edited by K. Bleuler, H. R. Petry and A. Reetz. VI, 626 pages. 1978.

Vol. 678: D. Dacunha-Castelle, H. Heyer et B. Roynette. Ecole d'Eté de Probabilités de Saint-Flour. VII-1977. Edité par P. L. Hennequin. IX, 379 pages. 1978.

Vol. 679: Numerical Treatment of Differential Equations in Applications, Proceedings, 1977. Edited by R. Ansorge and W. Törnig. IX, 163 pages. 1978.

Vol. 681: Séminaire de Théorie du Potentiel Paris, No. 3, Directeurs: M. Brelot, G. Choquet et J. Deny. Rédacteurs: F. Hirsch et G. Mokobodzki. VII, 294 pages. 1978.

Vol. 682: G. D. James, The Representation Theory of the Symmetric Groups. V, 156 pages. 1978.

Vol. 684: E. E. Rosinger, Distributions and Nonlinear Partial Differential Equations. XI, 146 pages. 1978.

Vol. 690: W. J. J. Rey, Robust Statistical Methods. VI. 128 pages. 1978.

Vol. 691: G. Viennot, Algèbres de Lie Libres et Monoïdes Libres. III, 124 pages. 1978.

Vol. 693: Hilbert Space Operators, Proceedings, 1977. Edited by J. M. Bachar Jr. and D. W. Hadwin. VIII, 184 pages. 1978.

Vol. 696: P. J. Feinsilver, Special Functions, Probability Semigroups, and Hamiltonian Flows. VI, 112 pages. 1978.

Vol. 702: Yuri N. Bibikov, Local Theory of Nonlinear Analytic Ordinary Differential Equations. IX, 147 pages. 1979.

Vol. 704: Computing Methods in Applied Sciences and Engineering, 1977, I. Proceedings, 1977. Edited by R. Glowinski and J. L. Lions. VI, 391 pages. 1979.

Vol. 710: Séminaire Bourbaki vol. 1977/78, Exposés 507-524. IV, 328 pages. 1979.

Vol. 711: Asymptotic Analysis. Edited by F. Verhulst. V, 240 pages. 1979.

Vol. 712: Equations Différentielles et Systèmes de Pfaff dans le Champ Complexe. Edité par R. Gérard et J.-P. Ramis. V, 364 pages. 1979.

Vol. 716: M. A. Scheunert, The Theory of Lie Superalgebras. X, 271 pages. 1979.

Vol. 720: E. Dubinsky, The Structure of Nuclear Fréchet Spaces. V, 187 pages. 1979.

Vol. 724: D. Griffeath, Additive and Cancellative Interacting Particle Systems. V, 108 pages. 1979.

Vol. 725: Algèbres d'Opérateurs. Proceedings, 1978. Edité par P. de la Harpe. VII, 309 pages. 1979.

Vol. 726: Y.-C. Wong, Schwartz Spaces, Nuclear Spaces and Tensor Products. VI, 418 pages. 1979.

Vol. 727: Y. Saito, Spectral Representations for Schrödinger Operators With Long-Range Potentials. V, 149 pages. 1979.

Vol. 728: Non-Commutative Harmonic Analysis. Proceedings, 1978. Edited by J. Carmona and M. Vergne. V, 244 pages. 1979.

Vol. 729: Ergodic Theory. Proceedings 1978. Edited by M. Denker and K. Jacobs. XII, 209 pages. 1979.

Vol. 730: Functional Differential Equations and Approximation of Fixed Points. Proceedings, 1978. Edited by H.-O. Peitgen and H.-O. Walther. XV, 503 pages. 1979.

Lecture Notes in Physics

Vol. 68: Y. V. Venkatesh, Energy Methods in Time-Varying System Stability and Instability Analyses. XII, 256 pages. 1977.

Vol. 69: K. Rohlfs, Lectures on Density Wave Theory. VI, 184 pages. 1977.

Vol. 70: Wave Propagation and Underwater Acoustics. Edited by J. Keller and J. Papadakis. VIII. 287 pages. 1977.

Vol. 71: Problems of Stellar Convection. Proceedings 1976. Edited by E. A. Spiegel and J. P. Zahn. VIII, 363 pages. 1977.

Vol. 72: Les instabilités hydrodynamiques en convection libre forcée et mixte. Edité par J. C. Legros et J. K. Platten. X, 202 pages. 1978.

Vol. 73: Invariant Wave Equations. Proceedings 1977. Edited by G. Velo and A. S. Wightman. VI, 416 pages. 1978.

Vol. 74: P. Collet and J.-P. Eckmann, A Renormalization Group Analysis of the Hierarchical Model in Statistical Mechanics. IV, 199 pages. 1978.

Vol. 75: Structure and Mechanisms of Turbulence I. Proceedings 1977. Edited by H. Fiedler. XX, 295 pages. 1978.

Vol. 76: Structure and Mechanisms of Turbulence II. Proceedings 1977. Edited by H. Fiedler. XX, 406 pages. 1978.

Vol. 77: Topics in Quantum Field Theory and Gauge Theories. Proceedings, Salamanca 1977. Edited by J. A. de Azcárraga. X, 378 pages 1978.

Vol. 78: Böhm, The Rigged Hilbert Space and Quantum Mechanics. IX, 70 pages. 1978.

Vol. 79: Group Theoretical Methods in Physics. Proceedings, 1977. Edited by P. Kramer and A. Rieckers. XVIII, 546 pages. 1978.

Vol. 80: Mathematical Problems in Theoretical Physics. Proceedings, 1977. Edited by G. Dell'Antonio, S. Doplicher and G. Jona-Lasinio. VI, 438 pages. 1978.

Vol. 81: MacGregor, The Nature of the Elementary Particle. XXII, 482 pages. 1978.

Vol. 82: Few Body Systems and Nuclear Forces I. Proceedings, 1978. Edited by H. Zingl, M. Haftel and H. Zankel. XIX, 442 pages. 1978.

Vol. 83: Experimental Methods in Heavy Ion Physics. Edited by K. Bethge. V, 251 pages. 1978.

Vol. 84: Stochastic Processes in Nonequilibrium Systems, Proceedings, 1978. Edited by L. Garrido, P. Seglar and P. J. Shepherd. XI, 355 pages. 1978

Vol. 85: Applied Inverse Problems. Edited by P. C. Sabatier. V, 425 pages. 1978.

Vol. 86: Few Body Systems and Electromagnetic Interaction. Proceedings 1978. Edited by C. Ciofi degli Atti and E. De Sanctis. VI, 352 pages. 1978.

Vol. 87: Few Body Systems and Nuclear Forces II, Proceedings, 1978. Edited by H. Zingl, M. Haftel, and H. Zankel. X, 545 pages. 1978.

Vol. 88: K. Hutter and A. A. F. van de Ven, Field Matter Interactions in Thermoelastic Solids. VIII, 231 pages. 1978.

Vol. 89: Microscopic Optical Potentials, Proceedings, 1978. Edited by H. V. von Geramb. XI, 481 pages. 1979.

Vol. 90: Sixth International Conference on Numerical Methods in Fluid Dynamics. Proceedings, 1978. Edited by H. Cabannes, M. Holt and V. Rusanov. VIII, 620 pages. 1979.

Vol. 91: Computing Methods in Applied Sciences and Engineering, 1977, II. Proceedings, 1977. Edited by R. Glowinski and J. L. Lions. VI, 359 pages. 1979.

Vol. 92: Nuclear Interactions. Proceedings, 1978. Edited by B. A. Robson. XXIV, 507 pages. 1979.

Vol. 93: Stochastic Behavior in Classical and Quantum Hamiltonian Systems. Proceedings, 1977. Edited by G. Casati and J. Ford. VI, 375 pages. 1979.

Vol. 94: Group Theoretical Methods in Physics. Proceedings, 1978. Edited by W. Beiglböck, A. Böhm and E. Takasugi. XIII, 540 pages. 1979.

Vol. 95: Quasi One-Dimensional Conductors I. Proceedings, 1978. Edited by S. Barišić, A. Bjeliš, J. R. Cooper and B. Leontić. X, 371 pages. 1979.

Vol. 96: Quasi One-Dimensional Conductors II. Proceedings 1978. Edited by S. Barišić, A. Bjeliš, J. R. Cooper and B. Leontić. XII, 461 pages. 1979.

Vol. 97: Hughston, Twistors and Particles. VIII, 153 pages. 1979.

Vol. 98: Nonlinear Problems in Theoretical Physics. Proceedings, 1978. Edited by A. F. Rañada. X, 216 pages. 1979.

Vol. 99: M. Drieschner, Voraussage – Wahrscheinlichkeit – Objekt. XI, 308 Seiten. 1979.

Vol. 100: Einstein Symposion Berlin. Proceedings 1979. Edited by H. Nelkowski et al. VIII, 550 pages. 1979.

Vol. 101: A. Martin-Löf, Statistical Mechanics and the Foundations of Thermodynamics. V, 120 pages. 1979.

Vol. 102: H. Hora, Nonlinear Plasma Dynamics at Laser Irradiation. VIII, 242 pages. 1979.

Vol. 103: P. A. Martin, Modèles en Mécanique Statistique des Processus Irréversibles. IV, 134 pages. 1979.

Vol. 104: Dynamical Critical Phenomena and Related Topics. Proceedings, 1979. Edited by Ch. P. Enz. XII, 390 pages. 1979.

Vol. 105: Dynamics and Instability of Fluid Interfaces. Proceedings, 1978. Edited by T. S. Sørensen. V, 315 pages. 1979.

Vol. 106: Feynman Path Integrals, Proceedings, 1978. Edited by S. Albeverio et al. XI, 451 pages. 1979.

Vol. 107: J. Kijowski, W. M. Tulczyjew, A Symplectic Framework for Field Theories. IV, 257 pages. 1979.

Vol. 108: Nuclear Physics with Electromagnetic Interactions. Proceedings, 1979. Edited by H. Arenhövel and D. Drechsel. IX, 509 pages. 1979.

Vol. 109: Physics of the Expanding Universe. Proceedings, 1978. Edited by M. Demiański. V, 210 pages. 1979.

Vol. 110: D. A. Park, Classical Dynamics and Its Quantum Analogues. VIII, 339 pages. 1979.

Vol. 111: H.-J. Schmidt, Axiomatic Characterization of Physical Geometry. V, 163 pages. 1979.

Vol. 112: Imaging Processes and Coherence in Physics. Proceedings, 1979. Edited by M. Schlenker et al. XIX, 577 pages. 1980.

This series reports new developments in physical research and teaching – quickly, informally and at a high level. The type of material considered for publication includes:

1. Preliminary drafts of original papers and monographs
2. Lectures on a new field or presentations of new angles in a classical field
3. Seminar work-outs
4. Reports of meetings, provided they are
    a) of exceptional interest and
    b) devoted to a single topic.

The timeliness of a manuscript is more important than its form, which may be unfinished or tentative. Thus, in some instances, proofs may be merely outlined and results presented which have been or will later be published elsewhere. If possible, a subject index should be included. Publication of Lecture Notes is intended as a service to the international physical community, in that a commercial publisher, Springer-Verlag, can offer a wide distribution of documents which would otherwise have a restricted readership. Once published and copyrighted, they may be documented in the scientific literature.

Manuscripts

Manuscripts should be no less than 100 and preferably no more than 500 pages in length.
They are reproduced by a photographic process and therefore must be typed with extreme care. Symbols not on the typewriter should be inserted by hand in indelible black ink. Corrections to the typescript should be made by pasting in the new text or painting out errors with white correction fluid. Authors receive 50 free copies and are free to use the material in other publications. The typescript is reduced slightly in size during reproduction; best results will not be obtained unless the text on any one page is kept within the overall limit of 18 x 26.5 cm (7 x 10½ inches). On request, the publisher will supply special paper with the typing area outlined.
Manuscripts in English, German or French should be sent to Prof. Dr. W. Beiglböck, Institut für Angewandte Mathematik, Im Neuenheimer Feld 5, 6900 Heidelberg/Germany, or directly to Springer-Verlag Heidelberg.

Springer-Verlag, Heidelberger Platz 3, D-1000 Berlin 33
Springer-Verlag, Neuenheimer Landstraße 28–30, D-6900 Heidelberg 1
Springer-Verlag, 175 Fifth Avenue, New York, NY 10010/USA

ISBN 3-540-09727-9
ISBN 0-387-09727-9